Cities and Nature

Series Editors

Peter Newman, Sustainability Policy Institute, Curtin University, Perth, WA, Australia
Cheryl Desha, School of Engineering and Built Environment, Griffith University, Nathan, QLD, Australia
Alessandro Sanches-Pereira, Instituto 17, São Paulo, São Paulo, Brazil

Cities and Nature fosters high-quality multi-disciplinary research addressing the interface between cities and the natural environment. It provides a valuable source of relevant knowledge for researchers, planners and policy-makers. The series welcomes empirically based, cutting-edge and theoretical research in urban geography, urban planning, environmental planning, urban ecology, regional science and economics. It publishes peer-reviewed edited and authored volumes on topics dealing with the urban and the environment nexus, including: spatial dynamics of urban built areas, urban and peri-urban agriculture, urban greening and green infrastructure, environmental planning, urban forests, urban ecology, regional dynamics and landscape fragmentation.

More information about this series at http://www.springer.com/series/10068

Bruce Dvorak
Editor

Ecoregional Green Roofs

Theory and Application in the Western USA and Canada

Springer

Editor
Bruce Dvorak
Department of Landscape Architecture and Urban Planning
305A Langford Architecture Center
Texas A&M University
College Station, TX, USA

ISSN 2520-8306 ISSN 2520-8314 (electronic)
Cities and Nature
ISBN 978-3-030-58394-1 ISBN 978-3-030-58395-8 (eBook)
https://doi.org/10.1007/978-3-030-58395-8

This Springer imprint is published by the registered company Springer Nature Switzerland AG
The registered company address is: Gewerbestrasse 11, 6330 Cham, Switzerland

Foreword

Today, more than at any time in history, we know that healthy people require fresh, clean air; pure, abundant water; safe, secure, weather-resistant shelter; nutrient-dense fresh food; proximity/access to natural landscapes; and habitat for birds, bats, butterflies, and a plethora of other organisms that are all part of a living ecology. We know that all of these and many other ecosystem services can be either enhanced or compromised by the surfaces of our buildings. We also know that a substantial portion of the existing building infrastructure throughout North America will be renovated or completely replaced/reconfigured over the next 50 years, and many urban areas will expand with new construction. Some project the total population to grow by 20% or approximately 75 million additional people on the continent during that time, and most will live, work, learn, and age in urban/peri-urban areas. Although it varies somewhat, building roofs typically occupy between 15 and 35% of cities, and those with higher population densities usually have less green space. It is hard not to conclude, then, that we would greatly benefit from the integration of living surfaces over most urban structures – *ecoregional green roofs*, the topic of this book.

Bruce Dvorak has a long history with living architecture and ecological practices (Associate Professor in the Department of Landscape Architecture and Urban Planning at Texas A&M University). In this book, he has compiled and crafted an extremely rich, detailed discussion on the current state of created/re-created living surfaces as a vehicle to illustrate the yet-unrealized potential for new life to be introduced and maintained over inert structures and conditions. Using a systematic approach, the book defines native vegetation and ecological landscape theory and practice, and summarizes the history and development of green roofs, emphasizing those adapted to local ecological contexts in Part I. In Part II, each chapter covers a particular ecoregion of the generally more arid portions of North America (west and south), including a detailed description, illustrations of a compendium of examples of green roofs, and observations from Bruce and eight co-authors, each with experience and knowledgeable about green roofs in their respective ecoregions. The book concludes in Part III with further observations that take the entire body of the green roof case studies into account, underscoring the need for further research, and celebrating the incredible opportunity we have to deploy practices that integrate technology with local, living ecologies.

Bruce lays the foundation for going beyond the notion of a vegetated roof to recreating living over structure landscapes, beginning with a discussion of North American grassland ecologies – setting the stage for us to think about roof space as part of a larger whole, rather than as an individual component. As Bruce describes in the first chapter, ecological, lightweight green roof and wall technology has evolved from early applications and publicity in Portland, Chicago, Toronto, and a handful of other pioneering cities to mainstream deployment in many locations throughout the continent. While much of the interest in living architecture (green roofs and walls) comes from a desire to achieve many of the more recognized attributes they offer (rainwater attenuation, energy efficiency/cooling, acoustic buffering, visual beauty, etc.), there has been a growing interest in the use of native plants on green roofs for various reasons. The performance and function of green roofs have advanced significantly in these areas in the past two decades, due in part to standards and metrics such as LEED, SITES, the Living Building Challenge, and most recently the Living Architecture Performance Tool (*LAPT, by the Green Infrastructure Foundation).* All promote biodiversity and native ecology as important potential functions a green roof <u>can</u> provide if properly designed, installed, and maintained, and in concert with other functions. This book provides a clear description of the implications of using native plants on a roof and gives direction on what to do (and not to do) to successfully establish and retain them over time.

In Chap. 2, Bruce and co-author Jennifer Bousselot, Ph.D. (Department of Horticulture and Landscape Architecture, Colorado State University*)*, further expand on the concept of ecoregional green roofs through a review of all of the various habitat elements that influence the health, vitality, morphology, and other characteristics of locally adapted native plants. In addition to soil substrate, rainfall regime, solar aspect, wind, and other factors, this includes relationships with other microorganisms in the soil, insects, birds, and other wildlife that are part of living ecosystems in natural landscapes. The discussion includes the human cultural relationship with native landscapes and the implications of adapting this essential facet of habitat on green roofs. The description of the case study approach then leads into the chapters in Part II.

Bruce has researched green roofs from the western ecoregions of the North American continent for the book's many case studies. These generally harsher, drier ecoregions illustrate and emphasize the need to carefully consider <u>all</u> of the design influences of these spaces as an integrated system over time (holism), not as isolated components of waterproofing, growing media, plants, and maintenance (reductionist thinking) as he reiterates throughout the book. The chapters in Part II describe the various West/Southwest continental ecoregions in collaboration with the eight chapter contributors, which provides a fantastic depth of local perspective. Each chapter then includes detailed case studies of a broad range of examples located in each ecoregion, describing each of them with the same metrics. Each case study includes information on the client/design/construction team, vegetation establishment methods and species, irrigation, maintenance protocols, observed wildlife, the most successful plants, and other data. The case studies also include perspectives from owners, designers, and others, and the author's observations. Some of the more helpful information in the book is included in Bruce's summaries, informed by two decades of experience, observation of dozens of green roofs around the world, and his test plots and trials.

At the end of each chapter, Bruce aggregates the trends and general consistencies with the native plant species that seem to do the best on roof surfaces. He makes additional summary observations and conclusions based upon the aggregation of each set of case studies. This book fills a wide knowledge gap by pairing descriptions of each defined ecoregion with a range of green roof applications that are deployed with native vegetation in various ways. His depth of research and data gained through personal visits, interviews with designers, building owners, researchers, and many others is extensive, and the insights he and his co-authors offer is extremely valuable. Pairing the examples with the ecoregions and the associated metrics and descriptions helps the reader to make the connection between all the elements of a functioning ecosystem, both at-grade and over-structure.

In the last chapter, Bruce and collaborator Lee R. Skabelund (Associate Professor in Landscape Architecture / Regional and Community Planning at Kansas State University) provide an overview and summary of the concept of ecological green roofs, informed by the body of this book's research. Some have criticized the use of green roof systems in dry/semi-arid climates that can require large volumes of water and/or expand the use of exotic species for rooftop vegetation (equating them to high-input lawns and gardens) and suggest green roofs are simply inappropriate for these climates. Bruce and Lee respond to this opinion with a resounding alternative – *adapt the technology/system to the local ecology*. While some degree of failure and learning curve is described in many of the case studies, overall, the potential for low-input, biodiverse, lush, and healthy native living surfaces is unequivocally demonstrated.

Bruce is uniquely suited to write this book, based upon his professional path up to this point – it is a culmination of the experience he has gained in both private/applied practice and in teaching/research, combined with his co-authors' perspectives and knowledge. It also benefits from the collective experience and wisdom of many others who have been part of his path through collaboration, research, and mutual passion for living architecture:

> Jeffrey L. Bruce has generously given so much of his brilliance with soil ecology, calibrated water delivery systems, and other practices essential to create living ecologies over-structure; Jeff passed away at the beginning of 2020, leaving a legacy of living architecture and ecological practice that substantially accelerated these ideas throughout the continent. Similarly, Paul Kephart has inspired Bruce and so many others since the very first eco-roofs in North America and continues to amaze with elegant, artful incorporation of living ecologies in all manner of places and contexts. Jeff and Paul are two of a handful of early pioneers and advocates of ecoregional living architecture practice, and their work is featured prominently in the case studies.

Other influences include Jim Patchett and Dr. Gerould Wilhelm, with whom Bruce worked in private practice while at Conservation Design Forum in the Chicago region, *(CDF, now ECT)*. Also pioneering an ecological approach, Jim's vision helped advance infiltration-based green infrastructure strategies in the Midwest and elsewhere. Jerry is the research director at Conservation Research Institute and co-author of *The Flora of the Chicago Region* (with Laura Rericha). He helped to nurture Bruce's botanical/ecological perspective and understanding of natural law and process. He unceasingly reiterates the dictum "*plants thrive in habitats to which they are adapted.*" His brilliance in ecoregional theory and the natural/cultural connections healthy native systems in North America rely upon have greatly helped inform the ecological perspective of this book.

As Bruce describes, he had the opportunity to help design and manage some of the first ecological green roofs in the country (*Chicago City Hall, Peggy Notebaert Nature Museum, among others*) while at CDF, and that experience was greatly enhanced through collaboration with Herbert Dreiseitl and his team. Bruce's first direct exposure to ecological green roofs was when he, Jerry, and several other CDF colleagues spent time with Herbert in his studio in Uberlingen, Germany, touring green roofs and other ecological design applications. Herbert is also a visionary; his amazing innovations in water-based design helped provide tangible examples of the potential for living systems to be applied and adapted in CDF's projects.

Bruce (and Jeff, Paul, and many, many others) has also benefitted enormously from a close-knit family of green roof/wall practitioners, academics, suppliers, and researchers. Steven Peck, founder of Green Roofs for Healthy Cities and the Green Infrastructure Foundation, has fostered this family for over 20 years. Steven and GRHC have supported/promoted the use of green roofs and walls for multiple benefits through training, conferences, courses, advocacy, research, publications, and collaboration. Bruce has been a part of these activities since the beginning and has been recognized by GRHC as one of the leading green roof researchers in the industry.

Why This Book Matters

The premise of this book is that when the local ecoregion of a place is the basis for the design, construction, and long-term maintenance of over-structure landscapes, ecosystem services are optimized on those surfaces. This should be self-evident. However, while the green roof and wall industry in North America has made significant strides in the past two decades, we are still in the infancy of these practices. Virtually all living surfaces (green roofs and walls) provide some significant benefits over inert waterproofing, but many do not realize their full potential. *Ecoregional Green Roofs: Theory and Application (in the Western U.S.A. and Canada)* is an important, well-researched, comprehensive guide to help progress rapidly towards high-performing, more authentic, more beneficial, and successful green roof/wall applications in the western ecoregions of North America and elsewhere.

This book will be useful to and enjoyed by many. It serves as a design tool with detailed botanical data and other information about ecological green roofs west of the 100th meridian in the U.S.A. and part of Canada. It is a textbook that describes the ecological history, theory, and the practice of living architecture. It is a compendium of ecological green roof case studies, documenting the common elements of successful green roofs – integrative design, understanding of local ecological and cultural context, proper maintenance, and stewardship calibrated for the specific landscape and client. This is in stark contrast with applying a standard recipe to every green roof installation as simply an alternative roofing material. It is a call for the urgent need for greater exploration, trials, and research for ecological surfaces in the face of rapidly changing climates and substantial investment in the renewal of our cities over the next 50 years and beyond. It is also a satisfying read that ties together often disparate concepts, uniting ecology, technology, and long-term maintenance/stewardship – essential aspects of sustainable site/landscape practice beyond ecological green roofs. The resulting vision makes very real some of the bold visions of future green neighborhoods, villages, and cities that are frankly available today.

Ecoregional Green Roofs: Theory and Application (in the Western U.S.A. and Canada) provides an invaluable resource to continue the path towards living urban places – how to optimize every surface for economic, ecological, and social performance. It underscores the glaring fact that the vast acreage over-structure (or at-grade with completely altered/capped soil) is simply too precious to squander on single-purpose materials and must be deployed with ecological surfaces for multiple purposes and benefits. Bruce and his co-authors share with us an optimism for the future, through clearly illustrating the potential to create new life on surfaces that were otherwise devoid of life – in other words, harnessing the infinite power of living systems to regenerate, renew, and generously provide for the health of people and place.

Board Chair, Green Infrastructure Foundation David Yocca
Balitmore, MD, USA

Acknowledgments

I would like to extend my deepest gratitude to the co-authors of this book (in order of case study chapters): Lee R. Skabelund, Dr. Jennifer Bousselot, Dr. Paul Coseo, Dr. Tom Woodfin, Dr. Philippa Drennan, Nancy Rottle, Dr. Olyssa Starry, and Daniel Roehr for their participation, collaborations, insights, and knowledge of green roofs and native landscapes. Their many contributions helped make this project possible and greatly improved its outcome. Also, for their contributions towards writing, editing, and their suggested improvements to the presentation of the material in this book was significant and much appreciated. Two graduate students also made important contributions to this book, Trevor Maciejewski and Dr. Tess Menotti. Trevor made first and second drafts of all of the ecoregion maps, and Tess worked through many iterations of spreadsheets to develop and refine the many plant lists and tables, final edits of ecoregion and climate maps, and assistance with the index. I would also like to thank Dr. Richard Sutton and Dr. Stephan Mondonca for their assistance in reviewing the manuscript, their feedback and insights are much appreciated. I also want to thank David Yocca for writing the foreword to this book. I also want to thank everyone at Springer, especially Juliana Pitanguy, Carmen Spelbos, Rathika Ramkumar, and Malini Arumugam for their guidance, feedback, and assistance with the making and publishing of this book, their contributions are much appreciated. I also appreciate the reviews and support from several unknown reviewers of this book through Springer, who gave their feedback and support for this project.

Contents

Editor and Contributors

About the Editor

Bruce Dvorak is an associate professor in the Department of Landscape Architecture & Urban Planning at Texas A&M University. He teaches sustainable site design, planning and construction, green roofs, and living walls. In 2009, Professor Dvorak established the Interdisciplinary Green Roof Research Group at Texas A&M University where he investigates research on green roofs and living walls.

Contributors

Jennifer Bousselot is an assistant professor in the Department of Horticulture and Landscape Architecture at Colorado State University. She teaches online courses in green roofs, urban horticulture, native plants, and the introductory horticulture course. Jen is on the Board of Directors of Green Roofs for Healthy Cities (GRHC), Associate Editor of the Journal of Living Architecture, and the co-leader of a Regional Center for Excellence in Living Architecture designated by GRHC.

Paul Coseo is an assistant professor in the Landscape Architecture Program at Arizona State University where he investigates how planning and design strategies such as green roofs can reduce extreme urban climates for more thermally resilient communities. His research and teaching emphasize an evidence-based design approach to expanding the use of green roofs in hot arid ecoregions by documenting the efficacy of design strategies that conserve water, provide native habitat, and reduce the production of unwanted heat.

Philippa Drennan is a professor in the Department of Biology at Loyola Marymount University in Los Angeles. She teaches vegetation ecology, field botany, and plant ecophysiology. Research interests in plant stress biology and restoration ecology are applied to green roofs, in particular to selecting native species for green roofs and their survival.

Daniel Roehr is an associate professor in the School of Architecture & Landscape Architecture at the University of British Columbia, Vancouver, Canada. Since 2007, he runs the research group Greenskins Lab. He co-authored the book *Living Roofs in Integrated Urban Water Systems* (Routledge 2015) and is currently writing the book *Sense-ible Design: Interacting with the Landscape for Designers* (Routledge 2021). Roehr was project architect of the award-winning Daimler-Chrysler Green Roof Project, Potsdamer Platz, Berlin, Germany (1995–2000) and in 2016 he received the Killam Teaching Prize from UBC.

Nancy D. Rottle is a professor in the Department of Landscape Architecture in the College of Built Environments at the University of Washington (UW), where she has taught since 2001. Informed by over 15 years of professional practice and subsequent studies in the USA, Europe, and Oceania, she directs the UW Green Futures Lab which undertakes research, design, and community engagement to explore and implement green infrastructure at all scales. She teaches and publishes on urban hydrology, public space theory and design, and green technologies, often taking a climate change adaptation and mitigation lens. While in professional practice, Professor Rottle was the lead landscape architect for the Cedar River Education Center and its living roof.

Lee R. Skabelund is an associate professor in the Department of Landscape Architecture and Regional & Community Planning at Kansas State University (KSU). His research focuses on several mixed-species green roofs and his teaching integrates ecology, ethics, and design. His primary professional interest is in creating livable communities for people while conserving and restoring healthy ecological systems.

Olyssa Starry is an associate professor at Portland State University where she teaches courses related to urban ecology and stormwater management in the Honors College. She studies and models the effects of green roof design and maintenance on stormwater management and biodiversity.

Tom Woodfin experimented with native prairie green roofs for the past 7 years in the southern Great Plains, has 30 years' teaching in landscape architecture, and extensive wilderness experience in the Intermontane West. He presently leads international study-abroad programs for landscape architecture and planning students in Germany and the Netherlands.

Part I
Background and Theory

Part I of this book outlines a background discussion regarding native vegetation, ecoregions, and their potential application to green roofs. Chapter 1 introduces ecoregional green roofs by discussing the development of plant communities, the history of modern green roofs, and the beginnings of ecoregional green roofs. Chapter 2 covers a theoretical understanding of what ecoregional green roofs are and why they are needed.

Chapter 1
Introduction to Ecoregional Green Roofs

Bruce Dvorak

Abstract This chapter introduces ecoregional green roofs by discussing the development of native plant communities, the history of modern green roofs, and some observations about ecoregional green roofs. It examines the development of the natural vegetation in the western U.S. and Canada and the kinds of plant communities that make up ecoregions appropriate for different forms of green roofs. The history of green roof origins and the development of ecoregional green roofs provide insight into the growth of the modern green roof industry in Europe and North America. Original intentions for green roofs can be misguided, as design decisions or maintenance practices can be out of line with the vegetation selected, or the microclimate of the roof. Several early examples of built ecoregional green roofs highlight successes and lessons learned. Although the conceptual framework laid out in Chap. 1 (and Chap. 2) can be applied anywhere, the climate characteristics for green roofs growing west of the 100th meridian provide background and rationale for the targeted regions of this book. Our knowledge and research literature is only beginning to include the analysis of ecoregional green roofs located in cities where plants experience prolonged exposure to heat and drought, or both.

Keywords Holocene · Prescribed burning · Indigenous people · Sod house · Germany · Extensive · Pilot project · Research · Ecoregion · History

1.1 Introduction

A major goal of this book is to introduce the concept of ecoregions, and how ecoregions can be useful for green roofs. It aims to provide diverse examples of how vegetation from various ecoregions can inspire designs and support a great diversity of ecosystem functions. This chapter establishes a basis for understanding the origins of green roofs and their longstanding correlation with native or adapted vegetation from local or adjacent ecoregions.

B. Dvorak (✉)
Department of Landscape Architecture and Urban Planning, 305A Langford Architecture Center, Texas A&M University, College Station, TX, USA
e-mail: bdvorak@tamu.edu

B. Dvorak (ed.), *Ecoregional Green Roofs*, Cities and Nature,
https://doi.org/10.1007/978-3-030-58395-8_1

Green roofs were first built in Europe over 300 years ago (Osmundson 1999; Grant 2006). Historically, green roofs were constructed on simple structures built with local materials and local vegetation. In their modern form (1980s onward), green roofs were developed with engineered and lightweight growing media and were planted with shallow-rooted-vegetation that grows in local, adjacent, or nearby ecoregions (Köhler 2006; Werthmann 2007).

Green roofs that are composed of a monoculture of exotic sedum is a relatively new and now popular planting concept in North America. These "exotic" green roofs were developed and popularized in the United States, where its vegetation may have little in common with the natural heritage of a place. A green roof planted with solely exotic vegetation may provide stormwater retention and microclimate benefits; however, they may not deliver a full spectrum of ecosystem services such as symbiotic associations with wildlife, preservation of native species, or provide a regional form of beauty (Leopold 1966; Lovell and Johnston 2009; Kowarik 2011; Cook-Patton and Bauerle 2012).

The following discussions and examples presume that non-native vegetation can provide valuable and worthwhile contributions to urban ecology. However, we also stress that it is equally important to preserve the native species, and ecosystem functions even in urbanized regions (Berry 2013). Green roofs that are planted with vegetation from the local ecoregions allow a richer and historically connected urban ecology to develop as a permanent way to bring nature into cities (Köhler 2006; Rowe 2015). This chapter covers a background discussion and theoretical foundation for justifying an ecoregional approach to green roofs.

1.1.1 The Rise of North American Grasslands during the Holocene Epoch

Over the past two million years (during the Pleistocene), and most recently, at least 20,000 years ago, mean sea levels were more than 100 meters (300 ft) lower than at present (Liss and Duce 2005; Gornitz 2009). During winter, sea ice in the Pacific Ocean covered the ocean surface as far south as present-day Los Angeles, California. In Alaska, however, contiguous areas of its interior experienced low amounts of snowfall. The exposed land formed a land bridge called Beringia that connected Asia and Europe to North America. Beringia allowed for an exchange of vegetation, animals, and humans from Asia and Europe to North America (Hopkins 1967; Wen et al. 2016).

The abundance, richness, and composition of plant communities across the North American continent 10,000 to 20,000 years ago was very different from present-day vegetation. Much of the northern half of the continent was covered in ice, while tundra and boreal forest habitats extended as far south as the 40th parallel, where Kansas and Nebraska share borders (Overpeck et al. 1992).

The rise of the latest ecosystems developed in North America during the Holocene, began after the Continental Ice Sheets started to retreat about 11,000 years ago (Graham 1999; Pielou 2008). As the ice receded and sea levels began to rise, the landscape biomes also slowly changed and adapted to the gradual warming of the earth (Overpeck et al. 1992). Over thousands of years, tundra biomes transitioned north as steppes (grasslands), woodlands and forests migrated from southern locations to the north. Biomes that existed much farther south began to migrate north (Webb III et al. 1983; Graham 1999) and evolved to form complex organizations of ecosystems. These habitats became adapted to the climate of geographic regions and began to stabilize about 4000 years ago (Brubaker 1988; Pielou 2008).

The plants and animals that formed the dominant ecosystems became naturally adapted to the region as the climate gradually warmed and glaciers retreated (Van Devender and Spaulding 1979; Graham 1999). Thus, over time, these locally adapted plant communities formed ecological regions or "ecoregions" that responded to the many fluxes and changes of climate, interacted with parent materials such as glacial till or bedrock in the formation of soils, slope exposure, elevation, and animal activities.

The science of ecoregions was introduced in the 1960s (Loucks 1962) and later developed in the 1980s through an evolving awareness of the sciences of ecology and geography, and the roles of climate and topographic influences on the development of vegetation (Bailey 1980; Bailey and Cushwa 1981; Omernik 1987). As the term "biome" is a general term for ecosystems at their climax stage (i.e. mature forest), ecoregion includes all phases of ecosystems including their climax stage (early through late-successional forest) (Bailey 2004). Examples of ecoregions in North America include tundra, alpine, prairie/steppe, scrubland, forest, woodland, desert, savanna, and others.

Humans arrived in North America no sooner than 20,000 or 30,000 years ago and may have had a role in the decline and elimination of megafauna which would have ecological consequences (Martin 1973; Goebel et al. 2008). As peoples from Eurasia migrated to North America during the several sequential exposures of the Beringia land bridge, the human influence widened (Goebel et al. 2008). At the beginning of the Holocene Epoch, the influence of humans on the landscape in North America was thought to be localized as the human population was small, compared to later in the Holocene (Overpeck et al. 1992; Vale 2013).

During the mid to later Holocene, the Native American population increased to at least nine million, but perhaps as high as 54 million from coast to coast prior to European settlement (Vale 2013). During the development of their cultural relationship with the land, they began to influence the composition, abundance, and organization of some ecosystems directly through the burning of the ground vegetation, planting of vegetation for food, medicine, and art (Anderson and Moratto 1996; Kimmerer and Lake 2001; Delcourt et al. 2004; Vale 2013).

The Native American cultures influenced the North American ecosystems for at least the previous 10,000 years. They would have lived with and influenced many kinds of ecosystems including prairies, savannas, woodlands, meadows, wetlands, and dense forests (Ricketts et al. 1999). In the southwestern United States, drought-tolerant and drought-resistant desert plant communities formed, including some shortgrass prairies (Van Devender and Spaulding 1979), which commonly had widely spaced vegetation that was often interspersed with bare rocks or soil.

Prairies formed where ground-fires were frequent, and meadows (not dependent on fire) formed on hillsides and openings of woodlands where woody vegetation would not grow due to excessive wind, salt spray or the persistence of shallow soils upon rocky outcrops or in locations with high elevations where trees and forests won't readily establish (Bailey 1980; Packard and Mutel 1997; Dunwiddie and Bakker 2011).

Expanded immigration and settlement in the nineteenth century, especially after the civil war, profoundly changed North American landscapes west of the 100th meridian. At the beginning of the 1800s, much of the West was in its Holocene condition including ancient anthropogenic influences by Native Americas (MacDougall 2003; Vale 2013). However, the Columbian exchange and the influence of the Spanish settlers had already been long underway as exotic and invasive plants (and some mammals) had already made a presence in parts of the Southwest (Crosby 2003). By the end of the 1800s, many of the prairie ecosystems in the Western United States and Canada were under private ownership. Much of the prairie had been plowed and fenced with barbed wire. Bison and other indigenous ungulates were largely diminished from the land and replaced by exotic livestock (e.g. cattle, pigs, chickens) imported and integrated into a new form of nascent industrial agriculture. This transformation fed a growing human population but it had profound negative environmental effects on the ecosystems (Cochrane 1979; Smith 1992; Ricketts et al. 1999; Goudie 2018).

New settlers in the West possessed little knowledge about prairies, native ecosystems, or how to live with the land (Libecap and Hansen 2002). The Native Americans knew much about the land, rivers, plants, and animals, but there was no written method in their culture that settlers could learn from. In the West, Native Americans already knew how to fish, raise cattle, grow crops, harvest plants and tend orchards, while European settlers learned as from anew, and it would be decades later that agroecology and ecological restoration would become a field of study to be learned, developed, appreciated, and applied (Packard and Mutel 1997; Young et al. 2005; Clewell and Aronson 2013).

As president, Thomas Jefferson (1801–1809) was eager to learn about the West including its native vegetation. Jefferson arranged an excursion to find an overland passage to the Pacific and learn about its native plants and animals. The Lewis and Clark expedition of 1804–1806 opened up a way for the westward expansion of European immigrants to North America and began the scientific exploration of the indigenous plants and animals of the West. Landscape painters visited the West to capture its character before its settlement, as it was unlike any landscape experienced by European immigrants. By the mid-nineteenth century, *plein-air* painters were aware of the changing conditions in the west (plowing, clear-cutting, mining, fur-trapping, etc.) and wanted to capture the essence or character of the West before it became fundamentally altered (Aikin 2000).

In the Willamette Valley in northwestern Oregon for example, over the previous 4000 to perhaps 10,000 years, the native peoples burned the valley grassland vegetation. The composition and character of its terrestrial vegetation were much different compared to the vegetation growing in the Willamette Valley today. The climax (mature) plant communities would have been more abundant and widespread, as barbed-wire fencing, clear-cut forestry, fire suppression, row crop agriculture, and grazing by exotic cattle, were not yet realized. The Native Americans maintained grasslands to encourage the persistence of wild game as a food source, covering their tracks and avoiding confrontations with warring tribes. They also planted native vegetation for food (e.g. camas, berries, vegetables) and cultural uses (Whitlock and Knox 2002).

During the initial settlement of the Oregon Territory by Europeans, writings and artistic renderings of these ecosystems were documented in various forms. The Hudson Bay Company, hired artist Paul Kane, to explore the western frontier, across Canada and the Oregon Territory. In 1847 Kane painted an image of the Willamette Valley near Portland (Fig. 1.1) capturing a landscape structure of open woods,

Fig. 1.1 Detail of *Wilhamet River from a Mountain*, Paul Kane, 1847. This image depicts a matrix of open prairies, woodlands, and forested habitats sprawled across the valley floor and hillsides during summer dormancy. The deliberate burning of the ground vegetation by Native Americans was done to provide open hunting grounds and forage desirable for wild game. The location of the painting is thought to be about 48 kilometers (30 miles) south of Oregon City, Oregon. The southern end of the valley (near Eugene) was thought to contain fewer woodlands and more prairie than shown in this figure (Boag 1992). (Courtesy of the Royal Ontario Museum)

prairies, and forests. The amount of open grassland is significant compared to the amount of natural grassland in the valley today. Although the painting included some obvious embellishments such as the deer, Paul's work was intended to be a true and accurate characterization of the landscapes he explored (Gehmacher 2014). Current research on the soils and other measures supports the idea that the Willamette Valley was historically vegetated with large areas of grasslands and open woods for the past millennia (Hulse and Gregory 2002; Whitlock and Knox 2002).

Many historic images, such as Kane's, were made of the unsettled landscapes of the West during the 1800s in California, Colorado, Texas, Utah, Arizona and, elsewhere. Part II of this book includes case studies with natural ecosystem descriptions, and these include historic paintings from the regions as a way to introduce the character of the ecoregions before the radical change and to demonstrate its structure, vegetation, and/or complexity.

Fig. 1.2 Magenta blooms of shooting star (*Dodecathion pulchellum*) plants grow streamside at the Kingston Prairie Preserve in Stayton, Oregon. It is the largest of the last remaining original prairie fragments in the Willamette Valley. Because of the shallow soils, underlying basalt rock, and wetlands, the land was not deemed suitable for agricultural uses. Thus, the private land trust that maintains the prairie (Greenbelt Land Trust) is preserving about 63 ha (155 acres) of the original prairie. With the study of plants native to grass-based ecosystems, meadow-based green roofs composed of species native to the Willamette Valley prairies might be a viable option for green roofs, even constructed wetland green roofs that treat wastewater (See Chaps. 7, 8 and 10). (Photo: Courtesy of the Greenbelt Land Trust, Corvallis, Oregon)

Unfortunately, the prairies and meadows of the Willamette Valley and western North America today are critically endangered ecosystems due to widespread displacement by ranching, agriculture, and urban development (Samson and Knopf 1996; Ricketts et al. 1999; Shorthouse 2010). Although the contiguous grasslands that once existed across the continent are gone, there are many small prairie remnants (Fig. 1.2) across the West that may serve as habitat templates, and as a way to identify potential native vegetation for green roofs (Lundholm 2006). Prairies, meadows, grasslands, and glades have already inspired planting design for green roofs, (Lundholm 2006; Dvorak and Volder 2010; Sutton et al. 2012) and have inspired a new generation of projects covered in this book.

1.1.2 North American Alpine, Barrens, and Rocky Outcrop Habitats of the West

Green roofs in their modern form present a new kind of niche, based upon the identification of plants and ecosystems with shallow soils such as barrens, rocky glades, alpine habitats, and places where succulents or other shallow-rooted-vegetation grows. As opposed to meadow-based green roofs, shallow extensive green roofs form the widest application to most buildings due to their lighter weight and lower cost. However, these kinds of green roofs generally sustain a less diverse mix of plant types. Due to the shallow soil depths, the top profile remains warmer than soils on the ground and as well, provide less moisture available for plants. This means that vegetation that is selected for shallow green roofs needs to be naturally adapted to similar conditions or have a permanent source of supplemental irrigation (VanWoert et al. 2005).

Mountainous terrain along the Pacific Rim distinctively defines the western edge of North America. A wide variety of ecosystems extends along the predominately east and west-facing mountain ranges. In a hypothetical cross-section from sea level to the continental divide, islands of ecosystems ring the mountains as precipitation generally increases in higher elevations. The Intermountain regions include hot and dry habitats in the valleys and gradual increases in moisture and density of vegetation with increasing elevation. However, where shallow soils persist, succulents and other drought-adapted vegetation grow.

There are at least 50 species of the succulent genera *Sedum* growing in North America, most of them are native to the Western U.S., and the native sedum rarely, if ever, form large masses of groundcover. They grow with other vegetation, as part of a plant community or ecosystem.

Alpine habitats are typically found above the timberline, in the higher elevations of the Cascade, Sierra, and Rocky Mountain ranges (Fig. 1.3). Alpine habitats have short summers, and typically have cooler diurnal summertime air temperatures than in the valleys below. In Europe, the alpine habitats contain several hundred species of *Sedum* and other alpine vegetation that has made a direct translation to green

Fig. 1.3 Sub-alpine habitat in Rocky Mountain National Park. Cushion plants, grasses, and sedges are common as well as forget-me-nots, phlox, daisy, columbine, and many more genera. Some plants growing in high elevation habitats with shallow soils have shallow roots and may grow in a variety of locations, and some have deep roots that find cracks in rocks to retain moisture and cool temperatures. Some alpine plants may not adapt to urban conditions of warmer and lower elevation ecoregions. (Photo: Bruce Dvorak, June 2018)

roofs in the cities of Europe's temperate climate. In western North America, the sometimes-extreme difference in air temperatures between alpine habitats, and the much warmer valleys below, may limit which species may adapt to green roofs. For this reason, research on plants of mountainous ecoregions is needed to better inform which plants might adapt to shallow-depth green roofs in a nearby city.

Coastal barrens and rocky outcrops are other landscape features where plants may be found that are suitable for green roofs with shallow substrates (Fig. 1.4). Coastal landscapes in the west include rocky habitats where vegetation meets the edge of the ocean, bay or, other habitats that have shallow and/or low-nutrient soils. Vegetation found growing in these conditions might adapt to extensive or semi-intensive green roofs. Vegetation growing on barrens can include grasses, succulents, mosses, lichens, and low-growing groundcovers. Stresses to these habitats can include wind, salt spray, and exposure to all elements. Plants naturally adapted to these conditions may be good candidates for green roofs in urban areas (Licht and Lundholm 2006).

Fig. 1.4 Pipers Lagoon on Vancouver Island, British Columbia, is a small peninsula that forms a lagoon. The remnant native Garry oak ecosystems intersect with plant communities that thrive in drought, wind, and nutrient-deprived soil conditions (**a**). Plants such as *Cladonia rangiferina* (**b**), *Sedum spathulifolium,* (**c**), and *Arctostaphylos uva-ursi* (**d**) all grow along the rocky outcrops. See Chaps. 7, 8, 9 and 10 for case studies along, the coastal ranges that make use of plants native to rocky outcrops. (Photos: Bruce Dvorak, August 2018)

1.2 Historic Overview of Vegetated Rooftops

The origins of green roofs began with sod-based green roofs. These developed at least three to four hundred years ago, but perhaps thousands of years ago in Scandinavia (Osmundson 1999; Grant 2006; Grancharov 2013). Practically, these early sod-based green roofs were designed to make use of local vegetation. Indigenous grass-based plant communities were the habitat template for the green roof vegetation (Lundholm 2006; Lundholm and Walker 2018). Sod roofs were planted with vegetation local to the ecoregion, and the roof deck was made from wood and bark from local trees (Grant 2007; Grancharov 2013).

From their earliest origins, green roofs reflected their indigenous landscapes. A green roof was a product of the local natural environment and provided natural insulation from the elements. The building and the land were synonymous with the character of the regional materials. Arable land was cleared of sod strips from a prairie or meadow. Stones from the same field were collected, sorted, and became the foundation of the buildings. Grass sod from a previously undeveloped site was cut and transported to the rooftop. The roofs were sloped, drained to the exterior of the structure, and contained thin layers of local soils clinging to the sod (Gates 1933; Carlson 1981). In temperate climates, precipitation persisted during the summer and the cool temperatures allowed the grass sod on the roofs to become established and remain green. Over time, this method of construction was perfected, but as new building technologies became popular, sod roof construction waned (Grant 2006).

In the twentieth century, grassed roofs were still constructed in Germany, Switzerland, Scandinavia, and England, but the frequency of their deployment was not significant enough to be considered a movement or a premeditated ecological intervention for future cities. The modern green roofs of today emerged out of central Europe, largely in dense cities such as Berlin, and elsewhere in Germany where major post-war reconstruction was taking place (Fig. 1.5), and the consciousness of the emerging environmental movement was gaining traction (Grant 2006; Werthmann 2007).

Beginning in the 1970s in Germany, green roofs became more organized in their design and construction. Green roof guidelines were published regarding their design, construction, and maintenance. Three kinds of green roofs were acknowledged including shallow extensive (succulent and shallow meadow roofs), semi-intensive (grassed roofs), and intensive green roofs or roof gardens (FLL 2008). These guidelines were to accompany the learned knowledge of an expert during the design process (Philippi 2005). The guidelines made allowances for the use of local materials for drainage and growing media, and local vegetation. After decades of trials, failures, and successes, a more reliable approach to green roof design became popular as green roofs became produced at larger scales and by an educated industry. During the second half of the twentieth century, there emerged a growth period for green roofs and continues from the 1990s onward (Koehler and Keeley 2005; Köhler 2006).

Beginning in the 1980s, lightweight green roof systems were developed in Europe. In Germany, extensive type green roofs (shallow lightweight substrates)

Fig. 1.5 Grass and wildflower green roofs located in Berlin (**a**) during the dormant season, and Bonn, Germany (**b**) during the growing season. (Photos: Bruce Dvorak)

were developed to mimic gravel-based and shallow-rooted plant communities found on hillsides such as the Alps. Open-graded lightweight media was developed with guidelines for growing native succulents and herbs. The non-profit German organization Forschungsgesellschaft Landschaftsentwicklung Landschaftsbau e. V. (FLL) developed guidelines for designing, building, and maintaining green roofs. It is a regulatory commission comprising a consortium of industry experts, university researchers, and seasoned professionals (FLL 2008). The *Guidelines for Growing Green Roofs* (FLL) were developed by a triad of researchers, green roof designers, and representatives of those that maintain green roofs. Low-slope roof decks became the standard with the introduction of internal drainage systems (Koehler and Keeley 2005). Most large low-sloped rooftops on buildings shed water not to the parapet at the edge, but in the central direction of the roof, with oversized drains. These drains take the water down through the interior of the building, hidden within the walls or structure of the building, and drain water below grade away from the building. Steeply sloped roofs, however, drain the substrate to the exterior of the building, and no water passes through the building (Weiler and Scholz-Barth 2009).

Because of new building materials and construction techniques developed at the beginning of the twentieth century, fewer buildings were designed to support the weight of floors and rooftops by walls. A new system was developed where steel and concrete beams and columns supported roof decks independent of the walls. Rooftops could now be designed to support minimal weights (Osmundson 1999). Strong roof decks supporting heavy gravel ballasted roofs were no longer necessary, as lightweight single-ply and other waterproofing materials emerged on the market. If green roofs were to be considered for new forms of architectural construction, then they needed to become shallow, lightweight, and to lie on low-sloped, internally drained roofs. The growing medium had to be deprived of silt or fine particulate matter to allow rapid through-flow of precipitation. Natural soils on flat green roofs could create problems such as silting and clogging of internally drained roofs (Dvorak 2011).

Tightly written specifications were developed for growing media for use on buildings with low-slope roofs and internal drainage systems, along with a suite of plants that thrive in those conditions. These early extensive green roofs were often a low-diversity mix of sedums native to the region but later became more diversified to sustain a nutrient balance (Köhler and Poll 2010). Thus, the modern form of extensive green roofs emerged (Fig. 1.6) (Werthmann 2007; Köhler and Poll 2010). Plantings typically consisted of succulents, bulbs, and shallow-rooted herbaceous vegetation native or adapted (naturalized) to the region or nearest mountains. The term "extensive" popularized in Germany intended a widespread application of low-maintenance green roofs covered with native or naturalized (adapted) vegetation (FLL 2008). Green roofs would largely sustain vegetation found in the region with few artificial means, although artificial watering and fertilizing was sometimes needed.

Invariably, vegetated rooftops became a part of the nature in the city, and in the early 2000's researchers in other locations of Europe also made significant contributions to the inclusion of biodiversity on green roofs (Brenneisen 2003; Gedge and

Fig. 1.6 Extensive green roofs in Berlin (**a**) during a rain event, and Frankfurt, Germany (**b**) in the late summer. These shallow type substrates typically maintain a low-diversity mix of sedums, succulents, and herbs native to Germany. (Photos: Bruce Dvorak)

Kadas 2005; Brenneisen 2006). Multiple benefits were recognized and incorporated into local policies such as credit for attenuation of stormwater, the reduction of flooding, and some provision for local plants and wildlife (Köhler et al. 2002; Keeley 2004; Werthmann 2007; Stovin et al. 2013).

1.3 The Growth of the Green Roof Industry in North America

The first wave of ecoregional green roofs in North America occurred during the settlement of the central plains by settlers from Scandinavia. Many plains settlers were not wealthy and the tracts of land they purchased were desolate places with few materials available to build a structure before the development of the railroad (Gates 1933; Dick 1937). Several kinds of sod (grass-covered) structures were built such as wood timber structures with sod roofs, and buildings constructed with sod for walls called a "soddy". These dwellings were constructed of layers of sod cut from the earth at about 10 cm (4 in) thick, dried out, and then stacked up to make walls. Soddies often included penetrations for windows, doors, and vent pipes for a fireplace. These grassed (sod) roofs were constructed of the same prairie vegetation that settlers cut out from the adjacent ground, and removed during the establishment of their dugouts and settlements.

From 1854 to 1890, homesteaders moved from the eastern United States to make settlements in rural lands of Kansas, Nebraska, eastern Oklahoma, and the Dakotas (Fig. 1.7). Many thousands of settlers built sod houses in methods learned from their homeland in Scandinavian countries, primarily Norway and Sweden. Custer County, located in western Nebraska for example, had over 8000 sod houses built in a wide variety of shapes, sizes, and materials (Kampinen 2008). In most cases, sod homes were temporary structures that were used while more permanent structures were

Fig. 1.7 Sod house complex (1887). In the foreground, the primitive sod dugout remains for cold storage, and the more advanced construction sod house with wood framing is in the background. The native prairie is visible behind the background. (Courtesy of PICRYL https://picryl.com/media/peter-m-barnes-dugout-near-clear-creek-custer-county-nebraska)

built; however, sometimes the structures built from sod were permanent and included some lumber, timber, or stone (Kampinen 2008). In Scandinavia, a shallow depth of sod (about 15–20 cm/6 in-8 in deep) and the cool climate would keep the vegetation on the roof green throughout the year. In the Great Plains, however, the sod would experience drought as the heat and lack of rainfall caused plant stress or dormancy.

In Canada, sod houses and structures were also built in the prairies from Saskatchewan to as far north as the Hudson Bay (Lemieux et al. 2011). Native peoples also used sod in some of their structures, but there was no known sharing of construction techniques between European settlers and the Native peoples. In Mexico, adobe houses were constructed, including soil on the roof (Van Wormer 2014). Although ruderal vegetation often invaded adobe roofs, it was intended to be removed. When this failed to happen, the vegetation on the adobe roof became a spontaneous kind of green roof seeded with wind-blown ruderal vegetation (Van Wormer 2014).

In the Great Plains, sod construction ended with the arrival of the railroad. Once stone, timber, and other materials became available for purchase, railroad routes and their supporting infrastructure supplied immigrants with long-term building materials, and vegetated rooftops in rural settings slowly faded into history. Although there are few original surviving structures today, there are reconstructed sod houses and sod roofed structures in various locations in the west that serve as a reminder of the ingenuity and necessity of the people at the time.

Although green roofs today are fundamentally different from these simple structures, today's green roofs still use some rudimentary principles. Sloped roof decks are still in use and make useful associations with many plant types for well-drained green roofs. The low-slope, two-percent (or less) roof deck construction popular in modern architecture makes for slow or poor draining conditions, which can be challenging for the establishment of succulents. Drought-tolerant succulent vegetation evolved on landscapes that typically have well-drained and sloped soils. Perhaps a second takeaway from these roofs is the idea that a roof can serve multiple purposes and functions including the use of native or naturalized vegetation. There are many case studies in Part II of this book that make use of grassed ecosystems, and a few of these replicate construction techniques of Old-World European green roofs. See Chaps. 6 and 10 for examples of Old-World technology still in use.

Far away from the desolate prairies, at the end of the 1800s and into the early 1900s, in urban areas roof gardens were being built in New York, Chicago, Seattle (Fig. 1.8), San Diego, Salt Lake City, Detroit, St. Louis, Pittsburgh and elsewhere. At one time, the United States had perhaps more intensive green roofs (roof gardens) than anywhere else (Osmundson 1999; Jarger 2008). The population of the United States was rapidly growing in its urban and industrialized cities. Much of the new building construction in the New World was innovative in its architecture, and rooftop gardens were becoming popular as places for entertainment and dining

Fig. 1.8 Detail of a historic postcard (ca. 1906) with a depiction of the roof garden built on the Lincoln Hotel in Seattle, Washington, prior to a catastrophic fire on April 7, 1920. This roof garden was built in 1899 and had maintained lawn, flowerbeds, shrubs, trees, and vines. It also had some of the best views from downtown to Elliott Bay, Mount Rainier, and the Olympic Mountains. (Courtesy of The Seattle Public Library, spl_pc_38004)

(Osmundson 1999). Modern air conditioning systems did not yet exist, and rooftops were the place to spend time to keep cool and relax during the summer heat. Rooftops provided the one location in the polluted cities where fresher air and cooler temperatures could be found (Werthmann 2007). Many of the park systems in place today did not exist at that time, so the roof gardens were often the greenest place and were one of the few connections to nature in the city.

Once air conditioning, heating, and ventilation systems became mainstream, rooftop gardens waned, and fewer were built or maintained in the United States. Some architects known to influence and establish the International Style of architecture, such as La Corbusier, explored with extensive type green roofs in Europe; however, the idea of including vegetation on rooftops was not a part of the new International Style of architecture that was popular in the United States. Regarding the modern American roof gardens, several technical issues such as reliable waterproof membranes, drainage, and heavy soils had not yet been refined, and interest in the technology lost its momentum in North America until its return again in the 1960s (isolated applications), and with greater interest begging in the late 1990s when architects began to encourage green roofs on buildings (Osmundson 1999; EarthPledge 2005; Jarger 2008).

In his seminal book *Roof Gardens: History, Design and Construction*, Theodore Osmundson wrote that in the 1960s when roof garden design was beginning to regain some interest, roof garden designers were "flying blind" (pg. 173–174), as there was little technical information available regarding the design of substrates and green roof systems (Osmundson 1999). A reliable design strategy for planting media and drainage materials on green roofs was largely undeveloped with no work or research to be referenced in the United States (Osmundson 1999).

During the re-emergence of the green roof industry in North America in the 2000s, much was borrowed and adapted from the European designs including substrate designs, waterproofing and drainage materials, and many of the same succulent species that grow native to the Alps or Asia. Extensive green roofs were developed across Canada first as an ecological response to urbanism, then later in the United States (Osmundson 1999; Peck et al. 1999). The industry group *Green Roofs for Healthy Cities* was very influential in developing and promoting standard practices, industry, education, and research on green roofs (Peck et al. 1999).

Editor of *Landscape Architecture Magazine*, Bill Thompson wrote, "An opportunity is emerging to introduce landscape architecture into another realm—the roofs of buildings—in a revolutionary way" (Thompson 1998). Bill Thompson was correct; however, many disciplines were involved: policymakers, insurance providers, roofers, horticulturists, ecologists, architects, engineers, biologists, and environmental engineers. The complexity of issues at hand are far-reaching, and integrated thinking was just emerging within the design professions (Van der Ryn and Cowan 2013).

In the United States, leaders in Portland, Oregon, Seattle, Washington, and Vancouver, British Columbia were influential in developing and promoting the extensive type of green roofs. In the Midwestern U.S., major projects such as the Ford Motor Rouge River Plant in Dearborn emerged, near the same time that Mayor

Richard M. Daley was seeking solutions to urban heat islands and visited Europe with his staff and began promoting green roofs in Chicago, Illinois (Laberge 2003). By the early 2000s, major cities across the eastern U.S. and Midwest adopted some form of extensive green roof pilot projects, research sites, and built projects (EarthPledge 2005; Peck 2012). Some exploration with native vegetation took place, but sedums, native or not, emerged as the workhorse of green roofs in the U.S. (Werthmann 2007).

The original connection to native and regional vegetation on green roofs, a standard practice in Europe, did not translate to the early development of green roofs in the North American markets (Luckett 2009; Dvorak and Volder 2010). There were some efforts to investigate native vegetation on green roofs beginning in the early part of the 2000s; however, the sudden interest in green roofs in the Mid-Atlantic coast and the Midwestern United States created immediate demand, and suppliers provided easily accessible materials for green roofs (Monterusso et al. 2005; Licht and Lundholm 2006; Martin and Hinckley 2007). Sedum-based green roofs (with exotics) quickly became in high demand and initially represented extensive green roofs in North America, but there was also a need for more comprehensive research (EarthPledge 2005; Getter and Rowe 2006; Cantor 2008). Exploration or research regarding native vegetation on green roofs was not a high priority for most green roof providers.

The development of sedum roofs, modular green roofs, and sedum-carpet roofs was easy to produce and made for easy installation (Luckett 2009; Peck 2010). Meanwhile, there developed a standard mix of exotic sedums (Table 1.1) that began to populate rooftops across North America, especially east of the 100th meridian (Dvorak and Volder 2010). These species of *Sedum* are hardy, adaptable, and perform well on extensive green roofs in the central U.S. and east coast where irrigation may be optional. These exotic plants became successful on green roofs in North America, and their uniform appearance helped to develop an image of a green roof as a "carpet" of green that connected with urban Americans (Jungels et al. 2013).

Table 1.1 Popular species of *Sedum* used on extensive green roofs east of the 100th meridian in North America

Species of *Sedum*	Nativity
S. acre	Europe, western and northern Asia
S. album	Europe, western and northern Asia
S. floriferum	Europe, western and northern Asia
S. kamtschaticum	Mongolia
S. hybridum 'Immergrunchen'	Siberia and Mongolia
S. 'Matrona'	Japan
S. middendorffianum var. *diffusum*	Siberia, China, Japan
S. reflexum or *S. rupestre*	Central and Western Europe
S. sexangulare	Central Europe
S. sieboldii	Japan
S. spurium	Armenia and Iran

Mainstream media outlets (television, magazines, online media) in the U.S. began covering green roofs and helped form public opinions, industry standards and aesthetic expectations as sedum roofs have been perceived more favorably over grassed roofs in some parts of the east coast (Peck 2008; Jungels et al. 2013).

The Stonecrop family (Crassulaceae) has at least 1500 species known worldwide. Many of these plants are native to temperate climates in the Northern Hemisphere, are adapted to well-drained low nutrient soils, and are drought- tolerant. Due to their tolerance of so many of these conditions often present on green roofs, *Sedums* are held in high regard and are often an excellent plant choice for green roofs, in the right climate, and roof condition. According to the USDA PLANT Database, there are at least 34 species of the genus *Sedum* that are native to North America (the US and Canada), and there are at least 11 *Sedum* species that have been introduced and have adapted (naturalized) in the landscape (USDA 2007). Some of these are frequently used on green roofs (Table 1.2). However, when subspecies of *Sedum* are included, there are at least 61 species and subspecies of *Sedum* native or introduced to North America (Table 1.2 and Table 1.3).

West of the 100th meridian, however; the semi-arid climates with long hot summers, and extreme surface and air temperatures are not a good match for *Sedums* from Europe's cool and temperate climates if irrigation is not used. Unpublished trials of green roofs in the western U.S. originally planted with some of these *Sedum* species proved unsuccessful. During the initial growth of the green roof industry in the western U.S., there was little research regarding plant viability in most ecological regions of North America. The central states of the U.S. and Pacific Northwest led with green roof plant research between 2000 and 2010; however, during that time there was limited research taking place in the semi-arid and arid west and southwest (Dvorak and Volder 2010).

North America has rich potential for the exploration of plants and or plant communities that could translate onto green roofs. Over 20,000 species of plants are native or naturalized in North America, and although many of them may not be

Table 1.2 Exotic species of *Sedum* that have been introduced (I) to North America, and have become adapted or naturalized on the land (USDA 2007). Bolded species are commonly used on green roofs in North America (Dvorak and Volder 2010)

	Botanical name	Status
1	***Sedum* acre L.**	I
2	***Sedum album* L.**	I
3	*Sedum dendroideum* Moc. & Sessé ex A. DC.	I
4	***Sedum diffusum* S. Watson**	I
5	*Sedum hispanicum* L.	I
6	*Sedum lineare* Thunb.	I
7	*Sedum mexicanum* Britton	I
8	***Sedum reflexum* L.**	I
9	***Sedum sexangulare* L.**	I
10	*Sedum sarmentosum* Bunge	I
11	***Sedum stoloniferum* S.G. Gmel.**	I

Table 1.3 Species and subspecies of *Sedum* native to North America (the U.S.A. and Canada). Bolded species have been trialed on some green roofs in North America (USDA PLANTS Database

1	*Sedum albomarginatum* R.T. Clausen	27	***Sedum oreganum*** **Nutt.**
2	*Sedum annuum* L.	28	*Sedum oreganum* Nutt. ssp. *oreganum*
3	*Sedum borschii* (R.T. Clausen) R.T. Clausen	29	*Sedum oreganum* Nutt. ssp. *tenue* R.T. Clausen
4	*Sedum citrinum* Zika	30	*Sedum oregonense* (S. Watson) M. Peck
5	*Sedum cockerellii* Britton	31	*Sedum paradisum* (Denton) Denton
6	*Sedum dasyphyllum* L.	32	***Sedum pulchellum*** Michx.
7	*Sedum debile* S. Watson	33	*Sedum pusillum* Michx.
8	***Sedum divergens*** S. Watson	34	*Sedum radiatum* S. Watson
9	*Sedum douglasii* Hook.	35	*Sedum radiatum* S. Watson ssp. *ciliosum* (Howell) R.T. Clausen
10	*Sedum eastwoodiae* (Britton) A. Berger	36	*Sedum radiatum* S. Watson ssp. *depauperatum* R.T. Clausen
11	*Sedum glaucophyllum* R.T. Clausen	37	*Sedum radiatum* S. Watson ssp. *radiatum*
12	*Sedum havardii* Rose	38	*Sedum rupicola* G.N. Jones
13	***Sedum lanceolatum*** **Torr.**	39	***Sedum spathulifolium*** Hook.
14	*Sedum lanceolatum* Torr. ssp. *lanceolatum nesioticum*	40	*Sedum spathulifolium* Hook. ssp. *pruinosum* (Britton) R.T. Clausen & Uhl
15	*Sedum lanceolatum* Torr. ssp. *subalpinum* (Blank.) R.T. Clausen	41	*Sedum spathulifolium* Hook. ssp. *purdyi* (Jeps.) R.T. Clausen
16	*Sedum lanceolatum* Torr. var. *rupicola* (G.N. Jones) C.L. Hitchc.	42	*Sedum spathulifolium* Hook. ssp. *spathulifolium*
17	*Sedum laxum* (Britton) A. Berger	43	*Sedum spathulifolium* Hook. ssp. *yosemitense* (Britton) R.T. Clausen
18	*Sedum laxum* (Britton) A. Berger ssp. *laxum*	44	*Sedum stenopetalum* Pursh
19	*Sedum leibergii* Britton	45	*Sedum stenopetalum* Pursh ssp. *monanthum* (Suksd.) R.T. Clausen
20	***Sedum moranense*** Kunth	46	*Sedum stenopetalum* Pursh ssp. *stenopetalum*
21	*Sedum moranii* R.T. Clausen	47	***Sedum ternatum*** Michx.
22	*Sedum nanifolium* Fröd.	48	*Sedum valens* Björk
23	*Sedum nevii* A. Gray	49	*Sedum villosum* L.
24	*Sedum nuttallianum* Raf.	50	*Sedum wrightii* A. Gray
25	*Sedum obtusatum* A. Gray		
26	*Sedum obtusatum* A. Gray ssp. *retusum* (Rose) R.T. Clausen		

appropriate for green roofs, few have been explored in long-term green roof trials with various ecological goals in mind (Dvorak and Volder 2010). A great diversity of ecoregions occur west of the 100th meridian and the green roof industry is just beginning to explore vegetation from those local ecoregions for potential use on green roofs. In Europe, green roofs make use of many forms and functions of plants including grasses, herbaceous perennials, annuals, succulents, sub-shrubs, bulbs, and corms. They also make use of native wetland vegetation to treat reclaimed water and perform like wetlands. This same broad spectrum of different forms and functions for plants is missing in the North American green roof market. Part II of this book begins to explore some of these ideas taking place in the West.

Rhodiola L. is another genus in the family Crassulaceae and has similar characteristics to *Sedum*. There are eight species of *Rhodiola* in North America, and little is known about their testing or application on green roofs. In Arizona, the southwest deserts have over 2000 species of plants native to the Sonoran Desert. There are over 75 genera and 800 species of plants native to the Anza Borrego Desert. California has over 7000 species of vegetation. There are over 500 species of plants native to the North America prairies. As there are at least 20,000 species of vascular plants native to North America, few have been tested on North American green roofs (Dvorak and Volder 2010). With the high temperatures and unreliable rainfall in the western United States and Canada, there is much room for exploration of new assemblages of plants and plant communities on green roofs.

1.4 The 100th Meridian

As a geologist, explorer, and researcher, John Wesley Powell advocated for the conservation of water and landscapes of the arid West of the United States. He worked hard to distinguish the importance of the 100th meridian as a gateway to the arid West and to disseminate knowledge about the lack of water available for domestic and agricultural uses. He informed the government that it should not encourage the rapid settlement of the west. Although the 100th meridian is only a line on a map, in the physical landscape it is that transition point where land east of the 100th is "humid" and land west of it is "dry" (Fig. 1.9) (Seager et al. 2018).

Powell lived his adult life on a mission to make known in Washington, D.C. these facts about land west of the 100th meridian, and their implications for the oncoming settlement (Powell 1879). Permanent settlements would need to address the fact that west of the 100th meridian, precipitation dropped off significantly from over 1000 mm (40 in) per year east of the 100th meridian to less than 500 mm (20 in) per year or less in much of the West. Thus, Powell argued that the climate, vegetation, and lack of consistent rainfall would become major issues for those making plans to settle the west (Powell 1879). The Great Plains and Desert Southwest were unlike any ecoregion that settlers from the cool temperate climates of Europe and the humid eastern U.S. had experienced. Although Powell delivered his message with passion and supported it with evidence from scientific research, he was unsuccessful in efforts to convince decision makers in Washington, D.C. The political motivation to settle the West quickly became official with the Homestead Act, beckoning thousands of Americans and European immigrants to settle the arid West (Seager et al. 2018). Many, if not most, of the immigrants, had little knowledge of how to farm, manage, or live on arid lands. Hence, the widespread building of reservoirs occurred across the West (Reisner 1993).

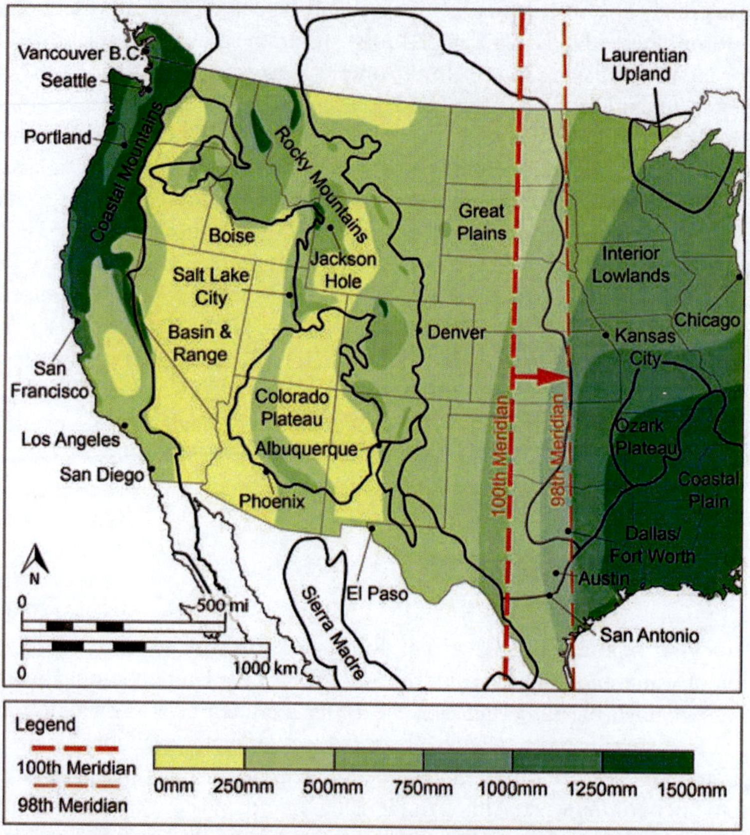

Source: National Oceanic and Atmospheric Association, National Climate Data Center

Fig. 1.9 Precipitation gradients at the 100th meridian and beyond. The red dashed line shows the location of the 100th meridian, as a transitional line between the humid east and semi-arid west. However, the highlighted vertical band east of the red lines show where there has been a shift in that arid/humid line since 1980 about 160 km (100 miles) to near the 98th meridian due to environmental impacts of human development (Seager et al. 2018). The bold black lines delineate physiographic region boundaries. (Graphic by Trevor Maciejewski, Tess Menotti, & Bruce Dvorak)

The Great Plains were settled, plowed, and populated in the late nineteenth century and early twentieth century. During the 1920s precipitation was abundant in the Great Plains and populations increased as the government encouraged the planting of crops such as wheat and corn. Then, in the mid-1930s, the kind of disaster that Powell had warned about hit the Great Plains. A series of droughts occurred in the Great Plains on and off for up to 8 years in some regions, and massive dust storms resulted from poor land management practices and the lack of conservation practices as topsoil blew off and over the Midwest. Dust storms could last up to a week or more, and dust even traveled as far as the East Coast and beyond. The Dust Bowl was one of the most catastrophic human-made environmental and economic disruptions on earth, and recent droughts in the 2000s indicate that the region remains vulnerable in the face of unsustainable management practices (Hornbeck 2012; McLeman et al. 2014). The primary causes of the Dust Bowl were the widespread uprooting of the prairies, poor land management practices, a lack of knowledge of prairie ecology, natural climate cycles, and drought (Libecap and Hansen 2002). Today, dust storms continue in the west, and permanent damage exists in some regions where natural vegetation is unable to establish due to loss of soil and salinization of the soil due to over-irrigation (Bolles et al. 2017; Tong et al. 2017). According to the Natural Resources Conservation Service, wind erosion remains the major source of sediment.

One may wonder if Powell and other voices of conservation at that time would have been more politically influential, perhaps the settlement and development of the West would have been more sustainable, and disasters avoided (Reisner 1993; Groenfeldt 2019). Combined with erosion and drought, availability of surface and groundwater west of the 100th Meridian remains a major issue, and will likely continue to be more so in the future. Many of the individuals who benefited economically from the land rush eventually experienced massive and long-term economic, environmental, and cultural losses. First World cultures and economies, therefore, need leaders that can demonstrate how conservation, preservation, and resilience are critical concepts to the development of the land, including cities (Beatley and Manning 1997; Berke and Conroy 2000; Jepson Jr. 2004; Groenfeldt 2019).

Scientists have demonstrated how the climate in North America has changed (albeit slowly) many times over millennia, but also how human influence has now accelerated these changes through the widespread use of unsustainable sources of water, energy, deforestation, and urbanization (Arnfield 2003; Cook et al. 2013; Tubiello et al. 2015). Planners, landscape architects, architects, developers, elected officials, and ethicists, must be challenged to respond to make way for an economy that counts the environmental and social costs of sterile forms of urbanization that were founded upon ideals generated from the Industrial Revolution (Bunker et al. 2007; Ahern 2013). They have an opportunity to prepare a way forward that is supported by research and the inclusion of resilient urban design that understands and respects ecosystem services as an integral part of green infrastructure including green roofs (Tzoulas et al. 2007; Pereira et al. 2010; Kowarik 2011; Weber 2013; Williams et al. 2014).

Precipitation east of the 100th meridian is more than 1000 mm (40 in) annually, and precipitation west of the line drops off to 500 mm (20 in) annually or less. In some parts of the Pacific Northwest annual precipitation is over 1500 mm (60 in); however, much of the precipitation occurs during the fall, winter, and spring, and summers can be very dry. Mountains in the interior montane regions form precipitation islands. The prevailing direction of the Jetstream from west to the east creates a moisture gradient from wet at the west coast mountains transitioning to dry at the mountain interior valleys and Great Plains, and wetter again near the 100th meridian where moist air from the Gulf of Mexico mixes with the Jetstream. Vegetation east of the 100th meridian generally receives consistent moisture throughout the growing season.

West of the 100th meridian in the interior mountain and desert southwest, many plants adapt to the arid conditions and high summer temperatures by retaining water in leaves and roots, or with summer dormancy. Much of the interior west and southwest receives its precipitation during the winter or intermittent events during the summer. Monsoon rains typically take place during July and August. Monsoons are regional events that produce intense but short-duration rain events across limited areas. Other factors also influence the distribution of vegetation west of the 100th meridian including wind, diurnal, and seasonal temperature fluctuations, reduced humidity, and heightened evapotranspiration (Lambrinos 2015).

Dramatic transitions in vegetative type and cover are visible on the landscape from near and beyond the 100th meridian. The natural range of the Tallgrass Prairie precedes the 100th meridian as it once stretched from San Antonio, Texas, in the south to Manitoba, Canada, in the north, and east to Chicago. Pioneers exploring the region were known to say that in some places the grass was as tall as a man on horseback (Gates 1933). West of the 100th meridian, the grasslands became progressively shorter in stature. The mixed-grass prairies extend beyond the 100th meridian but fade to the knee-high short grass prairies that extend to lower elevations of the Rocky Mountains and the interior west. Historically, in the lower elevations, semi-arid grasslands were dotted with many perennial and annual species as well as succulents (*Opuntia spp.*), and an abundance of low growing drought-tolerant woody vegetation such as sagebrush (*Artemisia tridentata)* and rabbitbrush (*Ericameria nauseosa)* (Gasch et al. 2016). Grasslands were the defining and prevailing ecosystem type across much of the land west of the 100th meridian, before the 1800s. (Vale 2013).

Much of the remainder of this book presents discussion and case studies that demonstrate how native and adapted (naturalized) vegetation can be used on green roofs in dry or drought-prone urban landscapes. The 100th meridian is a guideline for the application of theories and concepts presented in this book; however, any landscape where grasslands, meadows, glades, barrens, or rocky outcrops exist may find relevance. Another application is to locations where bi-modal or summer droughts are normal or frequent such as in the Pacific Northwest.

1.5 Green Roof Design Versus Reality

Green roofs are a simple idea. Their application, however, in cities where long summers and drought are frequent can be riddled with complex conditions that intertwine with urban policies and practices. Unlike planning a garden on the ground, there are important criteria that must be addressed on rooftops. Ignoring these critical criteria can result in the failure of plants, the roof deck, or the green roof system (Snodgrass and McIntyre 2010). Several of the critical criteria include working within the weight limits of the roof deck, designing a roof deck slope and drainage system, waterproofing, addressing soil or substrate stability, and nutrient requirements (Weiler and Scholz-Barth 2009). The selected plant material must adapt to the conditions of the green roof, or the green roof must be designed to support the intended vegetation (Snodgrass and Snodgrass 2006; Weiler and Scholz-Barth 2009). The persistence of a well-designed green roof often means that there is a "champion" or designated person that ensures the regular monitoring of the green roof. All green roofs, even those designed for low inputs, need some attention (Snodgrass and McIntyre 2010).

During the research and data gathering phase of this book, I visited 140 green roofs west of the 100th meridian. Many of the green roofs I visited had been re-planted at some point in time. Some green roofs needed only minor adjustments to the vegetation, while others were completely replanted. Some green roofs were not re-planted and were abandoned by a new and reluctant building owner. One of the most common reasons for replanting was due to either too little or too much irrigation water. I spoke with a number of the designers of green roofs and found that it was very common that the client or owner of the building that initiated the green roof no longer owned the building and passed the care of the green roof along to the care of others. It is these second, third, or fourth building owners that become the decision-makers of how and what happens to the green roof. My observation is that green roofs that are designed with attention to future building managers are the green roofs that persist over time. Contemplation of all of these observations and conservations led to a personal revelation: *green roofs should be designed with negligence in mind, but green roofs should not be neglected!*

My experiences as a green roof designer are shared here to inform about how ecoregional green roofs can be conceived of for urban rooftops with native vegetation, and I will begin to discuss some of the issues regarding design intentions and the sometimes unexpected requirements to maintain green roofs. Regarding vegetation, design intentions sometimes result in a different reality. There is no single way to design a green roof, but there are similar concepts shared across methods and means. The following three examples represent situations where there was little or no precedence (previous examples) from which to learn. The design, construction, and on-going maintenance of these projects informed my learning about vegetation for green roofs, green roofs as ecosystems, and the use of native vegetation on green roofs.

1.5.1 Chicago City Hall Urban Heat Island Initiative

Mayor Richard M. Daley (former mayor of Chicago) was the driver and motivation for a green roof on top of Chicago City Hall. Mayor Daley wanted to establish a green roof pilot project on the city hall building to test vegetation and learn about how green roofs might mitigate urban heat islands in Chicago (Laberge 2003). Weston Engineering was the prime consultant with the City of Chicago. Conservation Design Forum (CDF) was a member of a multidisciplinary team hired to design the green roof. I was employed at CDF and was the project manager and David Yocca was the project Principal. The Chicago City Hall green roofs were installed during

2000 and 2001 and were inspired by the natural environments once common throughout Chicago, the Tallgrass Prairie.

The concept for the Chicago City Hall green roof was to provide a setting where City staff could explore the use of native and adaptable non-native species (Laberge 2003). Since there was no green roof research in the region at the time, there was a great need for exploration of different forms of vegetation, varieties of species, and their adaption to different depths of green roof substrate. All of these variables came together in the design of the green roof. Roofscapes was the prime contractor responsible for detailing and installing the green roof system (Dvorak 2009).

Gravel hill prairies were once frequent in the Chicago region, before its settlement in the 1800s. A gravel hill prairie is a gently rolling hillside landscape feature, a remnant of glaciation (Fig. 1.10). Gravel hill prairies typically contain soils with a high percentage of gravel, sand, and rock, and are vegetated with drought-tolerant prairie vegetation. This plant community became a model (habitat template) for the Chicago City Hall green roof. The initial planting design for the Chicago City Hall green roof had plants disbursed in a sunburst pattern with plants grouped in color themes (Fig. 1.11). There was no intention for the pattern to be maintained over time, but it was a device to disburse an array of plants that may adapt to any specific location on the roof (Dvorak 2009).

Fig. 1.10 The Nachusa Grasslands is a 1537-ha (3800-acre) prairie preserve in north-central Illinois and is one of the largest Tallgrass prairie preserves in the state that is open to the public. This image shows one of many gravel hills on the property and has vegetation that was used on the Chicago City Hall green roof. The hillside has variations of solar aspect, slope, and exposure to wind and other elements. The artificial green roof mounds on Chicago City Hall were planted with prairie species, to see which plants might adapt. (Photo: Bruce Dvorak)

Fig. 1.11 Sunburst pattern can be seen in the white spray marks on top of the wind protection blanket. The biodegradable wind protection blanket (black fabric) was used to aide retention of the growing media, as the vegetation was immature. Shallow growing media depths prevent large-sized (mature) plants from being installed. (Photo: Bruce Dvorak, 2001)

The prairie vegetation adapted well to the extensive and semi-intensive substrates, both with and without irrigation. The sunburst pattern is long gone, and vegetation has self-selected according to microclimates on the roof (Fig. 1.12). Succulents have persisted in the shallowest well-drained and south-facing substrates, and herbs and grasses persist on the flat roof decks with slow draining substrates. The green roof receives maintenance each spring, summer, and fall. To avoid a build-up of thatch on the green roof, vegetation is cut down and hauled off to local compost processing facilities (Dvorak and Carroll 2008).

1.5.2 *Peggy Notebaert Nature Museum*

Chicago's Peggy Notebaert Nature Museum is located on the shores of Lake Michigan, at the intersection with the last remaining oak savanna dune in the city. The museum is known for a nationally recognized collection of live butterflies. Several years after it opened, the director and board members of the museum had the vision to develop a publicly accessible green roof on the nature museum. They determined that the Nature Museum green roofs should be visually and functionally different compared to the green roofs on the Chicago City Hall. Where Chicago

Fig. 1.12 Chicago City Hall prairie vegetation shown in the second growing season (June 2002). Many ecological characteristics of tallgrass prairie are present and seen above the soil line. City staff has periodically tested the growth media for nutrient loads. The natural decomposition taking place on the green roof has been sufficient, so staff has not fertilized the green roof. (Photo: Bruce Dvorak, July 2004)

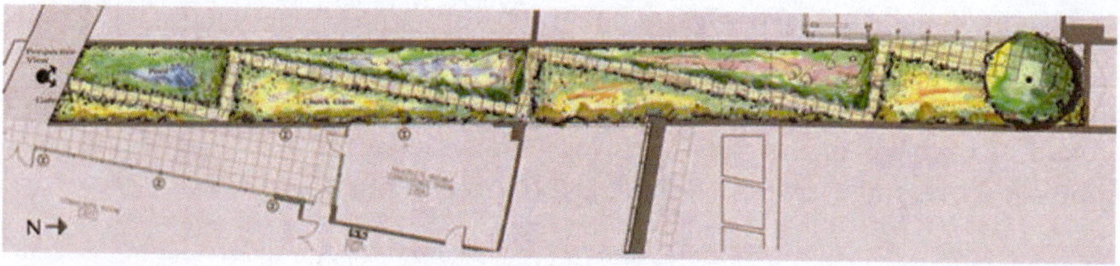

Fig. 1.13 Master plan for the green roof on the Peggy Notebaert Nature Museum. The zigzagged maintenance path was used as an organizing device to define different types of prairie vegetation. A wetland green roof is located to the left, shallow extensive prairie green roof in the middle section, semi-intensive green roof towards the right, and the last section is an intensive green roof with a tree. The depth of the substrate follows the allowable structural loads of the pre-existing roof deck. (Courtesy of Conservation Design Forum)

City Hall is a rectangular building with ample space with interior green roof habitats, the Peggy Notebaert Nature Museum building is a linear, blocky, and post-modern building. Where the Chicago City Hall green roofs are not accessible to the public (elevators do not access the roof), the board wanted the nature museum green roofs to be accessible and visible to the public. The green roof is located on the roof deck of an interior hallway, and is visible on the second floor, and is accessible for viewing. Together with colleagues at CDF, we developed the concept for the Nature Museum green roofs (Fig. 1.13). The green roof would be long and narrow with ascending depths of the substrate. The roof deck could support an extensive roof near the visible areas, and a semi-intensive green roof near the middle of the roof and an intensive green roof with a small tree at the far side (north) of the green roof. A narrow maintenance access path is the organizing element. The path is needed to allow infrequent access to the green roof and the upper museum rooftops (Dvorak 2003).

Near the public viewing area, some visual interest was needed. Discussions lead to the development of a wetland roof, with a small pool with recirculating water. Wetland green roofs are more frequent in Europe, and their contribution to the function of urban spaces is potentially significant. Phytoremediation is a process of removing nutrients or elements in water with plants. There are more than a few industrial uses of wetland green roofs to clean and improve water quality before discharge (EarthPledge 2005). Few wetland green roofs exist in North America, but it was known that the potential could be worth investigating. The first zone of the green roof was designed to retain water to a depth of up to 10 cm (4 in). Wetland vegetation native to the Chicago region was selected and planted on the roof and Chicago's first wetland roof was installed (Fig. 1.14, Fig. 1.15).

Fig. 1.14 Image shows how the roof drain system was elevated with a white PVC pipe to allow ponding of water in the substrate to a depth of about 10 cm (4 in). (Photo: Bruce Dvorak)

Fig. 1.15 The wetland habitat green roof located at the main viewing area at the Peggy Notebaert Nature Museum. (Photo: Courtesy of Conservation Design Forum)

1.5.3 Texas A&M University

As a landscape architect and a green roof designer that was in private practice in Chicago, when I relocated to central Texas, I found my transition from professional practice to green roof researcher and faculty member at Texas A&M University to be a challenge. Although I collaborated on the set up of green roof research in Chicago at the Notebaert Nature Museum and the Conservation Design Forum building in Elmhurst, Illinois, the climate and vegetation in east-central Texas were radically different from Chicago (Dvorak 2004; Dvorak and Volder 2013). No previous publications regarding plants existed for green roofs in Texas, and I was not yet familiar with the many ecoregions of Texas. In 2009, I first began working with native and exotic plants, to find vegetation that would grow without irrigation, since there was no water available on the roof, which served as my research facility (Dvorak and Volder 2013). By 2015, work with my colleagues had expanded from 12 green roof modules to include a larger roof area of 74 m^2 (800 ft^2) with irrigation.

After 3 years of exploring a single approach on two roofs, where we mixed succulents and herbs, we observed that we were either over-watering for succulents or under-watering for herbaceous vegetation. We then sorted modules into two separate gardens, based upon their moisture needs. The succulent plants were grown with minimal irrigation (few times a summer), and the herbaceous plot was irrigated every day the first year, then the next year we reduced irrigation to once every 4 days for about 20 minutes. We found that watering every day encouraged the rapid growth of unwanted volunteer plants such as bahiagrass (*Paspalum notatum*). Reducing irrigation to every 3 or 4 days eliminated bahiagrass as a dominant weed. Thus, two low-input approaches developed for two different kinds of vegetation on the research green roofs: a Succulent Roof and a Prairie Roof. Although exotic plants were already planted in each of the two green roofs, a new focus on vegetation from the ecoregion and adjacent ecoregions guided plant selection to favor natives. There was no effort to make a pristine habitat, as the goal was to explore taxa that may potentially adapt to the 10 cm-deep (4.5-in) green roof modules.

Succulent Roof

This plot is planted with succulents that are native to central or western Texas, boarding states, or to northern Mexico (Fig. 1.16). Plants trialed at Texas A&M University include *Agave colorata* x *parryi*, *Euphorbia antisyphilitica*, *Graptopetalum paraguayense* (native to Tamaulipas, Mexico), *Hesperaloe parviflora* (species, and the cultivar 'Brakelights'), *Hesperaloe parviflora* 'Perpa',

Fig. 1.16 The Succulent Roof shown here with vegetation native to Texas, New Mexico, Louisiana, Mexico, and southern Africa. This low-input roof receives water once or twice a year during extended periods of drought and high daytime temperatures. The research goal is to explore plants that are adapted to extensive green roofs with a low-watering approach (Photo: Bruce Dvorak)

Fig. 1.17 Students and faculty participated in the replanting of the original mix of succulents and herbs in the Prairie Roof. Some succulents remain, but many individual pre-grown plants were purchased and installed. Some seed was installed, but much of the seed did not establish. The reduction of irrigation eliminated bahiagrass as a dominant weed, but the drier environment allowed other invasive vegetation to appear. Our practice is to watch invasive species to see which ones may become aggressive, versus those that persist but are not aggressive. (Photo: Bruce Dvorak)

Opuntia cacanapa 'Ellisiana', *Talinum calycinum*, *Yucca louisianensis* 'Granted Wish', *Yucca filamentosa* 'Wilder's Wonderful', *Yucca filamentosa* (native to west Texas, New Mexico, and Louisiana).

Prairie Roof
This plot is planted with herbaceous grasses and forbs native to the Tallgrass Prairie that was once widespread in central Texas (Fig. 1.17). Plants were either installed as pre-grown plugs, or volunteers (self-seeded) (Table 1.4).

1.6 Summary and Overview

This chapter provided an overview of the development of ecoregions west of the 100th meridian, green roofs, and the long-term role of human culture in the alteration of plant communities in North America. An overview of the history of green roofs in Europe and North America outlines a long-term utilization of plants from

Table 1.4 Common and botanical names of 42 species observed on the Prairie Roof at Texas A&M University. Bolded vegetation indicates self-seeded. * = species volunteered in the yard of my residence and was transplanted to the green roof to see if they will survive with a low-watering approach. *Polypremum procumbens* has survived on the green roof 3 years without watering. *Habranthus tubispathus survived and bloomed the first year. Due to its erratic and short bloom time, it is not known if it has survived*

Common name	Botanical name
Common yarrow	*Achillea millefolium*
Autumn onion	*Allium stellatum*
Purple threeawn	*Aristida purpurea*
Butterfly milkweed	*Asclepias tuberosa*
Southern annual saltmarsh aster	***Aster exilis***
Flaxleaf whitetop aster	*Aster linearifolius*
Aromatic aster	*Aster oblongifolius*
Rooseveltweed	***Baccharis neglecta***
Blue wild indigo	*Baptisia australis*
Blue grama	*Bouteloua gracilis*
Bush's poppymallow	*Callirhoe bushii*
Scarlet Indian paintbrush	*Castilleja coccinea*
Small eyebane	***Chamaesyce nutans***
Canadian horseweed	***Conyza canadensis***
Largeflower tickseed	*Coreopsis grandiflora* 'Leading Lady Lauren'
Prairie lily	*Cooperia pedunculata*
Eastern purple coneflower	*Echinacea purpurea*
Button eryngo	*Eryngium yuccifolium*
Common boneset	*Eupatorium perfoliatum*
Graceful sandmat	***Euphorbia hypericifolia***
Indian blanket	*Gaillardia pulchella*
Rio Grande copperlily*	*Habranthus tubispathus*
Sneezeweed*	*Helenium amarum*
Zigzag iris	*Iris brevicaulis* Raf.
Texas barometer bush	*Leucophyllum frutescens*
Texas lupine	*Lupinus texensis*
Plains blackfoot	*Melampodium leucanthum*
Wild mint	*Mentha arvensis*
Carolina bristlemallow	***Modiola caroliniana***
Mexican feather grass	*Nassella (Stipa) tenuissima*
Pinkladies	*Oenothera speciose*
Slender yellow woodsorrel	***Oxalis dilleniid***
Common yellow oxalis	***Oxalis stricta***
Bahiagrass	***Paspalum notatum***
Turkey tangle fogfruit	*Phyla nodiflora*
Juniper leaf/rust weed*	*Polypremum procumbens*
Autumn sage	*Salvia greggii* 'Texas Wedding'
Canada goldenrod	*Solidago canadensis*
Indiangrass	*Sorghastrum nutans*
Largeflower fameflower	*Talinum calycinum*
Bluejacket	*Tradescantia ohiensis*
Texas vervain	***Verbena halei***

ecoregions on green roofs. The North American green roof movement from the early 2000s introduces a new form of green roof with the adoption of "exotic" non-native vegetation in urban regions. This new approach may have some positive ecosystem services; but they may not function to serve the historic ecosystems, plants, and animals of a region. Some early trials making use of native plants on green roofs reveal that native plants can be used on green roofs in North America, but there may be a need for more research and development of green roofs that make use of native vegetation.

Next, Chap. 2 covers details regarding the conceptual development of ecoregions, and ecological elements of green roofs and natural environments. Part II of this book covers case studies of green roofs west of the 100th meridian to demonstrate examples of vegetation from the ecoregions and green roofs that make use of vegetation from ecoregions. Chapter 3 covers green roofs found in the Tallgrass and Mixed-grass prairie ecoregions. Chapter 4 covers green roofs and vegetation of the Shortgrass prairie ecoregions. Chapter 5 covers green roofs and plant communities and vegetation of the Desert Southwest ecoregions. Chapter 6 covers green roofs and vegetation of the Intermontane ecoregions. Chapter 7 covers green roofs and vegetation of the California coastal ecoregions. Chapter 8 covers green roofs and vegetation of the Puget Lowlands ecoregions, and Chap. 9 covers green roofs and vegetation of the Willamette Valley ecoregions. Chapter 10 covers green roofs and vegetation growing in the Fraser Delta and Vancouver Island ecoregions. Part III consists of Chap. 11, a summary of observations from Part II, and future outlooks.

References

Ahern J (2013) Urban landscape sustainability and resilience: the promise and challenges of integrating ecology with urban planning and design. Landsc Ecol 28(6):1203–1212

Aikin RC (2000) Paintings of manifest destiny: mapping the nation. Am Art 14(3):79–89

Anderson MK, Moratto MJ (1996) Native American land-use practices and ecological impacts. In: Sierra Nevada ecosystem project: final report to Congress, 1996. University of California, Centers for Water and Wildland Resources Davis, pp 187–206

Arnfield AJ (2003) Two decades of urban climate research: a review of turbulence, exchanges of energy and water, and the urban heat island. Int J Climatol 23(1):1–26

Bailey RG (1980) Description of the ecoregions of the United States, vol 1391. US Department of Agriculture, Forest Service, Washington, DC

Bailey RG (2004) Identifying ecoregion boundaries. Environ Manag 34(1):S14–S26

Bailey RG, Cushwa CT (1981) Ecoregions of North America: after the classification of J.M. Crowley. UNT Libraries Government Documents Department, Fort Collins, Colorado

Beatley T, Manning K (1997) The ecology of place: planning for environment, economy, and community. Island Press, Washington, DC

Berke PR, Conroy MM (2000) Are we planning for sustainable development? An evaluation of 30 comprehensive plans. J Am Plan Assoc 66(1):21–33

Berry MM (2013) Thinking like a city: grounding social-ecological resilience in an urban land ethic. Idaho Law Rev 50:117

Boag PG (1992) Environment and experience: settlement culture in nineteenth-century Oregon. University of California Press Berkeley, Berkeley

Bolles K, Forman SL, Sweeney M (2017) Eolian processes and heterogeneous dust emissivity during the 1930s dust bowl drought and implications for projected 21st-century megadroughts. The Holocene 27(10):1578–1588

Brenneisen S (2003) The benefits of biodiversity from green roofs: Key design consequences. Paper presented at the 1st North American green roof conference: Greening rooftops for sustainable communities, Chicago, Illinois, 29–30 May

Brenneisen S (2006) Space for urban wildlife: designing green roofs as habitats in Switzerland. Urban Habitat 4(1):27–36

Brubaker LB (1988) Vegetation history and anticipating future vegetation change. Ecosystem management for parks and wilderness, pp 41–61

Bunker SG, Fisher WH, Gasson RA, Giljum S, Hakansson NT, Heyman J, Hughes JD, Jorgenson AK, Marks RB, Moore JW (2007) Rethinking environmental history: world-system history and global environmental change. Rowman Altamira, Lanham

Cantor S (2008) Green roofs in sustainable landscape design. Norton, New York

Carlson AW (1981) German-Russian houses in western North Dakota. Pioneer Am 13(2):49–60

Clewell AF, Aronson J (2013) Ecological restoration: principles, values, and structure of an emerging profession. Island Press, Washington, DC

Cochrane WW (1979) The development of American agriculture: a historical analysis. University of Minnesota Press, Minneapolis

Cook J, Nuccitelli D, Green SA, Richardson M, Winkler B, Painting R, Way R, Jacobs P, Skuce A (2013) Quantifying the consensus on anthropogenic global warming in the scientific literature. Environ Res Lett 8(2):024024

Cook-Patton SC, Bauerle TL (2012) Potential benefits of plant diversity on vegetated roofs: a literature review. J Environ Manag 106:85–92

Crosby AW (2003) The Columbian exchange: biological and cultural consequences of 1492, vol 2. Greenwood Publishing Group, Westport

Delcourt PA, Haccou P, Delcourt PA, Delcourt HR (2004) Prehistoric native Americans and ecological change: human ecosystems in eastern North America since the Pleistocene. Cambridge University Press, Cambridge

Dick E (1937) The sod-house frontier 1854–1890. D. Appleton-Century Company, New York

Dunwiddie PW, Bakker JD (2011) The future of restoration and management of prairie-oak ecosystems in the Pacific Northwest. Northwest Science 85(2):83–93

Dvorak B (2003) The greening of a nature museum–a demonstration project. Paper presented at the first annual greening rooftops for sustainable communities conference, Chicago, IL, 29–30 May

Dvorak B (2004) Green roof test plots: a green roof comparison project: the Illinois EPA-CDF green roof. Paper presented at the second annual greening rooftops for sustainable communities conference, Portland, Oregon, June 2–4

Dvorak B (2009) The Chicago City Hall green roof pilot project: a case study. Paper presented at the third international conference on smart and sustainable built environments. From building to promise building smartly in a changing climate, Delft, Netherlands, June 15–19, 2009

Dvorak B (2011) Comparative analysis of green roof guidelines and standards in Europe and North America. J Green Build 6(2):170–191. https://doi.org/10.3992/jgb.6.2.170

Dvorak B, Carroll K (2008) Chicago City Hall green roof: its evolving form and care. Paper presented at the sixth annual international rooftops for sustainable communities conference, Baltimore, MD, April 30–May 2

Dvorak B, Volder A (2010) Green roof vegetation for North American ecoregions: a literature review. Landsc Urban Plan 96(4):197–213

Dvorak BD, Volder A (2013) Plant establishment on unirrigated green roof modules in a subtropical climate. AoB Plants 5:pls049

EarthPledge (2005) Green roofs ecological design and construction. Schiffer Books, Atglen

FLL (2008) Guidelines for the planning, construction and maintenance of Green roofing, English edn. Forschungsgesellschaft Landschaftsentwicklung Landschaftsbau e. V, Bonn

Gasch CK, Huzurbazar SV, Stahl PD (2016) Description of vegetation and soil properties in sagebrush steppe following pipeline burial, reclamation, and recovery time. Geoderma 265:19–26. https://doi.org/10.1016/j.geoderma.2015.11.013

Gates DS (1933) The sod house. J Geogr 32(9):353–358

Gedge D, Kadas G (2005) Green roofs and biodiversity. Biologist 52(3):161–169

Gehmacher A (2014) Paul Kane: life & work. Art Canada Institute= Institut de l'art canadien, Toronto

Getter KL, Rowe DB (2006) The role of extensive green roofs in sustainable development. HortScience 41(5):1276–1285

Goebel T, Waters MR, O'Rourke DH (2008) The late Pleistocene dispersal of modern humans in the Americas. Science 319(5869):1497–1502

Gornitz V (2009) Sea level change, post-glacial. Encyclopedia of paleoclimatology and ancient environments. pp 887–893

Goudie AS (2018) Human impact on the natural environment. Wiley, Hoboken

Graham A (1999) Late Cretaceous and Cenozoic history of North American vegetation: north of Mexico. Oxford University Press on Demand, New York

Grancharov R (2013) Green roofs, history and the present. In: Proceedings in GVW-the 1st global virtual conference-workshop, April 2013. vol 1

Grant G (2006) Green roofs and façades, vol 70. IHS Bre Press, Bracknell

Grant G (2007) Extensive green roofs in London. Urban Habitat 4(1):1541–7115

Groenfeldt D (2019) Water ethics: a values approach to solving the water crisis. Routledge, New York

Hopkins DM (1967) The Bering land bridge, vol 3. Stanford University Press, Stanford

Hornbeck R (2012) The enduring impact of the American dust bowl: short-and long-run adjustments to environmental catastrophe. Am Econ Rev 102(4):1477–1507

Hulse D, Gregory S (2002) Willamette River Basin planning atlas: trajectories of environmental and ecological change. Oregon State University Press, Corvallis

Jarger E (2008) A pictorial of early roof top gardens in the United States. In: Greening rooftops for sustainable communities, Baltimore, MD, April 30–May 2 2008. The Cardinal Group, p 18

Jepson EJ Jr (2004) The adoption of sustainable development policies and techniques in US cities: how wide, how deep, and what role for planners? J Plan Educ Res 23(3):229–241

Jungels J, Rakow DA, Allred SB, Skelly SM (2013) Attitudes and aesthetic reactions toward green roofs in the Northeastern United States. Landsc Urban Plan 117:13–21

Kampinen AR (2008) The sod houses of Custer County. University of Georgia, Athens, Georgia, Nebraska

Keeley M (2004) Green roof incentives: tried and true techniques from Europe. In: Proceedings of the 2nd green roof conferenc: greening rooftops for sustainable communities, Toronto, June 2, 2004. The Cardinal Group, pp 119–120

Kimmerer RW, Lake FK (2001) The role of indigenous burning in land management. J For 99(11):36–41

Koehler M, Keeley M (2005) Green roof technology and policy development (Berlin). In: Green roofs: ecological design and construction. Schiffer Books, pp 108–112

Köhler M (2006) Long-term vegetation research on two extensive green roofs in Berlin. Urban Habitat 4(1):3–26

Köhler M, Poll PH (2010) Long-term performance of selected old Berlin greenroofs in comparison to younger extensive greenroofs in Berlin. Ecol Eng 36(5):722–729. https://doi.org/10.1016/j.ecoleng.2009.12.019

Köhler M, Schmidt M, Wilhelm Grimme F, Laar M, Lúcia de Assunção Paiva V, Tavares S (2002) Green roofs in temperate climates and in the hot-humid tropics–far beyond the aesthetics. Environ Manag Health 13(4):382–391

Kowarik I (2011) Novel urban ecosystems, biodiversity, and conservation. Environ Pollut 159(8–9):1974–1983. https://doi.org/10.1016/j.envpol.2011.02.022

Laberge K (2003) Urban Oasis: Chicago's City Hall green roof. Paper presented at the First North American green roof infrastructure conference: greening rooftops for sustainable communities, Chicago, IL, 5–05-03

Lambrinos JG (2015) Water through green roofs. In: Green roof ecosystems, Ecological studies. Springer, Cham, pp 81–105

Lemieux AM, Bhiry N, Desrosiers PM (2011) The geoarchaeology and traditional knowledge of winter sod houses in eastern Hudson Bay, Canadian Low Arctic. Geoarchaeology 26(4):479–500

Leopold A (1966) A sand county almanac: with essays on conservation from round river. Oxford University Press, Inc., New York

Libecap GD, Hansen ZK (2002) "Rain follows the plow" and dryfarming doctrine: the climate information problem and homestead failure in the upper great plains, 1890–1925. J Econ Hist 62(1):86–120

Licht J, Lundholm J (2006) Native coastal plants for northeastern extensive and semi-intensive Green roof trays: substrates, fabrics and plant selection. Paper presented at the fourth annual greening rooftops for sustainable communities conference, Boston, MA, May 11–12, 2006

Liss PS, Duce RA (2005) The sea surface and global change. Cambridge University Press, Cambridge

Loucks OL (1962) A forest classification for the maritime provinces. Proceedings of the Nova Scotian Institute of Science 25(Part 2):85–167

Lovell S, Johnston D (2009) Designing landscapes for performance based on emerging principles in landscape ecology. Ecol Soc 14(1)

Luckett K (2009) Green roof construction and maintenance. McGraw-Hill, New York

Lundholm JT (2006) Green roofs and facades: a habitat template approach. Urban Habitat 4(1):87–101

Lundholm JT, Walker EA (2018) Evaluating the habitat-template approach applied to green roofs. Urban Nat 1(1):39–51

MacDougall A (2003) Did native Americans influence the northward migration of plants during the Holocene? J Biogeogr 30(5):633–647

Martin PS (1973) The discovery of America: the first Americans may have swept the western hemisphere and decimated its fauna within 1000 years. Science 179(4077):969–974

Martin MA, Hinckley TM (2007) Native plant performance on a Seattle green roof. Paper presented at the fifth greening rooftops for sustainable communities conference, Minneapolis, Minnesota, April 29–May1, 2007

McLeman RA, Dupre J, Ford LB, Ford J, Gajewski K, Marchildon G (2014) What we learned from the dust bowl: lessons in science, policy, and adaptation. Popul Environ 35(4):417–440

Monterusso MA, Rowe BD, Rugh CL (2005) Establishment and persistence of *Sedum* spp. and native taxa for green roof applications. HortScience 40(2):391–396

Omernik JM (1987) Ecoregions of the conterminous United States. Ann Assoc Am Geogr 77(1):118–125

Osmundson T (1999) Roof gardens – history, design and construction. W.W. Norton & Company, New York

Overpeck JT, Webb RS, Webb T III (1992) Mapping eastern North American vegetation change of the past 18 ka: no-analogs and the future. Geology 20(12):1071–1074

Packard S, Mutel CF (1997) The tallgrass restoration handbook: for prairies, savannas and woodlands. Island Press, Washington, DC

Peck S (2008) Award winning green roof designs. Schiffer, Atglen

Peck S (2010) Green roof industry grows 16.1 percent in 2009 despite economic downturn. Green Roofs for Healthy Cities, Toronto

Peck SW (2012) The rise of living architecture. GreenRoofs for Healthy Cities, Toronto

Peck SW, Callaghan C, Kuhn ME, Bass B (1999) Greenbacks from green roofs: forging a new industry in Canada. Canada Mortgage & Housing Corporation, Ottawa

Pereira HM, Leadley PW, Proença V, Alkemade R, Scharlemann JP, Fernandez-Manjarrés JF, Araújo MB, Balvanera P, Biggs R, Cheung WW (2010) Scenarios for global biodiversity in the 21st century. Science 330(6010):1496–1501
Philippi PM (2005) Introduction to the German FLL-Guideline for the planning, execution, and upkeep of green-roof sites. Paper presented at the third annual greening rooftops for sustainable communities conference, Washington, DC, May 4–6, 2005
Pielou EC (2008) After the ice age: the return of life to glaciated North America. University of Chicago Press, Chicago
Powell JW (1879) Report on the lands of the arid region of the United States, with a more detailed account of the lands of Utah. Government Printing Office, Washington, DC
Reisner M (1993) Cadillac desert: the American West and its disappearing water. Penguin, New York
Ricketts TH, Dinerstein E, Olson DM, Eichbaum W, Loucks CJ, Kavanaugh K, Hedao P, Hurley P, DellaSalla D, Abell R (1999) Terrestrial ecoregions of North America: a conservation assessment, vol 1. Island Press, Washington, DC
Rowe B (2015) Long-term rooftop plant communities. In: Sutton RK (ed) Green roof ecosystems, vol 223. Springer, Cham, pp 311–332
Samson FB, Knopf FL (1996) Prairie conservation: preserving North America's most endangered ecosystem. Island Press, Washington, D.C
Seager R, Lis N, Feldman J, Ting M, Williams AP, Nakamura J, Liu H, Henderson N (2018) Whither the 100th Meridian? The once and future physical and human geography of America's arid–humid divide. Part I: the story so far. Earth Interact 22(5):1–22
Shorthouse JD (2010) Ecoregions of Canada's prairie grasslands. Arthropod Can Grassl 1:53–81
Smith DD (1992) Tallgrass prairie settlement: prelude to the demise of the tallgrass ecosystem. In: Proceedings of the twelfth North American prairie conference, Cedar Falls, IA, 1992. University of Northern Iowa, pp 195–199
Snodgrass EC, McIntyre L (2010) The green roof manual: a professional guide to design, installation, and maintenance. Timber Press, London
Snodgrass E, Snodgrass L (2006) Green roof plants. Timber Press, Portland
Stovin V, Poë S, Berretta C (2013) A modeling study of long term green roof retention performance. J Environ Manag 131(0):206–215. https://doi.org/10.1016/j.jenvman.2013.09.026
Sutton RK, Harrington JA, Skabelund L, MacDonagh P, Coffman RR, Koch G (2012) Prairie-based green roofs: literature, templates, and analogs. J Green Build 7(1):143–172. https://doi.org/10.3992/jgb.7.1.143
Thompson W (1998) Grass-roofs movement. Landscape architecture, vol 88. Magazine Publishers of America, Washington, DC
Tong DQ, Wang JX, Gill TE, Lei H, Wang B (2017) Intensified dust storm activity and Valley fever infection in the southwestern United States. Geophys Res Lett 44(9):4304–4312
Tubiello FN, Salvatore M, Ferrara AF, House J, Federici S, Rossi S, Biancalani R, Condor Golec RD, Jacobs H, Flammini A (2015) The contribution of agriculture, forestry and other land use activities to global warming, 1990–2012. Glob Chang Biol 21(7):2655–2660
Tzoulas K, Korpela K, Venn S, Yli-Pelkonen V, Kaźmierczak A, Niemela J, James P (2007) Promoting ecosystem and human health in urban areas using green infrastructure: a literature review. Landsc Urban Plan 81(3):167–178
USDA N (2007) The PLANTS database. USDA National Plant Data Center. https://plants.sc.egov.usda.gov/
Vale T (2013) Fire, native peoples, and the natural landscape. Island Press, Washington, DC
Van der Ryn S, Cowan S (2013) Ecological design. Island Press, Washington, DC
Van Devender TR, Spaulding WG (1979) Development of vegetation and climate in the southwestern United States. Science 204(4394):701–710
Van Wormer SR (2014) Mexican and American folk architectural traditions and adaptations at hedges: a late nineteenth–early twentieth-century mining camp in the California Desert. Calif Archaeology 6(1):95–118. https://doi.org/10.1179/1947461X14Z.00000000027

VanWoert ND, Rowe DB, Andresen JA, Rugh CL, Xiao L (2005) Watering regime and green roof substrate design affect Sedum plant growth. HortScience 40(3):659–664
Webb T III, Cushing EJ, Wright HE Jr (1983) Holocene changes in the vegetation of the Midwest. In: Late-quaternary environments of the United States, vol 2, pp 142–165
Weber C (2013) Ecosystem services provided by urban vegetation: a literature review. In: Urban Environment. Springer, Dordrecht, pp 119–131
Weiler SK, Scholz-Barth K (2009) Green roof systems: a guide to the planning, design, and construction of landscapes over structure. Wiley, Hoboken
Wen J, Nie ZL, Ickert-Bond SM (2016) Intercontinental disjunctions between eastern Asia and western North America in vascular plants highlight the biogeographic importance of the Bering land bridge from late Cretaceous to Neogene. J Syst Evol 54(5):469–490
Werthmann C (2007) Part I: Essay. In: Green roof: a case study: Michael Van Valkenburgh Associates' design for the headquarters of the American Society of Landscape Architects. Princeton Architectural Press, New York, p 159
Whitlock C, Knox MA (2002) Prehistoric burning in the Pacific northwest: human versus climatic influences. In: Fire, native peoples, and the natural landscape. Island Press, Washington, DC, pp 195–231
Williams NS, Lundholm J, Scott MacIvor J (2014) Do green roofs help urban biodiversity conservation? J Appl Ecol 51(6):1643–1649
Young TP, Petersen D, Clary J (2005) The ecology of restoration: historical links, emerging issues and unexplored realms. Ecol Lett 8(6):662–673

Chapter 2
Theoretical Development of Ecoregional Green Roofs

Bruce Dvorak and Jennifer Bousselot

Abstract This chapter investigates the theoretical background of environmental and ecological factors that can be used to inform the design of ecoregional green roofs. Ecoregions are defined by major or minor delineations of plant communities and their interactions with other resident or transient organisms. By observing and learning how native vegetation adapts and thrives in its natural settings, green roof researchers, educators, and designers can learn how to make decisions about the resourceful use of native vegetation on green roofs. This chapter discusses how green roofs must respond to environmental factors such as heat stress, drought, and varied slope and soil conditions, and how these factors can inform the design of green roofs with native vegetation. The chapter ends with a discussion regarding how ecoregions are defined in this book and are employed in the case studies in Part II of this book.

Keywords Ecoregion · Microclimate · Substrate · Native vegetation · Water · Maintenance · Biodiversity · Migration · Boundary · Anthromes

2.1 Environmental Factors for Ecoregional Green Roofs

2.1.1 Background

As human populations continue to grow exponentially, the expansion of urban development into natural environments means that biological diversity is threatened by fragmentation, loss of habitat, and the degradation of habitats as a result of

B. Dvorak (✉)
Department of Landscape Architecture and Urban Planning, 305A Langford Architecture Center, Texas A&M University, College Station, TX, USA
e-mail: bdvorak@tamu.edu

J. Bousselot
Department of Horticulture and Landscape Architecture, Colorado State University, Fort Collins, CO, USA

B. Dvorak (ed.), *Ecoregional Green Roofs*, Cities and Nature,
https://doi.org/10.1007/978-3-030-58395-8_2

pollution, urban heat islands, disturbances to surface and groundwater, and the introduction of exotic species (Rosenzweig 2003; Kowarik 2011). Although land-based conservation development practices have been conceived as a way to conserve native habitats within developments (Arendt 2013), most urbanized regions in the western U.S.A. and Canada have developed with little regard to the preservation and protection of the native vegetation within major metropolitan regions (e.g. Phoenix, AZ, Salt Lake City, UT, Los Angeles, CA, Vancouver, B.C.) (Ricketts 1999). If the original concept of a green roof is to create a regional ecology on the rooftop for a wide variety of benefits, then knowledge and a variety of examples of how this can be done, what it can look like, and how they can be maintained are needed.

Green roofs have the potential to bring native vegetation into urbanized regions where there is little remaining space on the landscape. Rooftops can account for 30% of impervious surfaces in urban watersheds (Carter and Jackson 2007). But what if green roofs, *by design*, are limited to standardized plantings of a single genus (i.e. *Sedum*)? From an ecological perspective, these generic green roofs may function as little more than a suburban lawn on a rooftop with minimal contribution to support the ecosystems that had previously endured for millennia. If contemporary culture has any responsibility to preserve some remnant population of its endangered native biodiversity for future generations, then green roofs are an efficacious tool to realize such goals, if they are intentionally designed to do so.

If green roofs (extensive, semi-intensive) are going to make significant contributions to buildings, sites, urban ecology, green infrastructure, and conservation biology, then they need to become developed much more in their capacity to support regional ecology, and not be conceived of as isolated generic patches or manicured gardens (Forman and Godron 1986; Lundholm and Richardson 2010; Kowarik 2011; Dvorak 2015b). If green roofs are going to be realized in support of ecosystem services (via policy), then ecoregional green roofs must be considered imperative and also include native vegetation (Dunnett 2015; Dunster and Coffman 2015; Rowe 2015; Sutton and Lambrinos 2015).

To learn about the characteristic ecology of a place, one must become familiar with its natural and cultural heritage (Leopold 1966; Packard and Mutel 1997). This means that green roof designers may need more environmental education and a spirit of self-learning. Being outdoors and observing natural landscapes is imperative to an understanding of how landscapes function. However, learning about principles of ecological restoration and habitat restoration may be equally important, and must become part of professional training (Bogner 1998; Melnick 2001; Tyrväinen et al. 2003; Cramer 2008). The following sections outline important factors that may prove critical to the design of a green roof through the observation of reference (native) ecosystems.

2.1.2 Climate and Microclimates

Perhaps no other environmental attribute contributes more to the potential success or failure of a green roof than climate, water being arguably the most important aspect (Lambrinos 2015). The occurrence and seasonal patterns of precipitation (rain, snow, fog), temperature norms and extremes, humidity, and diurnal atmospheric cycles influence which kind of plants grow and thrive in a particular region. The variation of climates west of the 100th meridian (Sect. 1.5) is great, and there are varieties of ways that vegetation responds to and is synergistic with climate (Lin et al. 2008; Reichstein et al. 2014). Each chapter in this book includes a synoptic discussion regarding climate, precipitation, temperature extremes, and natural vegetation.

Semi-arid and hot climates present challenges for plants in their acquisition of water and its efficient use. There are many drought adaption strategies that plants use to thrive, and take advantage of limited precipitation and hot diurnal temperatures (Puigdefábregas and Pugnaire 1999). Many of these plant adaptation strategies have developed over thousands of years. Some of the adaptions include minimizing water loss, developing ways to avoid drought stress, or tolerate water loss (Fig. 2.1) (Lambrinos 2015).

Different forms of plants have evolved tolerance strategies to survive the sometimes-brutal conditions of semi-arid and hot climates. For example, plants have adapted in their morphology by developing water efficiency mechanisms in their leaves, stems, or roots (Folsom 1995; MacDonagh and Shanstrom 2015). Some plants have evolved small narrow leaves, where others developed thickened or hairy leaves. Some developed succulent, shiny, or spiny leaves. Some plants have swollen stems, succulent stems, or developed stems as bulbs to conserve water. Root adaptations include shallow roots, taproots, thickened, succulent, or swollen roots (Folsom 1995). Figure 2.1 demonstrates these adaptions as demonstrated in vegetation native to conservation sites and ecoregional green roofs examined in this book.

Another type of adaptation within plant communities or ecoregions is drought avoidance. This can take place when ephemeral annual herbaceous vegetation adapts in a region. Since annual plants generally favor strategies such as rapid plant growth and seed production, they can avoid drought and take advantage of intermittent or erratic precipitation events. Annuals can fill roles in an ecosystem that perennial herbs may not. If a region experiences prolonged drought for example, where perennial vegetation may not survive, annuals may lie dormant until a favorable condition arrives, then flourish. Annuals have adapted to take advantage of winter, spring, and summer precipitation. The Desert Southwest, for example, is known for its abundance of annual wildflowers following winter or spring rains. Indian paintbrush (*Castilleja* sp.) is an example of a spring herbaceous fibrous-rooted annual plant that avoids drought. There is also drought avoidance with some perennial

Fig. 2.1 Morphological adaptations by plants growing in dry and hot climates west of the 100th meridian. Rocky Mountain juniper (**a**) has small scaly leaves. Rabbitbrush (**b**) has small light-colored leaves, and leaves of sagebrush (**c**) are light-colored and hairy and reflect light. Leaves of Oregon grape are thick, leathery, and shiny to conserve water (**d**), and Oregon stonecrop plants conserve water with thick and succulent leaves (**e**). Bulbous stems and fibrous roots of native allium (**f**) conserve water as does the thickened root and stems of prickly-pear (**g**). Prickly-pear also has small, spiny, and reflective leaves. It has shallow fibrous roots near the soil surface to absorb the moisture of light precipitation events, and it has deeper thickened roots and fibrous roots to harvest water deeper in the soil. (Graphic by Bruce Dvorak and Tess Menotti)

bulbs. Rain lilies, for example, can lie dormant for most of the year, and then emerge after a summer rain and complete its life cycle in just a few weeks (Dimmitt 2000). One species of rain lily, *Habranthus tubispathus*, was planted on a research green roof in south-central Texas and bloomed later that year (Table 1.4).

2.1.3 Microclimate

In natural landscapes, the microclimate of the land surface and immediate subsurface is one of the local environmental factors that can influence the presence or persistence of one plant over another at a specific location. Examples might be a drier environment near the top of a hill and a wetter environment at the bottom of a hill or warmer and drier southern-facing environment versus cooler and shadier north-facing environment in the Northern Hemisphere.

On green roof ecosystems, microclimatic conditions are more extreme than those at grade (Simmons 2015). The relatively shallow and lightweight substrate on green roofs do not buffer the environmental conditions of substrates in the same way as naturally occurring soil and rock conditions do at grade. Moisture conditions are more volatile, especially since the green roof substrate is typically well-drained and designed to have high pore space/air content. Substrate temperature has a wider diurnal variation on a green roof than at grade, mostly because the water content on a green roof is less compared to soils on the ground, and water maintains consistent temperatures. For example, during the winter, green roof substrates may be colder at night than the surrounding soils (Boivin et al. 2001), or warmer during the day compared to ground locations (Gaffin et al. 2009; Park et al. 2018).

Buildings are another factor that influences the microclimate of a green roof. Rooftop environments can cause stress for plants, as the microclimates on a roof can vary from the ambient climate at the ground level. Factors that can influence rooftop microclimates include solar orientation of roof deck, height above ground level, presence of surrounding facades such as brick, concrete, reflective glass, solar panels, and exposure to wind, heat reflectivity from dark building materials (Sect. 11.1.6) or negative effects of exhaust vents on rooftops (Sect. 11.1.6). Other factors can influence the microclimate of a roof due to accelerated or redirected winds from adjacent buildings, shade, or reflected light from adjacent buildings (Skabelund et al. 2015). Many rooftop microclimates produce more stressful conditions for plants than adjacent ground level landscapes do. Therefore, green roof designers must anticipate how these challenges affect the design and operation of green roofs (Jim and Tsang 2011; Lambrinos 2015; Simmons 2015).

2.1.4 Altitude

Altitude (topographic elevation) can have a profound influence on the local climate and the function and formation of ecosystems. In the Sonoran Desert, for example, mountains are ringed with bands of vegetation that have adapted to increasing levels of moisture and cooler air temperatures. Desert succulents and ephemeral annuals are common across the lower valley elevations of the Sonoran Desert, but succulent plants transition to mixed forests and ponderosa pine ecosystems in the upper elevations where temperatures are cooler and precipitation is more frequent. The

Willamette River in Oregon for example has an elevation of 1.2 m (4 ft) near Portland at the confluence with the Columbia River, and 131 m (430 ft) near Eugene at the south end of the Willamette Valley. The natural vegetation at the valley floor is prairie, woodland, and Garry oak ecosystems up to about 304 m (1000 ft) above the valley. Higher up in altitude, in the foothills of the Cascade Mountains, coastal temperate coniferous forests prevail. These abrupt changes in altitude are common west of the 100th meridian, and they have significant effects on the formation and persistence of plant communities.

One of the challenges for green roof designers is to identify native vegetation that is tolerant of the conditions that prevail on rooftops. A sub-alpine plant that is adapted to high altitude habitats, for example, may well be adapted to low-nutrient and well-drained soils, but may not be tolerant of high daytime temperatures that occur in the valley. Therefore, if the conditions that a plant is adapted to are not similar to conditions at the elevation of a green roof, then there may not be a good fit. Since plant communities at one altitude may have combinations of generalist and specialist plants, some plants may survive while others may not.

In some parts of the world, there has been a long record of plant trials to understand which plants can tolerate those conditions (Dunnett and Kingsbury 2004; Köhler 2006; Köhler and Poll 2010). With the green roof industry still relatively new to North America and other places, research regarding these efforts is still in its formative stages. Chapters 4 and 8 feature sub-alpine habitats and several green roofs in Chap. 4 demonstrate how sub-alpine vegetation has adapted to green roofs.

2.1.5 *Slopes and Microtopography*

In nature, the slope of the ground influences plant distribution in many ways, including the degree of slope steepness (angle/gradient), slope stability, level of flatness, drainage capacity, and potential for soil erosion. The steepness or flatness of a slope can have a profound effect on plant establishment, plant acclimation, and reproduction (Gibson et al. 2016). If a slope is too steep, some plants may not be capable of establishing. Likewise, a landscape with a slope that is nearly level, may exhibit traits that prohibit some plants from establishing, such as when too much moisture may be present for species that are adapted to well-drained soils. Landscapes west of the 100th meridian are distinctly defined by north to south aligned mountain ranges, valleys, and basins, and generally east and west-facing slopes. Slope direction and gradient are critical to the formation of microclimates. Generally, south and west-facing landscapes will accumulate more direct solar energy, becoming warmer and drier than cooler and wetter north- and east-facing slopes.

Knowledge of how plant communities become arranged by slope and aspect is critical to adapting vegetation for green roofs. A general good practice is to match the favorable microclimatic conditions of a plant, or plant community to the conditions of the roof. Likewise, ignoring the conditions of the plant community and rooftop could cause unfavorable conditions for the vegetation. Placing plants that are naturally adapted to well-drained soils and moderately sloped landscapes onto a flat and poorly drained north-facing slope on a green roof may prove difficult or unsuccessful (Palandino and Company 2006; Snodgrass and McIntyre 2010). In contrast, a moisture-loving plant community placed on a moderately sloped, and well-drained south-facing roof deck may require more frequent irrigation if the substrate is not of sufficient depth and moisture-holding capacity (VanWoert et al. 2005; Lambrinos 2015).

Some of the plants adapted to different moisture regimes that are defined by slope or aspect and include xeric or dry prairies, mesic or moist prairies, and wet or hydric prairies (Fig. 2.2). Plants that are members of xeric prairies thrive in dry habitats, where plants adapted to mesic prairies can tolerate some drought and some

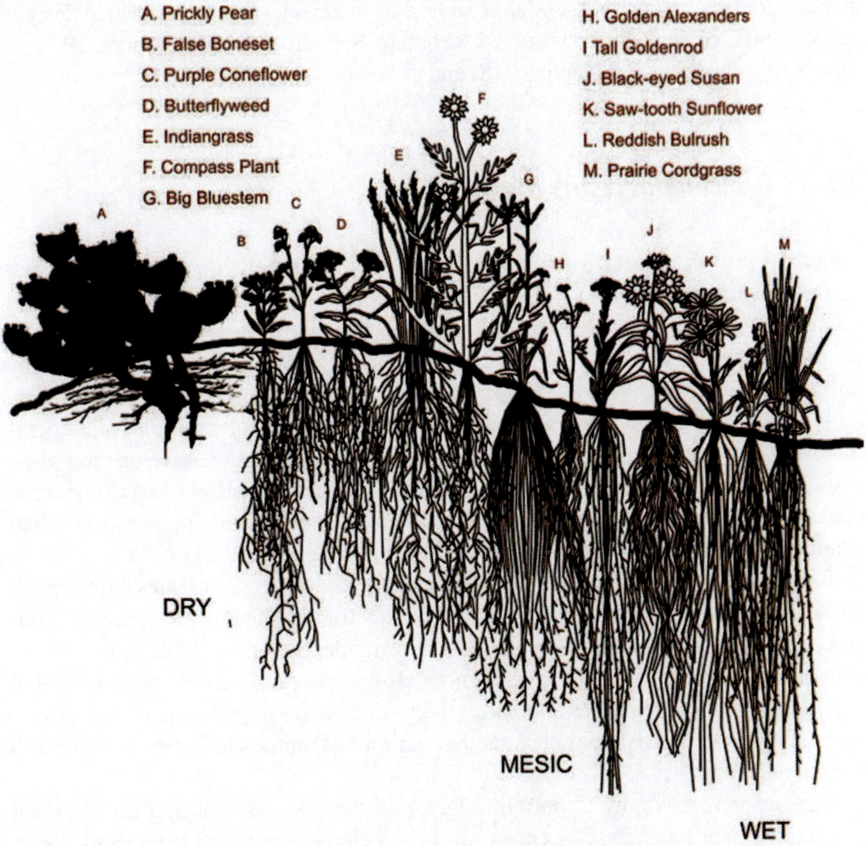

Fig. 2.2 A hypothetical cross-section of landscape with slope differentiation resulting in xeric (dry) prairie species that adapt to drier and warmer microclimates, species that adapt to mesic conditions with moderate or sometimes dry and sometimes wet conditions, and a wet prairie where soils are frequently saturated or wet. (Graphic by Bruce Dvorak and Tiantian Lyu)

flooding, but not too much exposure to the extremes. Plants that grow in anaerobic or hydric (wet) conditions function in specific ways and typically need constant saturation. Different plants adapt to the natural slope gradations of landscapes and can play various roles in a plant community.

Many green roofs are located on buildings that are designed with low-slope roof decks. Matching plants that naturally grow on undulating landscapes to flat roof decks may limit the variety or success of the vegetation. Plants that are native to mesic or wet conditions are naturally adapted to level and slow draining substrates; however, they are not yet frequently used on green roofs in North America. Most well-drained natural habitats are not flat or level. Therefore, there can be inherent contradictions to the practice of sourcing plants from native habitats to green roofs if the roof deck slope gradient and substrate design are not well-matched. To address these issues, designers can select native plants that can tolerate a variety of moisture conditions, elevate or slope roof decks with artificial means (lifts, or layers of polystyrene), or try a combination of approaches. If local codes allow, integrated designs can use flat roof decks as constructed wetlands. Several case studies in Part II demonstrate this concept (see Chaps. 7, 8, and 10).

2.1.6 Microtopography and Soils

Landscapes with natural topography rarely exhibit smooth or uniform surfaces. There are wide variations in soil profiles and exhibit microtopography. Untilled soil within a prairie or forest is often anything but uniform in its surface elevations. It can be bumpy, rugged, or irregular. In an undisturbed landscape, microtopography influences the arrangement of plants, drainage characteristics, soil character, distribution of micronutrients, and animal activity. In a prairie, for example, plant communities naturally adapt to the topographic distinctions in elevation and their associative moisture and nutrient regimes. It is in the summation of environmental factors, that influence which plants might adapt to the tops of slopes to those that adapt to the bottom of slopes (Schramm 1990; Bennie et al. 2008).

Some green roofs have sloped roof decks and/or undulating substrates (Brenneisen 2006). These variances in microtopography allow for diversity and a variety of habitats to develop on green roofs. If green roof designers intentionally design for these conditions, then rooftop habitats can more closely resemble their land-based analogs. Part II of this book covers many green roofs with sloped roof decks, or elevated substrates on a flat roof deck, and green roof substrates that vary in depth and are not uniform.

Landscape topography is only one factor of the environment that affects plant distribution in ecoregions. The composition and characteristics of the soil itself can greatly influence the survivability and growth of native plants. Native soils are vastly different from the soilless substrates used on green roofs. Because green roof substrates are designed with the desired benefits in mind, important characteristics of the native soils are sometimes overlooked. Soils can vary in the microbial community, micronutrient content and have unique physical and chemical properties that are challenging to replicate in engineered soils.

Native plants have evolved in native soils; therefore, it follows that for native plants to thrive on green roofs, it is ideal to emulate the characteristics of the native soils as much as possible. To that end, many green roofs in Europe include the use of native soils as a portion of the green roof substrate. It is not as common in North America, however, the Botanical Research Institute of Texas (BRIT) used up to 20% (by volume) of a local native soil when blending their green roof substrate (Best et al. 2015). They have subsequently found that over 90 species of native plants on the green roof, were not planted, and must have arrived as seed in the native soil that was mixed into the substrate.

2.1.7 Biodiversity and Ecoregional Green Roofs

On landscapes, it is well known that animals and microorganisms live in symbiosis with vegetation. Animals and insects can spread seeds, pollinate flowers, make the soil porous, live in the soil and decompose soil, participate in nutrient cycling, and become part of the soil after death. Microorganisms are vital to the life of the soil and plant life and are part of a plant community (McGuire et al. 2013; Best et al. 2015).

On green roofs, similar interactions between vegetation, animals, and microorganisms take place, but these interactions are just beginning to be understood (John et al. 2014; Buffam and Mitchell 2015; McGuire et al. 2015). The role of insects, wildlife, and microorganisms should be considered from the earliest development of a green roof and assumed into its long-term success (Brenneisen 2006). Most engineered substrates arrive at a rooftop in a near sterile condition, so there is a need for planning for soil amendments or other methods to accelerate these forms of interactions (Kephart 2005; McGuire et al. 2013).

There is great potential for biological diversity to be included on green roofs in the West, as there is a longstanding tradition for these provisions on green roofs in Europe and elsewhere (EarthPledge 2005; Brenneisen 2006; Cantor 2008; Dvorak 2015a; Madre et al. 2015). Some of the first green roofs that were designed to increase biodiversity in urban areas were developed and researched in Switzerland. Professor Stephan Brenneisen developed green roofs with varied substrate depths, rocks, branches, and growing media depths that were not uniform, but mixed on-site from local soils or materials. This pioneering work led to the design of green roofs to be designed specifically for greater biodiversity (Brenneisen 2003; Brenneisen 2006; Lundholm 2006; Dunster and Coffman 2015; Lundholm 2016).

Biodiverse green roofs may be designed formally, informally or they could be designed to mimic some particular habitat and have no intended geometric design. One of the main reasons for identifying a particular kind of plant, or plant community as a design goal, is that biologically diverse plant communities are typically more resilient and more ecologically valuable over time compared to monocultures (Kephart 2005; Cook-Patton and Bauerle 2012).

Research about how animals, insects, and microorganisms use green roofs is emerging (Gedge and Kadas 2004; Tonietto et al. 2011). There are many green roofs that have been designed to be biologically diverse in western parts of the U.S. and Canada, but very few studies of the biological aspects of green roofs are known or have been published (Hauth and Liptan 2003; Dvorak and Volder 2010).

It does not take long for insects such as bees, butterflies, moths, flies, and other flying insects to find their way onto a green roof (Coffman and Davis 2005; MacIvor and Ksiazek 2015). Insects can also crawl or find their way by being blown to a roof (Fig. 2.3) (Ksiazek 2014; MacIvor 2016). Nearby landscape features such as tall trees, water bodies, parks, ravines, or slopes that intersect or are near a green roof can bring these small but mobile creatures to rooftops. Birds can transport insects to a roof, and the insect may escape the bird and hide on a green roof. Grasshoppers, crickets, beetles, and other arthropods are frequently found on green roofs (Fig. 2.3) (Kadas 2006; Steck et al. 2015).

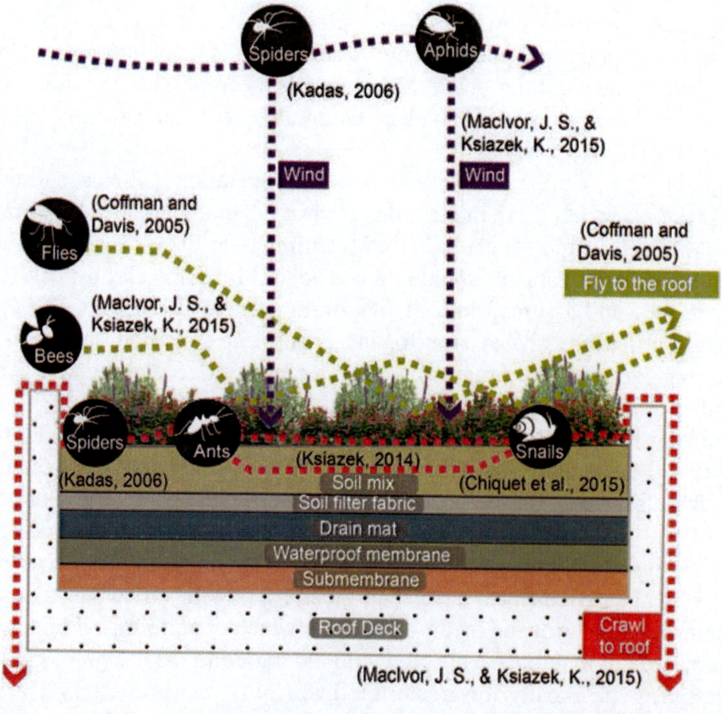

Fig. 2.3 Insects can find their way onto green roofs even during the first growing season. Most insects are beneficial to green roofs and cause little disturbance to the building systems. Insects arrive on green roofs through crawling, flying, being blown to the roof during strong wind events, or after being lifted or dropped by other animals, such as birds. (Graphic by Yinghui Chen and Bruce Dvorak)

Compared to conventional roofs (non-green roofs) arthropods and birds were more abundant on green roofs, and breeding was successful on green roofs in New York City (Partridge and Clark 2018). In a comparative study between the abundance and diversity of bees on green roofs compared to the ground level, green roofs had smaller and less diverse communities of pollinators compared to ground levels in Chicago, however, the ground level sites were much larger (10xs) than the green roofs sites, and had slightly more diverse vegetation. In Halifax, Nova Scotia, birds, insects, ants, flies, beetles, grasshoppers, moths, damselflies, lacewings, earwigs, Booklice, Springtails, were found in similar richness and abundance on green roofs and urban green spaces (MacIvor and Lundholm 2011a, b). Just as with natural systems, microorganisms play a vital role in ecosystems.

Managing green roofs for biodiversity is usually accomplished through directed and persistent observation of the green roof over time. Some of the most successful green roofs have active and informed maintenance managers, who monitor and read ecological conditions and make adjustments to the management of a green roof to become more resilient (Kephart 2005; Piana and Carlisle 2014; Dunster and Coffman 2015).

2.1.8 Species Migration and Ecoregional Green Roofs

The native plant communities provide habitat support for resident and transient (migrating) forms of wildlife. There are several major flyways for birds in North America: the Atlantic Flyway, Mississippi Flyway, Central Flyway, and Pacific Flyway. There are hundreds of bird species that make use of mostly natural habitats along flyways annually. In addition to mammals, birds, and butterflies, some insects migrate across ecoregions. Major population centers in the western U.S.A., Mexico, and Canada lie in the path of the migration routes. The historic native plant communities that covered the landscape before the nineteenth century served migrating species their migrations. With the successive loss of habitat each year, usable habitat for some local and some migrating species also shrink (Bender et al. 1998).

Migrating species have been observed to make use of green roofs, and in some cases, green roofs have been designed to include vegetation for migrating species (Brenneisen 2006; Coffman and Waite 2011; Partridge and Clark 2018). Most migrating species avoid urban spaces in place of natural habitats. However, green roofs designed for particular habitats have been used during migration for some species (Gedge and Kadas 2005; Fernandez-Canero and Gonzalez-Redondo 2010). Depending on location and time of year, birds use habitat for foraging, roosting, and nesting at their destinations. Many migrating species travel at night to avoid predation. This means that they use habitats during the daytime when many insects are active.

The following figures identify major flyways for songbirds, butterflies, and hummingbirds that intersect urban regions of North America (Figs. 2.4, 2.5, 2.6 and 2.7). Major urban centers are located as black dots to identify potential locations for the

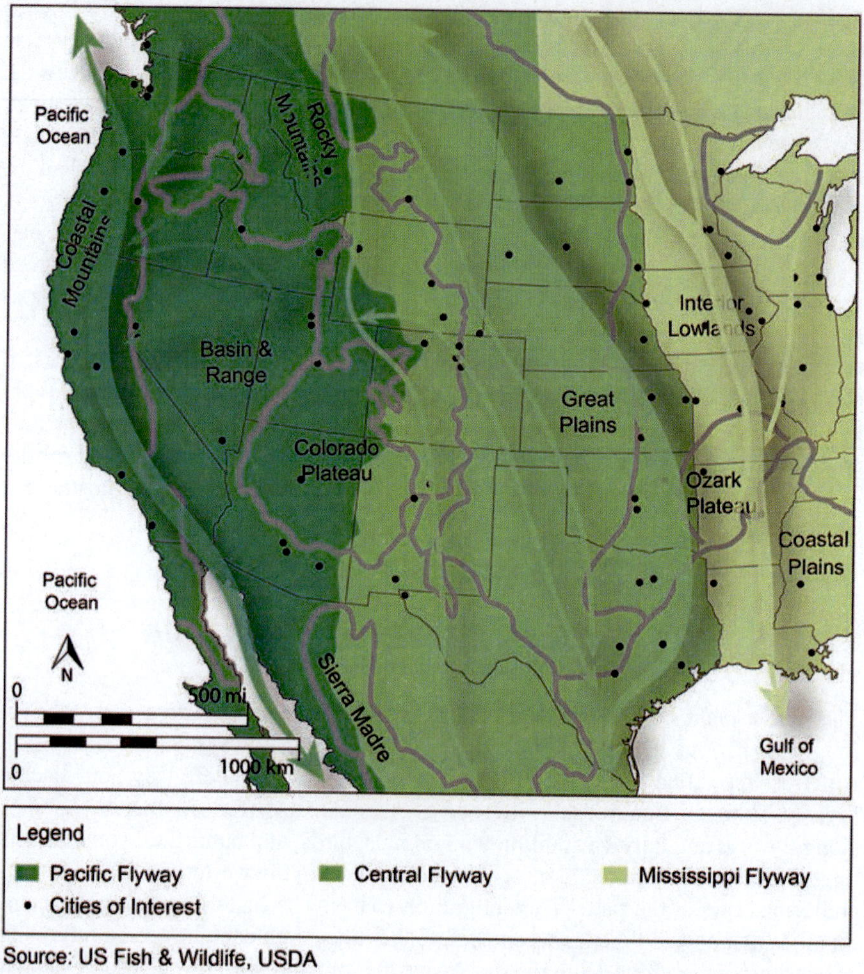

Fig. 2.4 Over 350 species of songbirds migrate between South and Central America to North American each year. Their particular destinations are unique to each species and form the Mississippi, Central, and Pacific flyways. Colored zones show limits of flyways and arrows show areas of concentration for some species. Black dots identify locations of primary and secondary urban population centers where ecoregional green roofs could be employed to provide habitat for species during their migrations. The bold grey lines delineate physiographic region boundaries. (Graphic by Trevor Maciejewski and Bruce Dvorak)

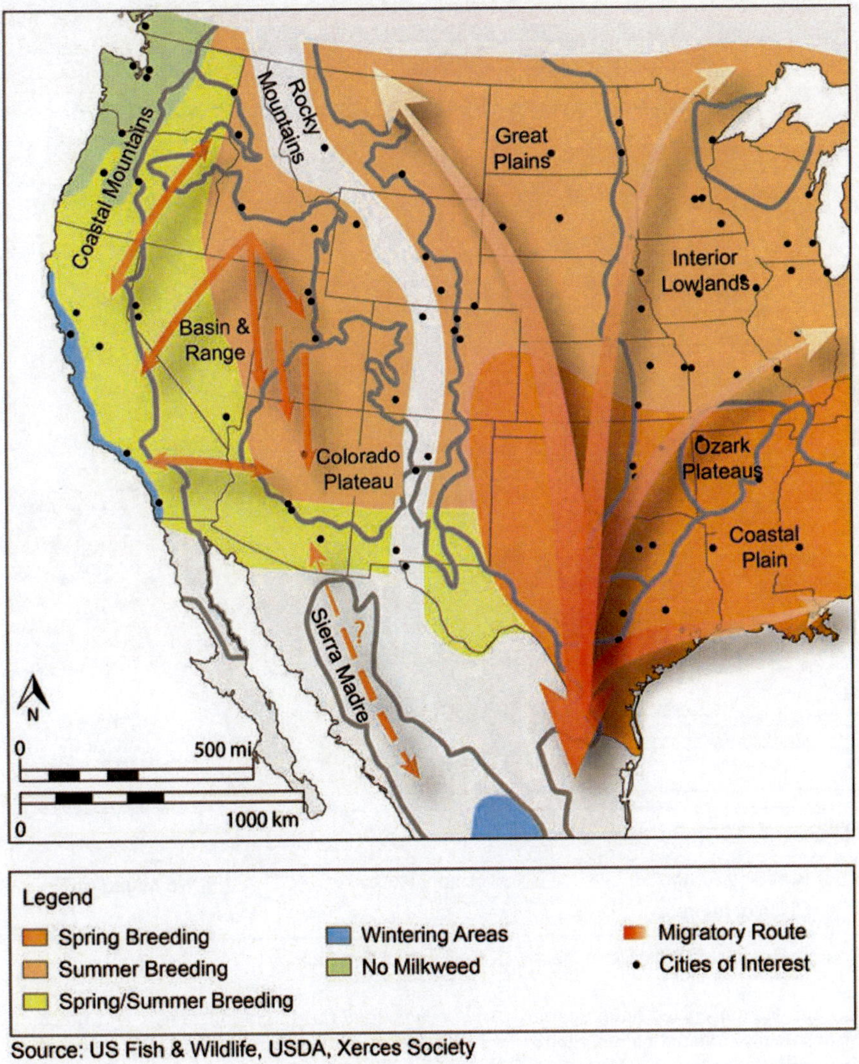

Fig. 2.5 Flyways of the monarch butterfly (*Danaus plexippus*). Colored zones delineate typical regions where swarms of monarch butterflies travel during migrations. Arrows show common migration routes. Black dots identify locations of primary and secondary urban population centers and bold grey lines delineate physiographic region boundaries. Since adult monarch butterflies typically live 2–6 weeks, it can take up to four generations of butterflies for swarms to migrate from south to north and return to their overwintering habitats. Ecoregional green roofs can be vegetated with milkweed and other plants to provide food and habitat for monarchs to reproduce, puddle, or rest. See Part II for a variety of examples of green roofs where monarchs have made use of green roofs. (Graphic by Trevor Maciejewski and Bruce Dvorak)

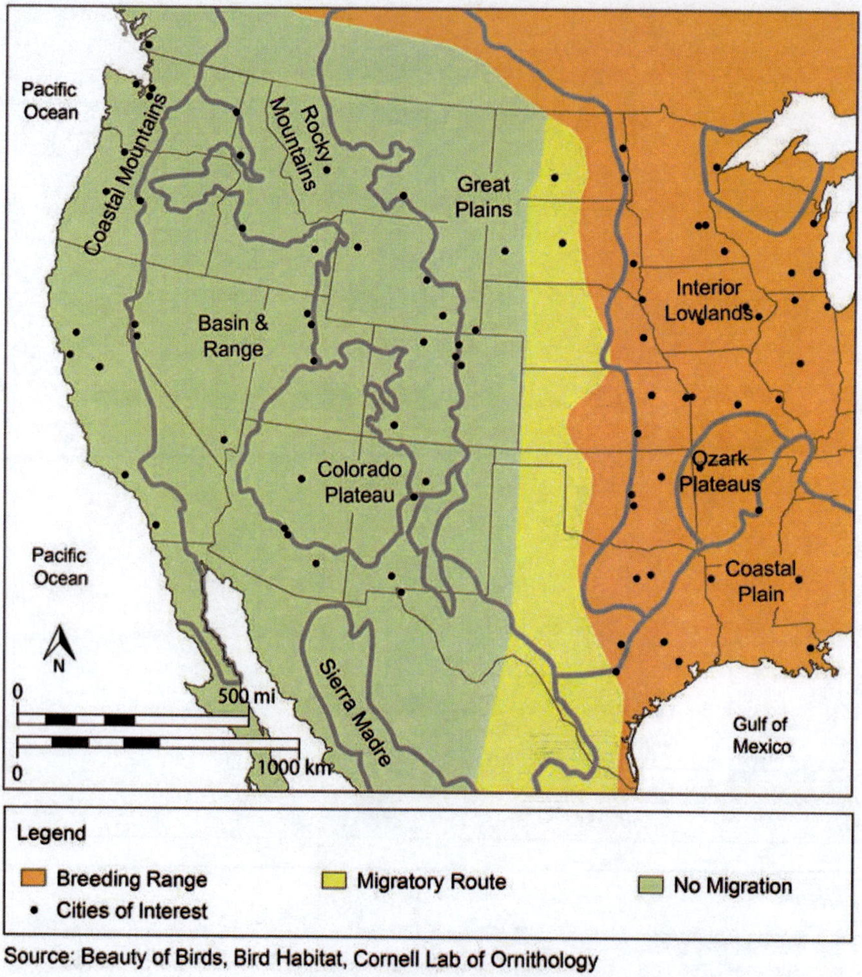

Fig. 2.6 Ruby-throated hummingbird (*Archilochus colubris*) migration ranges and breeding regions. Ruby-throated hummingbirds typically do not migrate west of the 100th meridian as they prefer the humid east. Many Ruby-throated hummingbirds have made use of green roofs in the ecoregional green roofs in Chapters 3 and 4. The bold grey lines delineate physiographic region boundaries. (Graphic by Trevor Maciejewski and Bruce Dvorak)

use of ecoregional green roofs for migrating species. If green roofs are designed as generic habitats with exotic plant species and don't address the specific needs of migrating species, then green roofs may never attain their ecological potential. Part II of this book presents many case studies where local and migrating species have used green roofs that include vegetation native to the ecoregion.

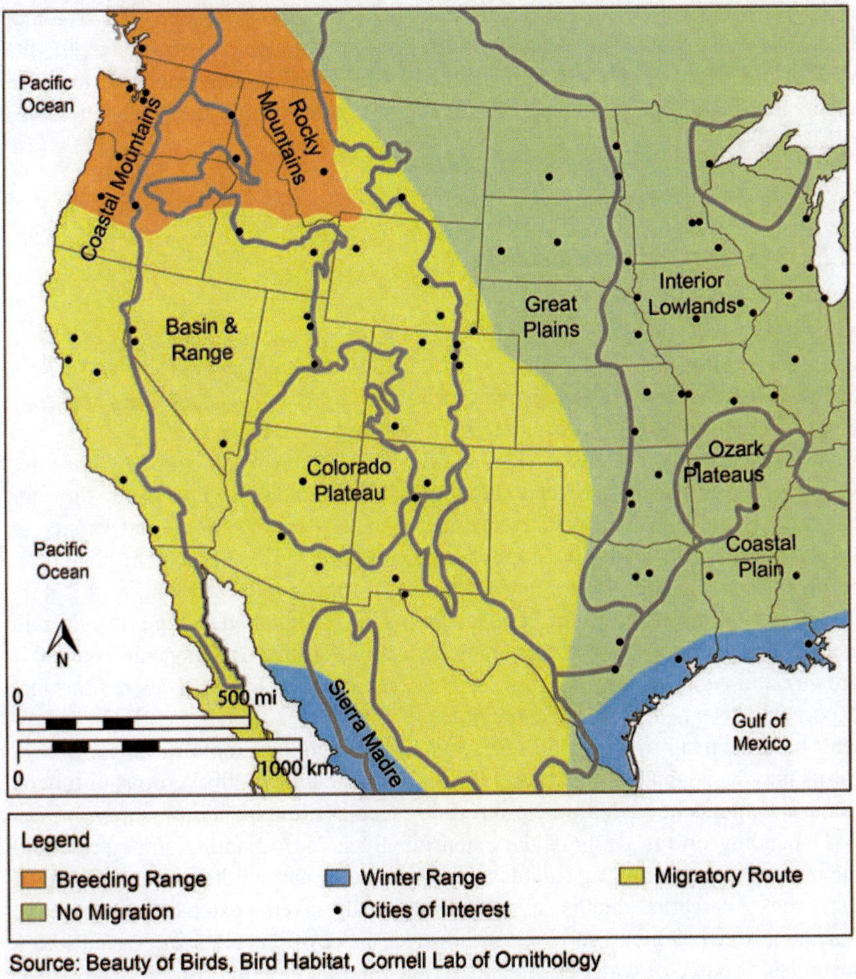

Fig. 2.7 Rufous hummingbird (*Selasphorus rufus*) migration ranges and regions. Rufous hummingbirds typically do not migrate east of the 100th meridian except during winter range along with the Gulf Coast habitats. See Chapters 5–10 for case studies of ecoregional green roofs where rufous hummingbirds have been observed on green roofs. (Graphic by Trevor Maciejewski and Bruce Dvorak)

2.2 Substrates and Fitness for Ecoregional Green Roofs

This section covers the adoption of substrates (growing media) for different classifications of green roofs. One of the most critical decisions in designing a green roof is specifying an appropriate blend and depth of substrates for particular forms of vegetation. The FLL Guidelines for Green Roofs specifies three categories of green roofs, each with its ascribed range of substrate depth: extensive, semi-intensive, and intensive green roofs. The categorical divisions are the depth of substrates, the fitness of plant forms, and levels of maintenance. These are considered universal in their categorical segregation and terminology (extensive, semi-intensive, intensive) (FLL 2008).

2.2.1 *Extensive Green Roofs*

Extensive green roofs are the shallowest kind of green roof. There are extensive green roofs with substrates as shallow as 2.5 cm (1 in) deep, and as thick as 12.7 cm (5 in). What characterizes extensive green roofs (by the FLL Guidelines and elsewhere) is their expectation for natural vegetation, low maintenance, and long-term survival of plant communities (FLL 2008). Due to their shallow depths, the substrates tend to warm up quicker and longer than the soil on the ground, and they dry out quickly as well. These characteristics tend to limit the forms and variety of vegetation that can naturally adapt to such conditions. Since succulent plants can retain moisture in their leaves and roots and can photosynthesize during the night (CAM cycle), they tend to thrive in shallow soils and take advantage of light rain events. However, most succulents are not tolerant of a slow draining subbase, and if grown on low-sloped rooftops may need special measures to survive in regions that experience heavy or persistent precipitation (Simmons 2015). Many desert succulents have deep tap root systems and retain moisture in their roots. A variety of root forms may be adaptable to extensive green roofs including bulbs, corms, or tubers. Some succulents have shallow fibrous roots, such as most species of *Sedum*.

Depending on the goals of the extensive green roof, planting strategies could include succulents only, succulents mixed with bulbs and wildflowers, additions of logs, rocks, or varied depths for more biologically diverse extensive green roofs. Extensive wetland green roofs exist, however, most of these are integrated into a sustainable source of water or integrated function of a building (EarthPledge 2005; Cantor 2008). Case studies in this book demonstrate a variety of options for extensive green roofs.

2.2.2 *Semi-Intensive Green Roofs*

Semi-intensive green roofs are intended to support larger forms of vegetation such as grasses, forbs, some groundcovers, and low-growing shrubs. Substrate depths typically begin at 15 cm (6 in) up to 30 cm (12 in) with the depth being selected by the kinds of vegetation intended to grow. It is not recommended to grow a meadow roof in less than 15 cm (6 in) depth of substrate, as the substrate temperatures can get too high for meadow grasses and forbs. Irrigation is recommended for most semi-intensive designs, depending upon species selected, slope orientation, gradient, and the climate and context of the site (Snodgrass and Snodgrass 2006; Dvorak and Volder 2010; Sutton et al. 2012).

Many case studies in this book demonstrate semi-intensive green roofs with a variety of substrate and maintenance designs. Although semi-intensive systems are heavier than extensive systems, the structural upgrade from an extensive to a semi-intensive system is typically nominal to the overall cost of new construction. Semi-intensive systems should be considered during the planning phases of a building project and considered for their multiple ecosystem services and life-cycle benefits (Licht and Lundholm 2006; Vacek et al. 2017).

2.2.3 Intensive Green Roofs

Intensive green roofs are roof gardens. These green roofs typically require substrates more than 30 cm deep (12 in), and higher forms of maintenance at a greater frequency. Most any kind of plant can be grown on rooftops or deck structures with the appropriate level of substrate depth and structural support (Weiler and Scholz-Barth 2009). Large trees have been planted on many intensive green roofs, but the appropriate structural, drainage, and watering support are necessary. Several intensive green roofs are highlighted in this book. However, due to the expense and variety of options, we limit coverage of intensive green roofs to a few to allow appropriate coverage of the more challenging and widespread use of extensive and semi-intensive green roofs.

2.2.4 Substrate Fitness for Ecoregional Green Roofs

Just as soil characteristics are vital to the development, support and persistence of ecosystems on the ground, substrates (green roof soils) are vital to the establishment and persistence of long-term vegetative cover on green roofs (Durhman et al. 2007; Köhler and Poll 2010; Brown and Lundholm 2015). Over time, plant community members can adapt to a green roof by moisture and nutrient gradients of the substrate (Köhler and Poll 2010; Nagase and Dunnett 2012). Natural soils and engineered soils have been used on green roofs, and each has its unique traits and should be used appropriately under the guidance of experienced green roof designers (Schrader and Böning 2006; Byerley 2014).

The FLL Guidelines for Green Roofs characterize essential qualities that make sustainable substrates for green roofs. Substrate stability over time (minimal slumping), moisture retention, moisture evacuation, nutrient management, and use of local resources are covered in the guidelines. The beauty of the guidelines is that they afford a substrate designer in any location a set of time-tested characteristics to begin designing a substrate; however, local research and expertise from those familiar with substrate design and guidelines should support local adoption of the baseline data in the guidelines (Philippi 2005; Dvorak 2011; ASTM E2777 142,014). The FLL guidelines can help prepare a substrate designer to use materials that are local to the region. They can provide guidance on the granulometric distribution of particle sizes in substrates so that it can retain structural stability, adequate moisture, drainage, and substrate air to moisture ratios (Beattie and Berghage 2008; ASTM E2777 142,014). The guidelines will likely need to be modified to adapt to local and microclimatic conditions (Williams et al. 2010).

We do not recommend that inexperienced designers fabricate green roof substrates or apply natural soils on green roofs without the advice of an experienced guide. Natural soils on rooftops can slump (shrink), clog drainages, become poorly drained, and not function the way they do on the ground (Dvorak 2011). However, some natural soils may be appropriate for some kinds of roof designs, and the addition of some natural soils and/or organisms such as mycorrhizal fungi can jump-start the biological community (Best et al. 2015).

On green roofs, some plants from prairie plant communities that have deep root systems have been found to adapt to shallow substrates, by growing laterally (Dvorak and Carroll 2008). Side-oats grama (*Bouteloua curtipendula*), little bluestem (*Schizachyrium scoparium*), and big bluestem (*Andropogon gerardi*) grow in shallow to deep soils in a variety of moisture conditions. These have been found to also grow on green roofs in deep extensive or semi-intensive green roofs throughout the Midwest, with various rates of irrigation on flat and sloped green roofs (Sutton et al. 2012). Providing vegetation that can adapt to varied moisture regimes can help make prairie-based green roofs resilient over time (Dvorak and Carroll 2008; Sutton et al. 2012; MacDonagh and Shanstrom 2015).

New directions in green roof design such as biodiverse roofs might favor substrates that are not uniform in depth or composition. Such substrates combine larger gravel or small stones in varying depths to a substrate to allow for a variety of microclimates and biodiversity (Gedge 2003; Brenneisen 2006). The pilot testing of substrates is critical to substrate success. Requiring installers to provide samples and adjust substrate mixtures to meet the specifications is necessary. Once a substrate is placed on a roof deck, it can be very expensive and difficult to remove and replace if its properties are not right for the selected vegetation (Osmundson 1999).

2.3 Maintenance, Water, and Aesthetics of Ecoregional Green Roofs

2.3.1 Maintenance

Expectations for green roof maintenance activities on green roofs can range from several visits a year, to weekly, or even daily maintenance visits for some high maintenance roof gardens (during some parts of the year). Green roofs with formal planting arrangements in geometric shapes or zones will require maintenance at much higher levels than biodiverse green roofs, or those designed to be natural in appearance. Many factors can influence levels of maintenance of green roofs including the application of too much or too little irrigation water, infestations of invasive plants or insects, disturbances such as wind or pecking by birds, poor nutrient management, and poor match between the vegetation, substrate and maintenance practices (Hauth and Liptan 2003; MacDonagh and Shanstrom 2015). The case studies in this book discuss a variety of maintenance practices, issues, and problems.

The successful design and maintenance of a green roof is as much an art as it is science (Cantor 2008; Luckett 2009; Snodgrass and McIntyre 2010). Green roof maintenance can become minimized when there is a synchronicity between the design of the green roof components, vegetation (climate and micro-climate adapted), maintenance regimen, and irrigation practices. However, using native plants on green roofs that do not exhibit characteristics that the plants naturally adapt to can result in high maintenance green roofs.

Knowledge and experience are needed for the appropriate selection of vegetation that will thrive on a green roof. Maintaining green roofs is a relatively new discipline and is quite different than maintaining landscapes on the ground. There are specialized skills, areas of knowledge, and execution. It is not uncommon for an excellent green roof design to be installed and established, and when an inexperienced maintenance crew takes over, it sees a decline or complete demise of the vegetation (Vijayaraghavan 2016). Likewise, appropriate maintenance practices performed throughout the growing season can yield a sustainable green roof (Cantor 2008).

2.3.2 *Water*

Precipitation west of the 100th meridian drops considerably in volume, frequency, and duration. Since green roof substrates dry out quickly and can warm up well beyond soils on the ground, supplemental water may be required on most green roofs in the West to sustain the vegetation (Van Mechelen et al. 2015). In areas with warm days and cool nights over the winter, plant survivability is greatly increased if winter watering is done every 3–4 weeks on warm days (Bousselot et al. 2010).

There are many choices for sourcing water for green roofs. Beyond using municipal potable water, more sustainable strategies include harvested rainwater from rooftops or parking lots, condensate from HVAC systems, use of greywater from inside buildings, use of blackwater from inside buildings, or reclaimed water from an industrial source (EarthPledge 2005; Cantor 2008; Lambrinos 2015). Given the scarcity and erratic nature of precipitation in many urban regions west of the 100th meridian, coupled with the potential benefits from green roofs (keep buildings cool and reduce urban heat islands), sustainable sources of water should be secured.

Most of the case studies in this book feature a variety of sustainable sources of water to irrigate green roofs including greywater, blackwater, and rainwater harvested from rooftops and pavements. Use of refuse water (greywater) from industrial processes that are clean and appropriate for green roofs has not been readily explored in North American green roofs, but it is worth investigating (Ouldboukhitine et al. 2014). One manufacturing company in Germany was able to pump refuse water from inside the building to a constructed wetland rooftop to filter and clean water. This creative use of water for the green roof eliminated an expense that the company paid for the city to clean the water (EarthPledge 2005).

2.3.3 Aesthetics

One of the purposes of this book is to provide visual examples of green roofs that adapt to dry, bimodal and semi-arid climates. Since monocultures of *Sedum* may work in some, but not any ecoregion, we present a variety of approaches from built projects. Due to weight limits on rooftops, some green roof applications may have few choices other than low growing succulents. However, as substrate depth increases, more variety of forms of vegetation are possible. Succulent green roofs will provide a carpet of green in many climates, but in the Western U.S.A., Canada, and parts of Mexico, tender *Sedums* native to cool and temperate climates may not survive. In these ecoregions, cacti and other drought-tolerant vegetation prevail. Annual wildflowers are an important part of an ecosystem in some desert ecoregions. Other ecoregions have precipitation evenly distributed throughout the year, and grasses and wildflowers may persist. The case studies in this book demonstrate a wide variety of aesthetic approaches.

One of the challenges of applying a universal concept such as a green roof in a wide variety of climates is that a single impression of what a green roof is or can be may be misleading, or create unrealistic expectations (Rayner et al. 2016). Building owners need to know what aesthetic options are realistic for their project, and examples of different green roof aesthetics are needed. Over time, most green roofs will be managed by many different building owners, and each owner will need to know how to care for the green roof and the intended aesthetic expectations. There are cases of building owners not being satisfied with their green roof for a variety of reasons. Most building owners and managers will not be experts in managing green roof systems and may need to be educated on how to care for the roof and what it should look like (Thurston 2017). Therefore, measures must be in place for the care of a green roof and the transfer of that knowledge will need to take place through multiple building owners, and different maintenance personnel. This may mean that training and education about how to maintain a particular kind of green roof may be necessary.

Sutton outlined three levels of aesthetic appreciation that can apply to green roofs: (1) enjoyable beauty (perceived through senses), (2) admirable beauty (culturally learned) and (3) ecological beauty (e.g. co-dependent species, local flora). What this means is that building owners may need to become aware of what kind of beauty they expect, can afford, and are willing to maintain. Sutton reiterates the importance of experience, ecosystem knowledge, participation, and maintenance of green roofs for building owners and users, and awareness of local landscape aesthetics (Sutton 2014, 2020).

2.4 Development of Ecoregions

"A descriptive term to denote regions recognized by vegetation in relation to these environmental criteria may be obtained by applying the prefix 'eco-'to the term 'region'. 'Ecoregion' is proposed therefore as the geographic unit within which ecological relationships between species and sites are essentially similar…" (Loucks 1962, pg. 91.)

The first use of the term ecoregion was used to delineate different classifications of forests for use in managing silviculture (Loucks 1962). Potential natural vegetation, biogeographic regions, ecosystems, and other terms were also in use, but the use of ecoregion over time became a way to define the geographic distribution and boundaries of related ecosystems (Table 2.1).

In 1976, Dr. Robert Bailey published a map of ecoregions to delineate potential natural vegetation of the United States called *Ecoregions of the United States*; which was later revised as *Ecoregions of North America* (Bailey 1983). These formative maps provoked further interest in the delineation of ecoregions during the 1980s, and onward (Bailey 1980, 1983; Omernik 1987). Over time, the term ecoregion has

Table 2.1 Reverse chronological summary of the development of the term ecoregion in North America

Development of the Term Ecoregion in North America	References
Various updates to delineations of ecoregions of North America for potential natural vegetation, agricultural uses, aquatic, conservation, etc.	Many authors (2000 to current)
A conservation assessment of terrestrial ecoregions of North America	Ricketts (1999)
Ecoregions map of North America updates, and a book that delineates ecoregions of oceans and continents	Bailey (1998b) and Bailey (1998a)
Maps and delineation of ecoregions of terrestrial and aquatic ecoregions for the United States. Updates include agricultural uses	Omernik (1987), Omernik (1995) and Omernik and Griffith (1991)
Publication of delineation of ecosystem regions of the 1976 map by R. G. Bailey	Bailey (1983)
Ecoregions of North America, map	Bailey and Cushwa (1981)
Map "ecoregions of the United States" was produced to delineate major ecosystem types of natural vegetation in the United States	Bailey (1976)
Article outlining the concept of "biophysical" regions. Uses term ecoregion, and map titled ecoregions of Canada is published	Crowley (1967)
Canadian forester uses the term "ecoregion" to delineate subdivisions of forests. The targeted function was silviculture	Loucks (1962)

developed to serve a broader range of uses, by various groups for various purposes. Ecoregions can define various kinds of vegetation such as native, exotic (cultivated, naturalized, or invasive), or a combination of different kinds of vegetation depending upon the purpose of the ecoregion function.

Ecoregions are typically defined by their associations such as terrestrial, freshwater, or marine habitats (Ricketts 1999; Bailey 2004). For this book, we focus on terrestrial ecoregions. An ecoregion typically comprises one or several ecosystems that can be identified at various scales (Omernik 1995; Bailey 1998b, 2004). Different groups and organizations define ecoregions for different purposes. For example, the USDA Forest Service identifies ecoregions based upon the management of the natural vegetation or potential natural vegetation (Bailey 1995), while the Environmental Protection Agency (EPA) identifies ecoregions based upon existing vegetation which could include natural vegetation, crops, and horticultural landscapes (Omernik 1995). Each organization uses different means and methods to define and classify ecoregions and their boundaries at various scales (Bryce et al. 1999).

The term Ecozones is used in Canada (Wiken 1986). The World Wide Fund for Nature uses its own system of classification for the conservation of wildlife (Morrison et al. 2009), and the Sierra Club has its own classification for similar purposes (Bailey 2002). The Commission for Environmental Cooperation publishes its own system to identify major ecosystems in North America in attempts to unify classifications in North America (Bailey 2002).

Because ecoregions can be discussed at various scales from continental at the largest scale to regional or site level, there is a complex hierarchy of nomenclature used to define ecoregion names. Classifications typically begin with a hierarchy from large to small-scale applications such as a domain, division, province, and section (USFS), or by regions I, II, II (EPA). Such systems use alternative names for ecoregions at different scales. For example, in the US Forest Service ecoregion classification system, the naming of prairie ecosystems includes a progression of terms such as Prairies, Prairie Parkland, and Mesquite-Buffalo grass at different categorical scales (Bailey 1997). The US EPA recognizes much of the same geographic area referenced by the US Forest Service system as the Western Corn Belt Plains ecoregion as it identifies the dominant existing vegetation, and not the historic natural vegetation (Omernik 1995). For this book, we define ecoregions to identify the natural vegetation.

2.4.1 Plant Nativity and Human Culture

One of the functions of ecoregions is to define the boundaries or ranges where native or potential native vegetation and naturally occurring ecosystems occur (Bailey 1983). A native plant is a plant that occurs naturally in the place where it evolved (Wasowski and Wasowski 2003; USDA 2019). Some native plants are found growing widespread across the continents such as common yarrow (*Achillea millefolium*), where others are endemic to an isolated location, such as Texas tubrose (*Manfreda maculosa*), which is endemic to several counties of south-central Texas and northeast Mexico. While the debate on the specific meaning of native plants for green roofs will continue, for the purposes of this book, we define a native plant for green roofs as one that is native to its ecoregion (Sect. 2.6).

In this book, we use several primary authorities to identify information about plants such as the Biota of North America *bonap.org*, or the United States Department of Agriculture (USDA) Natural Resources Conservation Service (USDA) PLANTS Database (*plants.sc.egov.usda.gov*) There are also other databases referenced, such as those maintained by state-based Native Plant Societies such as Calscape (*calscape.org*). We use these widely accessible sources to identify the character and nomenclature of vegetation in this book. Many of these databases identify the ecosystem or ecoregion of a plant, locations, and habitats of native plants, and whether a plant has been introduced (exotic), become naturalized (adapted) or is invasive (aggressive) (USDA 2019).

Native plant communities are complex organizations that are more than the sum of the independent elements such as living organisms, inert materials, or the dynamic ecosystem processes and cycles (Leopold 1966). They can become altered, disturbed, conserved, and managed through restoration (Harker 1999). Plant communities are constantly in flux, but where disturbances occur they can become unstable, and decline, or collapse. Steady-state and biologically rich plant communities take time to develop, and strategic management practices may be needed to achieve stability (Magurran et al. 2010; McDonald et al. 2016). Steady-state (stable) plant communities are a targeted goal for many habitat conservation and restoration groups, and reference ecosystems are important for observation and understanding of ecosystem stability (McDonald et al. 2016).

While ecoregions are an excellent way to delineate native ecosystems, the urban and peri-urban spaces where most green roofs are located are highly influenced by humans (Forman 2014; Goudie 2018). These anthropogenic biomes or 'anthromes' demonstrate the impact that sustained human presence has on native ecoregions. At the intersection of fairly undisturbed ecoregions and anthromes can yield a blending of the two ecosystem types (Martin et al. 2014). Additionally, some of the land set aside as native ecosystems (national and state parks, private preserves) were influenced by humans for thousands of years through the burning of the land, or other human-induced activities (Vale 2013). Thus, some native ecosystems may be dependent upon human activities such as burning the land (Schramm 1990; Packard and Mutel 1997; Holl et al. 2003). For example, Native American tribes intentionally burned the ground vegetation to maintain coastal prairies along the Olympic Peninsula in what later became part of the Olympic National Park in Washington state (Wray and Anderson 2003). The prairie was an integral part of the Native American economic system, culture, and way of living with the land on the peninsula (Wray and Anderson 2003). As the burning of the ground-level vegetation retained the prairie on some parts of the peninsula lowlands, the absence of burning fostered the prairie to become encroached by forests. Today, the practice of prescribed ground fires is becoming part of the fire management program to maintain the prairie ecosystems and some forests. Thus, some of the land inside Olympic National Park and other locations in the Pacific Northwest may appear to be natural as forested land (to the uninformed), but they might not be the historic ecosystem if lcft alone, without prescribed burns. Such plant communities may be dependent upon human intervention such as prescribed burning, reseeding or extracting aggressive invasives to retain or restore the plant communities (i.e. prairie) that were present for millennia (Storm and Shebitz 2006; Charnley et al. 2007; Anderson 2009).

Many examples exist in North America where habitat restoration in urban regions has been effective, especially where there is a longstanding culture of restoration activities, research, and outreach ensues (Schramm 1990; Packard and Mutel 1997; Young et al. 2005). There is a growing interest in the potential ability of green roofs to be used as a reconciliatory way to make provisions for local and migratory users of urban ecosystems, especially where natural habitats have already been significantly reduced or are no longer present (Rosenzweig 2003; Dunnett 2006; Coffman and Waite 2011; Francis and Lorimer 2011; Kowarik 2011; MacIvor and Lundhom 2011a, b). The 26 conservation site case studies in this book (Part II) demonstrate how various management activities are being used to conserve vegetation from many diverse ecoregions and how these sites can inform the design and management of green roofs.

2.5 Development of Ecoregional Green Roofs

In North America, much of the early research on the use of native plants on green roofs emerged from the disciplines of ecology, horticulture, engineering, and landscape architecture (Hauth and Liptan 2003; Livingston et al. 2004; Rowe et al. 2005; Lundholm 2006; Bousselot et al. 2010; Lundholm and Richardson 2010). Over time, this diverse group of participants began to develop a well-rounded argument to demonstrate how vegetation from ecoregions can be employed on green roofs in North America. This section outlines the theory, development, and practice of using vegetation native to ecoregions on green roofs.

Much of the academic research regarding the use of native vegetation on green roofs (in North America) has focused on the identification of potential plant taxa, plant survival, and plant health. Much of this work has taken place on small-scale experimental plots. There has been some applied research regarding the selection of plants based upon the natural ecosystems, either in separate designs, or integrated with introduced vegetation (Table 2.2).

As green roofs have become adopted into municipal ordinances and other regulatory uses, there has been a need for municipalities to address guidelines or regulations for green roofs, including vegetation (Table 2.3). Most of these documents provide examples for the use of native vegetation on green roofs; some provide plant species lists, while others discuss concepts such as biological diversity for green roofs. Although climate-adapted vegetation and biodiverse vegetation are recommended for use on green roofs, native vegetation is not required (Table 2.3).

Table 2.2 Reverse chronological summary of studies with a focus on native plants identified from Ecoregions for green roofs in North America

Development of Green Roof Research with Vegetation from the Ecoregion	References
Vegetation study based upon plants identified and selected through EPA ecoregions in New York state	Mandel et al. (2016)
Study on green roof plant trials with vegetation that is native to the central Great Plains	Sutton (2015b)
112 plant tax selected for green roofs growing in high elevation ecoregions were tested over five years. The study provides a benchmark for green roofs in shortgrass prairie and semi-arid ecoregions	Schneider et al. (2014)
This study investigates native and climate-adapted vegetation on three green roofs in Texas	Dvorak et al. (2013)
Studies of vegetation from the Pacific Northwest ecoregions on extensive green roofs	Schroll et al. (2011) and Martin and Hinckley (2007)
Plants native to shallow, rocky foothills evaluated *in situ* on the EPA region 8 headquarters green roof	Bousselot et al. (2010)
Conference presentation and paper regarding an ecological approach to green roofs, with case study example of pilot testing for the California Academy of Sciences green roof	Kephart (2005)
Native plant studies used on green roofs from coastal ecoregions in Canada (Nova Scotia)	Licht and Lundholm (2006)
Plant study regarding sedums and vegetation native to Michigan	Rowe et al. (2005)
Several papers provide results from North American projects that made use of native vegetation at the First Greening Rooftops for Sustainable Communities conference in Chicago, IL	Sharp (2003), Hauth and Liptan (2003), Shiner (2003), Pearce (2003) and Dvorak (2003)

Table 2.3 Reverse chronological development of references to green roofs regarding native vegetation, or local biodiversity discussed in guidelines, manuals, or other municipal publications

Development of Regulations for Ecoregional Green Roofs	References
The City of Denver's rules and regulations for the green building ordinance suggests using native plant lists provided by the Colorado Native Plant Society	COD (2019)
The City of San Francisco's guideline for green roof vegetation includes a discussion of native plants as well as direction for how to provide habitat on rooftops and includes case studies of green roofs with native vegetation	SFLRM (2015)
ASTM guideline for vegetated roofs includes the potential value of natural habitat on green roofs, "vegetative (green) roof systems may provide habitat to local and migratory wildlife" pg. 12	ASTM E2777 14 (2014)
The City of Toronto, Ontario, published a comprehensive guideline for the development of biodiverse green roofs, including the use of native vegetation and other elements to consider	Torrance et al. (2013)
FLL guidelines for green roofs define extensive green roofs as synonymous with natural vegetation, "…plants should be natives of Central Europe or plants which are fully integrated into that climate" (sect. 3.1.4 extensive greening, pg. 16)	FLL (2008)
ASTM standard guide for vegetative (green) roof systems (ASTM E2777–14) encourages "connections with wildlife corridors" pg. 2	ASTM E 2400 (2006)
Paper discusses major concepts of the German FLL guidelines for green roofs, first translation in English, by German authors	Philippi (2005)

There are other guidelines and regulations regarding green roofs published by North American organizations or municipalities, but the references cited in Table 2.3 make specific reference to the use of native vegetation. Due to the lack of research and industry support, there perhaps is not enough information yet regarding how native plants can be used on green roofs in North America.

There has been some theoretical development of the use of native vegetation on green roofs, and it has come from varied sources. Fundamental ideas such as reference or analog landscapes where native vegetation may be used on green roofs is not a new idea (Chap. 1). However, in North America, as the green roof industry arrived after 2000, a variety of approaches were used for planting green roofs: all exotic vegetation, hybrid of exotics with some natives, and all native vegetation. In publications, the development of theory and reasons why native vegetation may be valid for green roofs is still emerging (Table 2.4).

As outlined in the literature here (Tables 2.1, 2.2, 2.3 and 2.4), various terms and concepts are introduced regarding the use of native plants on green roofs. Terms such as habitat templates, biodiverse roofs, native green roofs, living roofs, ecoroofs, and other similar terms are used in some of these documents and convey similar concepts. These references associate common ideas with the term "ecoregion" for green roofs (Dvorak and Volder 2010; Schroll et al. 2011; Dvorak 2015b; Mandel et al. 2016; Catalano et al. 2018).

Table 2.4 Reverse chronological development of theory for ecoregional green roofs

Contribution to the Theoretical Development of Ecoregional Green Roofs	References
Paper explores 'ecomimicry' as an approach to ensure that biodiversity and locally important habitats are created on extensive green roofs	Nash et al. (2019)
Paper explores implications of guidelines and standards for green roofs in Europe in context to ecoregions	Catalano et al. (2018)
Green Roof Ecosystems was the first book (in English) to focus on green roofs as ecosystems. The edited chapters demonstrate how green roofs exhibit many traits of ecosystems, and there is a great need for rigorous research on green roof ecosystems. The book includes a chapter on ecoregional green roofs	Sutton (2015a) and Dvorak (2015b)
This paper reviews the use of native vegetation on green roofs. Knowledge and perceptions about native plants used on green roofs are examined. Outcomes of the study clarify a need for greater rigor and transparency when promoting the use of native plants on green roofs	Butler et al. (2012)
A literature review that covers different forms of vegetation used on green roofs in north American ecoregions and their provisions for wildlife	Dvorak and Volder (2010)
Fundamental paper that introduces the concept of referencing natural habitats as templates to source vegetation for green roofs	Lundholm (2006)
Pioneering paper that demonstrates how green roofs can be designed to become biologically diverse. Demonstrates concepts such as varying the substrate thickness and the inclusion of natural soils and materials (stones, branches) to construct a habitat for targeted or endangered species	Brenneisen (2006)
This paper presents key concepts in the early development of bio-diverse green roofs	Brenneisen (2003)

Although the term ecoregion is beginning to be associated with green roofs (Table 2.4), the term "eco-regional green roof" was first used in *Green Roof Ecosystems* (Sutton 2015a). It was defined as a green roof that "expands or profits the immediate urban or regional landscape ecology or ecosystem (Kowarik 2011; Forman 1995). An eco-regional green roof sustains local or regionally important plant species, plant communities, organisms, functions, or services (Lundholm 2005; Sutton et al. 2012; Dvorak and Volder 2010)." (Dvorak 2015a, b pp. 392–393).

We presume to make use of the above-cited definition to the application of ecoregional green roofs in this book. The spelling "ecoregion" is preferred in this book over the previous spelling "eco-region", due to its now widespread adoption. Furthermore, we have assessed that the 73 ecoregional green roof case studies in Part II fall into one of four categories ranged from most (1) to least (4) aligned with the adoption of native vegetation from an ecoregion:

1) A green roof where all of the intended vegetation is native to the site, or nearby;
2) A green roof where the intended vegetation is native to one or several adjacent ecoregions (i.e. vegetation from an alpine ecosystem that is planted on a green roof, in a valley below);
3) A green roof where the intended vegetation is native to various, but similar ecoregions of North America (west of the 100th meridian);
4) A green roof with vegetation from a combination of native (west of the 100th meridian) and non-native (introduced/exotic) vegetation. These climate-adapted green roofs feature native plants, but also include non-native plants.

These categories define a spectrum of applications of native plants on green roofs. Therefore, there are green roofs that use native vegetation exclusively, and those that make use of some exotic vegetation. The case studies in Part II cover all four categories, and we recognize that the green roof industry in North America is in the early adoption phase of a relatively new idea.

As a new idea, support from the research community is needed to provide feedback. However, the traditional approach to research may need to be expanded. In his book, *Ecoregion-Based Design for Sustainability*, Robert Bailey says, "Given this holistic view of nature, I see a need for evolutionary change in educational institutions to reduce compartmentalization of academic disciplines. Reductionism still holds sway in science" (Bailey 2007, pg. 17).

As such, there is a growing interest in the design of self-sustaining buildings, and many of these include green roofs as integral components through an integrated design process, which involves multiple professions and perspectives. Buildings can be designed to include renewable energy sources (solar panels with green roofs), water efficiency (harvesting roof water for irrigation), capturing and recycling wastewater (using green roofs to clean wastewater and irrigate), energy efficiency (reduce heat gain via green roofs), natural lighting, and more (Loftness 2019). If buildings of the future will become more like living systems, (i.e. *Living Building Challenge*), then there is a great deal of rigorous interdisciplinary research that will be needed. It seems that the foundational beginning point for this approach is a viewpoint that ecosystems can inform the design of buildings if they become integrated into landscape and ecosystem functions, and various points of view are included. As such, an integrated approach can produce self-sustaining buildings and green roofs can be vital components of integrated design (Kephart 2005; Brenneisen 2006; Lundholm 2006; Sutton 2015c).

Where green roof policies are in place, municipalities can encourage interdisciplinary engagement, more ecologically meaningful green roofs, and promote and educate about native or naturalized vegetation on green roofs (SFLRM 2015). However, many of the current green roof policies in place emphasize human-centered benefits, and attention to habitat or species complexity and diversity is not yet a priority (Peck and Goucher 2005). Pilot projects and publicly accessible examples of green roofs with native vegetation are needed for such an approach to thrive.

This book opens up a dialog for interdisciplinary learning and practice that will benefit from further development, exploration, and documentation. Researchers, designers, and educators will gain by learning how to observe and appreciate the ecological functions of green roofs from a diversity of perspectives.

2.6 Ecoregions of the Western United States and Canada

Our goal here in this book is to identify the natural terrestrial vegetation found growing in a region, and show the physiographic areas where ecosystems were historically present. We present a series of ecoregion maps (Fig. 2.8), and in successive chapters. They were inspired by several existing delineations, and terminologies used from various sources. We make a consistent terminology in ecoregions and physiographic areas to identify the natural vegetation in a consistent format.

Regarding context, we presume that ecoregions represent current day or potential natural vegetation. As there may be several terms for some of the plant communities identified, we default to consistency, when feasible. For example, while some sources use the term "steppe", we recognize prairie as a commonly recognized term for the grassland and wildflower vegetation. Three divisions of prairie are recognized based upon plant stature: Tallgrass Prairie, Mixed-grass Prairie, and Shortgrass Prairie. Some ecoregion classification systems use the term steppe to classify shortgrass species, but to be morphologically consistent we use the term, Shortgrass Prairie.

Since the physical geography of the landscape west of the 100th meridian is a primary organizing factor in determining vegetation, we use common physical geography terms of the land to organize ecoregions at the domain or providence level. This allows divisions of landscape vegetation that are influenced by major factors such as topographic elevation (altitude), temperature, humidity, and precipitation as a way to provide ecological context.

Twenty ecoregions are identified for the major physiographic areas in this book including Coastal Plains, Interior Lowlands, Great Plains, Rocky Mountains, Colorado Plateau, Basin and Range, Ozark Plateaus, and Coastal Mountains.

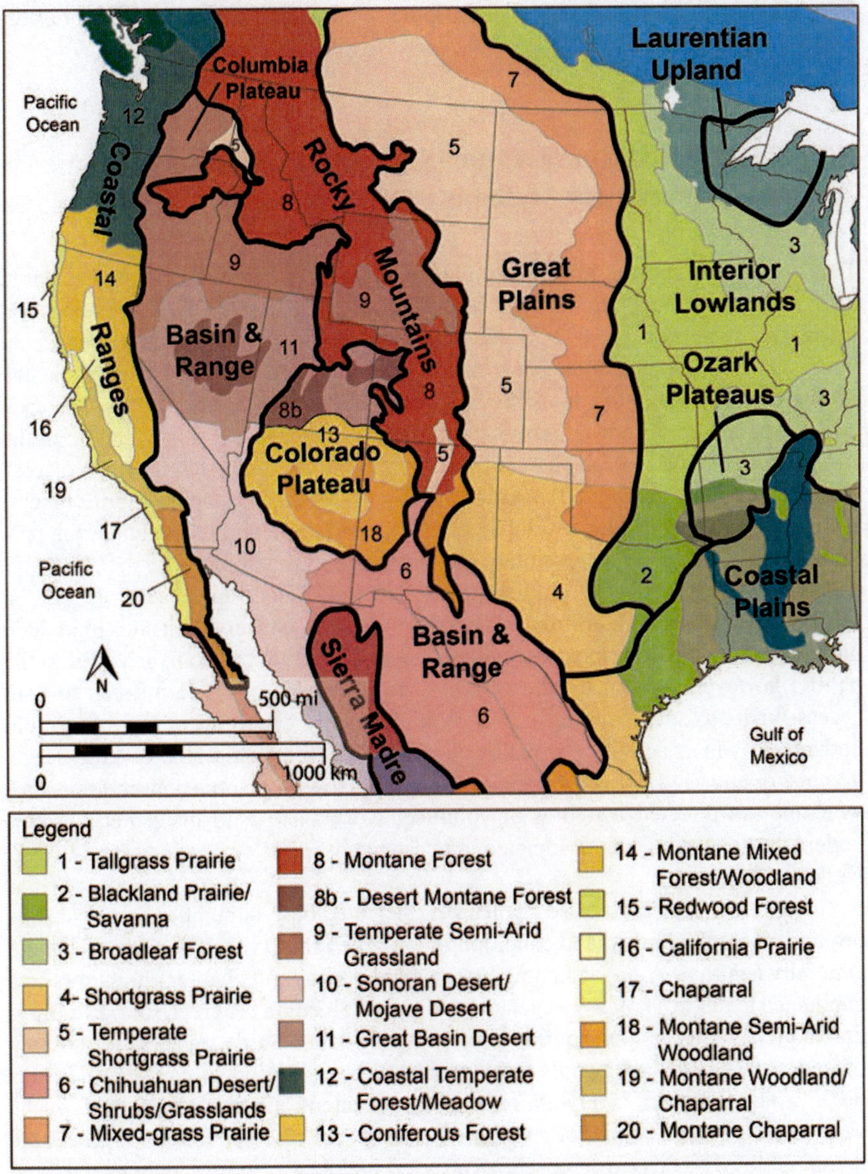

Fig. 2.8 North American ecoregions for potential green roof vegetation. This map identifies the major physiographic regions and their associative ecoregions in the central and western landscapes of North America (Hammond 1964). The designations identify the kind of natural vegetation that either currently or historically populated the region. Ecotones are transitional areas between ecoregions and some boundaries presented here are simplified. Ecoregion boundaries used in this book are largely influenced by the works of Bailey (1995), and Omernik (1995). Ecoregion names stem from various sources. (Graphic by Trevor Maciejewski and Bruce Dvorak)

Ecoregions are located within physiographic regions, and detailed descriptions are provided in each chapter in Section II.

2.7 Ecoregional Conservation Sites and Ecoregional Green Roof Case Study Methods

This book represents the first intentional collection of green roof case studies based upon ecoregional vegetation. In Part II, 26 conservation case studies of remnant and or restored natural sites, and 73 ecoregional green roofs are discussed. Remnant landscapes are located in and near major urban centers that exist throughout the ecoregions covered in this book. Several remnant landscapes are discussed in each case study chapter as a brief introduction to the characteristics of vegetation in the ecoregions covered in each chapter. Plant communities and different forms of vegetation currently growing in the region are discussed and presented. A brief natural and cultural history of the development of the landscape and its vegetation is provided as well as a brief discussion of the climate in the region.

Conservation site case studies were selected based upon several criteria. Accessibility was a high priority in determining which conservation sites to include in the chapters. Most of the conservation sites covered in this book are open to the public; however, a few sites are privately owned and require permission to gain access. Visits to sites covered in this book are highly encouraged. Visitors should understand which activities are allowed or not. Walking off trails, picking seeds, flowers, or any kind of disturbance is prohibited. However, some conservation sites welcome the directed activities of volunteer workers to help restore ecosystems under the direction of designated site stewards (Packard and Mutel 1997; Harker 1999).

Ecoregional green roof case studies were selected based upon their accessibility, proximity to a major city, and condition of the green roof (vegetation is present and generally healthy). A pre-screening by the main author and chapter co-authors was conducted to discern if native vegetation was maintained on the green roof. Buildings under private ownership required pre-arranged visits. Some owners of privately owned ecoregional green roofs were contacted but declined to participate in this book project. Owners of private residences are intentionally not identified in the case studies. Private offices or corporate buildings are also generally not accessible to the public. The private corporations covered in the case studies were generous to share information about their green roofs, to be photographed, and discussed for educational purposes, and to exemplify corporate social responsibility (Eichholtz et al. 2009; Dudley 2013). For these reasons, many of the conservation and green roof case studies included in Part II were selected because they are accessible to the public.

The main author (and some chapter co-authors) visited 140 green roofs (some had multiple green roofs) located in the western U.S.A. and Canada during 2018 and 2019. Not all of the green roofs that were visited were included in this book, as it was learned that a few of the green roofs were no longer maintained, and some green roofs did not have native vegetation. The ecoregional green roofs included in the book represent ecoregional green roofs, some more than others (Sect. 2.5). Some green roofs had goals other than a pure expression of native vegetation, and native plants were included, but the green roof was not intended to represent a pure expression of a native green roof. Because green roofs are in the early adoption phase in North America, the number of all native green roofs is just emerging.

2.7.1 Ecoregional Case Study Methods

The methods used to collect information about the case studies include site visits to each conservation site and a green roof, interviews, and online data searches. Interviews with current owners, designers, maintenance staff, or project representatives took place at the case study sites. Phone calls or contacts through e-mail replaced on-site interviews for some case studies. Information from published documents was also used to support information gathered during site visits. Post-occupancy observations include notable published findings, comments from designers or managers of green roofs. Preference was given to peer-reviewed publications; however, online web sites, databases, reports, and other media were included where necessary.

Green roof case studies cover an introduction to the project, identification of project goals, objectives, and approach to plant establishment. Discussions include maintenance practices, sources of irrigation water and irrigation use, and general care and upkeep for the green roof. Each case study includes a post-occupancy summary of what was learned. This could include observations by the owners, designers, researchers, regulators, and the authors' reflections.

Ecoregional green roof case study chapters:

- Ecoregion characteristics (natural and cultural history, climate summary)
- Ecoregional conservation sites
- Summary of published peer-reviewed green roof research within the ecoregion
- Conservation Case Studies:
  - Natural and cultural history
  - Identification of dominant ecosystems present
  - Overview of vegetation
- Ecoregional Green Roof Case Studies:
  - Project Introduction
  - Project Team
  - Overview and Objectives

  - Plant Establishment (forms of native vegetation installed on the green roof). Some green roofs have exotic vegetation, and may not be identified.
  - Irrigation
  - Maintenance
  - Observed Wildlife
  - Post-occupancy Observations

- Chapter summary includes a summation of some of the popular species used on ecoregional green roofs and an overview of how green roofs were used in case studies.

2.8 Summary

The collective amalgamation of conventional roofing (non-vegetated) and impervious surfaces throughout urbanized regions can result in significant ecosystem disturbances such as urban flooding, thermal and chemical pollution of runoff, loss of groundwater recharge and reduced evapotranspiration from the displaced vegetation, the development of urban heat islands, wasted energy in buildings, and the displacement of habitat that sustains local and migrating wildlife. Green roofs can minimize the ecological disturbances of building construction as they retain and delay rooftop stormwater runoff, keep air temperatures similar to vegetated landscapes, conserve energy in buildings, and potentially can support native biodiversity. When vegetation from local ecoregions is used, green roofs have the potential to mimic some of the same ecological benefits provided by their native analogs (reference landscapes) such as habitat patches or habitat corridors. However, knowledge of ecoregions, ecosystems, their structure, function, and evolved growing conditions for plants is needed to become a source of inspiration for green roofs. This chapter reviewed such knowledge as a background to understanding the underlying goals, objectives, and functions of ecoregional green roofs covered in Part II.

The harsh environments that can exist on rooftops, coupled with the unique, human-altered landscapes of urban spaces, can challenge the successful use of native plants on green roofs. This chapter outlined a background discussion regarding the application of green roofs when inspired by ecoregions. This includes environmental, ecological, microclimatic, and aesthetic factors for the design and use of ecoregional green roofs and some of the innovations that are taking place through integrated designs that allow resilient and reliable sources of water for green roofs. The theory and development of the concept of ecoregions and ecoregional green roofs were discussed. Ecoregions of North America were identified for the areas covered in this book, and for gathering information used in the case studies of conservation sites and green roofs. The methods used for the systematic gathering of information were described. This chapter provided a theoretical and background discussion for the 26 case studies of ecoregional conservation sites and 73 ecoregional green roof case studies presented in Part II of this book.

References

Anderson K (2009) The Ozette prairies of Olympic National Park: their former Indigenous uses and management. Pacific West Region, National Park Service, Port Angeles, Washington, DC

Arendt RG (2013) Growing greener: putting conservation into local plans and ordinances. Island Press, Washington, DC

ASTM E 2400 (2006) Standard guide for selection, installation, and maintenance of plants for green roof systems. ASTM International, West Conshohochen

ASTM E2777 14 (2014) Standard guide for vegetative (green) roof systems. ASTM, West Conshohocken

Bailey RG (1976) Ecoregions of the United States (map). USDA Forest Service, Intermountain Region, Ogden

Bailey RG (1980) Description of the ecoregions of the United States, vol 1391. US Department of Agriculture, Forest Service, Washington, DC

Bailey RG (1983) Delineation of ecosystem regions. Environ Manag 7(4):365–373

Bailey R (1995) Description of the ecoregions of the United States, 2nd. Edn. Rev. and Expanded. Misc. Publ, vol 1391. USDA, Washington, DC

Bailey RG (1997) Ecoregions of North America. U.S. Department of Agriculture, Forest Service, Washington, DC

Bailey RG (1998a) Ecoregions map of North America: explanatory note, vol 1548. US Department of Agriculture, Forest Service, Washington, DC

Bailey RG (1998b) Ecoregions: the ecosystem geography of the oceans and continents. Springer, New York

Bailey RG (2002) How land-management agencies, conservation organizations, and others use ecoregion maps. Ecoregion-Based Design for Sustainability: 133–151

Bailey RG (2004) Identifying ecoregion boundaries. Environ Manag 34(1):S14–S26

Bailey RG, Cushwa CT (1981) Ecoregions of North America: after the classification of J.M. Crowley. UNT Libraries Government Documents Department, Fort Collins

Beattie D, Berghage RD (2008) Designing green roof growing medium. In: ASTM & FLL standards workshop at the sixth annual international greening rooftops for sustainable communities conference, awards and trade show, Baltimore, MD, April 30, 2008. Green Roofs for Healthy Cities, pp 13–20

Bender DJ, Contreras TA, Fahrig L (1998) Habitat loss and population decline: a meta-analysis of the patch size effect. Ecology 79(2):517–533

Bennie J, Huntley B, Wiltshire A, Hill MO, Baxter R (2008) Slope, aspect and climate: spatially explicit and implicit models of topographic microclimate in chalk grassland. Ecol Model 216(1):47–59

Best BB, Swadek RK, Burgess TL (2015) Soil-based green roofs. In: Green roof ecosystems. Springer, Cham, pp 139–174

Bogner FX (1998) The influence of short-term outdoor ecology education on long-term variables of environmental perspective. J Environ Educ 29(4):17–29

Boivin M-A, Lamy M, Gosselin A, Dansereau B (2001) Effect of artificial substrate depth on freezing injury of six herbaceous perennials grown in a green roof system. Hort Technology 11(3):409–412

Bousselot JM, Klett JE, Koski RD (2010) Extensive green roof species evaluations using digital image analysis. HortScience 45(8):1288–1292

Brenneisen S (2003) The benefits of biodiversity from green roofs: key design consequences. Paper presented at the 1st North American Green Roof Conference: Greening Rooftops for Sustainable Communities, Chicago, Illinois, 29-30 May

Brenneisen S (2006) Space for urban wildlife: designing green roofs as habitats in Switzerland. Urban Habitats 4(1):27–36

Brown C, Lundholm J (2015) Microclimate and substrate depth influence green roof plant community dynamics. Landsc Urban Plan 143:134–142

Bryce SA, Omernik JM, Larsen DP (1999) Environmental review: ecoregions: a geographic framework to guide risk characterization and ecosystem management. Environ Pract 1(3):141–155
Buffam I, Mitchell ME (2015) Nutrient cycling in green roof ecosystems. In: Green roof ecosystems. Springer, Cham, pp 107–137
Butler C, Butler E, Orians CM (2012) Native plant enthusiasm reaches new heights: perceptions, evidence, and the future of green roofs. Urban For Urban Green 11(1):10. https://doi.org/10.1016/j.ufug.2011.11.002
Byerley B (2014) Personal communication on the plant diversity on the BRIT green roof
Cantor S (2008) Green roofs in sustainable landscape design. Norton, New York
Carter T, Jackson CR (2007) Vegetated roofs for stormwater management at multiple spatial scales. Landsc Urban Plan 80(1–2):84–94
Catalano C, Laudicina VA, Badalucco L, Guarino R (2018) Some European green roof norms and guidelines through the lens of biodiversity: do ecoregions and plant traits also matter? Ecol Eng 115:15–26
Charnley S, Fischer AP, Jones ET (2007) Integrating traditional and local ecological knowledge into forest biodiversity conservation in the Pacific Northwest. For Ecol Manag 246(1):14–28
COD (2019) City and county of Denver rules & regulations GoverningGreen building requirements. Department of Community Planning and. Development & Department of Public Health and Environment, Denver
Coffman RR, Davis G (2005) Insect and avian Fauna presence on the ford assembly plant Ecoroof. Paper presented at the Third Annual Greening Rooftops for Sustainable Communities Conference, Washington, D.C., May 4-6, 2005
Coffman RR, Waite T (2011) Vegetated roofs as reconciled habitats: rapid assays beyond mere species counts. Urban Habitats 6(1):10
Cook-Patton SC, Bauerle TL (2012) Potential benefits of plant diversity on vegetated roofs: a literature review. J Environ Manag 106:85–92
Cramer JR (2008) Reviving the connection between children and nature through service-learning restoration partnerships. Native Plants J 9(3):278–286
Crowley JM (1967) BIOGEOGRAPHY. The Canadian Geographer / Le Géographe Canadien 11(4):312–326. https://doi.org/10.1111/j.1541-0064.1967.tb00474.x
Dimmitt MA (2000) Plant ecology of the Sonoran Desert region. In: A natural history of the Sonora Desert. Desert Museum Press and University of California Press, Tucson/Berkeley, pp 129–151
Dudley B (2013) Google boom: Kirkland campus to double. The Seattle Times
Dunnett N (2006) Green roofs for biodiversity: reconciling aesthetics with ecology. Paper presented at the Fourth Annual Greening Rooftops for Sustainable Communities Conference, Boston, MA., May 11-12, 2006
Dunnett N (2015) Ruderal green roofs. In: Sutton RK (ed) Green roof ecosystems. Springer International, Cham, pp 233–255. https://doi.org/10.1007/978-3-319-14983-7_10
Dunnett N, Kingsbury N (2004) Planting green roofs and living walls. Timber Press, Portland
Dunster K, Coffman RR (2015) Placing green roofs in time and space: scale, recruitment, establishment, and regeneration. In: Sutton RK (ed) Green roof ecosystems. Springer, Cham, pp 357–390. https://doi.org/10.1007/978-3-319-14983-7_15
Durhman AK, Rowe BD, Rugh CL (2007) Effect of substrate depth on initial growth, coverage, and survival of 25 succulent green roof plant taxa. HortScience 42(3):588–595
Dvorak B (2003) The greening of a nature museum–a demonstration project. Paper presented at the First Annual Greening Rooftops for Sustainable Communities Conference, Chicago, IL, 29-30 May
Dvorak B (2011) Comparative analysis of green roof guidelines and standards in Europe and North America. J Green Build 6(2):170–191. https://doi.org/10.3992/jgb.6.2.170
Dvorak B (2015a) Conserving energy with biodiverse building skins: a review of literature. Paper presented at the Advanced Building Skins, 10th Annual Conference, Bern, Switzerland, November 3-5

Dvorak B (2015b) Eco-regional green roof case studies. In: Sutton RK (ed) Green roof ecosystems. Springer, Cham, pp 391–421
Dvorak B, Carroll K (2008) Chicago City hall green roof: its evolving form and care. Paper presented at the Sixth Annual International Rooftops for Sustainable Communities Conference, Baltimore, MD, April 30-May 2
Dvorak B, Volder A (2010) Green roof vegetation for north American ecoregions: a literature review. Landsc Urban Plan 96(4):197–213
Dvorak B, Byerley B, Volder A (2013) Plant species survival on three water conserving green roofs in a hot humid subtropical climate. J Living Archit 1(1):10
EarthPledge (2005) Green roofs ecological design and construction. Schiffer Books, Atglen
Eichholtz P, Kok N, Quigley JM (2009) Why do companies rent green? Real property and corporate social responsibility. Real Property and Corporate Social Responsibility (August 20, 2009) Program on Housing and Urban Policy Working Paper (W09-004):49
Fernandez-Canero R, Gonzalez-Redondo P (2010) Green roofs as a habitat for birds: a review. J Anim Vet Adv 9(15):2041–2052. https://doi.org/10.3923/javaa.2010.2041.2052
FLL (2008) Guidelines for the planning, construction and maintenance of green roofing, English edn. Forschungsgesellschaft Landschaftsentwicklung Landschaftsbau e. V, Bonn
Folsom DB (1995) Dry climate gardening with succulents. Pantheon Books, New York
Forman RT (2014) Urban ecology: science of cities. Cambridge University Press, Cambridge, UK
Forman RT, Godron M (1986) Landscape ecology. Wiley, New York
Francis RA, Lorimer J (2011) Urban reconciliation ecology: the potential of living roofs and walls. J Environ Manag 92(6):1429–1437. https://doi.org/10.1016/j.jenvman.2011.01.012
Gaffin SR, Khanbilvardi R, Rosenzweig C (2009) Development of a green roof environmental monitoring and meteorological network in New York City. Sensors 9(4):2647–2660
Gedge D (2003) From rubble to redstarts. Paper presented at the 1st Annual Greening Rooftops for Sustainable Communities, Chicago, IL, 29-30 May
Gedge D, Kadas G (2004) Bugs, bees and spiders: green roof Design for Rare Invertebrates. Paper presented at the Second Annual Greening Rooftops for Sustainable Communities Conference, Portland, Oregon, Jun 2-4, 2004
Gedge D, Kadas G (2005) Green roofs and biodiversity. Biologist 52(3):161–169
Gibson DJ, Donatelli JM, AbuGhazaleh A, Baer SG, Johnson LC (2016) Ecotypic variation in forage nutrient value of a dominant prairie grass across a precipitation gradient. Grassl Sci 62(4):233–242
Goudie AS (2018) Human impact on the natural environment. Wiley, Hoboken
Hammond E (1964) Classes of land-surface form in the United States. US Geological Survey
Harker D (1999) Landscape restoration handbook. CRC Press, Boca Raton
Hauth E, Liptan T (2003) Plant survival findings in the Pacific Northwest. Paper presented at the 1st North American Green Roof Infrastructure Conference: Greening Rooftops for Sustainable Communities, Chicago, IL, May 29-30
Holl KD, Crone EE, Schultz CB (2003) Landscape restoration: moving from generalities to methodologies. Bioscience 53(5):491–502
Jim CY, Tsang S (2011) Biophysical properties and thermal performance of an intensive green roof. Build Environ 46(6):1263–1274
John J, Lundholm J, Kernaghan G (2014) Colonization of green roof plants by mycorrhizal and root endophytic fungi. Ecol Eng 71:651–659
Kadas G (2006) Rare invertebrates colonizing green roofs in London. Urban Habitats 4(1):66–86
Kephart P (2005) Living architecture – an ecological approach. Paper presented at the Third Annual Greening Rooftops for Sustainable Communities Conference, Washington, D.C., May 4-6
Köhler M (2006) Long-term vegetation research on two extensive green roofs in Berlin. Urban Habitats 4(1):3–26
Köhler M, Poll PH (2010) Long-term performance of selected old Berlin greenroofs in comparison to younger extensive greenroofs in Berlin. Ecol Eng 36(5):722–729

Kowarik I (2011) Novel urban ecosystems, biodiversity, and conservation. Environ Pollut 159(8–9):1974–1983. https://doi.org/10.1016/j.envpol.2011.02.022

Ksiazek K (2014) The potential of green roofs to support urban biodiversity. In: MKhaldF JB (ed) Green cities in the world. Pronatur, Madrid, pp 101–124

Lambrinos JG (2015) Water through green roofs. In: Green roof ecosystems. Ecological Studies. Springer, Cham, pp 81–105

Leopold A (1966) A sand county almanac: with essays on conservation from round river. Oxford University Press, Inc., United States of America

Licht J, Lundholm J (2006) Native coastal plants for northeastern extensive and semi-intensive green roof trays: substrates, fabrics and plant selection. Paper presented at the Fourth Annual Greening Rooftops for Sustainable Communities Conference, Boston, MA, May 11-12, 2006

Lin BB, Perfecto I, Vandermeer J (2008) Synergies between agricultural intensification and climate change could create surprising vulnerabilities for crops. Bioscience 58(9):847–854

Livingston EH, Miller C, Lohr M (2004) Green roof design and implementation in Florida. Paper presented at the Second Annual Greening Rooftops for Sustainable Communities Conference, Portland, OR, June 2-4, 2004

Loftness V (2019) Environmental surfing: design for human health and environmental sustainability. In: ARCC 2019: Bridging Research & Practice, Ryerson University, Toronto, ON, 2019. ARRC, p 132

Loucks OL (1962) A forest classification for the maritime provinces. Proc Nova Scotian Inst Sci 25(Part 2):85–167

Luckett K (2009) Green roof construction and maintenance. McGraw-Hill, New York

Lundholm JT (2006) Green roofs and facades: a habitat template approach. Urban Habitats 4(1):87–101

Lundholm JT (2016) Spontaneous dynamics and wild design in green roofs. Israel J Ecol Evol 62(1–2):23–31

Lundholm JT, Richardson PJ (2010) MINI-REVIEW: habitat analogues for reconciliation ecology in urban and industrial environments. J Appl Ecol 47(5):966–975. https://doi.org/10.1111/j.1365-2664.2010.01857.x

MacDonagh LP, Shanstrom N (2015) Assembling prairie biome plants for Minnesota green roofs. In: Sutton RK (ed) Green roof ecosystems. Springer International, Cham, pp 257–283. https://doi.org/10.1007/978-3-319-14983-7_11

MacIvor JS, Ksiazek K (2015) Invertebrates on green roofs. In: Sutton RK (ed) Green roof ecosystems. Springer, Cham, pp 333–355

MacIvor JS, Lundholm J (2011a) Insect species composition and diversity on intensive green roofs and adjacent level-ground habitats. Urban Ecosyst 14(2):225–241. https://doi.org/10.1007/s11252-010-0149-0

MacIvor JS, Lundholm J (2011b) Reconciliation ecology opportunities reach new heights: Insect species composition and diversity on green roofs and adjacent ground-level habitat patches in an urban area In: America ESo (ed) 96the ESA Annual Meeting, Austin, Texas, August 7–12 2011. ESA, p 1

MacIvor JS (2016) Building height matters: nesting activity of bees and wasps on vegetated roofs. Israel J Ecol Evol 62(1–2):88–96

Madre F, Clergeau P, Machon N, Vergnes A (2015) Building biodiversity: vegetated façades as habitats for spider and beetle assemblages. Global Ecol Conserv 3(0):222–233. https://doi.org/10.1016/j.gecco.2014.11.016

Magurran AE, Baillie SR, Buckland ST, Dick JM, Elston DA, Scott EM, Smith RI, Somerfield PJ, Watt AD (2010) Long-term datasets in biodiversity research and monitoring: assessing change in ecological communities through time. Trends Ecol Evol 25(10):574–582

Mandel L, McCoy E, Liss T (2016) Reference community: adapting native plants to North American green roofs. J Green Build 11(4):15–36

Martin MA, Hinckley T (2007) Native plant performance on a Seattle green roof. Paper presented at the Fifth Greening Rooftops for Sustainable Communities Conference, Minneapolis, MN, April–1 May

Martin LJ, Quinn JE, Ellis EC, Shaw MR, Dorning MA, Hallett LM, Heller NE, Hobbs RJ, Kraft CE, Law E (2014) Conservation opportunities across the world's anthromes. Divers Distrib 20(7):745–755

McDonald T, Gann G, Jonson J, Dixon K (2016) International standards for the practice of ecological restoration–including principles and key concepts.(Society for Ecological Restoration: Washington, DC, USA.). Soil-Tec, Inc.,© Marcel Huijser, Bethanie Walder. Society for Ecological Restoration, Washington, DC

McGuire KL, Payne SG, Palmer MI, Gillikin CM, Keefe D, Kim SJ, Gedallovich SM, Discenza J, Rangamannar R, Koshner JA, Massmann AL, Orazi G, Essene A, Leff JW, Fierer N (2013) Digging the new York City skyline: soil fungal communities in green roofs and City parks. PLoS One 8(3):e58020. https://doi.org/10.1371/journal.pone.0058020

McGuire KL, Payne SG, Orazi G, Palmer MI (2015) Bacteria and fungi in green roof ecosystems. In: Green Roof Ecosystems. Springer, pp. 175–191

Melnick R (2001) Ecology and design: frameworks for learning. Island Press, Washington, DC

Morrison J, Loucks C, Long B, Wikramanayake E (2009) Landscape-scale spatial planning at WWF: a variety of approaches. Oryx 43(4):499–507

Nagase A, Dunnett N (2012) Amount of water runoff from different vegetation types on extensive green roofs: effects of plant species, diversity and plant structure. Landscape Urban Plan 104(3–4):356–363. https://doi.org/10.1016/j.landurbplan.2011.11.001

Nash C, Ciupala A, Gedge D, Lindsay R, Connop S (2019) An ecomimicry design approach for extensive green roofs. J Living Archit 6(1):62–81

Omernik JM (1987) Ecoregions of the conterminous United States. Ann Assoc Am Geogr 77(1):118–125

Omernik JM (1995) Ecoregions: a framework for managing ecosystems. In: The George Wright Forum. vol 1. JSTOR, pp 35–50

Omernik JM, Griffith G, E (1991) Ecological regions versus hydrologic units: frameworks for managing water quality. J Soil Water Conserv 46 (5):334–340

Osmundson T (1999) Roof gardens – history, design and construction. W.W. Norton & Company, New York

Ouldboukhitine S-E, Spolek G, Belarbi R (2014) Impact of plants transpiration, grey and clean water irrigation on the thermal resistance of green roofs. Ecol Eng 67:60–66

Packard S, Mutel CF (1997) The tallgrass restoration handbook: for prairies, savannas and woodlands. Island Press, Washington, DC

Palandino & Company (2006) King County green roof case study report. King County Department of Natural Resources & Parks, Seattle

Park J, Kim J-H, Dvorak B, Lee DK (2018) The role of green roofs on microclimate mitigation effect to local climates in summer. Int J Environ Res 12(5):671–679

Partridge DR, Clark JA (2018) Urban green roofs provide habitat for migrating and breeding birds and their arthropod prey. PLoS One 13(8)

Pearce K (2003) Toronto City hall–early research findings. Paper presented at the Annual Greening Rooftops for Sustainable Communities Conference, Chicago

Peck S, Goucher D (2005) Overview of North American policy development and the policy development process. Paper presented at the Third Annual Greening Rooftops for Sustainable Communities Conference, Washington D.C., May 4-6, 2005

Philippi PM (2005) Introduction to the German FLL-Guildeline for the planning, execution, and upkeep of green-roof sites. Paper presented at the Third Annual Greening Rooftops for Sustainable Communities Conference, Washington, D.C., May 4-6, 2005

Piana MR, Carlisle SC (2014) Green roofs over time: a spatially explicit method for studying green roof vegetative dynamics and performance. Cities Environ (CATE) 7(2):1

Puigdefábregas J, Pugnaire FI (1999) Plant survival in arid environments. In: FI Pugnaire, F Valladares (eds.) Handbook of functional plant ecology, pp 381–405

Rayner JP, Farrell C, Raynor KJ, Murphy SM, Williams NS (2016) Plant establishment on a green roof under extreme hot and dry conditions: the importance of leaf succulence in plant selection. Urban For Urban Green 15:6–14

Reichstein M, Bahn M, Mahecha MD, Kattge J, Baldocchi DD (2014) Linking plant and ecosystem functional biogeography. Proc Natl Acad Sci 111(38):13697–13702

Ricketts TH (1999) Terrestrial ecoregions of North America: a conservation assessment. Island Press, Washington, DC

Rosenzweig ML (2003) Reconciliation ecology and the future of species diversity. Oryx 37(02):194–205. https://doi.org/10.1017/S0030605303000371

Rowe B (2015) Long-term rooftop plant communities. In: Sutton RK (ed) Green roof ecosystems. Springer, Cham, pp 311–332. https://doi.org/10.1007/978-3-319-14983-7_13

Rowe B, Monterusso M, Rugh C (2005) Evaluation of Sedum species and Michigan native taxa for green roof applications. Paper presented at the Third Annual Greening Rooftops for Sustainable Communities Conference, Washington, DC., May 4-6, 2005

Schneider A, Fusco M, Bousselot J (2014) Observations on the survival of 112 plant taxa on a green roof in a semi-arid climate. J Living Archit 1(5):10–30. https://doi.org/10.46534/jliv.2014.02.01.010

Schrader S, Böning M (2006) Soil formation on green roofs and its contribution to urban biodiversity with emphasis on collembolans. Pedobiologia 50(4):347–356. https://doi.org/10.1016/j.pedobi.2006.06.003

Schramm P (1990) Prairie restoration: a twenty-five year perspective on establishment and management. In: Proceedings of the Twelfth North American Prairie Conference, 1990. University of Northern Iowa Cedar Fall, Iowa, pp 169–177

Schroll E, Lambrinos JG, Sandrock D (2011) An evaluation of plant selections and irrigation requirements for extensive green roofs in the Pacific northwestern United States. Hort Technol 21(3):314–322

SFLRM (2015) San Francisco living roof manual. City of San Francisco, San Francisco

Sharp R (2003) A coastal meadow in the sky: the Sechelt justice building. Paper presented at the Greening Rooftops for Sustainble Communities, Chicago, IL, 29-30 May

Shiner E (2003) Conservation architecture: endangered plants on an old slaugherhouse roof. Paper presented at the Greening Rooftops for Sustainable Communities, Chicago

Simmons MT (2015) Climates and microclimates: challenges for extensive green roof design in hot climates. In: Green roof ecosystems. Springer, Cham, pp 63–80

Skabelund LR, DiGiovanni K, Starry O (2015) Monitoring abiotic inputs and outputs. In: Green roof ecosystems. Springer, Cham, pp 27–62

Snodgrass EC, McIntyre L (2010) The green roof manual: a professional guide to design, installation, and maintenance. Timber Press, London

Snodgrass E, Snodgrass L (2006) Green roof plants. Timber Press, Portland

Steck A, Morgan S, Retzlaff W, Williams J (2015) Insect communities on green roofs that are close in proximity but vary in age and plant coverage. J Living Archit 2(1):1–11

Storm L, Shebitz D (2006) Evaluating the purpose, extent, and ecological restoration applications of indigenous burning practices in southwestern Washington. Ecol Restor 24(4):256–268

Sutton RK (2014) Aesthetics for green roofs and green walls. J Living Archit 1(2):1–20

Sutton RK (2015a) Green roof ecosystems, vol 223. Springer, Cham

Sutton RK (2015b) Green roof plant trials for the central Great Plains. J Living Archit 2(2):1–10

Sutton RK (2015c) Introduction to green roof ecosystems. In: Green roof ecosystems. Springer, Cham, pp 1–25

Sutton RK (2020) Reading the Nebraska landscape: an ecological aesthetic. In. CreateSpace Independent Publishing Platform, Moneee, p 16

Sutton RK, Lambrinos J (2015) Green roof ecosystems: summary and synthesis. In: Sutton RK (ed) Green roof ecosystems. Springer, Cham, pp 423–440. https://doi.org/10.1007/978-3-319-14983-7_17

Sutton RK, Harrington JA, Skabelund L, MacDonagh P, Coffman RR, Koch G (2012) Prairie-based green roofs: literature, templates, and analogs. J Green Build 7(1):143–172. https://doi.org/10.3992/jgb.7.1.143

Thurston R (2017) Defining and measuring green roof failure using a case study of incentivized industrial, commercial, and institutional vegetated roofs in Portland, Oregon. Evergreen State College

Tonietto R, Fant J, Ascher J, Ellis K, Larkin D (2011) A comparison of bee communities of Chicago green roofs, parks and prairies. Landscape Urban Plan 103(1):102–108. https://doi.org/10.1016/j.landurbplan.2011.07.004

Torrance S, Bass B, MacIvor J, McGlade T (2013) City of Toronto guidelines for biodiverse green roofs. Toronto City Planning, Toronto

Tyrväinen L, Silvennoinen H, Kolehmainen O (2003) Ecological and aesthetic values in urban forest management. Urban For Urban Green 1(3):135–149

USDA (2019) The PLANTS database. USDA, NRCS. https://plants.sc.egov.usda.gov. Accessed 2019

Vacek P, Struhala K, Matějka L (2017) Life-cycle study on semi intensive green roofs. J Clean Prod 154:203–213

Vale T (2013) Fire, native peoples, and the natural landscape. Island Press, Washington, DC

Van Mechelen C, Dutoit T, Hermy M (2015) Adapting green roof irrigation practices for a sustainable future: a review. Sustain Cities Soc 19:74–90

VanWoert ND, Rowe BD, Andresen JA, Rugh CL, Fernandez TR, Xiao L (2005) Green roof stormwater retention: effects of roof surface, slope, and media depth. J Environ Qual 34:1036–1044

Vijayaraghavan K (2016) Green roofs: a critical review on the role of components, benefits, limitations and trends. Renew Sust Energ Rev 57:740–752

Wasowski S, Wasowski A (2003) Native Texas plants: landscaping region by region, 2nd edn. Lone Star Books, Lanham

Weiler SK, Scholz-Barth K (2009) Green roof systems: a guide to the planning, design, and construction of landscapes over structure. Wiley, Hoboken

Wiken EB (1986) Terrestrial ecozones of Canada, vol 19. Ecological Land Classification Series. Environment Canada, Lands Directorate, Ottawa

Williams NSG, Rayner JP, Raynor KJ (2010) Green roofs for a wide brown land: opportunities and barriers for rooftop greening in Australia. Urban For Urban Green 9(3):245–251. https://doi.org/10.1016/j.ufug.2010.01.005

Wray J, Anderson MK (2003) Restoring Indian-set fires to prairie ecosystems on the Olympic peninsula. Ecol Restor 21(4):296–301

Young TP, Petersen D, Clary J (2005) The ecology of restoration: historical links, emerging issues and unexplored realms. Ecol Lett 8(6):662–673

Part II
Application: Ecoregional Green Roof Case Studies

Part II outlines case studies of conservation sites and ecoregional green roofs. From June 1, 2018 until October 30th, 2018 the lead author, Bruce Dvorak, traveled the western U.S. and Canada to visit over 140 green roofs. He met with the owners of these green roofs, designers, and those that maintain ecoregional green roofs. He also met with co-authors of this book to visit green roofs, and discuss the approach for this book. Part II of this book describes the findings of these visits and follow up investigations that took place. Each chapter focuses on an ecoregion or collection of ecoregions based upon a loose similarity of vegetation, climate geographical region, and dispersal of metropolitan areas. Each chapter consists of a brief overview of the ecoregion in its natural and cultural history, climate and ecoregional vegetation.

Chapter 3
Green Roofs in Tallgrass Prairie Ecoregions

Bruce Dvorak and Lee R. Skabelund

Abstract This chapter presents case studies of three tallgrass prairie conservation sites and nine ecoregional green roofs located in the Coastal Plains and Interior Lowlands of North America. Prairies were once one of the top ten ecologically rich ecosystems in the world with over 300 species of forbs and 70 species of grasses native to the tallgrass prairie. Today, less than one percent of tallgrass prairie remains across North America. On green roofs, however, 145 species of plants native to tallgrass prairies have been trialed on the ecoregional green roofs featured in this chapter located in Kansas, Nebraska, Texas, and Missouri. This chapter demonstrates how the diverse tallgrass prairie can inspire a variety of habitat types for green roofs including vegetation from dry, mesic, and wet prairie habitats on sloped and flat roof decks.

Keywords Prairie · Restoration · Graminoid · Forb · Biodiversity · Slope · Substrate · Drought · Rainwater harvesting · Maintenance

3.1 Tallgrass Ecoregion Characteristics

Prior to the mid-nineteenth century, tallgrass (Fig. 3.1), mixed-grass, and shortgrass prairies covered 60 million hectares (231,661 mile^2) across North America. Prairie ecosystems once covered the land from Manitoba in the north, to near the Mexican border and the Gulf Coast of Texas to the south. Not uniform in their distribution, there were subsets of prairie ecosystems that were influenced by climate, soils, and

B. Dvorak (✉)
Department of Landscape Architecture and Urban Planning, 305A Langford Architecture Center, Texas A&M University, College Station, TX, USA
e-mail: bdvorak@tamu.edu

L. R. Skabelund
Department of Landscape Architecture and Regional & Community Planning, Kansas State University, Manhattan, KS, USA
e-mail: lskab@k-state.edu

B. Dvorak (ed.), *Ecoregional Green Roofs*, Cities and Nature,
https://doi.org/10.1007/978-3-030-58395-8_3

Fig. 3.1 Detail of *Sunset of the prairies* (Albert Bierstadt, 1881–1882). In this painting, Bierstadt captures the distant views of a prairie landscape, which was often compared to a sea of grass. In the tallgrass prairie, warm-season grasses were typically less than knee-high in early summer. By late summer, prairie grasses could sometimes grow as tall as a rider on horseback. Bierstadt's painting at sunset illuminates the vegetation, and perhaps prophetically announces a glorious end to the reign of tallgrass prairie in North America at the dawn of the twentieth century. Less than 1% of the North American Tallgrass Prairie remains. (Courtesy of Wikiart.org https://uploads1.wikiart.org/images/albert-bierstadt/sunset-of-the-prairies.jpg!Large.jpg)

topography which made assemblages of dry, mesic, or wet habitats (Schramm 1990; Steinauer and Collins 1996).

Northern tallgrass prairies are subject to a continental climate where seasonal extremes such as sub-freezing temperatures and prolonged snowfall are common during the winter, and hot and dry conditions can persist during the summer. Central tallgrass prairies have a more moderate climate compared to northern prairies although they still experience climate extremes, while southern tallgrass prairies rarely experience freezing temperatures or snowfall, but frequently have extended periods of heat and drought during the summer. In the United States, the east-to-west expanse of prairie extended from western Indiana and Michigan in the north, and islands of prairies in Arkansas, Mississippi, Ohio, and Kentucky to the south. To the west, tallgrass prairie extended to eastern Nebraska, Kansas, Oklahoma, and Texas (Packard and Mutel 1997; Robertson et al. 1997; Chapman and Brewer 2008). In Canada, the tallgrass prairie was present in the southernmost sections of Saskatchewan and Manitoba (Robertson et al. 1997).

The average annual precipitation generally decreases east to west at 1000 mm in the eastern range of the tallgrass prairie to 750 mm of annual precipitation at the intersection of the mixed-grass prairie ecoregions and 380 mm annual precipitation in the shortgrass prairie ecoregion. Thus, the precipitation gradient influences the stature of vegetation growing in the region, giving rise to the terms high prairie and low prairie (Shively and Weaver 1939; Robertson et al. 1997).

Modern prairie ecosystems began forming in the mid-continent of North America from about 4000 to 7000 years ago. Continental ice was already retreating from the central United States, and glacial till, loess soils, and climate set conditions for biologically diverse grasslands to form (McMillan and Klippel 1981). Bison, elk, antelope, and Native American people lived with the prairies and influenced their composition and range of habitat (McMillan and Klippel 1981; Packard and Mutel 1997). Prairies, especially tallgrass prairies, co-evolved with humans in North America. Native Americans used fire for thousands of years to maintain prairies as a way to secure hunting grounds, medicinal plants, for food sources, clothing, and cultural uses (Kaiser et al. 1979; Samson and Knopf 1996; Kimmerer and Lake 2001; Kindscher 2009).

Some of the first explorers and settlers to visit and settle the tallgrass prairies traveled along with river trails, and across the land. Many recorded their experiences in writings, drawings, or paintings. Amos Andrew Parker wrote about his travels through the tallgrass prairies from Illinois to Texas. He described the tallgrass prairies as an Eden-like setting. He marveled that the beauty of the prairie was unlike any landscape he had ever seen (Parker 1836). Bottomland (wet) prairies were seen as some of the most difficult and horrid lands to travel through, whereas upland (dry) prairies were easier to travel through (Winsor 1987).

The southern tallgrass prairie was called the Blackland Prairie (Launchbaugh 1955; Srinath and Millington 2016). It received its name from the black soils that turned to muck each spring during rains. Blackland Prairies had many of the same species that grew in the tallgrass prairie further north; however, its composition favored taxa that could survive summer droughts and mild winters (White 2006). Early settlers to the Blackland Prairie remarked about its bountiful bloom season. Plein-air painter Robert Julian Onderdonk wrote, "San Antonio offers an inexhaustible field for the artist. In the spring, when the wild flowers are in bloom, it is riotous: every tint, every hue, every shade is present in the most lavish profusion, and even in the dead of summer, when one would imagine that any canvas could only convey the impression of intense heat, the possibilities of the landscape are still beyond comprehension" (Morseburg 2011).

Although prairies were North America's largest continuous ecosystem, the tallgrass prairie is now highly fragmented and critically endangered (Ricketts 1999). Since the implementation of the Homestead Act of 1862, the tallgrass prairie was encouraged to be plowed and replaced with agricultural uses (Anderson 2011). Iowa, for example, had about 80–85% of its land in tallgrass prairie ecosystems before the nineteenth century (Smith 1992, 1998). Although there were calls for conservation of Iowa's prairies during the time of great decline, less than 0.1% remains, primarily in private conservation sites (Smith 1992).

The epicenter of the tallgrass prairie ecoregion comprised Illinois, Iowa, eastern Nebraska and Kansas, western Missouri, and southern Minnesota (Fig. 3.2). Tallgrass prairies were usually treeless, ecologically rich, and contiguous across a variety of soil conditions, and many prairie species extended into partly shaded habitats of oak savannas. A few remaining large tracts of high-quality tallgrass

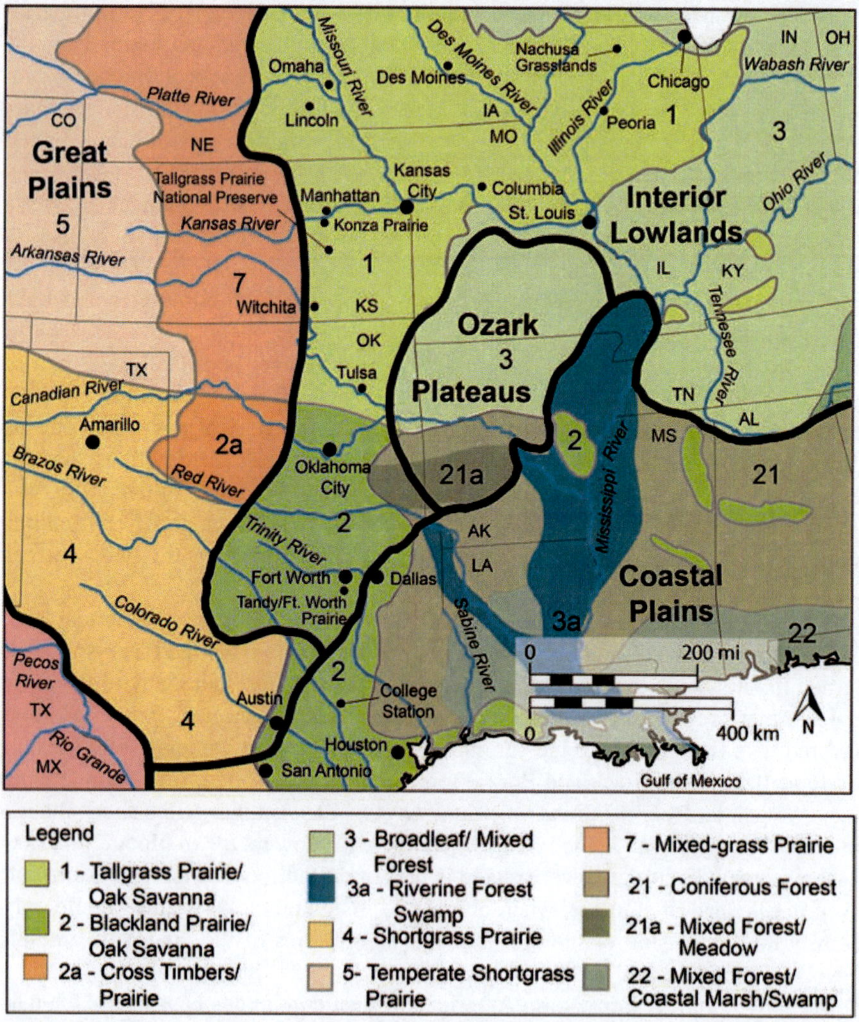

Fig. 3.2 Ecoregions and physiographic regions of the central U.S. states, and cities of interest covered in this chapter (Hammond 1964; Grace 2005; Bailey 1997; Smith and Butler 2011). Tallgrass Prairie (1) extended from near Chicago and beyond the map area to western Indiana and Michigan, southcentral Wisconsin, southwestern Minnesota, and south to Oklahoma. The Blackland Prairie (2) was located in the more humid and temperate climate of southern states and often coincided with oak savanna. Isolated islands of tallgrass prairie grew in parts of Kentucky, Ohio, and Michigan (not shown here), and was intermixed with savanna and open woodlands in Arkansas, Louisiana, Mississippi, and Alabama (Smith and Butler 2011). The Mixed-grass Prairie (7) included species found in the tallgrass and shortgrass prairies. (Graphic: Trevor Maciejewski & Bruce Dvorak)

prairie exist in the central U.S. (Sampson and Knopf 1994; Smith 1998). After many decades of work by conservationists, the 4409 hectare (10,894-acre) Tallgrass Prairie National Preserve, located in the Flint Hills of Kansas, was set aside as a national preserve in 2005 under a public/private partnership involving the National Park Service, The Nature Conservancy, and others (Licht 1997; Conard and Hess 1998).

Mixed-grass prairies grew west of the tallgrass prairies and were transitional between the tallgrass prairie and shortgrass prairie (Steinauer and Collins 1996). Mixed-grass prairies typically had higher diversity as they shared species of tallgrass and shortgrass prairies.

Today, the conservation of prairies is often managed with prescribed burning of the ground vegetation (Collins and Wallace 1990; Packard and Mutel 1997). Although many conservation sites are distant from urban centers, they inspire the creation and management of prairie-based green roofs (Sutton et al. 2012; Sutton 2013; Skabelund et al. 2014; MacDonagh and Shanstrom 2015; Blackmore 2019; Liu et al. 2019). By the early 2000s, tallgrass prairie vegetation increased in value in many urbanized areas as a part of green infrastructure planning and design activities (Beatley 2000; Benedict and McMahon 2002). This living green infrastructure includes habitat for pollinators and wildlife, improvements to stormwater management, and the creation of educational and functional green roof ecosystems (Harmel et al. 2006; Holman-Dodds 2007; Dvorak 2009; Li et al. 2010; Sutton et al. 2012; Van der Merwe et al. 2017).

Prairies are also conserved for their ecosystem functions as they are home to many species of insects, butterflies, bees, resident and migrating birds and waterfowl (Blair 1999; Bendel et al. 2018). Green roofs are also seen as an important part of broader efforts to conserve prairie species in urban and peri-urban areas (Benedict and McMahon 2002; Sutton 2015).

The climate varies greatly across the tallgrass ecoregions from north to south and east to west. Generally, summers are warm to hot (with high temperatures exceeding 35-40° C in July and August), with consistent precipitation across most months (although late July, August, and September tend to be drier than May, June, and early July in some ecoregions). Snowcover and freezing temperatures (i.e. -25° C) are common in northern ecoregions during winter, while freezing temperatures and snow are infrequent in the south (Fig. 3.3).

3.1.1 Vegetation of the Tallgrass Prairie Ecoregions

3.1.1.1 Grasses

In terms of biomass, grasses (graminoids) dominate the tallgrass prairie at about 80% of vegetation. Typically, 40–60 species of grasses can grow in some tallgrass prairies; however, only a few species typically dominate and include big bluestem (*Andropogon gerardii*), little bluestem (*Schizachyrium scoparium*), switchgrass

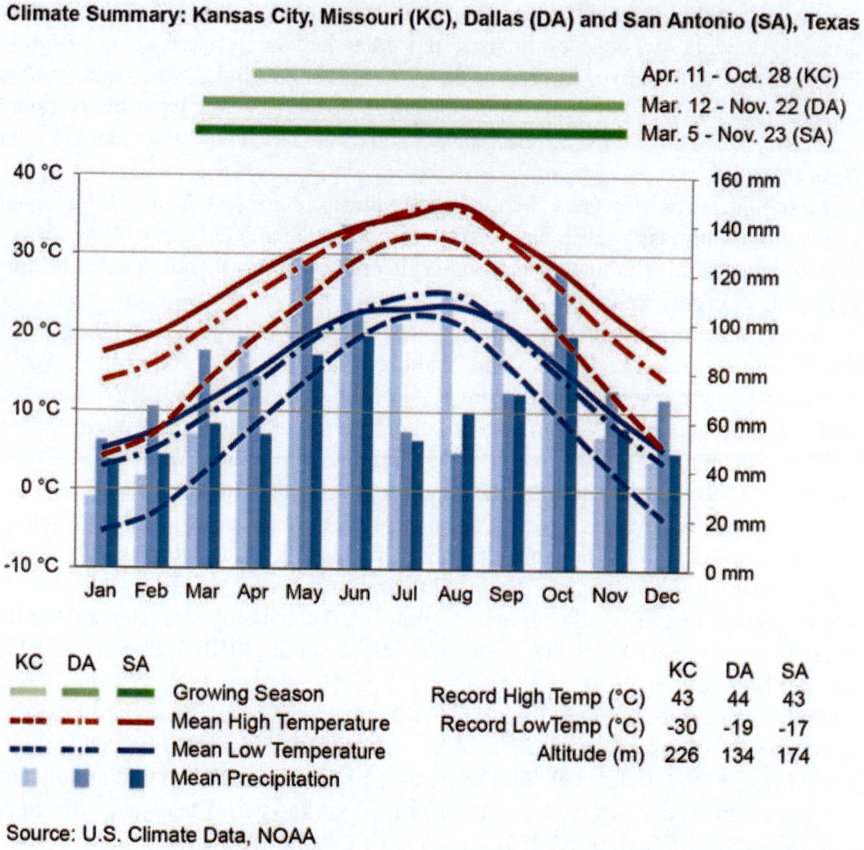

Fig. 3.3 Monthly precipitation and temperature averages for Kansas City, Missouri, and Dallas and San Antonio, Texas. Kansas City has the shorter growing season, however, it has more reliable summer precipitation compared to Dallas or San Antonio. Implications for vegetation are that in ecoregions where summer precipitation can be less frequent and combine with high temperatures, supplemental irrigation should be considered for any prairie green roof where full-coverage of vegetation is either required or highly desirable. (Graphic: Menotti and Dvorak 2020)

(*Panicum virgatum*) and Indian grass (*Sorghastrum nutans*). Other common grasses include prairie dropseed (*Sporobolus heterolepis*), sideoats grama (*Bouteloua curtipendula*), purple lovegrass (*Eragrostis spectabilis*), Virginia wild rye (*Elymus virginicus*), prairie Junegrass, (*Koeleria macrantha*), and in moist and shady conditions inland sea oats (*Chasmanthium latifolium*). Sedges and rushes also have a significant presence in wetter soil profiles (Schramm 1990; Williams 2010).

Invasive grasses *of* the tallgrass prairies include bahiagrass (*Paspalum notatum*), Caucasian bluestem (*Bothriochloa bladhii*), Johnson grass (*Sorghum halepense*), reed canary grass (*Phalaris arundinacea*), smooth brome (*Bromus inermis*), and many others. Invasive forbs include birdsfoot trefoil (*Lotus corniculatus*), black medic (*Medicago lupulina*) common mullein (*Verbascum thapsus*), dame's rocket (*Hesperis matronalis*), garlic mustard (*Alliaria petiolata*), ragweed (*Ambrosia artemisiifolia*) sweet clover (*Melilotus officinalis*), purple loosestrife (*Lythrum salicaria*), Canada thistle (*Cirsium arvense*), Queen Anne's lace (*Daucus carota*), and wild parsnip (*Pastinaca sativa*). These and many other early successional species (exotic or native) may need to be controlled during the establishment of a prairie green roof (Sutton et al. 2012; Dvorak 2015; MacDonagh and Shanstrom 2015; Skabelund et al. 2017).

3.1.1.2 Herbaceous Forbs

Forb diversity (wildflowers) varies across prairies depending upon the frequency of fire, grazing, or other disturbances. The diversity of the historic prairies likely had 100–200 species and up to 300 species commonly found on prairies depending upon many factors including size of land area, soil depth and type, and topographic features (Howe 1994). Some forbs require particular stability while others take advantage of disturbance. Some of the shorter forbs that may adapt to green roofs include aromatic aster (*Aster oblongifolius*), blazing-star (*Liatris aestivalis*), common yarrow (*Achillea millefolium*), firecracker penstemon (*Penstemon eatonii*), foxglove beardtongue (*Penstemon digitalis*), Mexican hat (*Ratibida columnifera*), Missouri goldenrod (*Solidago missouriensis*), prairie spiderwort (*Tradescantia occidentalis*), purple prairie clover (*Dalea purpurea*), tall blazing star (*Liatris aspera*), winecup or purple poppymallow (*Callirhoe involucrata*), wild bergamot (*Monarda fistulosa*), and wild onion (*Allium stellatum*) (Schramm 1990; Williams 2010). Many of the green roof case studies in this chapter feature these plants.

3.1.1.3 Succulents

Few succulent species persist in the eastern regions of tallgrass prairie ecosystems. In drier and warmer habitats, such as tops of ridges, south or west-facing slopes, or rocky outcrops, succulents can occupy such niches. Some of these include big-rooted prickly pear (*Opuntia macrorhiza*), brittle prickly pear (*Opuntia fragilis*), eastern prickly pear (*Opuntia humifusa*), large-flowered rock pink (*Phemeranthus calycinus*), three-leaved stonecrop (*Sedum ternatum*), widow's cross (*Sedum pulchellum*), and more frequent in southern states coastal stonecrop (*Lenophyllum texanum*), false aloe (*Manfreda maculosa*), Texas prickly pear, (*Opuntia engelmannii*), tulip pricklypear (*Opuntia phaeacantha*), and yellow stonecrop (*Sedum nuttallii*) (BONAP 2018).

Although succulents with thorns or spines can grow on green roofs with little maintenance or watering, their maintenance expecations should be well-addressed during the planning phase. Succulents such as cacti can grow along with grasses if the intended aesthetic is wild. However, if used in a garden setting on a green roof cacti can be problematic if the intent is to remove biomass around cacti year-to-year. The thorns or prickles may make maintenance and research activities more difficult and potentially hazardous.

3.1.2 *Conservation Site Case Studies (Arranged North to South)*

3.1.2.1 Nachusa Grasslands, Franklin Grove, Illinois

At 1538 hectares (3800 acres), the Nachusa Grasslands is one of the few large tallgrass prairies in the Midwest that is free and open to the public. The preserve consists of remnant prairie, woodland, and wetland habitats. Unique to the original 81 hectares (200 acres) is a variety of moisture gradient plant communities of sand prairie habitats that include wet, wet-mesic, mesic, dry-mesic, and dry prairie communities (Fig. 3.4). Much of the remaining land was added to the original conservation site after a former agricultural use. Therefore, a significant portion of total property is under active restoration. Research is underway regarding restoration of the prairie, and management of bison herds that live on portions of the site (Taft

Fig. 3.4 Pale purple coneflower (front and center), white indigo (center white blooms), and compass plant (tall yellow) bloom as part of the tallgrass plant community here. Purple coneflower grows on several green roofs in the ecoregion and is discussed in several green roof case studies in this chapter. (Photo: Bruce Dvorak)

2007; Blackburn 2018). One study measured mean diversity of the original sites with the restoration sites and found 71 native species present on the remnant sites and 66 species at the planted sites. The rolling topography, rocky outcrops, and sandy soils make this site a good study for vegetation that may translate to green roofs located in the upper Midwest. A few of the species growing at the grasslands include beardtongue (*Penstemon digitalis*), big bluestem (*Andropogon gerardii*), common milkweed (*Asclepias syriaca*), prairie dropseed (*Sporobolus heterolepis*), which grow on a number of the green roofs highlighted as case studies in this chapter.

3.1.2.2 Konza Prairie Biological Station, Manhattan, Kansas

Located in the Flint Hills region of north-central Kansas, the Konza Prairie is a 3487-hectare (8616-acre) tallgrass prairie research site is managed by Kansas State University in conjunction with The Nature Conservancy (Fig. 3.5). The Konza tallgrass prairie operates under a three-fold mission of long-term ecological research,

Fig. 3.5 *Echinacea angustifolia* and its associative plant community members grow on a hillside at the Konza Prairie. Several species of *Echinacea* grow at the Konza Prairie, and on green roofs in tallgrass ecoregions, such as the Memorial Stadium at Kansas State University in Manhattan, Kansas, Camp Young Judaea's Experiential Learning Center in Wimberley, Texas, and others. (Photo: Lee R. Skabelund, June 2009)

education, and prairie conservation. The prairie is managed through a planned set of strategies including prescribed burning at different times of the year and frequencies, and grazing by cattle or bison herds (Ernest and Towne 2020; KPBS 2020). After decades of research, the vascular flora of Konza Prairie has been determined to have at least 576 taxa growing on its hills, riparian zones, and wetlands. This represents 336 genera and 96 families including *Poaceae* (84), *Asteraceae* (79), *Fabaceae* (49), and *Cyperaceae* (33) as the most abundant; however, 16.7% of the total flora are non-native (Towne 2002).

Bison herds roam the prairie in designated and fenced zones. The herd size is set between 200–375 to prevent overgrazing. Thus, only 25% of vegetation is consumed by bison annually. Research regarding plant responses to fire frequency and grazing frequency demonstrates that frequent grazing by bison and less frequent burning sustains diverse vegetation and appropriate numbers of bison. Annual burns are found to reduce diversity; burning every several years allows more floristic diversity than annual burns (Knapp et al. 1999).

Research has aided an understanding of cattle stocking rates at Konza Prairie. These rates are set so that the bison grazing intensity removes about 25% of the annual above-ground plant production (KPBS 2020). In addition, research found that, "Plant diversity is greater in bison pastures than in cattle pastures. Forb numbers, especially annuals, are higher in bison pastures. Both bison and cattle primarily consume big bluestem, little bluestem and Indiangrass on the tallgrass prairie. Cattle remove about 46% of the annual net primary production of the grasses while bison remove about 54%" (KEEP 2020).

3.1.2.3 Fort Worth Prairie/Tandy Hills Natural Area

The Fort Worth Prairie was part of a large prairie ecosystem called the Grand Prairie. In north-central Texas, the tallgrass prairie historically extended in a swath, north to south, and was sandwiched between cross timber habitats and savanna. Northcentral Texas had a matrix of open prairie and woodlands and formed a very diverse group of ecoregions. At least 2000 species of plants belonging to the sunny, partly sunny, and shady prairie and wooded habitats (Diggs et al. 1999). The remaining fragments of Fort Worth Prairie are spread out north, south, and west of the City of Fort Worth, Texas (Diggs et al. 1999).

The vegetation of the Fort Worth Prairie (Fig. 3.6) includes the dominant grass little bluestem (*Schizachyrium scoparium*) which consists of about two-thirds of the vegetated cover. Also common to the prairie is sideoats grama (*Bouteloua curtipendula*), Indian grass (*Sorghastrum nutans*), tall dropseed (*Sporobolus compositus*), hairy grama (*Bouteloua hirsuta*), and big bluestem (*Andropogon gerardii*), which all grow on the Perot Museum of Science and Nature green roof as discussed in Sect. 3.3, (Diggs et al. 1999). These species also grow on other green roofs, as discussed in this chapter.

Fig. 3.6 Seen here, the Tandy Hills Natural Area is one of the last fragments of the Fort Worth Prairie and is located in the Fort Worth metro area, accessible to the public. Historically, little bluestem (above) was one of the dominant grasses in the region. The 65-hectare prairie (160-acres) has over 600 species of plants native to the barrens and hillside prairies (including prickly pear, prairie verbena, blanket flower, and coreopsis) all found nearby this portion of the prairie. Many plants of the Fort Worth Prairie have also been trialed on green roofs at the Botanical Research Institute of Texas, discussed in Sect. 3.3.6. Woody vegetation (background) often encroaches when the burning of the land is suppressed as a land management tool. Downtown Fort Worth can be seen in the distance. (Photo: Bruce Dvorak, June 2018)

3.2 Green Roof Research in the Tallgrass Ecoregions

Several peer-reviewed publications have reported on prairie-based green roof research in the ecoregion. Both native and non-native vegetation has been trialed across the region including Minneapolis, Minnesota, Lincoln, Nebraska, Manhattan, Kansas, Norman, Oklahoma, Fort Worth, Texas, College Station, Texas, Austin, Texas, and Houston, Texas (Dvorak and Carroll 2008; Sutton 2008; Sutton et al. 2012; Dvorak et al. 2013; Sutton 2013; Klein and Coffman 2015; MacDonagh and Shanstrom 2015; Simmons 2015; Liu et al. 2019). Although there has been active research across the tallgrass prairie, there is currently no single prairie green roof that has published long-term research (more than 10 years). Observational research has been undertaken on a small prairie and sedum green roof at Kansas State

Fig. 3.7 Native prairie plants grow with sedums on the campus of Kansas State University. Some of the early studies of native vegetation on campus, such as at this Seaton Hall Upper Green Roof research site, led the way for further expansion of green roofs on campus. See the case study in Sect. 3.3.4 which covers the largest native prairie green roof in Kansas. (Photo: Lee Skabelund, July 2009)

University (Fig. 3.7) since summer 2009, but 10-year findings still need to be reviewed and published.

Research on ecosystem services of green roofs in the Tallgrass Prairie Ecoregion has been recorded in Austin and College Station, Texas, Manhattan, Kansas, Norman, Oklahoma, and Chicago, Illinois. Investigations regarding temperature and microclimate conditions of irrigated and non-irrigated green roofs generally conclude that green roofs keep the rooftop surface near the same as the ambient temperature and conventional roofs exacerbate air temperatures by up to 38° C at the roof membrane (compared with irrigated green roof systems) and by up to 18° C for unirrigated systems at the surface and by up to 27.5° C beneath the membrane. Thus, conventional roofing in the Tallgrass Prairie ecoregion contributes to the buildup of the urban heat island effect, where green roofs keep roof surfaces near the ambient temperature (Dvorak and Volder 2013; Klein and Coffman 2015; Simmons 2015). Stormwater retention functions report a wide range of performance with some vendor-provided systems retaining 44% of annual precipitation and other systems retaining up to 88% of precipitation (Simmons et al. 2008; Volder and Dvorak 2013; Skabelund et al. 2015). Conventional roofing contributes to flooding in built-up regions, and green roofs mimic and aid in restoring characteristics of the natural hydrology.

3.3 Ecoregional Green Roof Case Studies (Arranged North to South)

3.3.1 Pioneers Park Nature Center, Lincoln, Nebraska

Part of a 270-hectare (668-acre) conservation site, the Pioneers Nature Center includes mixed-grass prairie, wetlands, and woodland ecosystems all connecting back to original and restored habitats. All of the lands were acquired from 1963 to 2005 through a series of acquisitions. Master plans for the nature center were developed in 1984, 1995, and 2001. These led to the programming of a new building to house environmental education. Constructed in 2007, the LEED-certified center includes natural daylighting, harvested rainwater, energy conservation practices, and a green roof located over the main hallway (Fig. 3.8). As the first building in Lincoln to be awarded LEED certification, the nature center draws in over 60,000 visitors annually. Although the green roof is visible from the ground level, there is no public access for viewing the green roof up close.

Fig. 3.8 Late spring image of the Pioneers Nature Center green roof. The vegetation shown here includes native grasses, and purple prairie clover (*Dalea purpurea*) ending its bloom cycle on the green roof (lower right) and on the prairie below (upper right). (Courtesy of Dr. Richard Sutton)

3.3.1.1 Project Team

Building Owner/Client: City of Lincoln Parks and Recreation Department
Green Roof Design Team Lead: Richard Sutton
Architect: The Architectural Partnership (Deb Hansen)
Landscape Architect: Richard Sutton
Installation Contractor: Richard Sutton, students, and volunteers
Maintenance Contractor: Nature center staff
Project completion: 2007
Green roof area: 83 m^2 (900 ft^2)

3.3.1.2 Overview and Objectives

The roof deck on this narrow green roof is flat with a 1% slope, has internal drains, and a 10-cm-deep (or 4-inch) substrate provided by American Hydrotech. From its initial conception, researchers at the University of Nebraska set up the green roof to test several different treatments of the substrate condition to investigate best practices to establish vegetation. Since engineered growth media is typically sterile when installed, researchers formulated a slurry made from water and sandy loam topsoil native to the prairie site below and was added to the substrate in designation sections to provide a natural source of nematodes and microbes. Also due to the shallow depth, Horta-Sorb® was added to aid retention of moisture to other sections of the substrate to compare to the substrate without any treatment. Due to the shallow depth of the substrate, vegetation for the green roof was selected from shortgrass and mixed-grass prairie ecosystems. Some of the vegetation selected for the green roof also grows in the prairies at the mixed-grass prairie below; thus, there is a blending of vegetation from several prairie ecoregions.

3.3.1.3 Plant Establishment

Flowering perennial vegetation was established with live plugs planted at 15–20 cm on center spacing. Grasses were established through seeding (Fig. 3.9).

Grasses & Sedges

blue grama (*Bouteloua gracilis*), hairy grama (*Bouteloua hirsuta*), little bluestem (*Schizachyrium scoparium*), plains muhly (*Muhlenbergia cuspidate*), prairie junegrass (*Koeleria macrantha*), sideoats grama (*Bouteloua curtipendula*), sun sedge (*Carex inops*)

Fig. 3.9 The mixed-grass and shortgrass prairie vegetation were established through seeding. This image shows the prairie vegetation in the late spring and blends into the native prairie beyond. This green roof is planted with vegetation native to locations further west than Lincoln, Nebraska as the design team wanted a low maintenance green roof (minimal watering and weeding) that had vegetation capable of adapting to the shallow substrate with minimal watering. Some of the same plant species also grow on the prairie hillside vegetation near the green roof. (Photo: Bruce Dvorak, June 2018)

Herbaceous Perennials

dotted gayfeather (*Liatris punctata*), Canada goldenrod (*Solidago altissima*), Fender's aster (*Symphyotrichum fendleri*), Ohio spiderwort (*Tradescantia ohiensis*), white prairie clover (*Dalea candida*), purple prairie clover (*Dalea purpurea*)

Shrubs

dwarf leadplant (*Amorpha canescens*), fringed sage (*Artemisia frigida*)

3.3.1.4 Irrigation

An 11,356-liter (3000-gallon) rainwater harvesting system feeds the irrigation system for gardens below, while the green roof receives watering by hand application from potable water sources. Nature Center staff keep track of natural rainfall and supplement by watering the green roof as needed. Application of supplemental watering generally begins in the spring, once low temperatures during the early morning are consistently above 21 °C (70 °F). When these conditions take place, weekly watering may be required. When precipitation takes place during the warm season (late spring, summer, and fall) irrigation can be reduced to once every 10 days.

3.3.1.5 Maintenance

On the prairie restoration sites, the burning of the land is used to maintain the native habitats. On the green roof, monthly maintenance of the green roof takes place by the Nature Center staff. Common invasive plants that are removed include tree seedlings, and herbs such as thistles, garlic mustard, dandelion, and leafy spurge. This green roof is periodically mowed if thatch begins to build (Sutton 2018).

3.3.1.6 Observed Wildlife

Bees, butterflies and birds frequent the green roofs. A pair of geese nested on the green roof for three consecutive years.

3.3.1.7 Best Performing Native Vegetation

Blue grama, sideoats gramma, prairie junegrass, spiderwort, dotted gayfeather, purple prairie clover, and white prairie clover, are top performers.

3.3.1.8 Post-Occupancy Observations

Publications

- Results from a one-year investigation into three methods of plant establishment demonstrated that more vigor was present from vegetation that was established in plots that received hydro-absorbent polymer gel application and the slurry with mycorrhizae inoculated substrate (Sutton 2008).

Authors' Reflections

- This project has the right mixture of components of a comprehensive pilot project. Initial relationships and collaborations were set up at early conceptions of the building between Nature Center staff and university researchers. Ongoing research is important to grow the green roof industry, so future research on the green roof could expand upon the foundation set by the initial design and prior investigations.

- Since access is not provided for viewing by visitors and the general public, any future expansion for the building should consider including a viewing exhibit to educate the public about prairie-based green roofs.

3.3.2 Kansas City Public Library, Kansas City, Missouri

During the early to mid-1800s, the landscapes near the confluence of the Kansas River with the Missouri River in eastern Kansas and western Missouri became an important and strategic place of settlement in the Tallgrass Prairie Ecoregion (Fig. 3.2). Historically, Kansas City was a riverine gateway that led out to the sod house frontier. Numerous sod houses were built in the treeless regions of western Missouri, Kansas, Nebraska, and Iowa as settlers traveled westward from Kansas City out into the Great Plains during the mid to late-1800s (Macbride 1928; Dick 1937).

The roof garden on the Kansas City Central Library was the first publicly accessible modern prairie-based green roof in Kansas City (Fig. 3.10). The original tenant of the building was the First National Bank. The building was

Fig. 3.10 A mixture of perennial wildflowers ring the edge of the roof garden. The white blooms of foxglove beardtongue (*Penstemon digitalis)* make a show in early June. The Library has maintained this low-maintenance prairie garden through private contracts since its installation in 2004. (Photo: Bruce Dvorak, June 2018)

purchased by the City and donations by Jonathan Kemper, president of the Kansas City Public Library Board of Trustees. Renovations were completed in 2004 to accommodate the relocation of the central library holdings and staff. The original roof of the four-story building was expanded with a roof terrace as a fifth floor was added as part of the library renovations. The modified roof deck was designed to include a roof garden with vegetation native to Kansas City prairies as a gesture towards the blending of old and new. The longevity of the green roof demonstrates the continuity of its original vegetation and is a testament to the feasibility, simplicity, and beauty of prairie plants on rooftops. The roof garden offers one of the finest views of downtown Kansas City and is accessible to the public.

3.3.2.1 Project Team

Building Owner/Client: City of Kansas City
Green Roof Design Team Lead: Conservation Design Forum
Landscape Architect: Conservation Design Forum (now Environmental Consulting & Technology, Inc.)
Installation Contractor: Roofscapes (green roof system)
Maintenance Contractor: private contracts
Project completion: 2004
Green roof area: 790 m^2 (8503 ft^2)

3.3.2.2 Overview and Objectives

The green roof was designed to be an extension of the interior lobby out onto the rooftop as a festive display of native grasses, forbs, bulbs and succulents, members of the tallgrass prairie. The library has made the roof terrace available to be rented for private or public events including outdoor movies projected onto a blank wall. A large-scale chessboard and pieces occupy a portion of the roof terrace, as well as benches and signage that interprets the roof garden as well as adjacent buildings past and present (that are or were viewable from the rooftop).

A mix of prairie grasses, forbs, and succulents was selected to depict a range of prairie vegetation. Long-lived species were selected for the interior planter beds and exterior planted buffers located beyond the safety rail (Fig. 3.10 & Fig. 3.12). One of the larger sections of the green roof is laid on a flat roof deck where plants native to mesic or slow-drained prairies were selected. A low stone wall with an elevated terraform behind it makes a dry habitat for cactus and other plants adapted to drier microclimates (Fig. 3.11). The FLL compliant substrate is 15 cm deep in the exterior prairie beds, and up to 35 cm deep where the trees were planted. A rigid polymer drainage board lays at the bottom layer with a separation fabric on top.

Fig. 3.11 The informal arrangement of vegetation allows plants to select their preferred habitat and create a natural aesthetic that contrasts the formal edging of the building and pavement. A low stone wall (left center) rises to support upland species such as prickly pear and buffalo grass, where vegetation below the wall includes plants adapted to moist conditions. The divide symbolizes the vertical limestone bluffs of the Missouri River as it cuts through downtown Kansas City. This view is from behind the window shades in the Missouri Valley Special Collections room on the fifth floor, where one can find several books on prairie-based sod roofs cited in this chapter. (Photo: Bruce Dvorak, June 2018)

3.3.2.3 Plant Establishment

Native herbaceous perennial vegetation was selected to provide color during the spring, summer, and fall (Table 3.1). Ornamental native grasses were selected to provide visual interest throughout the seasons, including winter dormancy. Vegetation was pre-grown in containers and installed into the substrate. Bulbs were planted in the substrate, and no intentional seeding or over-seeding has taken place. Several non-native early spring blooming bulbs were installed to add color, but are not listed here.

Annuals

crested prickle poppy (*Argemone intermedia*), Leavenworth's eryngo (*Eryngium leavenworthii*), tanseyleaf tansyaster (*Machaeranthera tanacetifolia*).

Table 3.1 Native herbaceous perennials on the Kansas City Public Library green roof

Common Name	Botanical Name
Whorled milkweed	*Asclepias verticillata*
White heath aster	*Aster ericoides*
Aromatic aster	*Aster oblongifolius*
Prairie goldenrod	*Aster ptarmicoides*
Western silver aster	*Aster sericeus*
Purple prairie clover	*Dalea purpurea*
Blacksamson echinacea	*Echinacea angustifolia*
Fewleaf sunflower	*Helianthus occidentalis*
Tall blazing star	*Liatris aspera*
Blackfoot daisy	*Melampodium leucanthum*
Wild bergamot	*Monarda fistulosa*
Firecracker penstemon	*Penstemon eatonii*
Foxglove beardtongue	*Penstemon digitalis*
Woolly paperflower	*Psilostrophe tagetina*
Mexican hat	*Ratibida columnifera*
Flat-top goldentop	*Solidago graminifolia*
Missouri goldenrod	*Solidago missouriensis*
Rocky Mountain zinnia	*Zinna grandiflora*

Bulbs

nodding onion (*Allium cernuum*; native to the central U.S.)

Grasses

blue grama (*Bouteloua gracilis*), little bluestem (*Schizachyrium scoparium*), sideoats grama (*Bouteloua curtipendula*)

Succulents

bigroot prickly pear (*Opuntia macrorhiza*), brittle prickly pear (*Opuntia fragilis*), cane cholla (*Opuntia imbricate*), plains prickly pear (*Opuntia polycantha*), Spinystar (*Coryphantha vivipara*), woodland stonecrop (*Sedum ternatum*; native to the eastern U.S.)

3.3.2.4 Irrigation

Irrigation water is sourced from potable water lines from inside the building and delivered in a subsurface drip irrigation system. Irrigation runs daily during the growing season.

3.3.2.5 Maintenance

The vegetation is maintained about three times per year with a visit during the spring, summer, and fall. The fall maintenance visit includes some deadheading and removal of vegetation top growth. Native grasses are left in place for winter interest. Other than the removal of three trees that were part of the original planting (but did not survive), there have been no issues reported with the vegetation growing on this roof, the green roof system, nor the waterproofing system.

3.3.2.6 Observed Wildlife

No formal investigations have taken place on the roof garden, but the staff that manages the library roof garden have observed bees, butterflies (Fig. 3.12), crickets, and other insects.

Fig. 3.12 Magenta blooms of wild bergamot (*Monarda fistulosa*) attract a yellow cloudless sulphur butterfly (*Phoebis sennae*) which can be seen in the center-right of the figure. Sulphur butterflies migrate similar to monarchs, although not as far south. This one found a resting place 15 meters (50 ft) above the downtown streetscape on prairie vegetation that once blanketed the bluffs of the Missouri River. (Photo: Bruce Dvorak, June 2018)

3.3.2.7 Best Performing Native Vegetation

All of the vegetation has adapted to the green roofs system except three trees, which were removed in 2017.

3.3.2.8 Post-Occupancy Observations

Author Reflections

- The contracted maintenance staff have maintained the original vegetation on the roof since 2004, with great success. The informal natural aesthetic allows the roof garden to be a low-maintenance green roof that also educates visitors about the historic vegetation native to the region.

3.3.3 TWA Former Headquarters, Kansas City, Missouri

Originally designed by architects Raymond E. Bales, Jr. and Morris Schechter and built in 1956, the massive modernist building was renovated in 2008 by architect El Dorado. Trans World Airlines (TWA) occupied the building as its headquarters until 1969 when it was left abandoned. By 2002, the site was added to the National Register of Historic Places. In 2005, investors began a process to renovate the building into a multi-office space. The fully restored building was renovated with modern updates to the exterior, interior, and with the addition of a rooftop terrace (Fig. 3.13). As a designated historic structure, the green roof was set back three meters from the edge so as to not be visible from below. Major updates to the roof included the transformation of what was a bituminous inaccessible rooftop into an accessible rooftop garden terrace. The terrace was programmed as an extension of an open utilitarian space indoors-to-outdoors. Uses include staff lounge, small group meeting space with overhead trellis, and provisions for outdoor barbeque cooking. The accessible pedestrian deck space (not vegetated) is covered with IPE wood tiles. Other rooftop updates included a replica fabrication of its iconic Moonliner II rocket, now located on the rooftop, and is partially visible from the roof garden.

3.3.3.1 Project Team

Building Owner/Client: Leased to Barkley Inc.
Green Roof Design Team Lead: Off the Grid (Maggie Riggs)
Architect: El Dorado Inc.
Structural Engineer: Norton and Schmidt

Fig. 3.13 The yellow blooms of lance-leaved coreopsis (*Coreopsis lanceolata*) and white blooms of foxglove beardtongue (*Penstemon digitalis*) dominate the green roof. However, initially planted in a bed of exotic sedum plants, the native vegetation was cultivated to grow in the extensive substrate in place of sedums. (Photo: Bruce Dvorak, June 2018)

Landscape Architect: Jeffrey L. Bruce and Company (irrigation design)
Installation Contractor: American Hydrotech, Off the Grid
Maintenance Contractor: Off the Grid (Sunflower.com)
Project completion: 2007
Green roof area: 2322 m^2 (25,000 ft^2) roof area, 1640 m^2 (18,000 ft^2) vegetated

3.3.3.2 Overview and Objectives

The green roofs were originally planted heavily with four species of exotic sedums in the shallow substrates (10 cm), and planted with taller native plants located on the outside perimeter. Because of the early success of the native vegetation, the plant palette was later expanded to include more diversity of native vegetation across the entire roofscapes (Fig. 3.14). In this case, the lead maintenance contractor (Maggie Riggs) saw the success of the initial planting and convinced the building owner that native vegetation could be expanded upon the entire roof, and

Fig. 3.14 This inaccessible section of the green roofs is located away from the terrace and has no designated paths. Much of the same vegetation that grows on the identically flat terrace roof also grows here; however, vegetation is allowed to compete and occupy its preferred microclimate. (Photo: Bruce Dvorak, June 2018)

the client was convinced that native plants could be appreciated more than the sedum monoculture. The evolution of the native vegetation brought more diversity and color to the rooftop and connects to the historic prairie vegetation of Kansas City.

3.3.3.3 Plant Establishment

Much of the herbaceous perennial vegetation was initially pre-grown in containers (Table 3.2). New plantings were achieved with additional plant material brought onto the roof, and now through infilling by seedling plants.

Grasses

little bluestem (*Schizachyrium scoparium*), prairie dropseed (*Sporobolus heterolepis*).

Table 3.2 Native herbaceous perennials on the TWA Former Headquarters green roof

Common Name	Botanical Name
American pasqueflower	*Anemone patens*
Wild red columbine	*Aquilegia canadensis*
Common milkweed	*Asclepias syriaca*
Butterfly milkweed	*Asclepias tuberosa*
Blue false indigo	*Baptisia australis*
Purple poppy mallow	*Callirhoe involucrata*
Lance-leaved coreopsis	*Coreopsis lanceolata*
Purple coneflower	*Echinacea purpurea*
Dotted blazing star	*Liatris punctata*
Dense blazing star	*Liatris spicata*
Beardtongue	*Penstemon digitalis* 'husker red'
Foxglove beardtongue	*Penstemon digitalis*
Aromatic aster	*Symphyotrichum oblongifolium*
Spiderwort	*Tradescantia virginiana*

3.3.3.4 Irrigation Schedule

Irrigation runs daily for 10 minutes at 5:00 am during the growing season. The irrigation system uses overhead spray rotors that are located on a head-to-head spacing.

3.3.3.5 Maintenance

Visits to the rooftop for maintenance vary from twice weekly to twice monthly during the active growing season so that an informal garden-like appearance is maintained near accessible patio areas. The owners desire the vegetation to grow in loose drifts of similar species. Unwanted volunteer vegetation is removed and composted on the rooftop, and later used to fertilize strawberries that were planted in select sections of the green roof. Mycelium and alfalfa pellets are periodically added to the soil to replenish the organic matter in the substrate where needed. No commercial fertilizers or herbicides are used on the roof. Manual cutting or removal of unwanted plants allows for the relocation of desired seedlings.

3.3.3.6 Observed Wildlife

The maintenance staff has observed bees, butterflies (including monarchs), birds, and several ducks that nested; however, hawks and humans ended the reign of ducks on this green roof.

3.3.3.7 Best Performing Native Vegetation

All of the vegetation listed above is thriving on the roof.

3.3.3.8 Post-Occupancy Observations

Authors' Reflections

- This green roof visually reads like a meticulously maintained "wild" garden. The minutia of edits by highly skilled maintenance staff has led current building owners to change their view about native vegetation; they now are convinced that native plants can be beautiful, vibrant, and desirable in a roof garden setting.

3.3.4 Memorial Stadium, Kansas State University, Manhattan, Kansas

The 17,500-seat historic stadium previously was the home field of the Kansas State University Wildcats football team, and track and field team events before 1967. In 2017, the renovated stadium was formally dedicated as the World War I Memorial Stadium, honoring the prior military history and training that took place in nearby Fort Riley, Kansas. Fort Riley was established in the mid-eighteenth century to aid settlers along the Oregon Trail and the Santa Fe Trail when the tallgrass prairie ecosystem was the dominant vegetative cover. Currently, Kansas State University seeks to demonstrate how green roofs can be built and maintained on campus with vegetation native to Kansas (Fig. 3.15).

The stadium has a bilaterally symmetrical layout, with green roofs ascending from the ground level up to about 12 meters (40 ft.) above the field. As a research site, ecosystem services are monitored by an interdisciplinary group that includes green roof design team members, along with university researchers, facilities staff, and students.

3.3.4.1 Project Team

Building Owner/Client: Kansas State University
Green Roof Design Team Lead: Jeffrey L. Bruce & Company LLC.
Architect: West – Gould Evans Associates; East – Ebert Mayo Design Group
Structural Engineer: West – Bob D. Campbell & Company; East – Orazem & Scalora Engineering
Landscape Architect: Jeffrey L. Bruce & Company LLC.
Installation Contractor: Blueville Nursery Inc. (green roof and irrigation)
Maintenance: Blueville Nursery Inc., university faculty, staff, and students
Project completion: The West Memorial Stadium green roof (WMS-GR) was planted and seeded in June and July of 2015, the East Memorial Stadium green roof (EMS-GR) was seeded and planted in March and April of 2016.
Green roof area: Each roof is about 2000 m^2 (21,525 ft^2), or 0.4 hectares (43,000 ft^2) total.

Fig. 3.15 The East Memorial Stadium green roof is shown with its native prairie vegetation during the early summer season. Warm-season grasses are short in stature as they are beginning to emerge, while forbs are in their rapid growth phase and some are in full bloom. Thus, many of the shorter stature forbs take advantage of the abundance of sunlight early in the growing season due to the low stature of the grasses. Taller warm-season forbs rise above grasses later in the summer. The yellow *Coreopsis tinctoria* was a late addition approved by the designer to provide additional first-year color to the east side seed mix. The upright (purple) *Verbena stricta,* likely arrived inadvertently as seed stowed away in the green roof substrate, or with live plants from the nursery. (Courtesy of Jeffrey Bruce Associates, 2016)

3.3.4.2 Overview and Objectives

Plantings were arranged to create prairie-like ecosystems on each green roof, which reflect the character of the Flint Hills Tallgrass Prairie Ecoregion. Vegetation selected for these green roofs was tallgrass prairie species native to the Great Plains and the central United States. The substrate consists of 12–15 cm (5–6 inches) of custom-blended growth media and is laid on top of insulated, terraced sub decking, which steps down at slopes of 20–22 degrees (36–40%). Both green roofs have sand-based substrates; however, the East Memorial Stadium green roof includes expanded shale (supplied by Buildex) to lighten its structural load. This east-side green roof had fewer native grasses at the time of installation due to grasses not being included in the seed mix.

An open structured polymeric geotextile erosion control mat and hydromulch were used in combination to retain substrates in place during intense but short duration summer storm events, but the long-lasting molten polymeric matrix makes pulling weeds with large roots difficult, and thus many small portions have been ripped and/or removed during weeding to allow desired species room to grow.

These two green roofs are seen as a long-term research site for green roofs in Kansas, with informal observational research initiated during the spring of 2015, and formal research (employing data collection along north-south transects and subsurface soil moisture and temperature sensor deployment on each green roof) initiated in June 2016.

3.3.4.3 Plant Establishment

All seeds and live plants were supplied by Applied Ecological Services' Taylor Creek Restoration Nurseries (AES) from Brodhead, Wisconsin, and Baldwin City, Kansas. Hydro-seeding of the native seed mix, followed by the planting of plugs within the geo-web cells on each green roof, was completed by Blueville Nursery staff. Initial planting consisted of 26 intentionally planted species on the WMS-GR while 22 species were planted on the EMS-GR (Table 3.3).

Bulbs

Atlantic camas/wild hyacinth (*Camassia scilloides*), autumn/prairie onion (*Allium stellatum*)

Table 3.3 Native herbaceous perennials on the Memorial Stadium green roofs

Common Name	Botanical Name
Common yarrow	*Achillea millefolium*
Butterfly milkweed	*Asclepias tuberosa*
Blue wild indigo	*Baptisia australis*
Plains coreopsis	*Coreopsis tinctoria*
Purple prairie clover	*Dalea purpurea*
Pale purple coneflower	*Echinacea pallida*
Tall blazing star	*Liatris aspera*
Prairie blazing star	*Liatris pycnostachya*
Wild bergamot	*Monarda fistulosa*
Stiff goldenrod	*Oligoneuron rigidum*
Cobaea beardtongue	*Penstemon cobaea*
Foxglove beardtongue	*Penstemon digitalis*
Mexican hat coneflower	*Ratibida columnifera*
Pinnate prairie coneflower	*Ratibida pinnata*
Azure blue sage	*Salvia azurea*
White heath aster	*Symphyotrichum ericoides*
New England aster	*Symphyotrichum novae-angliae*
Western silver aster	*Symphyotrichum sericeum*
Prairie spiderwort	*Tradescantia occidentalis*
Bluejacket	*Tradescantia ohiensis*

Grasses

blue grama (*Bouteloua gracilis*), Indian grass (*Sorghastrum nutans*), little bluestem (*Schizachyrium scoparium*), prairie dropseed (*Sporobolus heterolepis*), sideoats grama (*Bouteloua curtipendula*)

Note that big bluestem (*Andropogon gerardii*) was also unintentionally planted on the WMS-GR, as discovered during observations during 2018 and 2019 when it became clear that it was abundant on the WMS-GR.

Shrubs

white sagebrush (*Artemisia ludoviciana*)

3.3.4.4 Irrigation

Irrigation is sourced from potable water provided by the city and runs on a schedule that is adjusted during the growing season with feedback sent from precipitation sensors and the lead researcher. Water is typically set to be delivered overhead via adjustable rotors for 15–45 minutes per zone one to three times per week during the growing season between 4:00 and 7:00 am, depending on soil moisture levels monitored by K-State faculty and staff (Skabelund et al. 2016).

3.3.4.5 Maintenance

Maintenance activities have adapted over time as the vegetation has matured (Fig. 3.16). Vegetation was fertilized bi-annually the first 2 years using an organic fertilizer sprayed (via Blueville Nursery staff) on each green roof. Green roof vegetation is clipped annually by hand and/or mechanically trimmed/weed-whacked several times a year. In March or early April each year, dormant aboveground biomass, is clipped to 15–30 cm (6–12 inches) above the surface by KSU Grounds staff. Since the summer of 2016, volunteer assistance with weeding and clipping of tree/shrub seedlings and other weeds of concern has been provided and is expected to continue.

Due to the substrate, live plant storage, and irrigation practices (two times a day on the west green roof versus three times a day on the east green roof for most of the 2016 growing season), many agricultural weeds (including foxtail, pigweed, lambsquarters, ragweed, and marestail) were observed on the east green roof. On the west green roof, most of "weedy" or nuisance species (including wild sweet clover, marestail, and cottonwood) have been controlled.

In July 2017, researchers clipped or pulled roughly 10,000 marestail (*Conyza canadensis*) on the east green roof to keep these plants from forming and spreading seed. Despite this effort, thousands of additional marestail (a native plant found on disturbed sites) formed seedheads in early August on the east green roof. In

Fig. 3.16 (**a**) EMS-GR 26 Oct 2019, Students and a local community member assist with spot weeding of woody plants (primarily elms, cottonwood, and honeysuckle), wild sweet clover, and other readily visible and removable weeds on the Memorial Stadium green roofs in late October 2019. (**b**) WMS-GR 22 Oct 2019, sideoats grama, blue grama, little bluestem, and big bluestem were the dominant species observed along eight transects (four high and four low) in late June 2019 (Lee Skabelund, unpublished data). Big bluestem was not specified as part of the mix but is abundant so it came with either the native prairie seed or live plants. (**c**) EMS-GR 16 Dec 2019 in early winter. (**d**) blue wild indigo came into bloom early and was very abundant during the April 2017 dedication of the Memorial Stadium. In the background, the naturalized slope north of the EMS-GR is visible. Stairs allow both people and small mammals to access the green roofs. (Photos: Lee R. Skabelund)

mid-August and early September 2017, 8000 marestail plants were removed from the east green roof, with 10,000 more removed at the end of the 2018 growing season. Many plants of ragweed, tall and common aster, and Canada goldenrod have also been found on the east green roof, and ragweed and thistle plants are removed as soon as they are found. Thus, maintaining the roofs as a prairie will continue to occupy faculty, staff, and students.

3.3.4.6 Observed Wildlife

Many forms of wildlife have been observed on the Memorial Stadium green roofs including birds, butterflies, bees, ants, grasshoppers, and raptors such as hawks. Birds frequently drop seeds of many types on both green roofs. Many butterflies have been observed (hundreds during massive hatches), such as a painted lady

(*Vanessa cardui*) outbreak that occurred in 2017. Between May and September of 2017 and 2018, 18 species of butterflies were observed on the green roofs including monarch (*Danaus plexippus*), red admiral (*Vanessa atalanta*), variegated fritillary (*Euptoieta claudia*), and 15 other generalist butterfly species (Blackmore 2019, 72). Bee bowls were placed on the Memorial Stadium green roofs during the summer of 2019. Cotton rats were found to be burrowing into the substrate, while rabbits were also found to using the east side green roof. In October 2019, substrate samples were collected for an examination of nematode populations. The K-State research team includes landscape architects, ecologists, biologists, entomologists, soil scientists, and engineers. Future research will continue to examine flora and fauna on the two green roofs.

3.3.4.7 Best Performing Native Vegetation

All of the planted or seeded grass species have established, and most of the species of forbs have adapted. Strong perennial forb performers include purple prairie-clover, gray-headed coneflower, Mexican hat coneflower, blue wild indigo, spiderwort, sage, stiff goldenrod, tall blazing star, prairie blazing star, and yarrow. Top-performing perennial grasses include big bluestem, little bluestem, sideoats grama, blue grama, Indian grass, and prairie dropseed. Butterfly milkweed, planted as live plants on the West Memorial Stadium green roof, disappeared by 2018. Prairie onion (*Allium*) was abundant in 2019 on the two green roofs. If the university decides to irrigate less to save water, more drought-tolerant species (including adaptive ruderal species) will likely become more dominant over time.

3.3.4.8 Post-Occupancy Observations

Publications

- Skabelund, L.R., Knapp, M. Moore, T., van der Merwe, D., Bruce, J.L., Decker, A., Shrestha, P. et al. 2016. "Monitoring green roof dynamics on two large-scale prairie green roofs in the Flint Hills Eco-region with the aim of conserving potable water." Poster presented at the Future of Water in Kansas Conference, Manhattan, KS (Nov. 2016).
- Skabelund, L.R., A. Decker, T. Moore, P. Shrestha, J.L. Bruce. 2017. "Monitoring two large-scale prairie-like green roofs in Manhattan, Kansas." 15th Annual Cities Alive Green Roof and Wall Conference (Sep. 2017), Seattle, WA.
- Van der Merwe, D., L.R. Skabelund, P. Blackmore, A.J. Sharda. D. Bremer. 2017. "Characterizing green roof vegetation using color-infrared and thermal sensors." 15th Annual Cities Alive Green Roof and Wall Conference (Sep. 2017), Seattle, WA.

Authors' Reflections

- The Memorial Stadium green roofs stand as a creative example of how historic structures can be repurposed into a meaningful place that connects to its cultural and natural heritage together at a land-grant university.
- This green roof was designed and is maintained and monitored in ways that serve as a model pilot project by making strong alliances between native ecosystems, green roofs, construction, maintenance, and ongoing research.
- Steep sloped green roofs in hot continental climates need supplemental irrigation, and irrigation naturally increases coverage by a wide range of plant species, some of these undesirable due to their type, size, and form. As a result, ongoing maintenance of irrigation systems and management of vegetation are essential and must be anticipated and planned for.
- The ongoing research taking place on the Memorial Stadium green roofs makes it the largest tallgrass green roof ecosystem research site in North America. The research team has included the following Kansas State University faculty and students: Lee R. Skabelund, Mary Knapp, Dr. Mark Mayfield, Dr. Trisha Moore, Dr. Stacy Hutchinson, Dr. Deon Van der Merwe, Dr. Ajay Sharda, Dr. Dale Bremer, Dr. Dave Haukos, Dr. Brent Chamberlain, Dr. Brian Speisman, Dr. Tania Kim, Dr. Andrew Hope, Timothy Todd, Jeff Taylor, Pamela Blackmore, Ryan Peters, Allyssa Decker, Priyasha Shrestha, Kyle Koehler, Elizabeth Musoke, Lekhon Alam, Marcos Aleman, and others—with support from Jeff Bruce and Chuck Dixon.

3.3.5 *Perot Museum of Nature & Science, Dallas, Texas*

The mission of the Perot Museum of Nature and Science is to "inspire minds through nature and science." This one of a kind nature museum goes well beyond a simple display of natural objects, the entire building and site represent a rethinking of how learning, play, and physical setting can come together with a variety of functions, levels of engagement, and novel aesthetics. The museum leadership selected a multidisciplinary team to interpret its vision for a new building and site. The building and landscape are intertwined into a 14-story building that physically interprets the Texas landscape through green roofs (Fig. 3.17). Ledges, Vines, Drapery, Lichens are the four concepts used to integrate the building and site together. The entrance to the museum rises upward from the street level to the second floor, while the vegetation on the ground seamlessly transitions to external exhibits with water, sculptural frogs, ephemeral waterfalls, and lush vegetation that functions as bioswales and roof gardens. Limestone blocks placed on the ground are metaphorically embedded into the building façade, as layers of pre-cast concrete panels echo forms of limestone. The site's outdoor rooftop exhibits engage visitors to learn about five Texas ecologies: West Texas Rock Cap, Upland Prairie, Blackland Prairie, and East Texas Forests/Wetlands. Each of these ecosystems is highlighted on the museum's green roofs (Fig. 3.18).

Fig. 3.17 Little bluestem stabilizes a northeast-facing slope, where vegetation and stone slabs drain stormwater from the rooftop ecosystems down towards a waterfall ledge that delivers runoff to one of the two cisterns below. (Photo: Bruce Dvorak, June 2018)

3.3.5.1 Project Team

Building Owner/Client: Perot Museum of Nature & Science
Green Roof Design Team Lead: Talley Associates
Architect: Morphosis Architects (Thom Mayne)
Structural Engineer: Datum Engineers
Landscape Architect: Coy Talley & Associates
Project completion: December 2012
Green roof area: 0.4 hectares (1 acre)

3.3.5.2 Overview and Objectives

The multi-leveled and varied-slope building creates multiple habitats based upon moisture levels, sunlight, and reflected light (Fig. 3.18). The five ecosystems are placed within the micro-climate zones created by the building and site. The Cap Rock habitat which consists of Texas prickly pear which grows naturally in rocky outcrops and shallow soils occupies the southern and west-facing sections of the roof. The Upland Prairie covers the south, east, and northwest facing slopes of

Fig. 3.18 Images of the Ross Perot green roofs. (**a**) Shows the upland prairie roof sloping down towards the waterfall ledge that drains the roofs into a cistern below a rock filter bed. (**b**) Cap Rock habitat with Texas prickly pear and red yucca grows on the south and west-facing rooftops. (**c**) Big bluestem is seen as part of the wetland grassland exhibit seen from inside the ticket lobby. (**d**) Trees and shrubs of the East Texas exhibit are seen at the bottom slope which also has a small rooftop outdoor terrace. (Photos: Bruce Dvorak, June 2018)

the roof, while the Blackland Prairie wetland habitat lies at the lowest section of the roof. The big bluestem and wetland vegetation is seen from the ticket lobby, behind the windows.

3.3.5.3 Plant Establishment

Forms of plants include annuals, grasses (Table 3.4), herbaceous perennials (Table 3.5), shrubs, and succulents. All vegetation was pre-grown in containers.

Annuals

black-eyed Susan (*Rudbeckia hirta*), clasping coneflower (*Dracopis amplexicaulis*), cowpen daisy (*Verbesina encelioides*), drummond phlox (*Phlox drummondii*), golden-wave (*Coreopsis tinctoria*), lemon mint (*Monarda citriodora*)

Table 3.4 Native grasses on the Perot Museum of Nature & Science green roofs

Common Name	Botanical Name
Big bluestem	*Andropogon gerardii*
Bushy bluestem	*Andropogon glomeratus*
Cane bluestem	*Bothriochloa barbinodis*
Sideoats grama	*Bouteloua curtipendula*
Green sprangletop	*Bouteloua dactyloides*
Blue grama	*Bouteloua gracilis*
Inland sea oats	*Chasmanthium latifolium*
Prairie wildrye	*Elymus canadensis*
Horsetail reed	*Equisetum hyemale*
Sand lovegrass	*Eragrostis trichodes*
Texas cupgrass	*Eriochloa sericea*
Curly mesquite	*Hilaria belangeri*
Buffalograss	*Leptochloa dubia*
Pink muhly grass	*Muhlenbergia capillaris*
Upland switch grass	*Panicum virgatum*
Little bluestem	*Schizachyrium scoparium*
Indian grass	*Sorghastrum nutans*
Sand dropseed	*Sporobolus cryptandrus*
Grease grass	*Tridens flavus*

Table 3.5 Native herbaceous perennials on the Perot Museum of Nature & Science green roofs

Common Name	Botanical Name
Butterfly weed	*Asclepias tuberosa*
Winecup	*Callirhoe involucrata*
Lanceleaf coreopsis	*Coreopsis lanceolata*
Purple coneflower	*Echinacea purpurea*
Cutleaf daisy	*Erigeron compositus*
Gayfeather	*Liatris spicata*
Blackfoot daisy	*Melampodium leucanthum*
Pigeonberry	*Rivina humilis*
Scarlet sage	*Salvia coccinea*
Mealy blue sage	*Salvia farinacea*
Autumn sage	*Salvia greggii*
Spiderwort	*Tradescantia virginiana*
Wood violet	*Viola sp.*

Shrubs

coralberry (*Symphoricarpos orbiculatus*), mapleleaf viburnum (*Viburnum acerifolium*)

Succulents

red yucca (*Hesperaloe parviflora*), Texas prickly pear (*Opuntia engelmannii var. lindheimeri*)

3.3.5.4 Irrigation

The project captures rainwater from the parking lot, museum roofs, and condensate water from the air conditioning system. Two large cisterns retain water for the site where one 75,708-liter (20,000-gallon) tank holds water for flushing toilets and one 113,562-liter (30,000-gallon) tank is used to irrigate vegetation on the green roof and landscape. The sourcing of tanks can be switched if one tank happens to run dry. A backup connection to city water is in place if needed. The museum has never run out of water for irrigation or flushing. Water from the parking lot is filtered first through a bioswale, then screens, and a sand filter. Water from the rooftop exits the roof from a waterfall-like ledge of the building into a splash rock filter before entering the subgrade cistern.

3.3.5.5 Maintenance

The green roof was designed to receive minimal maintenance as the plant material was selected to match the microclimate of the building, and its arrangement is intended to emulate a natural habitat. The landscape near the parking lot up through the entrance including the bioswale receives routine maintenance. The irrigation system uses harvested water from the roofs, HVAC units, and bioswale from the parking lot and thus requires annual cleaning. Once a year, the museum shuts down for three days to deep clean the museum, including flushing out the cisterns.

3.3.5.6 Observed Wildlife

There is no formal assessment of wildlife available at the museum, but birds, insects, and lizards are commonly observed on the green roof.

3.3.5.7 Best Performing Native Vegetation

All of the vegetation is thriving in its designed ecosystem.

3.3.5.8 Post-Occupancy Observations

Authors' Reflections

- The green roofs are clearly perceived as roof gardens from the street level or from nearby adjacent buildings. However, perceived from the walking path from the parking lot up to the main entrance of the building or from views inside looking out, the vegetation is experienced as an integrated experience of the museum and its exhibits. The site and building are merged together and immerse visitors in views of green. This is important because the site is bounded on two sides by elevated freeways, tall buildings, and impervious surfaces. There is no nature in this urban location, except the ecoregional green roofs and bioswales.

- Each plant community was appropriately adapted and integrated into the many microclimates made from the building's dynamic forms. The vegetation has proven its hardiness to the habitats as the vegetation is all thriving, including the Rock Cap vegetation which is subjected to intense exposure from the south- and west-facing sun and wind.

3.3.6 Botanical Research Institute of Texas, Fort Worth, Texas

The Botanical Research Institute of Texas (BRIT) is committed to its mission to conserve plants through research, discoveries, education, and built examples. Designed to house a world-class collection of plants and books, the multi-functional building has more than its herbarium, library, and research offices, it has one of the first ecoregional green roofs built in the Fort Worth/Dallas metropolitan region.

The design process, construction methods and activities used to construct and maintain this ecoregional green roof all demonstrate prairie barrens green roofs. Inspired from the nearby Walnut and Goodland Limestone Prairie Barrens, which is part of the Fort Worth Prairie ecosystem, green roof test modules were built to pre-test vegetation. Results from preliminary investigations were used to integrate the knowledge of a custom-designed biodegradable green roof system that would integrate engineered soils with natural soils. This process began years before the early phases of the integrated design of the building and site.

The building and site are closely integrated with their function, performance, aesthetic, and institutional mission. Runoff from an unplanted rooftop on the herbarium collections building has solar panels and a white reflective roof and feeds cisterns used to provide supplemental watering of vegetation. Runoff from the site is minimized through the use of porous pavements, rain gardens, and the use of regionally adapted native vegetation. The green roofs on this LEED Platinum-certified building have received much coverage by media, and the research taking place at BRIT demonstrates that the green roof ecosystem services are functioning and performing well (Fig. 3.19). This means that without the ecoregional green roof, cisterns, and low-impact infrastructure, the building and site would waste energy, create urban heat islands, require a stormwater detention facility, and displace habitat for wildlife.

3.3.6.1 Project Team

Building Owner/Client: Botanical Research Institute of Texas (BRIT)
Green Roof Design Team Lead: Tony L. Burgess (Texas Christian University), Brooke Byerley Best (BRIT)
Architect: H3 Hardy Collaboration Architecture

Fig. 3.19 Two prairie ecosystems exist near the front entrance to BRIT, a grass-based Fort Worth prairie at the ground level (right), and succulent-based prairie barrens on green roofs located above BRIT offices. (Photo: Bruce Dvorak, June 2019)

Landscape Architect: Balmori Associates (landscape master plan)
Installation Contractor: American Hydrotech, Inc.
Maintenance: BRIT
Project completion: July 2010
Green roof area: 1083 m^2 (11,400 ft^2)

3.3.6.2 Overview and Objectives

Ecologists, botanists, landscape architects, and architects collaborated to envision prairie-based green roofs at a new BRIT facility. The process included four phases: 1) explore and describe the wild system, 2) create a model system, 3) compare the performance of the model system with the wild system, and 4) progressively refine the model for optimized desired functions (Best et al. 2015).

Whereas Blackland prairies grow in deep and rich soils, the limestone prairie barrens are an expression of prairie at the interface of exposed limestone rock where soils are shallow and the vegetation is adapted to often extreme environments during the summer heat, drought, and intense rainfalls, with below-freezing temperatures during the winter. Testing of many plant species in test modules allowed the top candidates to be incorporated into the green roofs. The installed green roof system consists of natural soils, engineered substrates, and biodegradable coconut fiber trays. These have a combined depth of 12.5 cm (5 inches) substrate made from natural soils with engineered soils to make the best of lightweight and free-draining materials. The natural soils had a seedbed, complete with micronutrients to help establish and sustain the green roof ecosystems.

3.3.6.3 Plant Establishment

Six native Texas species were pre-grown and transplanted installed into the modules. Remaining species were seeded onto the growth media surface to complete a total of 38 species applied to the green roof. Forms of vegetation include grasses, herbaceous perennials (successfully established forbs, and those that did not establish [Table 3.6]), and succulents.

Grasses

buffalograss (*Bouteloua dactyloides*), seep muhly (*Muhlenbergia reverchonii*), sideoats grama (*Bouteloua curtipendula var. curtipendula*), slim tridens (*Tridens muticus var. elongatus*)

Herbaceous Perennials (established)

blazing-star (*Liatris aestivalis*), frog fruit (*Phyla nodiflora*), greenthread (*Thelesperma filifolium*), Indian blanketflower (*Gaillardia pulchella*), pasture heliotrope (*Heliotropium tenellum*), stemmy hymenoxys (*Tetraneuris scaposa*), Texas bluebonnet (*Lupinus texensis*)

Succulents

desert prickly pear (*Opuntia phaeacantha*), fameflower (*Phemeranthus calycinus*), Missouri foxtail cactus (*Escobaria missouriensis*), pale yucca (*Yucca pallida*)

Table 3.6 Native herbaceous perennials trialed on the Botanical Research Institute of Texas green roofs (did not establish)

Common Name	Botanical Name
Antelope horns	*Asclepias asperula*
Winecup	*Callirhoe involucrate*
Texas bindweed	*Convolvulus equitans*
Polkadots	*Dyschoriste linearis*
Verbena	*Glandularia bipinnatifida*
Reverchon's false pennyroyal	*Hedeoma reverchonii*
Puffballs	*Marshallia caespitosa*
Michaux's stitchwort	*Minuartia michauxii*
Missouri evening-primrose	*Oenothera macrocarpa*

3.3.6.4 Irrigation

BRIT has explored and adjusted irrigation rates as vegetation moved from its plant establishment phase into its mature phase. Watering during the growing season was applied at 15.7 mm per week for the first 12 months, then intermittent to sparse thereafter. Researchers monitored soil moisture and plant performance to assess the watering needs of the desired plants.

3.3.6.5 Maintenance

As a research center, the green roofs receive a lot of observation, but they only receive maintenance when needed. The roofs have formed two distinct zones: prickly pear and yucca (Fig. 3.20). The prickly pear has become dominant on the roof, and cactus pads become dense and periodically need to be thinned back. The vegetation retains the substrate in place, as there has been no observed erosion of the substrate.

Fig. 3.20 Prickly pear (*Opuntia* sp.) has aggressively adapted and dominated this well-drained north-east sloping rooftop. Pale yucca (*Yucca pallida*) dominates in the top right of the rooftop, where a narrow band of gravel separates the two planting schemes. Native grasses are often allowed to dry out as supplemental irrigation occurs infrequently. (Photo: Bruce Dvorak, June 2018)

3.3.6.6 Observed Wildlife

In addition to bees, red wasps, butterflies, and other insects, mourning doves and ducks make use of the roof for feeding, cover, nesting, and, breeding.

3.3.6.7 Best Performing Native Vegetation

Of the 38 species initially installed desert prickly pear (*Opuntia phaeacantha*) and pale yucca (*Yucca pallida*) have become dominant on the roof along with *Bouteloua curtipendula*. The dominance of these species is greater than found at the barrens sites and the plant diversity is higher on the green roof as well. Investigations on microhabitats on the roof indicate that species are self-sorting based upon the moisture regimes of the sloped roof, as the top of slope favors species adapted to drier habitats and an assortment of species favors those that prefer more moisture near the toe of the roof slope (Dvorak et al. 2013; Best et al. 2015).

3.3.6.8 Post-Occupancy Observations

Publications

- BRIT maintains a current list of publications on their website for the living roof.

Authors' Reflections

- This green roof was designed and is maintained and monitored as an ideal model for a pilot project in any ecoregion as it makes strong alliances between potentially viable native ecosystems, green roofs, construction, maintenance, and ongoing research. The living roof website publishes current weather conditions on the green roof, current publications regarding the roof ecosystem, and is visible from the main entrance of the building below.
- Irrigation could be more frequent, however, BRIT staff are willing to have a partially brown roof as a way to observe vegetation dynamics while minimizing the use of supplemental water. This seems to fit well with BRIT's larger educational and research goals.

3.3.7 University of Texas at Austin's Dell Medical School, Austin, Texas

Located in downtown Austin, the University of Texas Dell Medical School overlooks a parking garage rooftop. The original roof covering was a ballasted gravel roof. After the Medical School leadership updated its mission to create a more healing environment for patients, staff and visitors, the gravel rooftop which is visible from offices and waiting areas inside, was identified as an ideal location to transform the highly visible space from grey to green (Fig. 3.21). The south-facing rooftop is located above the eighth floor of the parking garage and is adjacent to a riparian corridor. The exposed site is subject to full sun, wind, reflected light from the medical center, and within the influence of the invasive seed corridor in the riparian habitat below. The design team resolved to design a hardy green roof that is well adapted to the sometimes-stressful microclimate on the rooftop, but likely would not provide as much of a cooling benefit during summer months, and thus would reduce flowering potential, and have a different aesthetic.

3.3.7.1 Project Team

Building Owner/Client: University of Texas
Green Roof Design Team Lead: Ecological Research & Design
Architect: Page Southerland Page

Fig. 3.21 This green roof is dominated by succulent vegetation and is supplemented by the inclusion of some herbaceous plants all native to central Texas habitats that exhibit dry soils. This assemblage was pre-tested in modules at the Lady Bird Johnson Wildflower Center research plots. Vegetation in the front half of the image had unwanted marestail (*Conyza canadensis*) removed 2 weeks before the photo. The back half of the image shows marestail remaining, which was also removed several weeks later. (Photo: Bruce Dvorak, October 2019)

Landscape Architect: Sasaki Associates
Ecologist: Lady Bird Johnson Wildflower Center
Installation Contractor: Ecological Research & Design
Maintenance Contractor: Ecological Research & Design
Project completion: 2016
Green roof area: 1022 m^2 (11,000 ft^2)

3.3.7.2 Overview and Objectives

Ecological Research & Design at the Lady Bird Johnson Wildflower Center in Austin, Texas devised a green roof system that would adapt to the microclimate, require minimal maintenance, and not need ongoing irrigation. The semi-intensive green roof features a 30-cm (12-inch) custom-designed substrate that was pre-tested by staff at the Ecological Research & Design. The substrate is made from local and recycled resources such as crushed brick and local sources of organic matter and micronutrients (McCullough 2016).

3.3.7.3 Plant Establishment

Succulent vegetation was pre-grown and transplanted to the roof (Fig. 3.22). Other vegetation was seeded directly onto the green roof.

Annual

Texas bluebonnet (*Lupinus texensis*)

Herbaceous Perennials

Plains coreopsis (*Coreopsis tinctoria*)

Succulents

coastal stonecrop (*Lenophyllum texanum*), Manfreda (*Manfreda maculosa*), red yucca (*Hesperaloe parviflora*), spineless prickly pear (*Opuntia ellisiana*), twistleaf yucca (*Yucca pallida*)

Fig. 3.22 Heat and drought tolerant twistleaf yucca and spineless prickly pear persist on this semi-intensive ecoregional green roof with limited irrigation. Yellow blooms of coreopsis can be seen in the background. (Photo: Bruce Dvorak, August 2019)

3.3.7.4 Irrigation

The green roof received watering during the first several months of the plant establishment period. Due to the hardiness of plants selected and substrate depth and design, this green roof does not typically require irrigation, although a subgrade drip irrigation system is present to supplement during periods of drought as needed.

3.3.7.5 Maintenance

The green roof is maintained periodically when invasive vegetation is present. Due to the later addition of the green roof (previously gravel-balested roof), since the green roof was not designed into the original program for the Medical Center, maintenance visits take place early morning on weekends before visiting hours to avoid workers walking through office space and hallways with tools or bags of vegetation removed from the roof. Currently, the medical center is preparing plans for external access to accommodate easier access. Due to the flat roof deck below the substrate, moisture persists after rainfall events, and thus some invasive early invader species such as marestail, barnyard grass, crabgrass, and spurge compete for space.

3.3.7.6 Observed Wildlife

Hummingbirds are frequent visitors to the green roof, as are bees, butterflies, and ants.

3.3.7.7 Best Performing Native Vegetation

All of the vegetation has adapted to the green roofs. The manfreda is outcompeted by some of the tall invasive vegetation, which stunts its growth. In sunny locations, manfreda is thriving.

3.3.7.8 Post-Occupancy Observations

Authors' Reflections

- This green roof is a clear testament to the value of research and development of an ecoregional approach used to inform the green roof design process. Since all of the components and vegetation are local to central Texas, there was no need to import any part of the green roof substrate or vegetation.
- The simplicity of the planting design allows maintenance staff to easily recognize invasive vegetation. One maintenance challenge is that during persistent summertime wet periods, invasive vegetation can establish where drainage is slow, and moisture persists. The same vegetation on a different green roof with a sloped and shallower substrate (John Gains Park at the Mueller Development) has fewer invasive plants present. However, the sloped roof is not located adjacent to a riparian zone, and its moderate sloped substrate is well-drained and is irrigated during drought periods.
- Because the project is located in the Edwards Aquifer recharge zone, no pesticides or herbicides are allowed to be used. Thus, maintenance of the vegetation is best accomplished by maintenance staff that have an awareness of common aggressive species, and knowledge of how to make pro-active visits to prevent outbreaks. Maintenance of green roofs is critically important and is best left to companies that have the specialized knowledge to care for the green roof as an ecosystem.

3.3.8 Camp Young Judaea's Experiential Learning Center, Wimberley, Texas

Central Texas was once known as a landscape where Blackland Prairie and Post Oak Savanna were dominant ecosystems. Today, the Blackland Prairie is nearly extinct as rangeland plant species and invasive woody and herbaceous vegetation

Fig. 3.23 Camp Young Judaea's prairie green roof. Viewed from on top a small mechanical roof, this prairie roof demonstrates resilience, biodiversity, and the viability of prairie-based green roofs on shallow substrates with irrigation systems. (Courtesy of Ecological Research & Design August 2019)

are widespread. Camp Young Judea is a learning center operated for Jewish and Zionist youth. Located just west of Austin, Texas, the camp instills environmental values that connect to the local natural history of Texas. One of its new building additions was designed as a semi-intensive prairie green roof, that is viewed from an accessible roof terrace (Fig. 3.23). The Experiential Learning Center (ELC) features energy and water-saving elements including solar panels, cisterns to retain water for irrigation, and a prairie green roof.

3.3.8.1 Project Team

Building Owner/Client: Camp Young Judaea
Green Roof Design Team Lead: Ecological Research & Design (ERD)
Architect: Sanders Architecture
Ecologist: Ecological Research & Design
Installation Contractor: Ecological Research & Design
Maintenance: Camp Young Judaea and ERD

Project completion: 2013
Green roof area: 223m^2 (2400 ft^2)

3.3.8.2 Overview and Objectives

Nestled between a cluster of mature live oak trees (*Quercus virginiana*), the single-story multi-functional building was topped with a green roof to keep the building cool, retain stormwater, and display local biodiversity. Since the green roof is visible from the adjacent roof terrace, the camp leadership wanted the color and dynamic qualities and textures of the prairie vegetation to be close to camp visitors. The 25-cm-deep (10-inch) substrate is a custom-blended growth media made from materials local to central Texas. The roof deck slopes to the north at about a 6% gradient and has several drains located at the lowest parapet edge to trickle flow water below.

3.3.8.3 Plant Establishment

Forms of vegetation include annuals/biennials (Table 3.7), grasses, and herbaceous perennials (Table 3.8). All vegetation was established through seeding.

Grasses

big bluestem (*Andropogon gerardii* Vitman), blue grama (*Bouteloua gracilis*), buffalograss (*Bouteloua dactyloides*), little bluestem (*Schizachyrium scoparium*), sand lovegrass (*Eragrostis trichodes*), sideoats grama (*Bouteloua curtipendula*)

Table 3.7 Native annuals/biennials on the Camp Young Judaea's Experiential Learning Center green roofs

Common Name	Botanical Name
Huisache daisy	*Amblyolepis setigera*
Lazy daisy	*Aphanostephus skirrhobasis*
Annual winecup	*Callirhoe leiocarpa*
Partridge pea	*Chamaecrista fasciculata*
Plains coreopsis	*Coreopsis tinctoria*
Scrambled eggs	*Corydalis aurea*
Clasping coneflower	*Dracopis amplexicaulis*
Indian blanket	*Gaillardia pulchella*
Standing cypress	*Ipomopsis rubra*
Texas yellow star	*Lindheimera texana*
Bluebonnet	*Lupinus subcarnosus*
Lemon mint	*Monarda citriodora*
Clammyweed	*Polanisia* Raf.
Sleepy daisy	*Xanthisma texanum*

Table 3.8 Native herbaceous perennials on the Camp Young Judaea's Experiential Learning Center green roofs

Common Name	Botanical Name
Showy milkweed	*Asclepias speciosa*
Butterfly weed	*Asclepias tuberosa*
Winecup	*Callirhoe involucrata*
American basketflower	*Centaurea americana*
Golden wave	*Coreopsis intermedia*
Lanceleaf coreopsis	*Coreopsis lanceolata*
White prairie clover	*Dalea candida*
Purple prairie clover	*Dalea purpurea*
Illinois bundleflower	*Desmanthus illinoensis*
Narrow leaf coneflower	*Echinacea angustifolia*
Purple coneflower	*Echinacea purpurea*
Cutleaf daisy	*Erigeron compositu*
Texas bluebells	*Eustoma exaltatum* ssp. *russellianum*
Prairie verbena	*Glandularia bipinnatifida*
Gayfeather	*Liatris spicata*
Missouri primrose	*Oenothera macrocarpa*
Pink evening primrose	*Oenothera speciosa*
Prairie parsley	*Polytaenia nuttallii* DC.
Mexican hat	*Ratibida columnifera*
Prairie coneflower	*Ratibida columnifera*
Black-eyed Susan	*Rudbeckia hirta*
Scarlet sage	*Salvia coccinea*
Mealy blue sage	*Salvia farinacea*
Awnless bush sunflower	*Simsia calva*

3.3.8.4 Irrigation

The green roof vegetation is watered with a subgrade drip irrigation system. The water source is from the rooftop of an adjacent two-story building that has a metal roof. Runoff is conveyed directly into an above-ground metal cistern located next to the building. The 34 m^3 (9000-gallon) cistern is sized to hopefully never run dry. Irrigation is typically shut down from December through April.

3.3.8.5 Maintenance

The prairie roof has gone through several transitions. All the vegetation established during its initial growing season in 2013. During 2018, the irrigation system became non-functioning and was shut down. Consequently, the vegetation did not emerge in early 2019, and the surface of the green roof was bare of top growth in April. However, after the irrigation system was repaired in May 2019, and after several

Fig. 3.24 This vibrant green roof engulfs visitors with a diverse mixture of tallgrass Blackland Prairie vegetation. Seen here, the interior habitat grows a diversity of grasses with little bluestem (front middle) and big bluestem (center back) growing alongside the blooms of black-eyed Susan (yellow) and scarlet sage (red). (Photo: Bruce Dvorak, August 2019)

months with the irrigation running, the initial vegetation planted on the green roof emerged from dormancy and reestablished (Fig. 3.24).

The maintenance manual for the green roof includes laminated pages that show labeled photos of vegetation that was planted, along with common unwanted invasive plants. The maintenance manual also includes a seasonal schedule with timing and frequency of common maintenance activities such as weeding, deadheading, and fertilization.

3.3.8.6 Observed Wildlife

Dragonflies, grasshoppers, a diversity of bees, butterflies, and birds frequent the green roof.

3.3.8.7 Best Performing Native Vegetation

All of the vegetation is well-adapted to microclimate conditions on the green roof.

3.3.8.8 Post-Occupancy Observations

Authors' Reflections

- This green roof demonstrates the resilience of native vegetation once it is well-established. When the irrigation system was shut down for nearly a half-year, the vegetation remained dormant until the irrigation system was repaired. One of the results of the irrigation system being down is that little bluestem became more dominant on the roof than it might have otherwise been early on.
- This green roof also demonstrates how Blackland Prairie vegetation can grow on green roofs in central Texas, with irrigation.

3.3.9 Trinity University, Center for the Sciences and Innovation, San Antonio, Texas

Initiated from ambitions within the university, the Center for Sciences and Innovation was envisioned over several years to become a place where students and faculty can reconnect to the excitement of science, technology, and Texas' natural history. The design team worked with the university to program and plan how to best accomplish its goals. An ecoregional green roof became a focal point of the building. Located on top of the first floor, with second-floor access, and views to the garden from the third floor, the roof garden was designed to recall the prairie habitats that were once common throughout southcentral Texas. Kelly Lyons, Professor of Biology sums up the collective vision for the green roof. She says that it was designed to be "a place of contemplation to combat daily stress and nature deficit disorder, and increase awareness of the value of natural capital." Rialto Studio lead the green roof design process and articulated the space to function as an outdoor classroom, a native plant demonstration garden, and an urban wildlife habitat. O'Neil Ford Architects designed the LEED-certified building (Sierra 2019) (Fig. 3.25).

3.3.9.1 Project Team

Building Owner/Client: Trinity University
Green Roof Design Team Lead: Rialto Studio with Trinity University
Architect: Einhorn Yaffe Prescott (EYP)
Landscape Architect: Rialto Studio (site and green roof design)
Research Consultant: Dr. Kelly Lyons, Biology, Trinity University
Installation Contractor: Maldonado (green roof) Vaughn (General Contractor)
Maintenance Contractor: Trinity University facilities, students, and faculty
Project completion: February 2015
Green roof area: 303 m^2 (3200 ft^2)

Fig. 3.25 Green roof at the Trinity University, Center for the Sciences and Innovation. Its visually simple but diverse composition of plants is didactic, as it teaches visitors about native Texas grasses, wildflowers, succulents, and its resident and visiting wildlife. The roof garden occupies a prominent place central to the mission of the Center. (Courtesy of Rialto Studio)

3.3.9.2 Overview and Objectives

The planting design for the green roof was inspired by a vision to use all locally native low-stature vegetation. Plant selection heavily favors native grasses and succulents with the addition of native herbaceous perennial wildflowers to provide seasonal color interest. Designed to be a contemplative space for students, faculty, and staff to get away, the planting design consists of no formal patterns to be maintained, as the geometry of the accessible paths provide visual contrast (Fig. 3.26). The growing media is 30-cm-deep (about 12 inches).

3.3.9.3 Plant Establishment

Forms of vegetation on the roof garden include grasses (Table 3.9), herbaceous perennials (Table 3.10), and succulents. Vegetation was installed as pre-grown plugs or larger sized plants.

Succulents

red yucca (*Hesperaloe parviflora*), sacahuista (*Nolina Lindheimeriana*), twistleaf yucca (*Yucca pallida*)

Fig. 3.26 The native grasses grow to about one meter in height during the summer. Table and chairs were provided to convert the roof garden into an outdoor classroom and study space. Plant name tags are provided to educate visitors about the vegetation used in the garden. (Courtesy of Rialto Studio)

Table 3.9 Native grasses on the Trinity University, Center for the Sciences and Innovation green roofs

Common Name	Botanical Name
Big bluestem	*Andropogon gerardii*
Sideoats grama	*Bouteloua curtipendula*
Texas grama	*Bouteloua rigidiseta*
Inland sea oats	*Chasmanthium latifolium*
Texas cupgrass	*Eriochloa sericea*
Gulf muhly grass	*Muhlenbergia capillaris*
Lindheimer muhly grass	*Muhlenbergia lindheimeri*
Texas wintergrass	*Nasella leucotricha*
Switchgrass	*Panicum virgatum*
Little bluestem	*Schizachyrium scoparium*
Yellow Indiangrass	*Sorghastrum nutans*

3.3.9.4 Irrigation

Runoff from the roof garden is directed towards a subgrade cistern. Collected rainwater is used to irrigate the vegetation. The irrigation water not absorbed by plants drains back into the cistern for reuse. This circular design preserves valuable rainfall for the green roof ecosystem and ensures that city water which can include trace amounts of chlorine, fluoride, and other contaminants are not being introduced to

Table 3.10 Native herbaceous perennials on the Trinity University, Center for the Sciences and Innovation green roofs

Common Name	Botanical Name
Flame acanthus	*Anisacanthus quadrifidus var 'wrightii'*
Damianita	*Chrysactinia mexicana*
Blue mist flower	*Conoclinium greggii*
Turk's cap	*Malvaviscus arboreus 'drummondii'*
Blackfoot daisy	*Melampodium leucanthum*
Dwarf wax myrtle	*Myrica pusilla*
Bear grass	*Nolina texana*

plants. The quality of water in the cistern is periodically monitored to ensure it is suitable for use. The drip irrigation system is located at the substrate surface, is buried under cedar mulch (without fines), and is divided into six zones. Irrigation run times vary depending upon seasonal variables such as daytime and nighttime temperatures, and length of time between precipitation events.

3.3.9.5 Maintenance

Students maintain the green roof under the direction of faculty and staff. Since vegetation is not arranged in formal geometric design, seedlings of successful and desirable vegetation are allowed to take hold or be relocated. Seasonal activities include the removal of unwanted plants, trimming back of the herbaceous green roof vegetation at the end of the year, and applying organic forms of fertilizer when needed.

3.3.9.6 Observed Wildlife

Hummingbirds are frequently attracted to the blooms of Turk's cap and red yucca. Other frequent visitors include pollinators such as bees and butterflies during the growing season.

3.3.9.7 Best Performing Native Vegetation

All of the vegetation has thrived, except wax myrtle.

3.3.9.8 Post-Occupancy Observations

Research Faculty

- Most of the vegetation has thrived. Several species are struggling, including wax myrtle, Texas wintergrass, yellow Indiangrass, and Texas cupgrass. *These last two might not adapt well to the proportion of minerals in the substrate.*

- Several species were added to pre-test their adaptability and include silver bluestem, purple three awn (performed very well, and several volunteer plants emerged), and bluebonnets. Some species of native milkweeds were trialed and did not establish. Blackfoot daisy performed well; however, as a biennial, it can fade away if new plants do not establish from seed.

Authors' Reflections

- This project has an ideal set up as a pilot project. The design team included many points of view from the beginning, including university staff, administration, maintenance, and facilities representatives, and the long-term interest and involvement by the research faculty has endured. They all participated with the design team to help set up a vision, realistic expectations, and a commitment to succeed.
- This is one few green roofs where the vegetation includes locally native and some endemic vegetation.

3.4 Plants for the Prairie Ecoregion

There were 151 native taxa observed in the nine ecoregional case studies discussed in this chapter. One-hundred forty-five (145) are native to the Tallgrass Prairie ecoregion. Thirty-four (34) species occur more than once across the ecoregional green roofs including five (5) annuals/biennials, eight (8) grasses, 20 herbaceous perennials, and one (1) succulent. Ten (10) species occur in three or more of the ecoregional green roofs (Table 3.11).

As might be expected, grasses and herbaceous perennials were most frequently used. The four dominant kinds of grass most utilized (Table 3.11) on the green roofs

Table 3.11 Plant species that occur on three or more ecoregional green roofs in this chapter

Plant Type	Common Name	Botanical Name	A	B	C	D	E	F	G
Grasses	Sideoats grama	*Bouteloua curtipendula*	x		x	x	x	x	x
Grasses	Buffalograss	*Bouteloua dactyloides*					x	x	x
Grasses	Blue grama	*Bouteloua gracilis*	x		x	x		x	x
Grasses	Little bluestem	*Schizachyrium scoparium*	x	x	x	x		x	x
Herbaceous perennials	Butterfly weed	*Asclepias tuberosa*	x	x				x	x
Herbaceous perennials	Winecup	*Callirhoe involucrata*		x				x	x
Herbaceous perennials	Lanceleaf coreopsis	*Coreopsis lanceolata*		x				x	x
Herbaceous perennials	Purple prairie clover	*Dalea purpurea*	x		x				x
Herbaceous perennials	Purple coneflower	*Echinacea purpurea*		x				x	x
Herbaceous perennials	Mexican hat	*Ratibida columnifera*	x		x				x

Key = A (K-State MSGRs), B (TWA Headquarters Building), C (K.C. Central Library), D (Pioneers Nature Park), E (BRIT), F (Ross Perot Museum), G (Camp Young Judea).

in the chapter are also dominant grasses in the native prairies. Butterfly milkweed (*Asclepias tuberosa*) was observed on four of the case studies. Its ecological function was to supplement floral diversity on the green roof and to potentially attract migrating monarch butterflies. Butterfly milkweed performed best when it was established from seed on many of the green roofs, and when it was planted as a live plant *Asclepias tuberosa* disappeared from both of the K-State Memorial Stadium green roofs.

Compared with the most diverse conservation site, the green roofs were not as diverse as their native analogs. The Konza Prairie has 576 taxa growing on its hills, riparian zones, and wetlands. However, the Nachusa Grasslands has 71 native species, and much of the land was acquired and under restoration. The Konza Prairie has large portions of the original prairie and has more natural diversity. Only one portion of a green roof had wetland vegetation, the Ross Perot Museum of Science and Nature. Dry and mesic habitats were the most common kind of habitat designed. One dry green roof ecosystem (Dell Hospital) was designed to function with minimal irrigation but has an irrigation system if needed.

3.5 Summary

There has been a growing interest in implementing green roofs in the tallgrass ecoregions since Mayor Daley promoted green roofs in Chicago beginning around 2000. Several of the green roofs highlighted in this chapter were built before 2010, thus the endurance of the green roof ecosystems and their vegetation demonstrate that plants native to the tallgrass prairies can effectively establish and are resilient on green roofs with appropriate maintenance and irrigation practices. The green roofs have been developed for various purposes including to:

- connect visually with the surrounding environment, e.g., K-State Memorial Stadium, Perot Museum of Nature & Science, Kansas City Public Library;
- provide a learning tool to demonstrate green roofs, e.g., Pioneers Park Nature Center, Kansas City Public Library, Perot Museum of Nature & Science, Camp Young Judaea's Experiential Learning Center;
- provide an outdoor meeting space for workers, students, and or visitors, e.g., TWA Former Headquarters, Camp Young Judaea's Experiential Learning Center, Trinity University;
- benefit from the ecosystem services generally associated with green roofs, energy conservation, runoff amelioration, temperature and noise abatement e.g., Botanical Research Institute of Texas, Perot Museum of Nature & Science;
- function as a green roof research site, e.g., K-State Memorial Stadium, Botanical Research Institute of Texas, Trinity University Center for the Sciences and Innovation, Pioneers Park Nature Center;

- increase local habitat diversity, e.g., K-State Memorial Stadium, Botanical Research Institute of Texas, Camp Young Judaea's Experiential Learning Center, Perot Museum of Nature & Science;
- increase the aesthetic appeal of buildings especially where the function of the building itself may be associated with increased stress for visitors, e.g., hospitals (the University of Texas at Austin's Dell Medical School, Trinity University);
- provide seasonal attributes of prairie vegetation during the plant selection and ongoing maintenance at the K-State Memorial Stadium, TWA Former Headquarters, Camp Young Judaea's Experiential Learning Center.

There are several elements common to the green roofs described in this chapter. Most of the green roofs have a substrate depth of 12–15 cm (5–6 inches) which allowed grasses, annuals, and herbaceous perennials to establish. While built-up and layered (monolithic) forms of construction were the most popular, biodegradable modules (green roof trays) were successfully used on one project (BRIT).

Herbaceous grasses and perennial wildflowers were the dominant vegetation represented on the green roofs in this chapter. Succulents (of which there are few native to tallgrass prairies) were present but used as accents. Exceptions include the University of Texas at Austin's Dell Medical School, the BRIT green roof, and parts of the Ross Perot Museum green roof.

All of the green roofs have irrigation systems as drip systems buried in the substrate or placed on top. Overhead spray or capillary systems were also used. For most of the green roofs, harvested rainwater is used to maintain the vegetation. Most green roofs are irrigated during the late spring, summer, and autumn months only. Species for the most part reflect the particular ecosystems of the area, although horticultural cultivars and a few locally non-native species have been used.

Underlying their success is detailed planning, not only for their creation and installation but also for ongoing maintenance. The maintenance is either carried out by contracted specialists or transitioned to occupants of the building who are trained, or by university faculty and students. Attention to potentially invasive volunteer species is critical to maintaining the ecological integrity of these green roofs. A couple of the green roofs are maintained with a high level of knowledge of local plants, volunteer plants, and how to selectively remove or add vegetation during maintenance visits to retain a garden-like or naturalistic appearance.

Several green roofs have robust research taking place. This is important, as research is a key component in the feedback loop of design, maintenance, and persistence over time. Research has allowed key ecosystem services and vegetative plant communities to be identified. However, most of the major urban centers lacked connections to ongoing green roof research. For example, common maintenance activities include the strategic scattering of seed or the relocating and watering in live plants).

Acknowledgments We would like to thank the following individuals for their generosity in giving of time and sharing of information about their projects: staff at the Pioneers Nature Center, Maggie Riggs at Off The Grid Rooftop Gardens; Jeffery Bruce, David Stokes at Jeffrey L. Bruce & Company; Coy Talley at Talley Associates; Dr. Brooke Byerley Best with the Botanical Institute of Texas; John Hart Asher with Ecological Research & Design; Ron Waide, Sr. Property Manager Dell Medical Center; Bobby Eichholz at Rialto Studio; Dr. Kelly Lyons Trinity University; Pamela Blackmore, Kansas State University; and Dr. Richard Sutton, University of Nebraska.

Dedication We dedicate this chapter to Jeffrey L. Bruce for his superb and innovative work in green roof design and landscape architecture. Jeffrey often integrated native vegetation on green roofs in the Midwest, and throughout the western United States. His death in January 2020 is a great loss to his family, friends, firm, the design profession, and to the communities where he played such an important and energizing role.

References

Anderson HL (2011) That settles it: the debate and consequences of the homestead act of 1862. Hist Teach 45(1):117–137

Bailey RG (1997) Ecoregions of North America. U.S. Department of Agriculture, Forest Service,Washington, DC

Beatley T (2000) Preserving biodiversity: challenges for planners. J Am Plan Assoc 66(1):5–20

Bendel CR, Hovick TJ, Limb RF, Harmon JP (2018) Variation in grazing management practices supports diverse butterfly communities across grassland working landscapes. J Insect Conserv 22(1):99–111

Benedict MA, McMahon ET (2002) Green infrastructure: smart conservation for the 21st century. Renewable Resources Journal 20(3):12–17

Best BB, Swadek RK, Burgess TL (2015) Soil-based green roofs. In: Green roof ecosystems. Springer International, Cham, pp 139–174

Blackburn RC (2018) Diet and vegetation impacts of a reintroduced bison herd Northern Illinois University, DeKalb

Blackmore P (2019) Butterflies, tallgrass prairie, and green roofs. Kansas State University, Manhattan

Blair RB (1999) Birds and butterflies along an urban gradient: surrogate taxa for assessing biodiversity? Ecol Appl 9(1):164–170

BONAP (2018) The Biota of North America Program (BONAP), Taxonomic Data Center. The Biota of North America Program (BONAP). Taxonomic Data Center, Chapel Hill, NC

Chapman KA, Brewer R (2008) Prairie and savanna in southern lower Michigan: history, classification, ecology, vol 47. Michigan Botany Club, Ann Arbor

Collins SL, Wallace LL (1990) Fire in north American tallgrass prairies. University of Oklahoma Press, Norman

Conard R, Hess S (1998) Tallgrass Prairie National preserve legislative history, 1920–1996. Tallgrass Historians LC, National Park Service Midwest Support Office: 65, Iowa City

Dick E (1937) The sod-house frontier 1854–1890. D. Appleton-Century Company, New York

Diggs GM, Lipscomb BL, O'Kennon RJ, Mahler WF, Shinners LH (1999) Shinners' and Mahler's illustrated flora of north Central Texas. Botanical Research Institute of Texas, Fort Worth

Dvorak B (2009) The Chicago City Hall green roof pilot project: a case study. Paper presented at the Third International Conference on Smart and Sustainable Built Environments. from Building to Promise Building Smartly in a Changing Climate, Delft, Netherlands, June 15–19, 2009

Dvorak B (2015) Plant viability trials for shallow extensive green roofs in central Texas. In: International Low Impact Development Conference, Houston, Texas, January 19–21 2015

Dvorak B, Carroll K (2008) Chicago City Hall Green Roof: Its Evolving Form and Care. Paper presented at the Sixth Annual International Rooftops for Sustainable Communities Conference, Baltimore, MD, April 30–May 2

Dvorak B, Volder A (2013) Rooftop temperature reduction from unirrigated modular green roofs in south-Central Texas. Urban For Urban Green 12(1):28–35. https://doi.org/10.1016/j.ufug.2012.05.004

Dvorak B, Byerley B, Volder A (2013) Plant species survival on three water conserving green roofs in a hot humid subtropical climate. Journal of Living Architecture 1(1):10

Ernest T, Towne G (2020) How Bison Grazing Habits Affect Plant Composition. Tallgrass Gazette, vol Feb. Konza Prairie Research Natural Area, Manhattan, KS

Grace JB, Allain LK, Baldwin HQ, Billock AG, Eddleman,WR, Given AM, & Moss R (2005) Effects of prescribed fire in the coastal prairies of Texas U.S. Geological Survey, Reston, VA, p 46

Hammond E (1964) Classes of land-surface form in the United States. US Geological Survey

Harmel R, Richardson C, King K, Allen P (2006) Runoff and soil loss relationships for the Texas Blackland prairies ecoregion. J Hydrol 331(3–4):471–483

Holman-Dodds JK (2007) Towards greener stormwater management. Journal of Green Building 2(1):68–96

Howe HF (1994) Managing species diversity in tallgrass prairie: assumptions and implications. Conserv Biol 8(3):691–704

Kaiser PH, Berlinger SS, Fredrickson LH (1979) Response of blue-winged teal to range management on waterfowl production areas in southeastern South Dakota. Rangeland Ecology & Management/Journal of Range Management Archives 32(4):295–298

KEEP (2020) Bison on Konza Prairie. Konza Environmental Education Program. https://keep.konza.k-state.edu/prairieecology/bison.html. Accessed 8 Jan 2020

Kimmerer RW, Lake FK (2001) The role of indigenous burning in land management. J For 99(11):36–41

Kindscher K (2009) Edible plants of the prairie. Symphony in the Flint Hills Field Journal 9

Klein PM, Coffman R (2015) Establishment and performance of an experimental green roof under extreme climatic conditions. Sci Total Environ 512:82–93

Knapp AK, Blair JM, Briggs JM, Collins SL, Hartnett DC, Johnson LC, Towne EG (1999) The keystone role of bison in north American tallgrass prairie: Bison increase habitat heterogeneity and alter a broad array of plant, community, and ecosystem processes. Bioscience 49(1):39–50

KPBS (2020) Konza Prairie site description. K-State University. http://lter.konza.ksu.edu/konza-prairie-site-description. Accessed 3 July 2020

Launchbaugh JL (1955) Vegetational changes in the San Antonio prairie associated with grazing, retirement from grazing, and abandonment from cultivation. Ecol Monogr 25(1):39–57. https://doi.org/10.2307/1943213

Li M-H, Dvorak B, Sung CY (2010) Bioretention, low impact development, and stormwater management. In: Aitkenhead-Peterson J, and Astrid Volder (ed) Urban ecosystem ecology, vol 55. American Society of Agronomy pp 413–430

Licht DS (1997) Ecology and economics of the Great Plains, vol 10. U of Nebraska Press, Lincoln

Liu J, Shrestha P, Skabelund LR, Todd T, Decker A, Kirkham M (2019) Growth of prairie plants and sedums in different substrates on an experimental green roof in mid-continental USA. Sci Total Environ 697:134089

Macbride TH (1928) In cabins and sod-houses. State Historical Society of Iowa, Iowa City

MacDonagh LP, Shanstrom N (2015) Assembling prairie biome plants for Minnesota green roofs. In: Sutton RK (ed) Green roof ecosystems. Springer, Cham, pp 257–283. https://doi.org/10.1007/978-3-319-14983-7_11

McCullough A (2016) The green goes up. Landscapes

McMillan RB, Klippel WE (1981) Post-glacial environmental change and hunting-gathering societies of the southern prairie peninsula. J Archaeol Sci 8(3):215–245

Morseburg J (2011) Julian Onderdonk an illustrated biography. WordPress. https://julianonderdonk.wordpress.com/tag/julian-onderdonk-biography/. Accessed 25 Oct 2016

Packard S, Mutel CF (1997) The tallgrass restoration handbook: for prairies, savannas and woodlands. Island Press, Washington, DC

Parker AA (1836) Trip to the West and Texas: comprising a journey of eight thousand miles, through New-York, Michigan, Illinois, Missouri, Louisiana and Texas, in the autumn and winter of 1834–5. With a brief sketch of the Texian War. W. White, Concord
Ricketts TH (1999) Terrestrial ecoregions of North America: a conservation assessment. Island Press, Washington, DC
Robertson KR, Anderson RC, Schwartz MW (1997) The tallgrass prairie mosaic. In: Conservation in highly fragmented landscapes. Chapman & Hall, New York, pp 55–87
Sampson F, Knopf F (1994) Prairie conservation in North America. Other Publications in Wildlife Management:41
Samson FB, Knopf FL (1996) Prairie conservation: preserving North America's most endangered ecosystem. Island Press, Washington DC
Schramm P (1990) Prairie restoration: a twenty-five year perspective on establishment and management. In: Proceedings of the Twelfth North American Prairie Conference, 1990. University of Northern Iowa Cedar Fall, Iowa, pp 169–177
Shively SB, Weaver JE (1939) Amount of underground plant materials in different grassland climates
Sierra C (2019) When green equals green. Trinity University Campus News, San Antonio
Simmons MT (2015) Climates and microclimates: challenges for extensive green roof design in hot climates. In: Green roof ecosystems. Springer, Cham, pp 63–80
Simmons M, Gardiner B, Windhager S, Tinsley J (2008) Green roofs are not created equal: the hydrologic and thermal performance of six different extensive green roofs and reflective and non-reflective roofs in a sub-tropical climate. Urban Ecosyst 11(4):339–348
Skabelund LR, Blocksome C, Hamehkasi M, Jin Kim H, Knapp M, Brokesh D (2014) Semi-arid green roof research 2009–2014: resilience of native species. Paper presented at the Cities Alive 2014 12th Annual Green Roof & Wall Conference, Nashville, TN, November 12–15, 2014
Skabelund LR, DiGiovanni K, Starry O (2015) Monitoring abiotic inputs and outputs. In: Green roof ecosystems. Springer, Cham, pp 27–62
Skabelund LR, Knapp M, Moore T, van der Merwe D, Bruce JL, Decker A, Shrestha P (2016) Monitoring green roof dynamics on two large-scale prairie green roofs in the Flint Hills Ecoregion with the aim of conserving potable water. Poster presented at the Future of Water in Kansas Conference. Manhattan
Skabelund LR, Decker A, Moore T, Shrestha P, Bruce JL (2017) Monitoring two large-scale prairie-like green roofs in Manhattan, Kansas. In: CitiesAlive: 15th Annual Green Roof and Wall Conference 2017 Conference Proceedings, Seattle, WA, September 18–21 2017. The Cardinal Group, p 30
Smith DD (1992) Tallgrass prairie settlement: prelude to the demise of the tallgrass ecosystem. In: Proceedings of the twelfth North American prairie conference, Cedar Falls, IA. University of Northern Iowa, pp 195–199
Smith DD (1998) Iowa prairie: original extent and loss, preservation and recovery attempts. Journal of the Iowa Academy of Science: JIAS 105(3):94–108
Smith D, Butler B (2011) Map of the mid-continent tallgrass prairie. Tallgrass prairie (map by D. Smith and B. Butler)
Srinath I, Millington A (2016) Evaluating the potential of the original Texas land survey for mapping historical land and vegetation cover. Land 5(1):4
Steinauer EM, Collins SL (1996) Prairie ecology: the tallgrass prairie. Prairie conservation. Island Press, Washington, DC
Sutton RK (2008) Media Modifications for Native Plant Assemblages on Green Roofs. Paper presented at the Sixth Annual International Rooftops for Sustainable Communities Conference, Baltimore, MD, April 30–May 2
Sutton RK (2013) Seeding green roofs with native grasses. Journal of Living Architecture 1(1):15–35
Sutton RK (2015) Green roof ecosystems, vol 223. Springer, Cham
Sutton RK (2018) Maintenance of the Pioneer nature center green roof

Sutton RK, Harrington JA, Skabelund L, MacDonagh P, Coffman RR, Koch G (2012) Prairie-based green roofs: literature, templates, and analogs. Journal of Green Building 7(1):143–172. https://doi.org/10.3992/jgb.7.1.143

Taft JB (2007) Results from 2006 resample of permanent vegetation sampling transects at Nachusa Grasslands: changes five years after baseline sampling. INHS Division of Biodiversity and Ecological Entomology (DBEE); INHS Section …, Champaign

Towne EG (2002) Vascular plants of Konza Prairie Biological Station: an annotated checklist of species in a Kansas tallgrass prairie. SIDA, Contributions to Botany: 269–294

Van der Merwe D, Skabelund LR, Sharda A, Blackmore P, Bremer D (2017) Towards characterizing green roof vegetation using color-infrared and thermal sensors. In: Proceedings of the CitiesAlive 15th Annual Green Roof and Wall Conference, Seattle, WA, September 18–21 2017. pp 18–21

Volder A, Dvorak B (2013) Event size, substrate water content and vegetation affect storm water retention efficiency of an un-irrigated extensive green roof system in Central Texas. Sustain Cities Soc 10(0):59–64. https://doi.org/10.1016/j.scs.2013.05.005

White M (2006) Prairie time: a Blackland portrait, vol 10. Texas A&M University Press, College Station

Williams D (2010) The Tallgrass Prairie Center guide to seed and seedling identification in the upper Midwest. University of Iowa Press, Iowa City

Winsor RA (1987) Environmental imagery of the wet prairie of east Central Illinois, 1820–1920. J Hist Geogr 13(4):375–397

Chapter 4
Green Roofs in Shortgrass Prairie Ecoregions

Bruce Dvorak and Jennifer Bousselot

Abstract This chapter examines four case studies of conservation sites and nine green roofs located in the Shortgrass Prairie and Rocky Mountain ecoregions. The shortgrass prairie once covered the Great Plains from the Texas Panhandle in the south to Alberta, Canada, in the north, and from western Kansas in the east to Montana in the west. With 300–400 mm of annual precipitation, the shortgrass prairie vegetation is much shorter in stature than the mixed-grass or tallgrass prairies, however, it is ecologically significant. Topographic relief has a dramatic effect on plant diversity, as the natural vegetation of open plains tend to be simple compositions where drainages and transitions to dry habitats can be highly diverse. The Rocky Mountains form a massive range with many ecoregions including mountain meadows and alpine plant communities. Ecoregional green roofs located in Laramie, Wyoming, Boulder, Colorado, Denver, Colorado, and the Rocky Mountains represent 347 taxa in total, 274 of which are native to the ecoregions in this chapter.

Keywords Drought · Succulent · Diversity · Grass · Residential · Urban redevelopment · Winter dormancy · Resilience

4.1 Ecoregion Characteristics

The Great Plains covers a large geographic area that encompasses the southern half of the central provinces in Canada, the Dakotas, eastern Montana, Wyoming, and Colorado, the western half of Nebraska, Kansas, Oklahoma, and Texas. The

B. Dvorak (✉)
Department of Landscape Architecture and Urban Planning, 305A Langford Architecture Center, Texas A&M University, College Station, TX, USA
e-mail: bdvorak@tamu.edu

J. Bousselot
Department of Horticulture and Landscape Architecture, Colorado State University, Fort Collins, CO, USA
e-mail: Jennifer.Bousselot@colostate.edu

B. Dvorak (ed.), *Ecoregional Green Roofs*, Cities and Nature,
https://doi.org/10.1007/978-3-030-58395-8_4

shortgrass prairies were once part of the second-largest contiguous ecosystems in North America. Today ranching, industrial agriculture practices, and urbanization have largely rendered the shortgrass prairie as a critically endangered ecosystem (Lauenroth et al. 2008; Cook et al. 2019).

The central plains are defined by several distinct topographic regions: the Alberta plain in Canada, the Missouri Plateau in southern Canada and the United States, the High Plains, and the Interior Lowlands. The topographic elevation of each region steps downward from the northern altitudes at 1524–1829 m (5000–6000 ft) down to 457–609 m (1500–2000 ft) at the Central Lowlands in Nebraska to central Texas. Each plain is defined by climate, vegetation, and the unique cultural influences that have shaped the ecoregions.

At the beginning of the Holocene, the Great Plains were influenced by decades-long drought periods, where dunes formed and steppe vegetation declined, and then wetter periods persisted where shortgrass (treeless) ecosystems expanded, like much of the region today. Biomass generally follows higher amounts of precipitation toward the 100th meridian and declines west of the 100th meridian (Forman et al. 2001).

In addition to the influences of wind and precipitation, Native Americans frequently followed and hunted bison in the region. As grazers, bison were a major source of disturbance and regeneration of vegetation across the Great Plains (Fig. 4.1) (Vinton and Collins 1997). The presence of fire was less of a factor in the region compared to the impact it had on mixed-grass and tallgrass prairies. However, fire was used by Native American tribes as a tool in shortgrass prairies to burn the vegetation, and as a method of hunting, by forcing bison to move or congregate. Fire had effects of clearing the land of woody vegetation growing in isolated areas, returning nutrients to the soil, and generating patches of early successional vegetation (Scheintaub et al. 2009; Augustine et al. 2010).

The vegetation of the shortgrass prairie before its settlement consisted of grasses (graminoids), herbaceous wildflowers (forbs), annuals, mosses, succulents, subshrubs, shrubs, trees, and mosses. The shortgrass prairie is dominated by both sod-forming grasses and bunchgrasses, as both are common across the diverse landscapes of the Great Plains. Plant diversity on remnant plots of shortgrass prairie can range from 38 to 52 species, or more. However, some shortgrass prairies are more diverse than tallgrass prairies, especially where they intersect at drier ecotones. There are also plant communities that are sorted by moisture gradients, based upon the presence of soil moisture found along upland, mid-slope and lowland topography (Singh et al. 1996; Brockway et al. 2002).

Succulents and forbs are often intermixed with grasses and forbs in shortgrass prairies. They typically grow on the warmer and drier south and west-facing slopes. They also grow in transitional landscapes where prairies are interrupted by bluffs, mountains, or locations with shallow, rocky, or nutrient-deprived soils (Carter et al. 2018). Aquatic and wetland vegetation in the Great Plains, especially the northern section in the potholes region of the Dakotas has a diversity of over 500 species of vascular plants (Larson 1993).

As wetland vegetation can play significant functional roles for green roofs (and building system functions), their inclusion in the discussion here refers to their

Fig. 4.1 George Catlin was one of the most prolific and early painters of the American West, before its settlement. *Buffalo Bulls in a Wallow* (1837), by Catlin, captures a scene where he experienced a phenomenon where bison cool themselves by digging into a wet or moist place in a prairie. The mud and moisture cool the large animals in the heat of the day. Catlin observed that these wallows later become vegetated with diverse early successional herbage making for colorful circles of wildflowers that are like mysterious circles dotting the prairie (Catlin 1841). The bison and shortgrass prairie are shown here extending uninterrupted far into the horizon of the Great Plains in Montana. (Courtesy of Creative Commons https://upload.wikimedia.org/wikipedia/commons/2/27/George_Catlin_-_Buffalo_Bulls_in_a_Wallow_-_1985.66.425_-_Smithsonian_American_Art_Museum.jpg)

potential use and could be explored on green roofs in the Great Plains. Although not considered associated with grasslands, ferns and their allies grow in protected habitats in the Great Plains, and could function on shady green roofs where irrigation is provided (Hazlett 2004). Some regions of the Great Plains contain high diversity and rare plants, as there are 388 of 3211 plants native to Colorado that are considered rare (Hazlett 2004; Ackerfield 2015).

Remnant shortgrass prairie sites in some parts of the Great Plains are subject to invasive plants such as annual brome grasses, cheatgrass (*Bromus tectorum*), and Japanese brome (*Bromus japonicus*) (Ashton et al. 2016). Some introduced plants such as Kentucky bluegrass, and other grasses and forbs can invade remnant prairie sites, thus outcompete native grasses where prescribed burns no longer take place (Ditomaso et al. 2006; DeKeyser et al. 2015). Some common invasive forbs include bindweed (*Convolvulus arvense*), Canadian thistle (*Cirsium arvense*), yellow star thistle (*Centaurea solstitalis*), and many others (Hazlett 2004).

The climate across the region is influenced by a complex dynamic of fluctuating influences of plains, mountains, and air currents. The alignment of the jet stream oscillates seasonally to direct air from three main sources: polar air masses from the north, mountain air masses from the Pacific, and humid air masses from the Gulf of Mexico. Depending upon seasonal patterns, climate and weather events influence the vegetation patterns across the Great Plains (Anderson 1990). The climate of cities that lie at the foot of the Front Range experience differences in temperature, precipitation, and snowfall, based upon their altitude (Fig. 4.2).

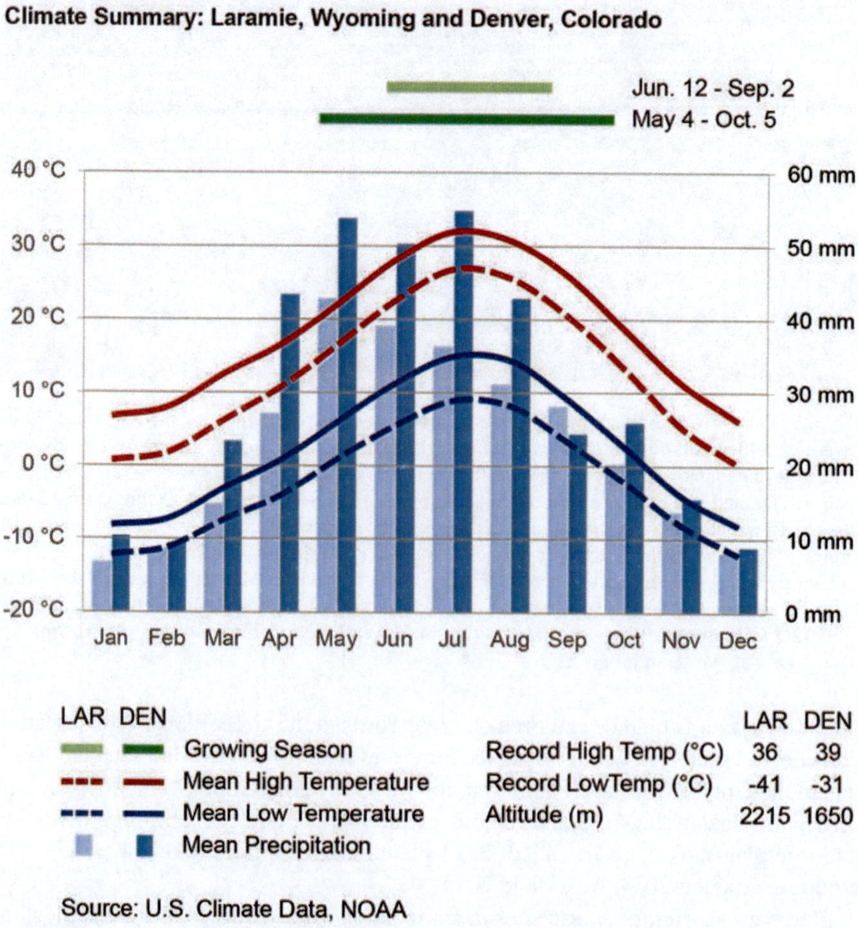

Fig. 4.2 Laramie, Wyoming averages 300 mm (12 in) of precipitation and 163 cm (64 in) of snow annually. Denver, Colorado averages 395 mm (16 in) of precipitation and 140 cm (55 in) of snow annually. Record high/low temperatures for Laramie are 36 °C/−41 °C, and 39 °C/−31 °C for Denver. Laramie is located at an altitude of 2215 m (7165 ft) and Denver is located at an altitude of 1650 m (5279 ft) (Graphic: Tess Menotti and Bruce Dvorak)

4.1.1 *Vegetation in the Shortgrass Prairie and Rocky Mountain Ecoregions*

Vegetation in the ecoregions includes grasses, herbaceous perennials, succulents, mosses, and many other forms. The major ecoregion delineations are generally based upon altitude, temperature, and precipitation, which are influenced by the presence of the Rocky Mountains (Fig. 4.3).

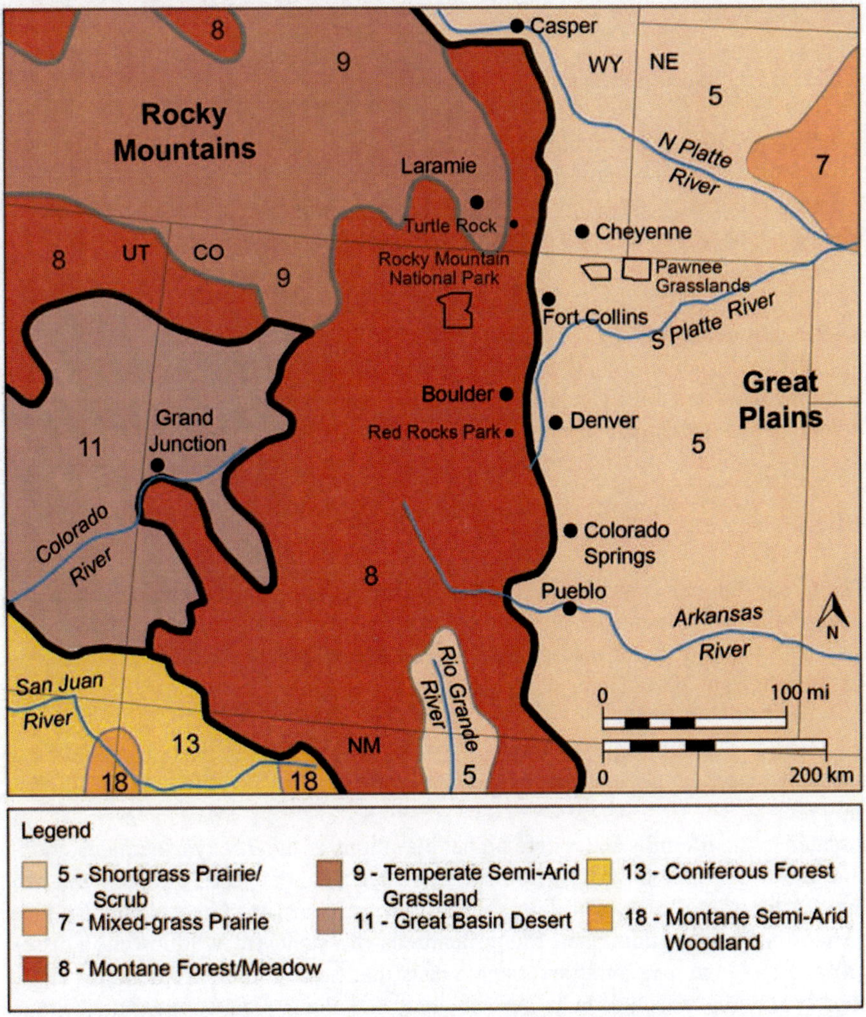

Fig. 4.3 Ecoregions covered in this chapter include Temperate Shortgrass Prairie (5), Temperate Semi-Arid Grasslands (9), and Montane Forest/Meadow (8) (Hammond 1964; Bailey 1997). Abrupt changes in ecoregions occur with topographic change which results in many ecotones. (Graphic: Trevor Maciejewski & Bruce Dvorak)

4.1.1.1 Grasses (Graminoids)

Dominant grasses include blue grama (*Bouteloua gracilis*), buffalograss (*Buchloe dactyloides*) little bluestem (*Schizachyrium scoparium*), purple three-awn (*Aristida purpurea*), sand dropseed (*Sporobolus cryptandrus*) sideoats grama grass (*Bouteloua curtipendula*), and western wheatgrass (*Pascopyrum smithii*) grow on drier habitat in shortgrass prairie (Brockway et al. 2002). Also present are many species in these genera: *Aristida, Bouteloua*, *Elymus*, *Dragrostis*, *Mulehbergia*, *Panicum*, *Sporobolus*, and in wetter areas *Carex* and *Juncus* (Hazlett 2004).

4.1.1.2 Herbaceous Perennials (Forbs)

Some of the larger plant families of herbaceous wildflowers include Bean (*Amorpha, Astragalus, Dalea),* Composite or Sunflower family (*Aster*, *Erigeron*, *Solidago*), Evening-primrose (*Oenothera*), Figwort (*Penstemon*), Knotweed, Milkweed, Mustard, Goosefoot, Mint (*Mondarda*), Phlox, and Rose (Hazlett 2004).

4.1.1.3 Succulents

Some of the common succulents in the drier habitats of the Great Plains include *Echinocereus*, *Opuntia, Phemeranthus*, *Sedum,* and *Yucca* (Hazlett 2004).

4.1.1.4 Mosses

The Great Plains has at least 333 species of mosses, 118 genera, and 39 families (Churchill 1976). As mosses are used on green roofs in Europe and elsewhere, there may be a future role for mosses in the Great Plains, where supplemental water is available (Grant 2006; Heim et al. 2014).

4.1.1.5 Montane Forests and Meadows

Located in the foothills and mid to higher elevations of the Rocky Mountains, montane meadows exist in areas with disturbed sites, shallow soils, or above the tree line. Many of the same plant families that are present in the Great Plains are also present but are not dominant. Many hundreds of species of wildflowers, grasses grow in montane meadows, and many plants that favor cooler temperatures and a shorter growing season form unique subalpine and alpine plant communities. Some of the species that grow in higher elevations include candytuft (*Iberis sempervirens*), pasque flower (*Pulsatilla patens*), rock clematis (*Clematis columbiana*), snow buttercup (*Ranunculus adoneus*), and western spring beauty (*Claytonia lanceolata*).

4.1.2 *Ecoregional Conservation Site Case Studies (Arranged East to West)*

4.1.2.1 Pawnee National Grasslands, Colorado

Not far from the borders of Wyoming and Nebraska in northcentral Colorado, the Pawnee National Grasslands lies in the heart of the High Plains or Colorado Piedmont (Fig. 4.4). These grasslands remain as one of the few places one can visit remnant shortgrass prairie with unobstructed views of grasslands in its native condition. Although the entire 1456 hectares (193,060 acres) of grasslands were first dedicated to be left in its natural state when the preserve was created, the mostly private lands of the grasslands now function to produce wind, oil and gas energy. Wind turbines sprawl across the grasslands, as do fracking stations. Below the shortgrass prairie, hydraulic fracking infrastructure marks the land, as does the steady flow of trucks transporting gas and oil from the National Grasslands to urban centers. Ranching by cattle also takes place on the private holdings of the grasslands, which affects the plant community dynamics. A portion of land where the buttes lie is preserved from the effects of wind turbines, fracking, grazing, and other

Fig. 4.4 The twin buttes are shown aligned in tandem here, with the smaller butte located behind the larger butte seen here. Yucca (*Yucca glauca*) grows on rolling well-drained topography with shortgrass prairie vegetation. (Photo: Bruce Dvorak, June 2018)

disturbances. Regardless, the habitats near the escarpment and buttes are similar to what the vegetation of the Great Plains was like for the previous thousands of years.

Surveys of vegetation have reported 521 plant taxa on the grasslands with about 22% of those as introduced species. Native plants include 406 taxa and 74% percent are perennials (grasses & forbs), 20% annual species, and 6% biennials. Much of the diversity is located in the riparian corridors as the upland prairies have lower diversity and a more predictable mixture of drought-tolerant vegetation (Hazlett 1998). Dominant vegetation includes blue grama (*Bouteloua gracilis*), and buffalo grass (*Bouchloe dactyloides*), which are commonly associated with thread-leaf sedge (*Carex filifolia)*, ring muhly (*Muhlenbergia torreyi)*, western wheatgrass (*Pascopyrum smithii*), and the prickly pear cactus (*Opuntia polyacantha*) (Badaracco 1972). Little bluestem (*Schizachyrium scoparium*) can be found in arroyos. Common shrubs include fourwing saltbush (*Atriplex canescens*), yucca (*Yucca glauca*), and Rocky Mountain juniper (*Juniperus scopulorum*). Near the edges of escarpments and rocky outcrops, several drought-tolerant plants grow including Indian ricegrass (*Oryzopsis hymenoides*), prickly poppy (*Argemone*), prickly pear cactus (*Opuntia* spp.) and veiny dock (*Rumex venosus*) among others. Some of the semi-arid vegetation that grows at the Pawnee Grasslands also grows on green roofs in the case studies, including Sect. 4.3.8, the Denver Botanic Gardens.

4.1.2.2 Red Rocks Park, Denver, Colorado

Located immediately west of Denver, Colorado, Red Rocks Park is a 277 hectare (686-acre) preserve where 2 million-year-old rocks and shortgrass prairie intersect the foothills of the Rocky Mountains. Famous for its natural amphitheater that was first popularized in the early 1900s when musical performances fist took place. The surrounding landscape is a diverse habitat that has easy accessibility with hiking trails (Fig. 4.5).

The vegetation in Red Rocks Park is dominated by grasses and forbs, and a few low growing shrubs local to the ecotone of the Great Plains and foothills of the Rockies. There are also succulents, annuals, and some trees on the preserve. Dominate grasses include blue grama (*Bouteloua gracilis*), side oats grama (*Bouteloua curtipendula*), buffalograss (*Buchloe dactyloides*), thickspike wheatgrass (*Elymus lanceolatus*), needle and thread grass (*Hesperostipa comata*), prairie Junegrass (*Koeleria macrantha*) and little bluestem (*Schizachyrium scoparium*) (KCROS 2011). Common forbs include yarrow (*Achillea lanulosa*), harebell (*Campanula rotundifolia*), Indian paintbrush (*Castilleja integra*), sulphurflower buckwheat (*Eriogonum umbellatum*), wild gernanium (*Geranium caespitosum*), hairy goldenaster (*Hetereotheca villosa*), gayfeather (*Liatris punctata*), prairie coneflower (*Ratibida columnifera*) (Turner and Turner 2009). Shrubs in Red Rocks Park are dominated by mountain mahogany (*Cercocarpus montanus*), rubber rabbitbrush (*Ericameria nauseosa*), Rocky Mountain juniper (*Juniperus scopulorum*), golden current (*Ribes aureum*) and three-leaf sumac (*Rhus trilobata*). A number of

Fig. 4.5 Drought tolerant native vegetation such as these forbs and grasses were once widespread across the foothills of the Front Range. Low growing shrubs and yucca thrive on this well-drained slope. The iconic red rock formations at the amphitheater can be seen in the upper right hand of the photo. (Photo: Bruce Dvorak, June 2018)

plants found growing at Red Rocks Park have been trialed on serveral of the green roof case studies in this chapter.

4.1.2.3 Vedauwoo Recreation Area (Turtle Rock)

As part of the Medicine Bow National Forest, the Vedauwoo Recreation Area near Laramie, Wyoming, covers some outstanding scenic landscapes that include forests, meadows, and ancient rocky outcrops. The granite rock formations are 1.4 billion years old. The pink Sherman granite punctures up through the ground as part of the Laramie Mountains and forms a variety of ecological niches. Vegetation and plant communities adapted to shallow and well-drained soils to support these conditions (Figs. 4.6 and 4.7).

Some of the dominant wildflowers include blanket flower (*Gaillardia aristata*), daisy fleabane (*Erigeron* ssp.), flowery phlox (*Phlox multiflora*), goldenrod (*Solidago* ssp.), groundsel (*Senecio* ssp.), Hood's phlox (*Phlox hoodii*), northern Idaho biscuitroot (*Lomatium orientale*), pasque flower (*Pulsatilla patens*), sand lily (*Leucocrinum montanum*), showy locoweed (*Oxytropis splendens*), slender

Fig. 4.6 Lewis flax (*Linum lewisii*) growing with purple locoweed (*Oxytropis lambertii*) on a south-facing slope with gravely soil. This gently undulating meadow lies at the base of Turtle Rock (mountain), and transitions into a pine savanna in the background of the scene shown here. (Photo: Bruce Dvorak June 2018)

wildparsley (*Musineon tenuifolium*), tufted milkvetch (*Astragalus spatulanthus*), Wyoming kittentail (*Besseya wyomingensis*) and Wyoming townsend daisy (*Townsendia alpigena*).

A few species are unique to the area including Wyoming locoweed (*Oxytropis nana*), which grows only in Wyoming. Succulent plants include purple pincushion (*Coryphantha vivipara*), and spearleaf stonecrop (*Sedum lanceolatum*) which also grows on the green roofs at the Berry Prairie green roof and the Mordecai Children's Garden green roof at the Denver Botanic Gardens (Sect. 4.3).

4.1.2.4 Rocky Mountain National Park, Colorado

Extending from Alberta, Canada, to New Mexico, the Rocky Mountains form an island in the sky that is covered with a variety of ecoregions and forms of vegetation growing in montane forests, montane meadows, upland, subalpine, alpine meadows and tundra habitats. One of the main features for visitors to explore at Rocky Mountain National Park is the road to the summit. The Alpine Visitor Center sits at 3595 m (11,796 ft) and the access road travels through many habitats that have

Fig. 4.7 Lichens, moss, grasses, and *Sedum lanceolatum* (yellow bloom) grow in the shallow soils of gently sloped rocky outcrop near a summit of one of the ancient rock formations at the Vedauwoo Recreation Area. Microclimates such as this can be similar to those on green roofs, and habitats such as these are an ideal location to study vegetation for extensive green roofs. (Photo: Bruce Dvorak, June 2018)

vegetation that may translate to green roofs. Not all of the alpine vegetation thrives at lower elevations (Fig. 4.8), so appropriate investigations into the individual plant requirements for alpine plants is necessary.

Over 900 taxa of plants grow in the park. Some of the larger families of plants include Asteraceae with 155 known species, the Figwort (Scrophulariaceae) family has 41 species which include penstemons and the mustard family (Brassicaceae) has 37 species. There are 101 kinds of grass and 66 sedges growing within the park boundary, including black alpine sedge (*Carex nigricans*) which grows in moist habitats (Terrell 2012).

Some of the perennial wildflowers that grow in the high-altitude habitats include, pasque flower (*Pulsatilla patens*), rock clematis (*Clematis columbiana*), snow buttercup (*Ranunculus adoneus*), western spring beauty (*Claytonia lanceolata*).

Succulents also grow in some of the high-altitude environments (Fig. 4.8) and include ball cactus (*Pediocactus simpsonii*), spearleaf stonecrop (*Sedum lanceolatum*), and the plains pricklypear *(Opuntia polyacantha)* grows on dry and sunny montane slopes.

Fig. 4.8 Alpine vegetation growing amongst rocks and gravely soil at about 3200 meters (10,500 feet). The short stature plant growing over the rock with pink flowers is moss campion (*Silene acaulis*), and the short plant with taller lavender flowers is dwarf clover (*Trifolium nanum*) which both grow on the green roof at the Berry Biodiversity Center in Laramie, Wyoming (see Sect. 4.3). (Photo: Bruce Dvorak June 2018)

Some of the common low growing shrubs and groundcovers native to montane environments include golden currants (*Ribes aureum*) gooseberries (*Ribes grossularia)*, Kinnikinnick (*Arctostaphylos uva-ursi*) grape holly (*Mahonia repens*), wax currants (*Ribes cereum*) and willows (*Salix* spp.) (RMPN 2018).

With the abundance of wildflowers growing in the ecoregions, pollinators abound. Butterfly diversity is high, with at least 141 confirmed species, and many of them prefer vegetation that is unique to different altitudes in the park. Thus, conservation of butterfly habitat is strongly linked to the conservation of butterflies (Mason 2019). For this reason, conservation of meadows, grasslands, and similar habitats where plant and animal relationships are co-dependent, the persistence of flora and fauna relationships in urban areas where they were once abundant is vital to their survival. Designers of green roofs have much to learn from conservation sites, ecological and botanical research, and the application of this knowledge in practice (Dunnett and Kingsbury 2004; Benjamin et al. 2013). Section 4.3 covers the green roofs at the Mordecai Children's Garden at the Denver Botanic Garden and feature plants from sub-alpine and alpine habitats.

4.2 Research in the Ecoregion

Colorado native plants, in general, are excellent candidates for use on green roofs due to their drought tolerance, ability to survive in harsh climates, and the fact that they have evolved in alkaline, low nutrient soils. Five Colorado native plants and one non-native were evaluated for use on extensive green roofs – four of the five native plants thrived (Bousselot et al. 2010).

Some of the first publications in the ecoregion regarding vegetation research for green roofs in the Great Plains included publications from plant trials at the Denver Botanic Gardens from 2007 to 2012. Native and some non-native drought-tolerant plants were selected to be trialed by their potential rate of establishment, adaption to environmental stresses, commercial availability, and aesthetic value. Results indicate that taxa can be grouped into categories of "perish, survive, and thrive". To date, 112 plant taxa have been trialed on this low water green roof. Twenty percent of the creeping forbs thrived, 18% survived and 62% perished (Schneider et al. 2014). Other publications regarding plants for green roofs in dry locations focus on alternatives to sedums and covers a number of taxa used on several green roofs in Denver (Fusco 2013), and a conference proceeding regarding the use of native flora on Colorado green roofs (Bousselot 2016).

Research on green roofs is emerging across the ecoregions of the Great Plains, and its populated ecotones along the Front Range of the Rocky Mountains in Colorado and Wyoming. Ecosystem services that have been studied include rooftop microclimates, stormwater retention, and some investigation into plants for green roofs.

Evapotranspiration from a green roof helps stabilize rooftop temperatures and keeps them closer to ambient temperature more effectively than conventional rock ballast rooftops. This is how green roofs help mitigate the urban heat island. In the high elevation, high solar radiation environment of Denver, Colorado, the rock ballast 'control' roof emitted over twice as much net thermal radiation over the summer as the green roof at the Environmental Protection Agency Region 8 Headquarters (Slabe and Bousselot 2013).

Some of the first research regarding stormwater retention in the region includes runoff data from 20 storms (2011–2014), where green roofs at the Denver Botanic Gardens' Mordecai Children's Garden had a runoff coefficient of 0.27 (representing a 73% reduction in runoff volume) and 15 storms (2011–2013) the control gravel ballasted roof coefficient was 0.54 (representing a 46% reduction in runoff volume) (Piza 2015).

Stormwater quality is an increasingly important environmental requirement, even in the semi-arid states of Colorado and Wyoming. This is in part because the continental divide runs through both states and the headwaters of some of the most important rivers in North America begin there. Green roofs are known for their ability to filter stormwater. However, they can also act as a sink for particulates and nutrients that are deposited via wind and rain, especially if materials that are used in filters are added to the substrate. A naturally occurring material called zeolite, which is often used in commercial water filters, can be added to the substrate to boost plant health and capture nutrients. Bousselot et al. (2012) found that too much zeolite may not improve the survivability of all species on the EPA Region 8 Headquarters rooftop (Bousselot et al. 2012). But lower percentages added as a topdressing can add nutrients to the substrate and improve the filtration capability of the green roof.

Water is the limiting factor for plant growth on the high, dry, and windy front ranges of Colorado and Wyoming. Green roofs are especially prone to drought considering the shallow, well-drained substrate and the fact that most have an artificial drainage layer under the substrate. A baseline was established of how long green roof plants in green roof substrate survived when irrigation was withheld (Bousselot et al. 2011). Substrate moisture declined to zero at about 18 days after irrigation. Succulent plants had viable foliage five times longer than herbaceous plants during the 5-month greenhouse trial. And succulents were nearly twice as likely to revive when irrigation was again provided (Bousselot et al. 2011).

In Colorado, there is an increased interest in the synergy between green roofs and technology. Solar panels were installed on one section of the EPA Region 8 Headquarters green roof. An analysis was performed on the temperature variation and plant survivability under the protection of the solar panels versus the exposed area. The temperature was moderated in both winter and summer, substrate moisture was higher and overall plant survivability was higher near the solar panels (Bousselot et al. 2017).

4.3 Ecoregional Green Roof Case Studies (Arranged North to South)

4.3.1 Berry Biodiversity Conservation Center, Laramie, Wyoming

Laramie is at an altitude of 2184 m (7165 ft), at the intersection of several ecoregions including the valley basin prairie vegetation, foothills of the Rocky Mountains, and sub-alpine and alpine ecoregions. This semi-intensive green roof on the University of Wyoming campus is affectionately called the "Berry Prairie". There is no other green roof like the Berry Prairie in North America. Entrance to the green roof is free, accessible to the public, it is a research site, a biological preserve, and an informal garden in its appearance (Fig. 4.9). The green roof sits above the Vertebrate Collections of the museum below. This green roof is very diverse and most colorful during the growing season. The design accomplishes a blend of program functions as a place for learning about plants native to various habitats in the local mountains and prairies and teaches about the aesthetics of a natural garden on a university campus.

The green roof has gone through two designs and installations. The original design was also a prairie roof which was built in 2011. The first green roof had a

Fig. 4.9 Since the museum roof deck is flat below the green roof, the low-profile substrate is slightly elevated above the roof deck with lifts of polystyrene to fabricate an appropriate microclimate favorable to the vegetation. The varied-slope design sustains plants that are native to well-drained conditions in the landscape and provides variation in the solar aspect. Grasses, forbs, and succulents that like well-drained and sunny conditions are thriving here on the green roof. Light-colored gravel mulch was added to reduce heat gain and reflect conditions of the natural site, such as those discussed in Figs. 4.6 and 4.7. (Photo: Bruce Dvorak, June 2018)

uniform substrate depth and it was laid directly on the low-sloped roof deck. After several issues with the waterproofing system, it was determined that the entire membrane needed to be replaced. The second design (featured here) included a local landscape architect that designed the green roof with input from 22 faculty and staff on campus including participants from the departments of Botany, Plant Sciences, Art and Sciences, Mathematics, Zoology, and the Berry Biodiversity Center. This second design intended to mimic local native habitats.

4.3.1.1 Project Team

Building Owner/Client: Berry Biodiversity Institute, University of Wyoming
Green Roof Design Team Lead: Charlotte Belton
Architect: Malone Belton Abel PC
Landscape Architect: Allison Fleury, PLA, Inside Out Landscape Architecture
Installation Contractor: Arcon Construction
Project completion: Originally completed September 2011, redone summer 2014
Green roof area: 330 m^2 (3,600 ft^2)

Fig. 4.10 Native Indian paintbrush (*Castilleja*) grows with native beardtongue (*Penstemon*) and many other native forbs that grow together in the Foothills Garden on the Berry Prairie green roof. (Photo: Bruce Dvorak, June 2018)

4.3.1.2 Overview and Objectives

This garden was designed to be used to educate the public about native plants. As an outdoor classroom and teaching space for K-16 students interested in any aspect of botany and ecology. It was designed to be used as a research site to investigate the practicality of using plants native to the Rocky Mountain ecoregions on green roofs (Fig. 4.10). The species represent valley basin vegetation at the front (south end) of the garden, foothills species at the center, and alpine species are at the far (north) end of the garden.

4.3.1.3 Plant Establishment

As a planting strategy, plant types were intermixed to avoid monocultures (Wanous 2011). There was also a slight elevation or mounding of the substrate that accomplished several diverse microclimates. This was done to mimic the natural drainage and sun/shade variations to better match the needs of the plants selected to grow on the roof. None of the plants selected for the green roof grow in the wild on low, two-percent slopes. Likewise, the designers sought to elevate and drain the substrate to better match the drainage, slope, and aspect characteristics of the plants found in the ecoregion.

Plants were arranged into zones on the roof garden according to their preferred microclimates found in natural habitats. Many of the plants were established as 5 cm (2 in) pots, a few were larger; the grasses were installed as plugs. Forms of plants include bulbs, grasses, sedges, herbaceous perennials (Table 4.1), cacti, shrubs, dwarf conifers. Some annual plants were established through seeding.

Table 4.1 Native herbaceous perennials on the Berry Prairie green roof

Common name	Botanical name
Yarrow	*Achillea millefolium*
Pacific anemone	*Anemone multifidi*
Pussytoes	*Antennaria parvifolia*
Laramie columbine	*Aquilegia laramiensis*
Colorado blue columbine	*Aquilegia coerulea*
Rocky Mountain blue columbine	*Aquilegia saximontana*
Short's milkvetch	*Astragalus shortianus*
Sundrops	Calylophus serrulatus
Harebell	*Campanula rotundifolia*
Clustered field sedge	*Carex praegracilis*
Mouse-ear chickweed	*Cerastium sp.*
Fireweed	*Chaemerion angustifolium (Epilobium)*
Rock clematis	*Clematis columbiana*
Sugarbowl	*Clematis hirsutissima v. scottii*
Yellow bee plant	*Cleome lutea*
Rocky Mountain bee plant	*Cleome serrulata*
Purple prairie clover	*Dalea purpurea*
Sierra shootingstar	*Dodecatheon jeffreyi*
Cusick's shootingstar	*Dodecatheon pulchellum (Primula pauciflora* v. *cusickii)*
Fewseed draba	*Draba oligosperma*
Hooker's sandwort	*Eremogone hookeri*
Cutleaf fleabane	*Erigeron compositus*
Rockslide fleabane	*Erigeron leiomerus*
Desert yellow fleabane	*Erigeron linearis*
Featherleaf fleabane	*Erigeron pinnatisectus*
Onestem fleabane	*Erigeron simplex*
Showy fleabane	*Erigeron speciosa*
James' buckwheat	*Eriogonum jamesii*
Cushion buckwheat	*Eriogonum ovalifolium*
Sulfur-flower buckwheat	*Eriogonum umbellatum*
Western wallflower	*Erysimum asperum*
Blanketflower	*Gaillardia aristata*
Parry's gentian	*Gentiana parryi* 'Hybrid'
Prairie smoke	*Geum triflorum*
Broom snakeweed	*Gutierrezia sarothrae*
Bracted alumroot	*Heuchera bracteata*

(continued)

Table 4.1 (continued)

Common name	Botanical name
Scarlet gilia	*Ipomopsis aggregata*
Blue flag	*Iris missouriensis*
Winterfat	*Krascheninnikovia lanata*
Alpine bladderpod	*Lesquerella alpina*
Bitterroot	*Lewisia rediviva*
Spotted gayfeather	*Liatris punctata*
Blue flax	*Linum lewisii*
Tenpetal blazingstar	*Mentzelia decapetala*
Wild candytuft	*Noccaea fendleri*
Clustered broomrape	*Orobanche fasciculata*
Purple locoweed	*Oxytropis lambertii*
Nuttall's oxytrope	*Oxytropis multiceps*
Stalkpod locoweed	*Oxytropis podocarpa*
Showy locoweed	*Oxytropis splendens*
Parry's lousewort	*Pedicularis parryi v. parryi*
Broadbeard beardtongue	*Penstemon angustifolius*
Sand penstemon	*Penstemon arenicola*
Fuzzytongue beardtongue	*Penstemon eriantherus*
Smooth beardtongue	*Penstemon glaber v. alpinus*
Larchleaf beardtongue	*Penstemon laricifolius v. exilifolius*
Waxleaf beardtongue	*Penstemon nitidus*
Littleflower beardtongue	*Penstemon procerus v. tolmiei*
Rocky Mountain beardtongue	*Penstemon strictus*
Front Range beardtongue	*Penstemon virens*
Dwarf phlox	*Phlox condensata*
Spiny phlox	*Phlox hoodia*
Kelsey's phlox 'Lemhi Purple'	*Phlox kelseyi*
Flowery phlox	*Phlox multiflora*
Sharpleaf twinpod	*Physaria acutifolia*
Devils gate twinpod	*Physaria eburniflora*
Sticky polemonium	*Polemonium viscosum*
Pasque flower	*Pulsatilla patens*
Moss campion	*Silene acaulis*
Blue-eyed grass	*Sisyrinchium idahoense*
Scarlet globemallow	*Sphaeralcea coccinea*
Rock tansy	*Sphaeromeria capitate*
James's telesonix	*Telesonix jamesii*
Stemless four-nerve daisy	*Tetraneuris acaulis*
Graylocks four-nerve daisy	*Tetraneuris grandiflora*
Golden banner	*Thermopsis montana*
Large-flower Townsend-daisy	*Townsendia grandiflora*
Hooker's Townsend daisy	*Townsendia hookeri*
Alpine clover	*Trifolium dasyphyllum*
Parry's clover	*Trifolium parryi*
Grassy deathcamas	*Zigadenus venenosus* v. *gramineus*

Native Bulbs

Geyer's onion (*Allium geyeri*) and wood lily (*Lilium philadelphicum*).

Native Grasses/Sedges

Blue grama (*Bouteloua gracilis*), Indian ricegrass (*Oryzopsis hymenoides*), little bluestem (*Schizachyrium scoparium*), needle and thread grass (*Stipa comate*), needleleaf sedge (*Carex duriuscula*), prairie dropseed (*Sporobolus heterolepis*), prairie junegrass (*Koeleria macrantha*), Sandberg's bluegrass (*Poa secunda*), side-oats grama (*Bouteloua curtipedula*), and Western wheatgrass (*Pascopyrum* (*Elymus*) *smithii*).

Native Succulents

Mountain ball cactus (*Pediocactus simpsonii*), plains pricklypear (*Opuntia polyacantha*), soapweed yucca (*Yucca glauca*), spearleaf stonecrop (*Sedum lanceolatum*), and spiny star cactus (*Coryphantha vivipara*).

Native Annual Wildflowers

Splitleaf Indian paintbrush (*Castilleja sulphurea*), and Wyoming paintbrush (*Castilleja linariifolia*).

Native Shrubs

Birdfoot sage (*Artemisia pedatifida*), black sage (*Artemisia nova*), fringed sage (*Artemisia frigida*), sand cherry (*Prunus pumila* v. *bessyi*), silver sage (*Artemisia cana*), and Wyoming big sage (*Artemisia tridentata wyomingensis*).

Native Dwarf Conifers

Limber pine (*Pinus flexilis*), pinyon pine (*Pinus edulis*), Rocky Mountain bristlecone pine (*Pinus aristata*; native to the southwestern U.S.)

4.3.1.4 Irrigation

In 2017, the subgrade drip irrigation system operated 15 min every other day. In 2018, run times were reduced to 5 min every other day during the growing season.

4.3.1.5 Maintenance

Maintenance activities consist of removing unwanted invasive plants and removing plants that have proven to be too aggressive or too dominant. These activities take place on an ongoing basis through the growing season (May–September). Some of the species have proven to be quite aggressive, requiring a lot of effort to control. Plants from other gardens in the vicinity have established on the green roof (e.g., yarrow, Maximillian sunflower) and are difficult to remove. Other common invasive plants such as grasses (Poaceae), dandelions, and willowherb find their way onto the roof garden. High standing vegetation in winter 2017 attracted rodents. Staff is hopeful that reduced watering will mitigate some or all of these problems in the future.

4.3.1.6 Observed Wildlife

Many insects frequent the green roof including pollen wasps, bumblebees, leafcutter bees, butterflies, earthworms, cottontails, crows, and hummingbirds. Since the green roof has a ground connection, there has been evidence of mice in the roof meadow.

4.3.1.7 Best Performing Native Vegetation

Many species are performing well. Some of the most successful and showy plants include *Penstemon virens*, *P. eriantherus*, *P. glaber*, *Carex praegracilis*, *Draba oligosperma*, *Bouteloua curtipendula*, *Geum triflorum*, *Telesonix jamesii*.

Other more common plants that have performed well include *Allium geyeri*, *Campanula rotundifolia*, most members of the Asteraceae family (yarrow), many members of the Brassicaceae (western wallflower, candytuft), Cactaceae (*Coryphantha vivipara*, *Opuntia polyacantha*, *Pediocactus simpsonii*), Fabaceae family, Plantaginaceae, Poaceae and Ranunculaceae families.

4.3.1.8 Post-occupancy Observations

Research Staff

- Native plants can be used successfully on the green roof, but they must be selected with knowledge of their capacity to thrive, and not become too aggressive or dominant.
- These plants could not thrive without irrigation. However, finding the appropriate watering rate has been difficult, especially given the different demands of basin plants vs. alpine plants.

- Our native species have done better than expected with the commercial planting medium, which is unlike the native soils.
- The substrate was mulched with light-colored pea gravel to conserve water and keep the soil cooler. Prior to the application of the mulch, the dark color of the planting medium got hot in the summer sun!
- Dwarf conifers are not suitable for this garden.

Authors' Reflections

- This project serves as an ideal kind of pilot project for ecoregional green roofs. Intentional partnerships were established between the design team, researchers, and university representatives. The design process was commendable, as it led to an outstanding ecoregional green roof. The collaborations continue through its maintenance and ongoing research.
- The first design included a uniform substrate depth on a flat roof deck, which formed a mesic condition across the entire roof. For the second design, the substrate topography was designed to replicate the substrate microclimates of the plants in the wild. For example, low rising mounds elevate the shallow substrate above the flat roof deck to make a variety of moisture and solar conditions. Plants were selected to match their native condition, thus drier species were included in the drier substrates near the tops of mounds (i.e. native *Sedum*) and moisture-loving plants were located near the roof deck drains (i.e. native *Iris missouriensis*).

4.3.2 Forsyth Ridge Residence, Colorado

Designed to be integral to the aesthetic and functions of this mountainside private residence, this montane meadow-based green roof was designed to blend into the natural setting (Fig. 4.11). The site is located at an altitude of 2123 m (7000 ft), which is in a montane forest and meadow ecoregion. Initially planted with a limited range of plant species that grow on the natural meadows of the property, the roof now shares a very similar composition to the meadow and valley vegetation below, due to the voluntary nature of some species that thrive in the adjacent meadows.

4.3.2.1 Project Team

Building Owner/Client: Private residence
Green Roof Design Team Lead: Andy Creath and Mark Fusco
Installation Contractor: Green Roofs of Colorado
Project completion: 2009
Green roof area: 134 m^2 (1,445 ft^2)

Fig. 4.11 Grasses and wildflowers native to this montane meadow also thrive on this semi-intensive green roof on a private residence above Boulder, Colorado. The dark-colored adjacent south-facing wall creates a warmer and drier condition near the wall, and thus the composition of the vegetation near the wall prefers the warmer and drier habitat. The taller vegetation away from the wall adapts to the cooler microclimate. (Photo: Bruce Dvorak, June 2018)

4.3.2.2 Overview and Objectives

The green roof was designed to be low maintenance and easily accessed from a ladder. The substrate was a custom mix with some native soil and laid at a depth of 15–17 cm (6–7 in). Vegetation was selected to reflect the montane meadow vegetation located on the property (Figs. 4.12 and 4.13).

4.3.2.3 Plant Establishment

Plants were installed from seed and plugs.

Grasses

Big bluegrass (*Poa ampla*), blue wildrye (*Elymus glaucus*), mountain bromegrass (*Bromus marginatus* Nees), prairie junegrass (*Koeleria macrantha*), Rocky Mountain fescue (*Festuca saximontana*), sandberg bluegrass (*Poa sandbergii*),

Fig. 4.12 Wildflowers bloom in mid-June including yellow (*Gaillardia aristata*), white (*Oxytropis sericea*), and violet (*Oxytropis lambertii*). The roof deck gently slopes from the center (near the stone chimney) to lower at the edges. Thus, positive drainage is achieved at the exterior edges of the green roof, and the gentle slope mimics the ground condition, where similar vegetation thrives on gently undulating topography. (Photo: Bruce Dvorak, June 2018)

slender wheatgrass (*Elymus trachycaulus*), streambank wheatgrass (*Elymus lanceolatus*), tufted hairgrass (*Deschampsia cespitosa*).

Herbaceous Perennials

Blackeyed Susan (*Rudbeckia hirta*), blue flax (*Linum lewisii*), fringed sage (*Atemisia frigida*), gaillardia (*Gaillardia aristata*), scarlet gilia (*Ipomopsis aggregata*).

4.3.2.4 Irrigation

The green roof is hand-watered on an 'as needed' basis.

4.3.2.5 Maintenance

The green roof is maintained about once or twice annually. Its low maintenance is due to a simple planting design where plants were randomly distributed (via

Fig. 4.13 Montane meadow vegetation located on the property directly below the green roof. Similar species such as the yellow *Gaillardia aristata* (foreground) is simultaneously blooming on the ground and the green roof (Fig. 4.12). Many other ground-to-roof species associations exist but have not been formally studied. However, the complexity of the habitat shown here is only partially replicated on the green roof, as the microtopography, natural stone, and ground connections don't exist on the green roof. (Photo: Bruce Dvorak, June 2018)

seeding) onto the green roof where plants establish and compete. Since plants were not installed as plugs, or in drifts or geometric patterns, they are free to establish in their preferred microclimate, and the green roof blends into the natural habitat.

4.3.2.6 Observed Wildlife

Pollinators such as bees and butterflies frequent the roof.

4.3.2.7 Best Performing Native Vegetation

All of the intended vegetation has adapted to the roof, and some desirable volunteer plants have established as well.

4.3.2.8 Post-occupancy Observations

Authors' Reflections

- This green roof demonstrates a very similar translation of the vegetation found growing on the montane meadow, also growing on the green roof. Thus, the direct application of a habitat template was highly successful due to the detailed articulation of design elements such as roof deck slope, seeding method, substrate depth and composition, and plant species composition.
- Although the vegetative composition is similar between the two meadow habitats (roof/ground), this project demonstrates that a green roof habitat can never fully replace a native habitat on a site. The lack of a ground connection, complex microclimates involving stone, soil microflora, temperature variations, animal activity limit the green roof to a valuable but limited replacement of the ground habitat displaced by the building.

4.3.3 Residence 1, Boulder, Colorado

This simple green roof was integrated into the Boulder residence as a way to provide views of native meadow vegetation from inside the residence on top of the garage. Although the green roof is not directly accessible from inside the residence, the owners enjoy periodic visits to the roof via a ladder to maintain the green roof (Fig. 4.14).

4.3.3.1 Project Team

Building Owner/Client: Private residence
Green Roof Design Team Lead: Andy Creath
Installation Contractor: Green Roofs of Colorado
Project completion: 2018
Green roof area: 37 m^2 (400 ft^2)

4.3.3.2 Overview and Objectives

The goal for this semi-intensive green roof was to provide a roof garden that is viewable from the main residence second floor. The 15 cm (6 in) depth of substrate is a pre-tested and custom-blended growth media that is covered with a light layer of pumice mulch. Most of the vegetation is native to Colorado or farther east. Plants were arranged into single-species drifts to achieve the appearance of drifts in a natural setting (Fig. 4.15).

Fig. 4.14 View from an adjacent residence onto the garage roof of a private residence in Boulder, Colorado. This new neighborhood in Boulder is adjacent to older and smaller lot properties. The green roof is one of few meadow habitats in the city. (Photo: Bruce Dvorak)

4.3.3.3 Plant Establishment

Vegetation was pre-grown in containers.

Grasses

Big bluestem (*Andropogon gerardi*), Ruby Muhly Grass "undaunted" (*Muhlenbergia reverchonii*), and non-native blue fescue (*Festuca glauca*).

Herbaceous Perennials

Gayfeather (*Liatris punctata*), white sagebrush (*Artemisia ludoviciana*), and non-native yarrow (*Achillea* 'Coronation Gold').

4.3.3.4 Irrigation

The green roof has a subsurface drip irrigation system. Residents apply water as needed.

Fig. 4.15 Loose drifts of non-native fernleaf yarrow (yellow blooms), and white sagebrush (silver foliage) are shown here about 1 month after installation. The angular edges of the dark pumice lock together to protect the lightweight substrate located beneath from potential high wind gusts. Once mature, the vegetation will maintain a similar function. (Photo: Bruce Dvorak, June 2018)

4.3.3.5 Maintenance

The residents enjoy maintaining the roof as needed, about once a month during the growing season.

4.3.3.6 Observed Wildlife

Pollinators such as bees and butterflies frequent the roof.

4.3.3.7 Best Performing Native Vegetation

All of the vegetation is established and is performing as expected.

4.3.3.8 Post-occupancy Observations

Authors' Reflections

- This green roof demonstrates how a mix of dry and mesic meadow vegetation can grow on a slow draining low-sloped roof deck.
- As a small garden isolated from the ground, visits by birds and butterflies can be observed from inside the residence. The dense spacing of homes in this neighborhood does not allow for an open meadow due to accelerated land values. Most residences make use of their limited ground space such as a patio or some other active space. In this case, a meadow-roof is meant to replace the herbaceous vegetation displaced by the buildings.

4.3.4 Residence 2, Boulder, Colorado

This accessible rooftop garden was integrated into this hillside residence as a way to provide owners a rooftop garden view of the neighborhood and adjacent hillside. With no open space at the ground level, this urban lot allows the owners to maintain a small garden. On this roof garden, native plants share space with a few herbs, strawberries, and tomatoes. The green roof also functions to keep the top level of the residence cool, and as a place to hang out (Fig. 4.16).

4.3.4.1 Project Team

Building Owner/Client: Private residence
Green Roof Design Team Lead: Andy Creath
Installation Contractor: Green Roofs of Colorado
Project completion: 2018
Green roof area: 45 m^2 (484 ft^2)

4.3.4.2 Overview and Objectives

The design theme was to use local hardy vegetation around the border edges of the green roof. Opportunities for residents to plant strawberries and tomatoes are intended to be located in small patches near the edge of the accessible pathway. The substrate is 15.25 cm deep (6 in).

4.3.4.3 Plant Establishment

All of the vegetation was pre-grown in containers, lifted to the roof, and planted as a meadow-like habitat (Fig. 4.17).

Fig. 4.16 A narrow path leads to a small patio set back from a row of rooftop photovoltaic panels. With a limited narrow configuration, a solid parapet wall would create dense shade and block views. The visible pass-through design of the screened wire fence makes this green roof a sunny location, and perhaps more accessible to pollinators. (Photo: Bruce Dvorak, June 2018)

Fig. 4.17 A small patch of native grasses and prairie coneflower is designed to attract local pollinators. (Photo: Bruce Dvorak, June 2018)

Grasses

Blue grama (*Bouteloua gracilis*), blue grama 'Blonde ambition' (*B. gracilis* 'Blonde Ambition').

Herbaceous Perennials

Wild blue flax (*Linum lewisii*), prairie coneflower (*Ratibida columnifera*), red hummingbird mint (*Agastache* 'Coronado Red'), and non-native coral bells (*Heuchera sanguinea splendens*), coral bells (*Heuchera micrantha* 'Purple Palace').

Succulents

Non-native dragons blood sedum (*Sedum spurium* 'Dragons Blood') was planted at the perimeter to maintain vegetation where wind scour may take place.

4.3.4.4 Irrigation

The green roof has an overhead spray irrigation system. Residents water the vegetation as needed.

4.3.4.5 Maintenance

The roof is maintained by the residents on an 'as needed' basis.

4.3.4.6 Observed Wildlife

Bees and other pollinators frequent the roof.

4.3.4.7 Best Performing Native Vegetation

All of the vegetation has adapted to the rooftop environment.

4.3.4.8 Post-occupancy Observations

Authors' Reflections

- This green roof is designed to respond to the microclimates around the open, sunny, and shady locations near and around facades.

- The attention to detail in this roof garden makes it work. The collaborations between the architect and green roof design team accomplished important decisions together. The parapet screen and roofline protection should make this green roof viable for a long time.

4.3.5 *Community College of Denver, Denver, Colorado*

The Confluence Building on the campus of the Community College of Denver is a LEED Gold-certified building that houses classrooms, offices, and a student services center. It was the first addition to the campus since 1967 and was intentionally designed to update the campus with a new identity and modern amenities for students under a sustainable solution. The open concept of the building allows for natural daylighting into the building and an accessible roof deck allows views to the mountains and downtown Denver. The rooftop houses a semi-intensive mixed-grass meadow that represents the vegetation of the nearby ecoregions (Fig. 4.18). The green roof is located on the third floor, is accessible to the public, and is visible from classrooms located on the fourth floor.

Fig. 4.18 Precisely articulated quadrangles of grass and gravel are assembled to make a visually vibrant roof. The gravel zones are a design element meant to provide some visual order and structure to the green roof and blend the rooftop design with the quadrant designs on the campus below. (Photo: Bruce Dvorak, June 2018)

4.3.5.1 Project Team

Building Owner/Client: Community College of Denver
Green Roof Design Team Lead: Green Roofs of Colorado & Studio INSITE
Architect: BORA
Landscape Architect: Studio INSITE
Installation Contractor: Green Roofs of Colorado
Project completion: 2013
Green roof area: 1,580 m^2 (17,000 $ft.^2$)

4.3.5.2 Overview and Objectives

The community college administrators were looking for a green roof that was accessible to the public, could be used for research, was visible from inside the building, had space for movable furniture and met needs for safety. The design team's solution included a mixed-grass meadow with native vegetation (Fig. 4.19). The substrate is 15.25 cm deep (6 in) and is topped with light-colored gravel.

Fig. 4.19 Native grasses grow near the edge of a planter with a built-in bench. (Photo: Bruce Dvorak, June 2018)

4.3.5.3 Plant Establishment

Vegetation was established though pre-grown plants in containers.

Grasses

Blue grama grass (*Bouteloua gracilis*), blue grama 'Blonde ambition' (*B. gracilis* 'Blonde Ambition'), sideoats grama (*Bouteloua curtipendula*).

Perennials and Groundcovers

Rocky Mountain penstemon (*Penstemon strictus*), white sagewort (*Artemesia ludoviciana*).

4.3.5.4 Irrigation

Irrigation runs during the growing season to maintain vegetation so that it is green throughout the summer, including periods of drought.

4.3.5.5 Maintenance

Typical maintenance activities include removing unwanted vegetation, replacing vegetation as budget allows, keeping gravel areas clear, and deadheading grasses and other perennial vegetation as necessary.

4.3.5.6 Observed Wildlife

Birds, bees and butterflies. No formal observations are taking place.

4.3.5.7 Best Performing Native Vegetation

The native grasses are thriving and reseeding. Some of the perennial herbs have declined over time.

4.3.5.8 Post-occupancy Observations

Authors' Reflections

- This green roof was designed and installed about 5 years before the Denver Green Building Ordinance that passed into law on November 7, 2017. This green roof stands as one of the few predominantly grassland green roofs in a publicly accessible space in the Denver metro area. This green roof is a prime example of a strong design that was implemented on public infrastructure but has yet to become maintained and used as a pilot project for education and outreach. An increase in the maintenance budget might allow the green roof to be replanted where needed and developed to support much-needed research on plants, ecosystem services, and maintenance needs of green roofs along the Front Range.
- In the summer of 2019, big bluestem (*Andropogon gerardii*) spontaneously arrived on the green roof, which commonly grows with the other grasses that were planted.

4.3.6 Zeppelin Station, Denver, Colorado

Zeppelin Station is a new mixed-use building that sits alongside an active freight rail line and is within walking distance of a light rail stop. Part of a new live/work TAXI district, the mixed-use building houses office space, a great hall, restaurants, and shopping. It is located in a former industrial zone located just north of downtown Denver. The multi-block area is being transformed into a new live-work-play district for residents of the new River North (RiNo) neighborhood. Multi-level green roofs extend the office space out onto patio terraces planted with native vegetation (Fig. 4.20) (McKnight 2018; ZS 2018).

4.3.6.1 Project Team

Building Owner/Client: Zeppelin Development
Green Roof Design Team Lead: Wenk Associates
Architect: Dynia Architects
Landscape Architect: Wenk Associates
Installation Contractor: Landtech
Project completion: 2018
Green roof area: 362 m^2 (3,900 ft^2)

4.3.6.2 Overview and Objectives

The rooftop vegetation was designed to bring biophilic properties into this work environment and to provide shade when the terrace is occupied. Native vegetation consists of trees, shrubs, grasses, and wildflowers which are compacted into small

Fig. 4.20 Retractable folding glass walls extend the office spaces on three levels. With the retractable walls open, the plants become part of the workspace. (Courtesy of Stephan Werk)

yet visually dynamic spaces (Fig. 4.21). The commercially blended substrate varies in depth from 20-cm-deep (8 in) up to 45-cm-deep (18 in) substrates to support shrubs and trees.

4.3.6.3 Plant Establishment

Plants were installed into a variable depth commercially available blended growth media, as pre-grown plants in containers.

Deciduous Trees/Shrubs

Autumn brilliance serviceberry (*Amelanchier* x *grandiflora*), smooth sumac (*Rhus glabra*).

Grasses

Prairie dropseed (*Sporobolus heterolepis*), tufted hairgrass (*Deschampsia cespitosa*).

Herbaceous Perennials

Fringed sage (*Artemisia frigida*), purple prairie clover (*Dalea purpurea*).

Fig. 4.21 Office space walls can open up to the terrace. Bike parking is located on the terrace space as is seen in the lower left. Native trees (Autumn Brilliance Serviceberry), shrubs (Smooth Sumac), grasses, and wildflowers grow in the variable depth substrates. Vines planted into the above terrace are beginning to grow onto the wire screen that intends to make a green façade above the window-doors as vegetation matures. (Photo: Bruce Dvorak, June 2018)

Vines

Virginia creeper (*Parthenocissus quinquefolia*), and yellow trumpet vine (*Campsis radicans*), however not considered native by local authorities (Ackerfield 2015).

4.3.6.4 Irrigation

The green roofs are irrigated with subsurface irrigation when plants are actively growing.

4.3.6.5 Maintenance

The green roofs are in their establishment period. Typical maintenance activities include the removal of unwanted vegetation, replacing mulch, and inspecting drains and irrigation.

4.3.6.6 Observed Wildlife

Observation of wildlife is not formally reported.

4.3.6.7 Best Performing Native Vegetation

All of the vegetation is currently adapting to the green roofs.

4.3.6.8 Post-occupancy Observations

Authors' Reflections

- Creative development solutions such as Zeppelin Station should become the norm. People want and need a connection to nature in their working environments throughout the day. In this case, the roof terraces allow nature to be present from inside many offices. The integrated design allows people to bike to work, walk, or take public transportation. Restaurants, shopping, and entertainment all take place at the station. The days of large single-use office parks that are isolated from other daily needs may become less popular over time as a new generation of alternative planning and integrated designs stand the test of time.
- The planting design of the roof terraces takes into account the views from multiple perspectives from inside adjacent office spaces. A sense of privacy and an extension of personal space are achieved for offices in opposing directions. This was accomplished through the use of careful consideration and placement of trees, shrubs, and vines.

4.3.7 Flight Building, Denver, Colorado

Located in Denver's River North (RiNo) Arts District, the Flight Building is part of the TAXI mixed-use community. It is a live/work district and demonstrates how offices of the new millennium can engage their regional heritage and the needs of entrepreneurs. Designed to provide views from every office space and most worker spaces, these grassland green roofs provide a regional connection to vegetation and landforms that once dominated the plains. Boa Technologies houses the first floor, where *The Nature Conservancy* and other organizations in need of a small to moderately sized office space are located in one of the 36 suites located on the second floor. Photovoltaic panels also compete for space on the roof to generate up to 20% of the building's energy needs (Fig. 4.22). Occupants can bike to work and park in one of the 100 spaces provided for bikes. The Flight building is on target to receive LEED Platinum certification (Malone 2019).

Fig. 4.22 This aerial view shows the layout of the two grassed roofs that flank the 36 offices. Photovoltaic panels share roof space on the tops of the first and second floors. These supply a renewable source of energy for 20% of internal energy demand. Rain gardens and biofiltration swales are located at the ground level to complete the stormwater management system. Short bridges connect the walkway over the biofiltration swale and can be seen perpendicular to the path (top of photo) on the ground lead to patios of private residences. (Courtesy of Stephan Werk)

4.3.7.1 Project Team

Building Owner/Client: Zeppelin Development
Green Roof Design Team Lead: Wenk Associates
Architect: Dynia Architects
Landscape Architect: Wenk Associates
Installation Contractor: Landtech
Project completion: 2018
Green roof area: 1,207 m^2 (13,000 ft^2)

4.3.7.2 Overview and Objectives

The green roofs were intended to be a simple design statement made with six species of native grasses. The grasses were planted into substrates that are either laid in flat horizontal planes or gently sloped (Fig. 4.23). The mounded landforms are intended to reflect dunes found on the eastern plains. They are a simple design statement made with only a few species of grasses and a few wildflowers. A variable depth substrate (20–30 cm/8–12 in) of lightweight media (American Hydrotech)

Fig. 4.23 View of the mounded grassed green roof from a small outdoor terrace at the Flight Building. Offices located on the second-floor sprawl along the green roof and share the view. A second vegetated mound (dubbed "the island") with trees is located to the left side of the roof photo beyond the photovoltaic panels. (Courtesy of Wenk Associates)

was placed either on the flat deck or on polystyrene lifts. The lifts were used to create the landforms to elevate vegetation for visual and ecological interest. Views from offices looking out onto the roofs, and from the first floor below (through skylights) can see grasses moving with the wind, and various microclimates are made from the mounded substrate. Substrate depths on the section with trees is up to 1 m deep in tree wells.

4.3.7.3 Plant Establishment

Vegetation was both seeded and installed as pre-grown plugs and containers.

Grasses

Big bluestem (*Andropogon gerardii*), blue grama (*Bouteloua gracilis*), Indiangrass (*Sorghastrum nutans*), slender wheatgrass (*Elymus trachycaulus* ssp. *trachycaulus*), switchgrass (*Panicum virgatum*), little bluestem (*Schizachyrium scoparium* 'The Blues').

Herbaceous Perennials

Apache sunset hyssop (*Agastache rupestris* 'apache sunset'), goldenrod (*Solidago* x 'golden baby'), purple prairie clover (*Dalea purpurea*).

Woody Vegetation

Dwarf blue rabbitbrush (*Ericameria nauseosa* var. *nauseosa*), fourwing saltbush (*Atriplex canescens*), soapweed (*Yucca glauca*), sunburst common honeylocust (*Gleditsia triacanthos inermis* 'Sunburst'), Virginia creeper (*Parthenocissus quinquefolia*).

4.3.7.4 Irrigation

The green roofs are irrigated with subsurface irrigation when plants are actively growing.

4.3.7.5 Maintenance

The vegetation is allowed to grow to its natural height. Maintenance activities include the removal of unwanted vegetation and periodic inspection of drains and irrigation systems. Irrigation pipes failed at one section of one green roof and was temporarily shut down. Some of the vegetation became dormant and a few plants were replaced after the irrigation was repaired.

4.3.7.6 Observed Wildlife

There is no formal observation of wildlife taking place on the green roofs. Birds and other insects frequent the green roofs.

4.3.7.7 Best Performing Native Vegetation

All of the grasses and forbs have established and are performing well.

4.3.7.8 Post-occupancy Observations

Publications

- While the green roofs on the Flight Building withstood no significant damage during a violent hailstorm May 2017, exposed white roofs on the Flight Building were significantly damaged but the green roofs were not. 1.5 billion dollars in hail damage took place on traditional roofs in the Denver area from the same storm (Rebchook 2017).

Design Team's Reflections

- Infill seeding of the grassed areas between where container plants were installed helps fill in the gaps and helps cover the visual impacts of the lightweight soil.
- Irrigation was a key component to monitor through the establishment period as well as varying times through the seasons and should be monitored on a regular frequency.

Authors' Reflections

- This grassed roof is currently one of the largest grassed green roofs in Denver and is one of the few green roofs that mimic the vegetation type once-dominant long the Front Range below the foothills of the Rocky Mountains.
- Amazingly, the cost for the green roofs added only about 1% to the overall cost to construct the building (Rebchook 2017). Meanwhile, the less expensive white reflective roofs required entire replacement after hail damage, they convey accelerated and polluted runoff down to the ground during rain storms, heat the microclimate above the roof during the summer, and provide little to no value for native local or migratory wildlife.
- Since green roofs are known to benefit and prolong photovoltaic systems, perhaps a future replacement of the white reflective roofs could include a green roof.

4.3.8 Denver Botanic Gardens, Denver, Colorado

The City of Denver's first green roof (on a city-owned building) is located at the Denver Botanic Gardens, first master-planned by landscape architect Garrett Eckbo and architects Victor Hornbein and Ed White, Jr. This demonstration green roof displays intensive and extensive areas planted with primarily succulent and drought-tolerant vegetation. At over 100 species, it is the largest rooftop collection of plants from cold temperate climates around the world. This was the first ecoregional green roof in Denver that was designed to demonstrate how local, regional, and some exotic succulent vegetation can thrive in Denver's climate. This rooftop retrofit was adapted onto a 1950s multi-purpose building with a bistro below the green roof and a conservatory adjacent to the green roof (Fig. 4.24), all of which are part of the original garden facilities (Krishnan 2019).

Fig. 4.24 A low irrigation management approach is used on this green roof to minimize the use of potable water and keep invasive self-sown seeds from germinating. A few self-sown plants are visible here (vertical shoots of scarlet gilia [*Ipomopsis aggregata*]), and remain as part of a low maintenance approach. Hand weeding between spiny succulents can be labor-intensive, and this reduced watering approach minimizes weeding activity. (Photo: Bruce Dvorak, June 2018)

4.3.8.1 Project Team

Building Owner/Client: Denver Botanic Gardens
Green Roof Consultant: Mark Fusco, Denver Botanic Gardens
Architect: Victor Hornbein and Ed White, Jr.
Structural Engineer: Sam McGlammery
Landscape Architect: Civitas Inc. (pro-bono participation)
Installation Contractor: Green Roofs of Colorado
Project completion: 2007
Green roof area: 195 m^2 (2100 ft^2)

4.3.8.2 Overview and Objectives

Civitas, working with Denver Botanic Garden Senior Horticulturist Mark Fusco, designed this green roof to be publicly accessible, interpretive, and educational. Substrate depth ranges 10–30 cm (4–12 in) in a custom blend designed by former Senior Horticulturalist at the Denver Botanic Gardens, Mark Fusco.

4.3.8.3 Plant Establishment

Plant taxa were selected for their rate of establishment, environmental tolerances, commercial availability, aesthetic value, and potential to add quality to local wildlife habitat (Schneider et al. 2014). Forms of plants include grasses, herbaceous perennials (Table 4.2) shrubs (Table 4.3), and succulents (Table 4.4).

Grasses

Alpine fescue (*Festuca brachyphylla*), beargrass (*Nolina microcarpa*; native to the southwestern U.S.), broomsedge bluestem (*Andropogon virginicus*; native to eastern U.S. and California), bullgrass (*Muhlenbergia emersleyi*; native to the

Table 4.2 Native herbaceous perennials on the Denver Botanic Gardens green roof

Common name	Botanical name
Dwarf pussytoes	*Antennaria parvifolia* 'McClintock'
Desert marigold	*Baileya multiradiata*
Smooth rockcress	*Braya alpina*
Rock clematis	*Clematis columbiana* var. *tenuiloba*
Pretty draba	*Draba streptocarpa*
Hooker's mountain-avens	*Dryas octopetala* var. *hookeriana*
Tall fleabane	*Erigeron elatior*
Rockslide yellow fleabane	*Erigeron leiomerus*
Early bluetop fleabane	*Erigeron vetensis*
Sulphur flower buckwheat	*Eriogonum umbellatum* var. *aureum*
Alpine wallflower	*Erysimum capitatum* var. *purshii*
Pennyroyal	*Hedeoma* clone
Jones' false goldenaster	*Heterotheca jonesii*
Scarlet gilia	*Ipomopsis aggregata*
Grass widow	*Olsynium biflorum*
Broadbeard beardtongue	*Penstemon angustifolius*
Mat penstemon	*Penstemon caespitosus*
Wasatch beardtongue	*Penstemon cyananthus*
Fendler's penstemon	*Penstemon fendleri*
Graham beardtongue	*Penstemon grahamii*
Colorado narrowleaf beardtongue	*Penstemon linarioides* ssp. *coloradensis*
Pineleaf penstemon	*Penstemon pinifolius*
California bluebells	*Phacelia campanularia*
Cleft phlox	*Phlox bifida* 'Betty Blake'
Front range twinpod	*Physaria bellii*
Arctic cinquefoil	*Potentilla hyparctica*
Mountain desert sage	*Salvia pachyphylla*
Heartleaf twistflower	*Streptanthus cordatus*
Colorado greenthread	*Thelesperma ambiguum*
Townsend's daisy	*Townsendia eximia*

Table 4.3 Native shrubs on the Denver Botanic Gardens green roof

Common name	Botanical name
Dwarf false indigo	*Amorpha fruticosa* 'Nana'
Greenleaf manzanita	*Arctostaphylos patula*
Wormwood	*Artemisia* spp.
Shadescale sagebrush	*Atriplex confertifolia*
Hairy mountain mahogany	*Cercocarpus breviflorus*
Desert sweet	*Chamaebatiaria millefolium*
Desert willow	*Chilopsis linearis*
Rubber rabbitbrush	*Ericameria nauseosa* ssp. *nauseosa* var. *nauseosa*
Wright's fendlerbush	*Fendlera rupicola* var. *wrightii*
Mat rockspirea	*Petrophytum caespitosum*
Autumn sage	*Salvia greggii*

Table 4.4 Native succulents on the Denver Botanic Gardens green roof

Common name	Botanical name
Parry's agave	*Agave parryi*
Fendler's hedgehog cactus	*Echinocereus fendleri*
Kingcup cactus	*Echinocereus triglochidiatus*
White Sands kingcup cactus	*Echinocereus triglochidiatus* white sands strain
Nylon hedgehog cactus	*Echinocereus viridiflorus*
Missouri foxtail cactus	*Escobaria missouriensis*
Spinystar	*Escobaria vivipara*
Red yucca, hummingbird yucca	*Hesperaloe parviflora*
Golden pricklypear	*Opuntia aurea*
Tulip pricklypear	*Opuntia phaeacantha*
Plains pricklypear	*Opuntia polyacantha*
Mountain ball cactus	*Pediocactus simpsonii*
Largeflower flameflower	*Phemeranthus calycinus*

southwestern U.S.), mountain muhly (*Muhlenbergia montana*), muttongrass (*Poa fendleriana*), switchgrass (*Panicum virgatum*).

4.3.8.4 Irrigation

The green roof irrigation has zones of drip and overhead watering applied during the growing season. Varying amounts of water are applied; 25 mm (1 in) or less of supplemental irrigation per month, at about 6.35 mm (0.25 in) each irrigation cycle. The watering rate reflects a minimal watering approach, and thus this green roof may be irrigated only 6–10 times per growing season.

Fig. 4.25 Prickly pear cactus and agave grow in a desert-like setting, where plant spacing is generous and light-colored stone, and gravel of the substrate helps to reflect sunlight and retain moisture. (Photo: Bruce Dvorak, June 2018)

4.3.8.5 Maintenance

Invasive plants are removed monthly, as the current management plan seeks to explore a low-maintenance approach (Fig. 4.25). This roof is not maintained as frequently as the Mordecai Children's Garden, as it is intended to model a low maintenance approach.

4.3.8.6 Observed Wildlife

There is no formal recording of wildlife on this green roof, however typical pollinators such as bees and butterflies visit the roof.

4.3.8.7 Best Performing Native Vegetation

Cacti, agave, and other succulents thrive on this low water demonstration roof. See the publication by Schneider et al. (2014) for a detailed report of plants that thrived, survived, or perished (Schneider et al. 2014).

4.3.8.8 Post-occupancy Observations

Authors' Reflections

- This time-tested and mature green roof should help to inform the green roof industry along the Front Range as to a real solution for low water use and low maintenance green roofs.
- As harvested rainwater is currently not an easy option in Colorado, this low maintenance approach may shed light on how heat and cold tolerant vegetation can be used on green roofs.
- The variety of vegetation native to local, adjacent, and far-reaching ecoregions, helps to inform adjacent ecoregions with semi-arid locations.

4.3.9 Denver Botanic Gardens' Mordecai Children's Garden, Denver, Colorado

As part of a master plan update completed by Denver Botanic Garden leadership and staff under the direction of Tryba Architects, this extraordinary roof garden sits on top of the third-floor roof deck of a 300-space parking garage (Fig. 4.26). The

Fig. 4.26 One of the artificial mounds with engineered soils, natural stone, and the fabricated rock façade in the background. (Photo: Bruce Dvorak, June 2018)

garden is designed for children of all ages (and adults) to explore vegetative communities and plants that belong to six different ecosystems. The garden features over 600 taxa of primarily native plants. The entrance to the garden leads visitors from the parking garage elevator, though ticketing and into a tunnel-like entrance. At the end of the short tunnel is the welcome plaza with a panoramic view of mounded gardens. The sub-alpine and alpine habitats site on top of a flat roof deck that was modified by the addition of many layers of polystyrene stacked up to form a series of mounded habitats. The polystyrene elevates the intensive substrates to create the effect of many microclimates and habitats for plant collections, and as a way to organize the exhibits.

4.3.9.1 Project Team

Building Owner/Client: Denver Botanic Garden
Green Roof Design Team Lead: Mundus Bishop, American Hydrotech
Architect: Tryba Architects
Landscape Architect: Mundus Bishop
Installation Contractor: Green Roofs of Colorado
Project completion: 2010
Green roof area: 4040 m^2 (1 acre)

4.3.9.2 Overview and Objectives

This children's garden was designed for small children 12 months to 8 years old, with programmed activities and is easy to access. However, children and adults of all ages can engage with the garden. Six habitats are zoned on the roof garden including alpine, sub-alpine forest, montane forest, plains grasslands, riparian, and montane shrubland. Substrate depths vary from 15–20 cm (6–8 in) where low-growing vegetation occurs, and up to 45–60 cm (18–24 in) where shrubs and small trees are located. Stone, boulders, and gravel mulch is used throughout to make a variety of habitats and is meant to provide an illusion of mountain environments and habitats of the Great Plains. As a botanic garden, its mission and purpose are to explore with new plants, educate, and maintain a constant state of beauty.

4.3.9.3 Plant Establishment

Plants were installed as pre-grown specimens in containers, bare-root, or bulbs. The garden staff re-plant parts of the garden as new plants are trialed, routine maintenance takes place, and small portions of the garden feature seasonal plants. Exotic plants have been trialed on the green roof as well, but we only present native plants in this case study and some of their cultivated forms. The following list is provided by the DBG staff. Forms of plants include annuals, grasses, herbaceous perennials (Tables 4.5 and 4.6), shrubs (Table 4.7), succulents, and trees.

Annuals

American thorow wax (*Bupleurum americanum*).

Table 4.5 Native herbaceous perennials on the Mordecai Children's Garden green roof

Common name	Botanical name
Common yarrow	*Achillea millefolium*
Jones' bluestar	*Amsonia jonesii*
Western pearly everlasting	*Anaphalis margaritacea*
Pacific anemone	*Anemone multifida*
Pacific anemone	*Anemone multifida var. multifida*
Rosy pussytoes	*Antennaria rosea*
Golden columbine	*Aquilegia chrysantha*
Colorado blue columbine	*Aquilegia coerulea*
Western columbine	*Aquilegia formosa*
Crested pricklypoppy	*Argemone polyanthemos*
Thrift seapink	*Armeria maritima*
Spider milkweed	*Asclepias asperula*
Butterfly milkweed	*Asclepias tuberosa*
Alpine aster	*Aster alpinus*
Rimrock milkvetch	*Astragalus desperatus*
Lyreleaf greeneyes	*Berlandiera lyrata*
Gunnison's mariposa lily	*Calochortus gunnisonii*
Yellow sundrops	*Calylophus serrulatus*
Bluebell bellflower	*Campanula rotundifolia*
Trumpet creeper	*Campsis radicans*
Wholeleaf Indian paintbrush	*Castilleja integra*
Purple prairie clover	*Dalea purpurea*
Blacksamson echinacea	*Echinacea angustifolia*
Ballhead sandwort	*Eremogone congesta*
Cutleaf daisy	*Erigeron compositus*
Spreading fleabane	*Erigeron divergens*
Rockslidef fleabane	*Erigeron leiomerus*
Aspen fleabane	*Erigeron speciosus*
Rambling fleabane	*Erigeron vagus*
Matted buckwheat	*Eriogonum caespitosum*
Parsnipflower buckwheat	*Eriogonum heracleoides*
James' buckwheat	*Eriogonum jamesii*
Sulphur-flower buckwheat	*Eriogonum umbellatum*
Sanddune wallflower	*Erysimum capitatum*

(continued)

Table 4.5 (continued)

Common name	Botanical name
Blanketflower	*Gaillardia aristata*
Northern bedstraw	*Galium boreale*
Prairie smoke	*Geum triflorum*
Gray aster	*Herrickia glauca*
Oceanspray	*Holodiscus discolor*
Flaxflowered ipomopsis	*Ipomopsis longiflora*
Pagosa ipomopsis	*Ipomopsis polyantha*
Roundleaf bladderpod	*Lesquerella ovalifolia* ssp. *ovalifolia*
Bladderpod	*Lesquerella* sp.
Wood lily	*Lilium philadelphicum*
Granite prickly phlox	*Linanthus pungens*
Lewis flax	*Linum lewisii*
Great blue lobelia	*Lobelia siphilitica*
Silvery lupine	*Lupinus argenteus*
Hoary tansyaster	*Machaeranthera canescens*
Bigelow's tansyaster	*Machaeranthera bigelovii*
Plains blackfoot	*Melampodium leucanthum*
Prairie bluebells	*Mertensia lanceolata*
Spotted evening primrose	*Oenothera canescens*
Purple locoweed	*Oxytropis lambertii*
Nuttall's oxytrope	*Oxytropis multiceps*
Gilia beardtongue	*Penstemon ambiguus*
Broadbeard beardtongue	*Penstemon angustifolius*
Beardlip penstemon	*Penstemon barbatus*
Cobaea beardtongue	*Penstemon cobaea*
Dusty beardtongue	*Penstemon comarrhenus*
Crandall's beardtongue	*Penstemon crandallii*
Firecracker penstemon	*Penstemon eatonii*
Brandegee's penstemon	*Penstemon glaber var. brandegeei*
Large beardtongue	*Penstemon grandiflorus*
Griffin's beardtongue	*Penstemon griffinii*
Harbour's beardtongue	*Penstemon harbourii*
Toadflax penstemon	*Penstemon linarioides*
Toadflax penstemon	*Penstemon linarioides* ssp. *coloradoensis*
Waxleaf penstemon	*Penstemon nitidus*
Palmer's penstemon	*Penstemon palmeri*
Littleflower penstemon	*Penstemon procerus*
Ramaley's beardtongue	*Penstemon ramaleyi*
Sidebells penstemon	*Penstemon secundiflorus*
Rocky Mountain penstemon	*Penstemon strictus*
Oneside penstemon	*Penstemon unilateralis*
Watson's penstemon	*Penstemon watsonii*

(continued)

Table 4.5 (continued)

Common name	Botanical name
Whipple's penstemon	*Penstemon whippleanus*
Largeflower fameflower	*Phemeranthus calycinus*
Sunbright	*Phemeranthus parviflorus*
Front range twinpod	*Physaria bellii*
Parasol bladderpod	*Physaria subumbellata*
Alpine bluegrass	*Poa alpina*
Silverweed cinquefoil	*Potentilla anserina*
Bigflower cinquefoil	*Potentilla fissa*
Rocky Mountain cinquefoil	*Potentilla rubricaulis*
Colorado cinquefoil	*Potentilla subjuga*
Eastern pasqueflower	*Pulsatilla patens*
Cutleaf anemone	*Pulsatilla patens* ssp. *multifida*
Azure blue sage	*Salvia azurea*
Lanceleaf figwort	*Scrophularia lanceolata*
Sticky skullcap	*Scutellaria resinosa*
Compassplant	*Silphium laciniatum*
Rocky Mountain goldenrod	*Solidago multiradiata*
Mt. Albert goldenrod	*Solidago simplex*
Scarlet globemallow	*Sphaeralcea coccinea*
Munro's globemallow	*Sphaeralcea munroana*
Stemless mock goldenweed	*Stenotus acaulis*
James' telesonix	*Telesonix jamesii*
Tall Townsend daisy	*Townsendia eximia*
Rocky Mountain zinnia	*Zinnia grandiflora*
Meadow zizia	*Zizia aptera*

Grasses

Alpine fescue (*Festuca brachyphylla*), finestem needlegrass (*Nassella tenuissima*; native to the southwestern U.S.), Idaho fescue (*Festuca cf. idahoensis*), little bluestem (*Schizachyrium scoparium* 'The Blues'), mountain muhly (*Muhlenbergia montana*), prairie dropseed (*Sporobolus heterolepis*), prairie Junegrass (*Koeleria macrantha*).

Succulents

Pacific stonecrop (*Sedum divergens*; native to Pacific coast of North America), red yucca (*Hesperaloe parviflora* 'Brakelights'; cultivar of native of TX), redflower false yucca (*Hesperaloe parviflora*; native of TX), Spanish bayonet (*Yucca harrimaniae*).

Table 4.6 Native herbaceous perennials on the Mordecai Children's Garden green roof – *Native-Adjacent Ecoregions and Cultivars*

Common name	Botanical name
Western yarrow	*Achillea tomentosa* 'Maynard's Gold'
McClintock's pussytoes	*Antennaria* 'McClintock'
Golden columbine	*Aquilegia chrysantha* 'Denver Gold'
Remembrance columbine	*Aquilegia* 'Remembrance' 'Swan Violet & White'
Thrift seapink	*Armeria maritima* 'Splendens'
Field sagewort	*Artemisia campestris*
Prairie sagewort	*Artemisia frigida*
White sagebrush	*Artemisia ludoviciana* 'Silver King'
White sagebrush	*Artemisia ludoviciana* 'Valerie Finnis'
White sagebrush	*Artemisia* 'Powis Castle'
Crescent milkvetch	*Astragalus amphioxys* 'San Felipe'
Bellflower	*Campanula* 'Hilltop Snow'
Eastern purple coneflower	*Echinacea purpurea*
Eastern purple coneflower	*Echinacea purpurea* 'Magnus'
Garrett's firechalice	*Epilobium canum* ssp. *garrettii* 'Orange Carpet'
Cutleaf daisy	*Erigeron compositus* ''Pink'
Cutleaf daisy	*Erigeron compositus* 'Red Desert'
Hairy false goldenaster	*Heterotheca villosa* x *jonesii* 'Goldhill'
Lupine	*Lupinus* 'The Governor'
Beardlip penstemon	*Penstemon barbatus* 'Coral Baby'
Scarlet bugler	*Penstemon barbatus* 'Prairie Dusk'
Large beardtongue	*Penstemon grandiflorus* 'Prairie Jewel'
Bluemat penstemon	*Penstemon linarioides* ssp. *coloradoensis* 'Silverton'
Rocky Mountain penstemon	*Penstemon strictus* 'Bandera'
Kelsey's phlox	*Phlox kelseyi* 'Lemhi Midnight'
Kelsey's phlox	*Phlox kelseyi* 'Lemhi Purple'
Azure blue sage	*Salvia azurea* 'Nekan'
Sticky skullcap	*Scutellaria resinosa* 'Smoky Hills'
White heath aster	*Symphyotrichum ericoides* 'Pink Cloud'

Trees– Native and Cultivars

Blue spruce (*Picea pungens*), bristlecone pine (*Pinus aristate*), limber pine, (*Pinus flexilis* 'Spider'), limber pine (*Pinus flexilis* J. Morris), limber pine (*Pinus flexilis*), thinleaf alder (*Alnus incana* ssp. *tenuifolia*), Utah juniper (*Juniperus osteosperma*), whitebark pine (*Pinus albicaulis*), ponderosa pine (*Pinus ponderosa* 'Janet Joy').

4.3.9.4 Irrigation

Irrigation runs about 30 min three times per week during the growing season. On days when temperatures rise above 35 °C (95 °F), irrigation is sometimes run for a short duration to cool off the plants.

Table 4.7 Native shrubs on the Mordecai Children's Garden green roof

Common name	Botanical name
Cliff fendlerbrush	*Fendlera rupicola*
Creeping barberry	*Berberis repens*
drummond's false pennyroyal	*Hedeoma drummondii*
Dwarf false indigo	*Amorpha nana*
Fivepetal cliffbush	*Jamesia americana*
Leadplant	*Amorpha canescens*
Little-leaf mountain mahogany	*Cercocarpus intricatus*
Manzanita	*Arctostaphylos* x *coloradensis* 'Mock Bearberry'
Manzanita	*Arctostaphylos* x *coloradensis* 'Cascade'
Manzanita	*Arctostaphylos* x *coloradensis* 'Chieftain'
Oregon grape	*Berberis aquifolium*
Rubber rabbitbrush	*Ericameria nauseosa* ssp. *nauseosa* var. *nauseosa*
Saskatoon serviceberry	*Amelanchier alnifolia*
Shortfruit willow	*Salix brachycarpa*
Desert princesplume	*Stanleya pinnata*
Shrubby cinquefoil	*Dasiphora fruticosa*
Silver sagebrush	*Artemisia cana*
Silverberry	*Elaeagnus commutata*
Stretchberry	*Forestiera neomexicana*

4.3.9.5 Maintenance

As a botanic garden, the green roof receives high maintenance to maintain a refined garden appearance (Fig. 4.27). This means that the green roof often receives daily maintenance during the active growing season, depending upon environmental conditions. This means that if invasive seeds are actively establishing in the area, the roof may need more maintenance. If there are few plants in the area producing seed, then maintenance may be less frequent.

4.3.9.6 Observed Wildlife

Researcher and volunteer-based observations take place and are currently being gathered for research at Colorado State University. Preliminary data suggest the gardens are frequently used by pollinators, both native solitary pollinators (especially in the early season) and European honey bees (all season long).

4.3.9.7 Best Performing Native Vegetation

Most of the plants originally installed on the green roof are still thriving on this irrigated and well-maintained green roof. Many notable plant families that are thriving on the green roof are Asteraceae (especially *Achillea* sp., *Antennaria* sp.,

Fig. 4.27 One of the many display gardens at the Mordecai Children's Garden. Displays demonstrate mostly low growing vegetation with light-colored expanded aggregate that acts as a mulch to retain moisture and reduce plant competition when new plants are installed. (Photo: Bruce Dvorak, June 2018)

Artemisia sp., *Ericameria* sp., *Erigeron* sp., etc.), the Lamiaceae (especially *Salvia* sp.), the Plantaginaceae (especially *Penstemon* sp. Fig. 4.28), and the Polygonaceae (especially the *Eriogonum* sp.).

4.3.9.8 Post-occupancy Observations

Authors' Reflections

- The programmatic aspects of this green roof are ideal and could be replicated elsewhere as it has all of the elements of an integrated design: it is planted with a wide mix of native and adapted plants, the staff that maintains the garden are highly knowledgeable about green roofs and plants for green roofs, research is connected to local universities, the garden is accessible to the public, and the green roof demonstrates multiple plant communities and substrate microclimates, and the garden exhibits a variety of garden esthetics.
- This is one of the most amazing ecoregion-inspired green roofs in North America. Although many of these plants are naturally adapted to higher altitude environ-

Fig. 4.28 *Penstemon digitalis* 'Husker Red' (shown here) is one of the 48 species of *Penstemon* growing on the Mordecai Children's Garden green roofs. This one grows along with companion plants in a semi-intensive substrate 15–20 cm (6–8 in). (Photo: Bruce Dvorak, June 2018)

ments in the Rocky Mountains, the plants trialed here demonstrate the adaptability and tolerance of this vegetation.

- Green roof designers, researchers, and suppliers might explore some of the vegetation growing in these gardens in shallow substrates, to see how they might play a role with extensive or semi-intensive green roofs in Denver to meet the demands of the Green Buildings Ordinance (Updated).

4.4 Plants for the Shortgrass Prairie Ecoregion

The plant diversity on the ecoregional green roofs in this chapter is the highest across all the ecoregions covered in this book. Across the case study sites in this chapter, there are 347 taxa in total, 274 of which are native to the ecoregions in this chapter (Table 4.8). Of those native to the ecoregions in this chapter, 54 species occur more than once across the case studies in this chapter. Of those occurring more than once, 11 are grasses/sedges, 33 are herbaceous perennials, 5 are shrubs, 2 are trees, and 3 are succulents.

Table 4.8 Thirteen species occur in three or more of the ecoregional green roof case study sites in the chapter

Plant type	Common name	Botanical name	A	B	C	D	E	F	G	H	I
Grasses/Sedges	Blue grama	*Bouteloua gracilis*	x	x	x	x	x				
Grasses/Sedges	Prairie Junegrass	*Koeleria macrantha*				x		x	x		
Grasses/Sedges	Prairie dropseed	*Sporobolus heterolepis*				x			x		x
Herbaceous Perennials	Pussytoes	*Antennaria parviflora*				x			x	x	
Herbaceous Perennials	Purple prairie clover	*Dalea purpurea*				x			x		x
Herbaceous Perennials	Sulphur-flower buckwheat	*Eriogonum umbellatum*				x			x	x	
Herbaceous Perennials	Blanketflower	*Gaillardia aristata*				x		x	x		
Herbaceous Perennials	Scarlet gilia	*Ipomopsis aggregata*				x		x		x	
Herbaceous Perennials	Lewis flax	*Linum lewisii*		x		x		x	x		
Herbaceous Perennials	Broadbeard beardtongue	*Penstemon angustifolius*				x			x	x	
Herbaceous Perennials	Rocky Mountain beardtongue	*Penstemon strictus*			x	x			x		
Herbaceous Perennials	Largeflower fameflower	*Phemeranthus calycinus*							x	x	x
Shrubs	Prairie sagewort	*Artemisia frigida*				x			x		x

Key = A (Residence Boulder #1), B (Residence Boulder #2), C (Community College of Denver), D (Berry Prairie), E (Flight Building), F (Forsyth Ridge), G (DBG Mordecai), H (DBG), I (Zeppelin Station)

4.5 Summary

The major population centers in Colorado and Wyoming are at the intersection of shortgrass prairie and foothill ecoregions of the Rocky Mountains. Therefore, the native plants that thrive in the dry, shallow soils in parts of those two ecosystems are present and well-represented on the green roof case studies in this chapter. The shortgrass prairie ecoregion, which is often composed of 70% grasses and 30% wildflowers in conservation sites, has been abundantly surveyed for species that are used on the green roofs profiled in this chapter (e.g. *Buchloe dactyloides*, *Dalea purpurea*, *Gaillardia aristata*, etc.). Additionally, the drought-tolerant groundcover species of the foothills areas are also found abundantly on these green roofs (e.g. *Antennaria parvifolia*, *Bouteloua gracilis*, *Eriogonum umbellatum*, *Sedum lanceolatum*, etc.).

The aesthetic styles and maintenance practices of green roofs in this region vary significantly – from formally planted and carefully tended green roofs to those that are fairly low input as far as maintenance and irrigation. Depth of substrate also varies significantly with artificial berms to lift the substrate above a flat roof deck being a major feature of some of the green roofs featured in this chapter. Precipitation is a major limiting factor in this ecoregion, which practitioners often address by including irrigation in every green roof design, as an inexpensive way to ensure that the green roof can receive adequate water in times of drought.

Some of the highlights of this chapter include:

- Multiple examples of shortgrass prairie vegetation on deep extensive and semi-intensive green roofs;
- Several examples of subalpine and alpine vegetation from the Rocky Mountains that grow in the lower elevations with some irrigation;
- Colorado currently does not allow greywater or harvested water for use on green roofs. Since alternative sources of water for green roofs are viable elsewhere in the West, the green roof industry could potentially become more sustainable if these alternative sources of water became trialed, studied, and readily adopted where appropriate;
- Various maintenance practices and aesthetics are possible with native vegetation including high maintenance gardens and low maintenance, low-water-use green roofs;
- Exotic sedums are used on green roofs in the ecoregion (not covered here); however, the high level of diversity of the native flora, and regional flora of montane meadows, subalpine and alpine ecosystem is largely unexplored, outside the case studies in this chapter. With policies driving green roofs in Denver, the variety of vegetation and assemblages demonstrated here should be explored as viable options for green roofs in the region.

Acknowledgments We would like to thank the following individuals for their time and sharing of knowledge regarding their respective projects: Andy Creath with Green Roofs of Colorado; Dr. Dorothy Tuthill with the University of Wyoming; William Wenk, Kristin Danford, and J.C. Culwell, with Wenk Associates; Stephan Werk and Chris T LeFebvre with Dynia Architects; Micheal Guidi at the Denver Botanic Garden; and various staff at Zeppelin Development. We would also like to thank the owners of several private residences who allowed us access to their green roofs and were willing to share knowledge about their green roofs.

References

Ackerfield J (2015) Flora of Colorado. BRIT Press, Fort Worth

Anderson RC (1990) The historic role of fire in the North American grassland. In: Fire in North American tallgrass prairies. University of Oklahoma Press, Norman

Ashton IW, Symstad AJ, Davis CJ, Swanson DJ (2016) Preserving prairies: understanding temporal and spatial patterns of invasive annual bromes in the Northern Great Plains. Ecosphere 7(8):e01438. https://doi.org/10.1002/ecs2.1438

Augustine DJ, Derner JD, Milchunas DG (2010) Prescribed fire, grazing, and herbaceous plant production in shortgrass steppe. Rangel Ecol Manag 63(3):317–323

Badaracco RJ (1972) An interpretation resource analysis of Pawnee Buttes. Colorado Colorado State University, Fort Collins

Bailey RG (1997) Ecoregions of North America. U.S. Department of Agriculture, Forest Service, Washington, DC

Benjamin LL, Dakin K, Pantiel M (2013) The professional design guide to green roofs. Timber Press, Portland

Bousselot JM (2016) Conservation of Colorado flora on rooftops. In: Turner J (ed) Aquilegia annual conference issue 2016 annual conference 2016; forty years of change: plants, people, places, vol 2. Colorado Native Plant Society

Bousselot JM, Klett JE, Koski RD (2010) Extensive green roof species evaluations using digital image analysis. Hort Science 45(8):1288–1292. https://doi.org/10.21273/HORTSCI.45.8.1288

Bousselot JM, Klett JE, Koski RD (2011) Moisture content of extensive green roof substrate and growth response of 15 temperate plant species during dry down. HortScience 46(3):518–522

Bousselot JM, Klett JE, Koski RD (2012) Evaluating a natural zeolite as an amendment for extensive green roof substrate. J Environ Hortic 30(4):201–206

Bousselot J, Slabe T, Klett J, Koski R (2017) Photovoltaic array influences the growth of green roof plants. J Living Archit 4(3):9–18

Brockway DG, Gatewood RG, Paris RB (2002) Restoring fire as an ecological process in shortgrass prairie ecosystems: initial effects of prescribed burning during the dormant and growing seasons. J Environ Manag 65(2):135–152

Carter JL, Carter MA, Stevens DJ, Bousselot JM (2018) Common Southwestern native plants: an identification guide. Colorado Native Plant Society, Ft. Collins

Catlin G (1841) Letters and notes on the manners, customs, and condition of the North American Indians, vol 1. Wiley and Putnam, New York

Churchill SP (1976) Mosses of the Great Plains: introduction and catalogue. Prairie Nat 8:44–57

Cook T, Dinerstein E, Simms P, Chaplin S, Smith S, Carney K (2019) Western short grasslands. World Wildlife Fund. https://www.worldwildlife.org/ecoregions/na0815. Accessed 2 Oct 2019

DeKeyser ES, Dennhardt LA, Hendrickson J (2015) Kentucky bluegrass (Poa pratensis) invasion in the northern Great Plains: a story of rapid dominance in an endangered ecosystem. Invasive Plant Sci Manage 8(3):255–261

Ditomaso JM, Brooks ML, Allen EB, Minnich R, Rice PM, Kyser GB (2006) Control of invasive weeds with prescribed burning. Weed Technol 20(2):535–548

Dunnett N, Kingsbury N (2004) Planting green roofs and living walls. Timber Press, Portland

Forman SL, Oglesby R, Webb RS (2001) Temporal and spatial patterns of Holocene dune activity on the Great Plains of North America: megadroughts and climate links. Glob Planet Chang 29(1–2):1–29

Fusco M (2013) Green roofs: dry in the sky. Pacific Horticulture, October

Grant G (2006) Green roofs and façades, vol 70. IHS Bre Press, Bracknell

Hammond E (1964) Classes of land-surface form in the United States. US Geological Survey

Hazlett DL (1998) Vascular plant species of the Pawnee National Grassland. General Technical Report RMRS-GTR-17 Fort Collins, CO: US Department of Agriculture, Forest Service, Rocky Mountain Research Station 17:26

Hazlett DL (2004) Vascular plant species of the Comanche National Grassland in southeastern Colorado. General Technical Report RMRS-GTR-130, vol 130. Forest Service, Rocky Mountain Research Station, Fort Collins, CO: US Department of Agriculture

Heim A, Lundholm J, Philip L (2014) The impact of mosses on the growth of neighboring vascular plants, substrate temperature and evapotranspiration on an extensive green roof. Urban Ecosyst 17(4):1119–1133

KCROS (2011) Ken-Caryl ranch open space plant list. Ken-Caryl Ranch Open Space, Denver

Krishnan S (2019) Green Roof–a year-and-a-half later. Denver Botanic Garden. http://navigate.botanicgardens.org. Accessed 2 Nov 2019

Larson GE (1993) Aquatic and wetland vascular plants of the northern Great Plains. General Technical Report RM-238 Fort Collins, CO: US Department of Agriculture, Forest Service, Rocky Mountain Forest and Range Experiment Station 238:681

Lauenroth WK, Burke IC, Morgan JA (2008) The shortgrass steppe: the region and research sites. In: Ecology of the shortgrass steppe a long-term perspective. Oxford University Press, New York, pp 1–13

Malone D (2019) Denver office building features 13,000 sf green roof

Mason S (2019) NPS.govPark homelearn/about the park/nature/animals/butterflies. National Park Service. https://www.nps.gov/deva/learn/nature/butterflies.htm. Accessed 15 June 2019

McKnight J (2018) Zeppelin Station market and offices by Dynia Architects overlooks rail tracks in Denver, August 1

Piza H (2015) Water quality summary report of the Denver Botanic Gardens green roof. Denver

Rebchook J (2017) Green roofs supported by Zeppelin

RMPN (2018) Montane ecosystem. NPS.gov. https://www.nps.gov/romo/learn/nature/montane_ecosystem.htm. Accessed 13 Jan 2020

Scheintaub M, Derner J, Kelly E, Knapp A (2009) Response of the shortgrass steppe plant community to fire. J Arid Environ 73(12):1136–1143

Schneider A, Fusco M, Bousselot J (2014) Observations on the survival of 112 plant taxa on a green roof in a semi-arid climate. J Living Archit 1(5):10–30. https://doi.org/10.46534/jliv.2014.02.01.010

Singh J, Bourgeron P, Lauenroth W (1996) Plant species richness and species-area relations in a shortgrass steppe in Colorado. J Veg Sci 7(5):645–650

Slabe T, Bousselot J (2013) Implications of the Stefan-Boltzmann Law for green roofs. Paper presented at the CitiesAlive, the Green Roof and Wall Conference, San Francisco, October 23–26

Terrell T (2012) Common wildflower families. NPS. https://www.nps.gov/romo/common_wildflower_families.htm. Accessed 15 June 2019

Turner J, Turner C (2009) Wildflowers of Red Rocks Park. Rabbitbrush Publishing, Golden

Vinton MA, Collins SL (1997) Landscape gradients and habitat structure in native grasslands of the central Great Plains. In: Ecology and conservation of Great Plains vertebrates. Springer, New York, pp 3–19

Wanous B (2011) How many thousands? Biodiversity Institute. removed. Accessed July 26, 2019

ZS (2018) Zeppelin Station Dynia Architects. Architect Magazine:2

Chapter 5
Green Roofs in Desert Southwest Ecoregions

Bruce Dvorak and Paul Coseo

Abstract This chapter presents case studies of four conservation sites and eight green roofs located in the Desert Southwest and montane semi-arid ecoregions. Although precipitation in the deserts is generally limited to 100–300 mm annually in lower elevations, the region has one of the world's most diverse desert vegetation. The monsoon season arrives in the summer, providing relief from high air and soil temperatures, and it is the primary season when plants grow and reproduce. Bird and reptile diversity are also highly rich and abundant. Ecoregional green roofs located in Texas, New Mexico, Arizona, and southeastern California demonstrate how 106 taxa native to the desert and montane meadow ecoregions can be employed on green roofs.

Keywords Drought · Desert · Succulent · Montane meadow · Monsoon · Bare soil · Heat · Irrigation · Annuals · Summer dormancy

5.1 Ecoregion Characteristics

This chapter covers conservation sites and green roofs located in the Desert Southwest ecoregions covering the Sonoran Desert, Colorado Desert, Mojave Desert, Chihuahuan Desert, and montane ecoregions near urban centers. Although precipitation is limited to 100–300 mm (3.9–11.8 in) in the lower elevation deserts, and 500–700 mm (19.7–27.6 in) in montane regions, the Desert Southwest has one

B. Dvorak (✉)
Department of Landscape Architecture and Urban Planning, 305A Langford Architecture Center, Texas A&M University, College Station, TX, USA
e-mail: bdvorak@tamu.edu

P. Coseo
Landscape Architecture Program, Arizona State University, Tempe, AZ, USA
e-mail: Paul.Coseo@asu.edu

B. Dvorak (ed.), *Ecoregional Green Roofs*, Cities and Nature,
https://doi.org/10.1007/978-3-030-58395-8_5

of the world's most diverse desert vegetation. The Sonoran Desert, for example, has over 2000 species of plants, 550 species of vertebrates, and thousands of invertebrate species and is the most biologically rich desert in the world (Dimmitt 2000). The major physiographic regions covered in this chapter include Basin and Range and the Colorado Plateau.

During the early Holocene, the climate in the Basin and Range region, where the Sonoran Desert is now, was cooler and wetter as Juniper-Joshua tree woodlands were dominant at higher elevations, and oaks and grasses were abundant in the valleys (Davis and Shafer 1992). About 4000 years ago, the climate stabilized to near its current conditions, and saguaro (Fig. 5.1), and brittlebush plant communities became dominant (Van Devender 1987).

Before the nineteenth century, the vegetation was not free from human influence, as our Native American communities inhabited the southwest, and employed the desert and mountain meadow vegetation in their ways of living over the past 10,000 years. Native Americans developed a complex and reciprocal relationship with the land. These communities lived and moved seasonally between montane meadows and lower valley desert habitats for food, shelter, trade, and spiritual practices.

Fig. 5.1 *Camelback Mountain* by Marjorie Thomas (1925). Pictured is a *Lower Colorado River Ecosystem* dominated by the tall saguaro cactus (*Carnegiea gigantean*), the creosote bush (*Larrea tridentata*), and the low-growing triangle-leaf bursage (*Ambrosia deltoidea*). The saguaro plant community typically includes succulents, woody, and herbaceous vegetation. This painting is characteristic of the desert vegetation in the Phoenix area before settlement. (Courtesy of The Picerne Collection of Arizona Landmark Art)

Montane meadows are high elevation ecosystems that are dominated by grasses, forbs, desert scrub vegetation, and some trees. On the Colorado Plateau, the Chacoan people lived in accord with the ecosystems of the high semi-desert habitats, climate, and environments. These communities planned the arrangement of their living environments by taking advantage of wind, sun, shade, and air circulation patterns (Baxter 1982). They arranged multiple living environments, at different altitudes that were interconnected networks of settlements that were used seasonally. Their culture engaged trade that coincided with their seasonal movements between the valleys, mountains, and high plateaus. Over thousands of years, their cultural practices altered riparian and other landscapes near present-day Phoenix.

As recently as 1450, the Hohokam civilization thrived in the lowest, hottest deserts in the Salt-Gila river basins in what is now the Phoenix metropolitan area (Crown 1990). It was the Hohokam's knowledge of desert ecology that allowed their communities to adapt to the lower desert environments. Their knowledge of the seasonal, yearly, and decadal patterns of precipitation, temperatures, plant and wildlife, and most importantly surface and groundwater availability. Success was dependent on their capacity to modify the hydrology of the Salt River and its wash tributaries to spread water over large stretches of the Salt-Gila River valley through an elaborate agricultural terrace and canal system to support more reliable agricultural practices (Wienhold 2013) (Fig. 5.2). Their manipulation of the landscape had many ecological consequences. Widening of the riparian zone (riparianization) extended the functions of riparian ecosystems far beyond the natural riparian zone. Riparian zones in desert environments are some of the most important biologically productive and rich areas in deserts. Thus, the presence of cottonwoods, willows, and other riparian species restructured hydrological and ecological processes and functions on a wider scale – creating an oasis out of what would have been lower desert creosote (*Larrea tridentata*) and bursage (*Ambrosia sp.*).

Parts of the once vast Hohokam system are still preserved, as in this agricultural terrace at the Beverly Canyon trailhead in the City of Phoenix (Fig. 5.2). The remnants of the Hohokam water transport system provided the foundation of today's cities in and around Phoenix. The Desert Suthwest has many such humanized landscapes across the large area covered in this chapter, however, there were also large areas that had minimal human interaction. When exploring the natural vegetation of the desert conservation sites, it is important to know the extent that human influence had on the persistence of the dominant ecosystems.

Over the last several centuries, the Desert Southwest was explored and settled by the Spanish, then other Euro-American colonizers ventured west for opportunity, access to land, and natural resources. The first European settlers were challenged by very different environmental conditions than they had known. However, through their reuse of indigenous water systems, new water diversions, and new advancements in agriculture, European immigrants were able to transform desert ecosystems into highly altered ecosystems dominated by introduced plants and animals, and large-scale engineering projects to secure water.

Today, a highly diverse mixture of people with varied backgrounds and cultural landscape traditions influence the valley vegetation. Within metropolitan regions of

Fig. 5.2 Images of the remnant desert scrub vegetation at the Beverly Canyon preserve outside Phoenix, Arizona during the dormant season. This site preserves terraforming remnants and terrace structures built by Hohokam communities that once lived in the region before 1450. Soil and rocks were arranged to divert and capture runoff at the toe of slopes across the southwest to grow crops such as species of *Agave* plants (century plant). Agave was grown and harvested for food, medicine, and fiber for tools. Because of this long history, humans have played a vital role in the formation of the Desert Southwest ecosystems. (Photos: Paul Coseo, Google Earth)

Phoenix, much of the original desert scrub and semi-arid grassland habitat has been displaced by urban development; however, outside of urban areas, much of these habitats remain, some of it in conservation sites. Much of the land in the valleys have been converted into rangeland and agricultural crop production over the twentieth century. Fragments of grassland and desert habitats are preserved in national parks, national monuments, state parks, and city and private preserves.

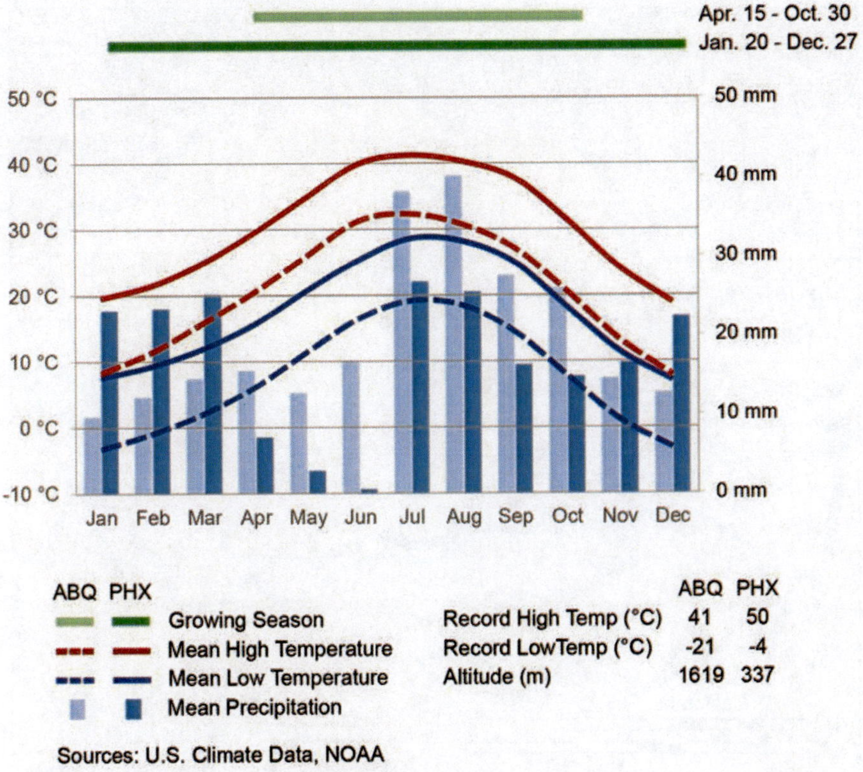

Fig. 5.3 Climate characteristics of representative urban regions in the lower desert southwest elevations with Phoenix (PHX) at about 337 m (1000 ft), and higher elevations of Albuquerque (ABQ) at about 1619 m (5300 ft) altitude. Climates of the lower desert ecoregions exhibit a sharp decline in moisture before the monsoon season beginning in July, which is amplified at lower elevations as seen in Phoenix's June average precipitation of only (0.04 in). Higher elevation ecoregions experience more consistent precipitation and a monsoon season, which lasts from June 15–September 30 (CLIMAS 2018). (Graphic: Tess Menotti, Bruce Dvorak)

Regarding the current climate across the region, the annual precipitation and temperature vary greatly across the ecoregions influenced predominantly by altitude and seasonal climate influences (Fig. 5.3). Monsoon rains are common across the entire southwest and consist of the majority of annual precipitation during the spring and summer. Monsoon rains are more reliable and consistent at higher elevations. At lower elevations, storms are spotty and unreliable, producing torrential downpours in one location, while leaving areas dry only a kilometer away. The scattered nature of summer precipitation is amplified in the lower altitudes. In the Phoenix metro area, for example, the sporadic distribution of rainfall for the 2018 Monsoon precipitation season had 25–52 mm (1.00–1.99 in) of rainfall in the valley, (June 15–September 30) whereas mountainous areas north of Phoenix had upwards of 406–533 mm of rain (16–21 in) during the same period (Flood Warning Branch 2018).

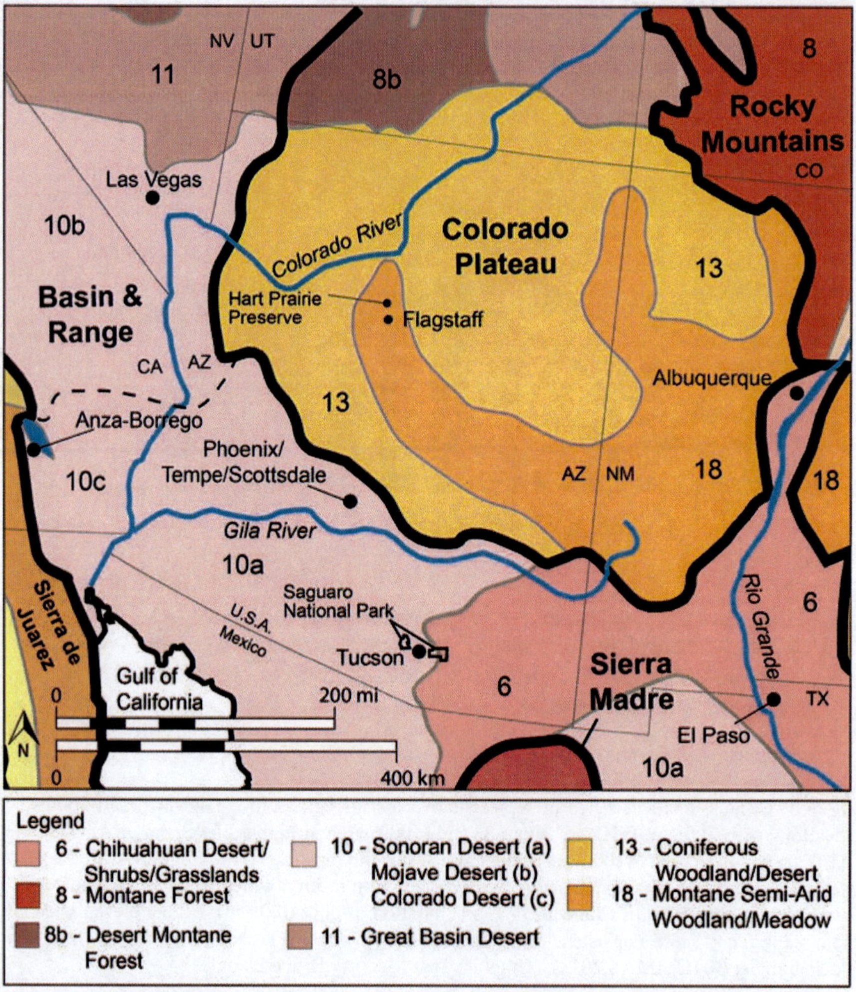

Fig. 5.4 Southwest Desert ecoregions (based on Bailey 1997) covered in this chapter include the Sonoran Desert (10a) the Mojave Desert (10b), the Colorado Desert (10c), the Chihuahuan Desert (6) and Montane Semi-Arid Woodlands, which includes montane meadows (18). Physiographic regions (Hammond 1964) where urban centers are located include the Basin and Range regions and the Colorado Plateau. Adjacent regions include the Rocky Mountains, Sierra Madre Mountains, and the Sierra de Juarez Mountains. (Graphic: Trevor Maciejewski & Bruce Dvorak)

The Desert Southwest is ecologically diverse (Fig. 5.4), as many of Earth's biomes are present either in the desert valleys, high plateaus, or mountain environments (Dimmitt et al. 2015). Of these types, the majority of urban areas lie in two ecoregions that are separated by altitude, montane meadows and valley deserts. Desert valley vegetation grows at altitudes below 1000 m (3000 ft). The valley vegetation is dominated by a variety of succulents, cacti, and associative vegetation.

Montane meadows are part of montane woodlands which are located at elevations above 1500 m (5000 ft).

5.1.1 Vegetation of the Desert Southwest Ecoregions

Natural vegetation in the region is diverse and plants that grow in arid regions take advantage of specialized methods to avoid or endure drought, heat stress, and frost. Leaf morphology is a common trait where specialized adaptations take place for plants to survive heat and drought stress. Some plant species have evolved leaves that are spiny, small, light-colored such as white or silver, fuzzy or fleshy. Chapter 2 has a more extensive discussion of the specialized adaptations of plants in hot, arid ecoregions (Sect. 2.1). Due to the extreme temperatures, long growing season and exposure to sunlight, horizontal or spreading forms of vegetation are uncommon, while upright forms of vegetation are common.

Perennial grasses and forbs avoid heat and frost by dormancy. Thus, individual plants persist to provide consistency in the habitat and or microclimate, nutrient cycling, and can serve specialist organisms when co-dependency occurs, such as between many penstemon species and pollination through visits by hummingbirds (Waser 1978). Succulents store water in their leaves and roots and endure drought and heat stress by minimizing growth during extreme heat and drought events. Woody vegetation in the region typically attains water through a combination of deep and shallow root systems, and survives drought and heat stress through dormancy. These plants also serve as nurse plants for many cacti. Nurse plants provide shade and cover for seedlings until cacti are more mature.

5.1.1.1 Valley Deserts

At the lowest elevations, the hot dry valley deserts are dominated by drought and heat tolerant cacti and shrubs communities such as creosote bush (*Larrea tridentata*), triangle-leaf bursage (*Ambrosia deltoidea*), and brittlebush (*Encelia farinosa*).

Succulent vegetation common to the valley floors includes cholla (*Cylindropuntia sp.*), barrel cactus (*Ferocactus wislizeni*), Engelmann's prickly-pear (*Opuntia engelmannii*), tulip prickly-pear (*Opuntia phaeacantha*) and the iconic saguaro (*Carnegiea gigantea*). Other common succulents include species of *Agave*, *Yucca*, *Sotol*, and *Nolina*. Altitude and its associative cooler temperatures, frost, and freezing keep many of the tender (not cold hardy) desert succulents growing below 1000 m (3000 ft), as frost and freezing temperatures are infrequent in the lower elevations of the Sonoran Desert.

Cool seasonal winter annuals (wildflowers) take advantage of precipitation when it occurs by seed germination if sufficient winter precipitation events take place. For instance, the most abundant blooming wildflower seasons in the lower deserts are triggered by winter rain patterns that start with October storms that bring roughly 25.4 mm (1 in) of rain, followed by monthly 25.4 mm (1 in) rainfalls until February (Dimmitt et al. 2015). Annual herbaceous wildflowers typically bloom longer than perennial vegetation and produce and drop seeds after they perish. Thus, annual vegetation is a critical element to the persistence of other desert forms of vegetation by attracting pollinators, covering exposed soil, and cycling of nutrients. Wildflower blooms vary according to rainfall patterns; thus, some areas may not have blooms where other areas where precipitation is present may have them in abundance. Wildflower blooms do not happen annually at any specific location but occur over time on decadal cycles. Seeds from these large blooms are stored for years in desert soils and a critical resource for supporting small mammals.

The diversity of wildflowers is great across the region. Some of the more common perennial wildflowers include common yarrow (*Achillea millefolium*), Western pearly everlasting (*Anaphalis margaritacea*), desert marigold (*Baileya multiradiata*), chocolate flower (*Berlandiera lyrate*). Some of the common annual wildflowers include desert chicory (*Rafinesquina neomexicana*), Indian paintbrush (*Castilleja integra*), winecup clarkia (*Clarkia purpurea*), blanket flower (*Gaillardia pulchella*), Arizona poppy (*Kallstroemia grandiflora*) and many more.

5.1.1.2 Montane Meadows

At the higher elevations, montane meadows are present in pine savannas, between the pine forests, and near exposed sites such as rocky outcrops. Meadows are dominated by vegetation in the grass (Poaceae), rush (Juncaceae), and sedge (Cyperaceae) families, along with a range of annual and perennial herbaceous plants. Although the Desert Southwest is known for its cacti, the islands of increased precipitation and cooler temperatures in the higher altitudes sustain herbaceous vegetation. Montane meadows were once much more prominent across the entire region. Suppression of ground fires has led to a reduction of meadow habitats as invasive woody vegetation is overtaking habitat that was once open meadow (Sect. 5.1.1).

Many genera and species of perennial herbaceous plants grow in the montane meadows. Some of the common genera of the mountain region include *Aster, Allium*, *Aquilegia*, *Astragalus*, Geranium, *Iris*, *Lupinus*, *Mirabilis*, *Penstemon*, *Ratibida*, *Solidago*, *Tradescantia* and many more.

Succulents growing in the high plateau include some of the generalist succulent species found in the valley below (*Agave*, *Opuntia*), plus some plants that thrive in cooler montane climates such as orange purslane (*Portulaca suffrutenscens*) and spearleaf stonecrop (*Sedum lanceolatum*).

Some common annual herbaceous vegetation of montane meadows includes *Astragalus, Baileya*, *Castilleja*, *Erigonum*, *Gaillardia*, *Linanthus*, *Oenothera*, and many more.

5.1.2 Ecoregional Conservation Site Case Studies (Arranged North to South)

5.1.2.1 Hart Prairie Preserve Nature Conservancy, Flagstaff, Arizona

The Hart Prairie Preserve is a 99-hectare (245-acre), high elevation (2590 m or 8500 ft) montane meadow (prairie), and Bebb willow preserve located in the San Francisco Mountain range northwest of Flagstaff, Arizona (Fig. 5.5), which receives about 59 cm (23 in) of precipitation annually. The preserve is owned and managed by the Nature Conservancy of Arizona. The unique habitat maintained on the preserve is the world's largest Bebb willow habitat and is one of the few remaining Bebb habitats in the world. Willows were culturally important for Native Americans (basket weaving) and the Bebb provides habitat for snowshoe hares, deer, elk, and moose (browse Bebb willow) (Favorite 2002). The buds, shoots, and catkins are eaten by birds, beaver, and small mammals. However, the preserve maintains a remnant hillside prairie habitat, also a rare habitat to find in the southwest (Conservancy 2010).

Fig. 5.5 Prairie restoration site at the Hart Prairie Preserve during summer. The montane meadow covers the slopes of hills and the bottomlands of the preserve. Prescribed burns are used to maintain the native fibrous-rooted vegetation, and clears out invasive high-water demanding woody species. In the distant background at the lower elevation, the Bebb willow preserve is seen, and coniferous forests are seen in the upland background. (Photo: Bruce Dvorak July 2018)

Before European settlement, the property was bordered by an open ponderosa pine savanna and grassland. Due to the suppression of low-intensity ground-level fires that once frequented the area, and the introduction of grazing by exotic cattle, the conifer forest has become denser and shadier. Consequentially, groundwater levels became reduced due to more groundwater use by the increase of conifers. Thus, the transition from savanna to forest began to threaten the prairie and Bebb willow habitat (Conservancy 2010). Though restoration work by the Nature Conservancy, the prairie, and Bebb willow habitats are also one of the few remaining high-quality prairies in the mountains of the Desert Southwest that represent vegetative communities once present in the region before permanent settlement (DeWald and Springer 2006).

Some of the wildlife on Hart prairie include elk, deer, porcupine, prairie dogs, over 120 species of birds, 37 species of butterflies, and over 250 species of plants. The land is maintained by Nature Conservancy staff, volunteers, and student workers. Maintenance activities include the removal of invasive species, seed dispersal, prescribed burns, and techniques to reduce overgrazing by elk and deer (Fig. 5.6).

Fig. 5.6 Property adjacent to the Hart Prairie Preserve (upslope of wire fence) is not subject to prescribed burns or prairie restoration management and is subject to grazing by cattle. Invasive tree seedlings can be seen spreading across the hill slope, uphill from the Hart Nature Preserve (foreground). The grassland at the preserve is maintained through appropriate site stewardship activities which include prescribed burns, and periodic removal of invasive species. Landscapes in the region, such as those uphill from the fence, may eventually transition into a coniferous woodland, from successional growth, such as those seen near the top of the hill. (Photo: Bruce Dvorak, July 2018)

No cattle graze the site, and the elk living in the region are introduced Rocky Mountain Elk that are controlled in size to prevent overgrazing. The indigenous Merriam's Elk was hunted to the point of extinction in the early twentieth century, and Rocky Mountain Elk was imported from Yellowstone National Park into Arizona to repopulate Arizona with elk. Herbivory by wildlife is the main threat to willow habitat restoration efforts. Restoration of the upland prairie is important to the recovery process of the groundwater levels and the delivery of baseflow groundwater, which supports the Bebb willow habitat located in the lower valley drainages (DeWald and Springer 2006; Conservancy 2010).

The Nature Conservancy maintains an active list of plants found on the preserve on their website as well as lists of observed wildlife dependent upon that habitat. Grassland habitats are viable ecosystems for green roofs. However, a grass-based green roof may require irrigation. The Hart preserve may serve as an excellent reference habitat for meadow-based green roofs located in the montane grassland ecoregions. Although access is limited, there are regularly scheduled outings for those who want to visit the preserve, and special visits can be made by reservation. Some of the plants found here have also been trialed on green roofs in the case studies (Sect. 5.3.4). Maintenance activities on a roof approximately replicate the effect of burning or grazing such as deadheading vegetation and removal of problematic invasive species.

5.1.2.2 Desert Botanical Garden, Phoenix, Arizona

The Desert Botanical Gardens (DBG) was started in 1939 by conservationists looking to conserve the Sonoran Desert plant communities. Located on 57 hectares (140 acres) of a preserved desert in Phoenix, AZ, at 390 m (1280 ft), it serves as a key resource for researchers, practitioners, and is open to the general public (Fig. 5.7). The DBG has over 4300 plant species including 379 rare or endangered species. Their activities help address the growing concern over degradation and extinction of desert plant species. The International Union for Conservation of Nature recognizes the cactus family as the fifth most threatened group of organisms (DBG 2020) and the DBG provides international leadership in desert ecosystems through its: Hazel Hare Center for Plant Science, Living Collection; Rare and Endangered Collection, and The Herbarium. DBG biologists are critical partners in advancing an ecoregional approach to green roofs in the Desert Southwest.

Biologists at the DBG, have learned that there are several important lessons to be considered for regionally appropriate green roofs. When choosing plants, they recommend the following advice. First, seeds may establish and perform better over time (for some species) than already established plants. The location of the plants is key in any environment but highly critical in hot arid places for solar incidence, which includes slope and aspect. The Lower Colorado River Valley is the ecological division of the Sonoran Desert and native plants in this location are adapted to resource-poor conditions. In the Lower Sonoran Desert root zone, look for plants that will do well together based on soil depth. Some comprehensive reference books

Fig. 5.7 The DBG has several plant collections that may be a good study for green roofs. Botanists at the DBG recommend trialing a variety of plants from their *Agave* (left) and *Oputina* (lower right) collections. Sotol (Dasylirion), is seen in the background (upper right) and may be appropriate for roof gardens that can support larger plants. More than half of the ecoregional green roof case studies in this chapter make use of desert succulents. (Photo: Bruce Dvorak, October 2018)

for the ecoregion vegetation are the *Vegetation and flora of Sonoran Desert*, by Forrest Shreve and Ira Wiggins, specifically the *Arizona Uplands* and *Lower Colorado Valley* chapters (Shreve and Wiggins 1964).

5.1.2.3 Saguaro National Park, Pima County, Arizona

Located outside the city limits of Tucson, Saguaro National Park is a conservation site established to preserve high-quality desert valley ecosystems of the Sonoran Desert (Fig. 5.8). The park is located in two separate landholdings, the larger and higher elevation Rincon Mountain district (elevations from 813 m/2641 ft to 2557 m/8336 ft), and the Tucson Mountain District (elevations from 664 m/2178 ft to 1428 m/4685 ft). Each district is distinct in its ecology and biodiversity which defined by unique plant communities. There are six biotic communities between the two districts which include desert scrub, desert grassland, oak woodland, pine-oak woodland, pine forest, and mixed conifer forest.

Regarding vegetation, the Tucson District has 71 families, 260 genera, 417 species, and 424 taxa. The Rincon Mountain District has 115 families, 543 genera,

Fig. 5.8 Eroded butte and alluvial fan geologic feature with Arizona Upland ecosystem communities. Vegetation growing in the alluvial fan (foreground) includes saguaro, creosote, cholla, bursage, ocotillo, opuntia, and their associative species. These ecosystems are common throughout Saguaro National Park and could be trialed on green roofs and roof gardens. (Photo: Bruce Dvorak, October 2018)

1138 species, and 1163 taxa. Conservation status includes several rare and sensitive plant species that are present in the park, and there are no threatened or endangered plants. There are currently at least 80 non-native plant species present in the park, and some are aggressive. Conservation practices are in process to reduce the influence of buffelgrass, which includes about 90% of the invasive vegetation (Buckley 2011). Ground fires are not natural to the valley vegetation, however, above the valley, ground fire is used to manage the upland grassland and pine ecosystems. The adaption of vegetation native to the Sonoran Desert in the Tucson valley is just beginning to be explored for use on green roofs (Rosenzweig 2016).

5.1.2.4 Anza-Borrego Desert, San Diego County, California

Located in the Lower Colorado River watershed of southeast California, the Anza-Borrego Desert State Park located in the Colorado Desert (Fig. 5.9) is home to over 70 genera of plants and over 800 species growing at elevations from near sea level to over 1829 m (6000 ft). The Colorado Desert is part of the Sonoran Desert, and its

Fig. 5.9 Anza-Borrego Desert State Park near the Visitor's Center. Bare soil and wide plant spacing are common across the desert as plants compete for soil moisture and nutrients. The microclimates between larger upright vegetation allow for shade-dependent species to co-exist. The Visitor Center landscape and rooftop have vegetation common to the desert identified near the building (Sect. 5.3.8). On the volunteer-based Anza-Borrego Desert website (iNaturalist) has hundreds of plants and animals identified across the desert preserve and can be used to identify vegetation. (Photo: Bruce Dvorak, October 2018)

vegetation is dependent upon precipitation from annual winter rains and summer monsoon storms. Much of the 127 mm (5 in) of the annual rainfall occurs during the winter and rejuvenates annual wildflower blooms. Although summertime temperatures in the desert can exceed those of the Death Valley, the desert succulents and some woody vegetation grow throughout the region. Plant cover and spacing are not contiguous cover and are often determined by available soil moisture. It is common to observe open or exposed soil without any year-round plant cover. Ground elevation, slope, aspect, microclimate, and soil characteristics influence plant distribution and regrowth. Few invasive plant species threaten the region, but Sahara mustard (*Brassica tournefortii*) is particularly harmful to the growth and establishment of some wildflower species, as it grows during the same timeframe as native wildflowers.

Anza-Borrego Desert State Park is accessible through many miles of trails, and seasonal guided hikes take place during the spring. Many of the desert succulents growing in the Colorado Desert also grow in the other American Deserts. Some of the succulent plant species are also found growing on the west side of the California coastal mountains such as cholla (Cylindropuntia sp.), prickly pear (Opuntia sp.), and others common to the coastal dunes and other low nutrient and well-drained locations.

5.2 Research in the Ecoregion

Although green roof research is relatively new to the region, there has been some publication and non-published experimental vegetation trials in the Desert Southwest. According to the Web of Science, journals or conferences have published 1126 "green roof" articles. Of those "green roof" articles, 30 were "arid" articles, and, eight were "desert" articles. However, after reviewing the abstracts from the eight "green roof desert" articles, only four were about green roofs in deserts. This section covers some peer-reviewed publications and some unpublished investigations, to present some preliminary results.

In Las Vegas, researchers at the University of Nevada, Las Vegas ran an experiment with 12 different plant species (Milburn et al. 2011). Of those species, ice plant (*Malephora luteola*), desert daisy (*Chrysactinia Mexicana*), bush morning glory (*Convolvulus cneorum*), and desert spoon (*Dasylirion wheeleri*) were tested for thermal performance. They reported that the green roofs reduced daytime roof temperatures compared to a white reflected roof, while at night the green roof was warmer than the same white roof. Plant performance was not reported.

Recent work at Arizona State University (ASU) in Tempe, AZ used a case-control research design to examine the plant, stormwater, and thermal performance of desert-adapted green roof systems (Fig. 5.10). The plots were located on a roof lab space on the Tempe, AZ ASU campus. The green roofs consisted of three 1.8 m × 2.4 m (6 ft × 8 ft) green roof plots to test against a conventional 1.8 m × 2.4 m (6 ft × 8 ft) white roof control plot. The green roof plots were made of twelve 61 cm × 61 cm × 20 cm (2 ft × 2 ft × 8 in deep) GreenGrid® modules by Weston Solutions. For plant species performance, they tested four regionally native (to the Sonoran Desert) and five non-native plant species. Each module was planted with nine plants or one of each of the nine different species. In total, they planted 36 replicates of each species or a total of 324 plants.

From July–December 2015, the roof received 146.6 mm (5.77 in) of rainfall. This compares to the 50-year (1948–1998) yearly average of 201.9 mm (7.95 in) (Western Regional Climate Center n.d.). Monsoon thunderstorms are highly sporadic in the Phoenix Valley. Three storms dropped 59 mm (2.33 in) of rain during 1 week in late August (August 27, 29, and 31, 2015). That week of rain represents 40% of the rain that fell from July–December 2015. During those three rainstorms, the green roof retained at least 49% more runoff than a conventional white roof. This compares to an average of 56% runoff retention from 12 other green roof studies as reported by Gregoire and Clausen (2011). Out of the nine species planted, only four species had 100% survival rates: Brittlebush (*Encelia farinosa*); Outback Sunrise (*Eremophila maculata 'Outback Sunrise'*); Myoporum (*Myoporum parvifolium*); and Texas prickly pear (*Opuntia engelmannii*). *Bougainvillea sp.* had a 16% survival rate, while *Chrysactinia mexicana*, *Lantana montevidensis*, *Justicia spicigera*, and *Rosmarinus officinalis* 'Huntington Carpet' had 0% survival rates at the end of the 6 months. Both *Encelia farinosa* and *Opuntia engelmannii* are native to the Sonoran Desert and increase roof biodiversity providing habitat benefits for wildlife, and can provide shade to rooftops.

Fig. 5.10 Green roof plant trial plots on the roof of the Design School at Arizona State University. (Photo: Paul Coseo)

A follow-up study from ASU has early indications for green roof thermal performance. The same modular plots were replanted with three different xeric plant mixes. The "shrub" plot was planted with desert milkweed (*Asclepias subulata*), Outback sunrise (*Eremophila maculata* 'Outback Sunrise'), globe mallow (*Sphaeralcea ambigua*), golden eye (*Viguiera parishii*), Sideoats grama (*Bouteloua curtipendula*). The "succulent plot" was planted with firestick euphorbia (*Euphorbia tirucalli* 'Rosea'), elephant food (*Portulacaria afra*), and rainbow bush (*Portulacaria afra* 'Variegata'). The "Sonoran plot" was planted with more ecoregional-appropriate species such as Texas prickly pear (*Opuntia engelmannii*), brittlebush (*Encelia farinosa*), and pink fairyduster (*Calliandra eriophylla*). The researchers varied the irrigation per bed with the shrub bed receiving the most supplemental water. The best-performing plants were *Eremophila maculata* 'Outback Sunrise' (high water requirements), *Euphorbia tirucalli "Rosea"* (moderate water requirements), *Portulacaria afra* (moderate water requirements), and *Opuntia engelmannii* (low water requirements).

Similar to the Milburn et al. (2011) study, the research team also found that the green roof systems were all (Sailor et al. 2012) warmer than the white roof at night and all cooler during the day. Yet on average, over the entire testing round, the shrub plot (−2.36 °C ± 1.68) and the succulent plot (−2.64 °C ± 2.34) green roofs were on average cooler than air temperatures, whereas the white roof system was more variable and only −0.60 °C (±5.16) on average cooler than air temperatures. This finding is consistent with previous computer simulations and is an encouraging sign for green roof development in hot arid environments. It may provide some evidence for the dampening effect soil and planted surfaces may have on heating and cooling cycles. However, more nuanced and sophisticated types of analysis must be conducted on these data to decipher more meaningful conclusions.

Some research has begun to explore the ecosystem services of green roofs in the Desert Southwest. Sailor et al. (2012) simulated the energy cost savings of a green roof compared to a conventional roof in Phoenix, AZ (Sailor et al. 2012). They showed that the green roof provided a net cost energy savings because it reduced energy consumption during peak hours (1 pm to 7 pm) when electricity rates were the most expensive. Although some green roof projects in the Desert Southwest have been installed and monitored (e.g. the Tempe Transit Center by Lerum and Thakare (2005), the Biosphere 2 green roof study in Tucson, and a study out of the University of Nevada, Las Vegas by Milburn and colleagues (Milburn et al. 2011)), none of these were published in peer-reviewed journals. This lack of knowledge on green roof performance in hot, arid conditions limits the ability to understand their performance on a range of ecosystem services.

5.3 Ecoregional Green Roof Case Studies (Arranged East to West)

5.3.1 Biology Building, University of Texas, El Paso, Texas

From its conception, the rooftop on the Biology Building at the University of Texas, El Paso was designed as a platform for research (Fig. 5.11). The campus was moving toward green and energy-saving design alternatives, and Ed Soltero, the director of UTEP's Office and Planning and Construction set into motion an opportunity for its first green roof on campus. The green roof emerged as a collaborative project between UTEP's Planning and Construction, The Department of Biology, and the College of Engineering. The purpose of the green roof was to establish a place for interdisciplinary green roof research regarding energy savings, plants, and watering techniques (Scott Nabers 2009).

5.3.1.1 Project Team

Building Owner/Client: the University of Texas, El Paso, Texas
Green Roof Design Team Lead: Department of Biology and Environmental Sciences faculty/students
Architect: Ed Soltero, AIA
Landscape Architect: Gutierrez Landscape Associates

Fig. 5.11 Green roof research plots on the campus of the University of Texas, El Paso. The green roof features drought-tolerant herbaceous plants and succulents that thrive with limited watering. (Photo: Bruce Dvorak, October 2018)

Installation Contractor: Columbia Green Technologies
Maintenance Contractor: Faculty/students/staff
Project completion: 2009
Green roof area: 850 m^2 (9156 ft^2)

5.3.1.2 Overview and Objectives

The overall goal was to explore green roofs as a tool to reduce rooftop temperatures, cut back energy demands during the summer, explore opportunities to lower carbon emissions, and make provisions for wildlife. During its conception, it was estimated that the green roof might contribute net energy savings between 20% and 30% for the building. The modular green roofs were set up to explore different planting options within the Columbia Green Technologies® modular system as research plots. Growing medium depth was set at 18–20 cm (7–8 in) of the vendor-provided substrate. Vegetation initially included many plants native to El Paso. Climate adapted exotic species have also been investigated on the roof over time (Stanley and Stanley 2010).

5.3.1.3 Plant Establishment

A majority of the vegetation was seeded onto the substrate, and many species have also re-seeded to sustain vegetative cover over time (Fig. 5.12). Some individual plants such as cacti were pre-grown in containers and planted into the substrate. A variety of forms of plants include grasses, annuals, herbaceous perennials (Table 5.1), shrubs, and succulents.

Fig. 5.12 Research plots include some native annuals such as *Gaillardia pulchella* which blooms and re-seed on the green roof modules each year. (Photo: Bruce Dvorak, October 2018)

Table 5.1 Herbaceous Perennials grown on the green roof

Common Name	Botanical Name
Yarrow, woolly	*Achillea tomentosa*
Milkweed, tropical	*Asclepias curassavica*
Chocolate flower	*Berlandiera lyrata*
Mistflower, Gregg's, palmleaf throughwort	*Conoclinium greggii*
Coreopsis, tickseed	*Coreopsis sp.*
Evening primrose, pale	*Oenothera pallida*
Evening primrose, Mexican	*Oenothera speciosa*
Penstemon, Rocky Mountain	*Penstemon strictus*
Mexican hat, coneflower	*Ratibida columnifera*
Globemallow, copper	*Sphaeralcea angustifolia*
Scarlet betony	*Stachys coccinea*
Dogweed	*Thymophylla pentachaeta*

Grass

Blue grama (*Bouteloua gracilis*).

Annuals

Blanketflower (*Gaillardia pulchella*); fleabane (*Erigeron sp.*); marigold, desert (*Baileya multiradiata*); tansyaster (*Dieteria canescens*; native to the western U.S.).

Shrubs

Texas creeping-oxeye (*Wedelia acapulcensis* var. *hispida*), turpentine bush (*Ericameria laricifolia*).

Succulents

Beavertail cactus, (*Opuntia basilaris)*, which is native to the Sonoran Desert (not Chihuahuan), cholla *(Opuntia imbricate*), native to west Texas and red yucca (*hesperaloe parviflora*) native to central Texas (did not thrive).

5.3.1.4 Irrigation

The original overhead spray irrigation was damaged during a winter freeze in 2011. A new zoned drip irrigation system was installed along with some new plants. The irrigation is an automated irrigation system that runs daily during the growing season.

5.3.1.5 Maintenance

Faculty and students make periodic visits to the green roof to maintain vegetation. Since many of the modules are planted with single species, it is easy to determine which plants remain and which can be removed.

5.3.1.6 Observed Wildlife

Butterflies, bees, hummingbirds, other birds all frequent the green roof.

5.3.1.7 Best Performing Native Vegetation

The species listed above are thriving on the green roofs, except where noted.

5.3.1.8 Post-occupancy Observations

Research Team

- After several years of plant trials, the following plants did not survive. It is not clear why the plants did not survive:
  - *Herbaceous Perennials: Echinacea purpurea, Engelmannia peristenia, Tetraneuris acaulis, Melampodium leucanthum, Oenothera caespitosa, Coreopsis grandiflora, Penstemon digitalis, Penstemon havardii, Zephyranthes drummondii, Salvia greggii, Salvia farinacea, Calylophus berlandieri 'Tucson', Oenothera lindheimeri*
  - *Herbaceous* Annuals: *Phacelia campanularia*; *Eschscholzia californica, Castilleja exserta, Simsia calva*

Authors' Reflections

- The green roof is large enough to capture important environmental data; however, there is no funding for long-term research to take place. Currently, there are individual studies by graduate students working with faculty to capture points in time.
- Future research could benefit from a new irrigation system that is set up with different zones that would allow different watering rates. It may be worthwhile to retest some of the species that did not thrive to see if they perform better with less or more water.
- Multi-partner efforts (industry/municipal) could expand the application of the research, as preliminary results show much promise for multiple ecosystem services from green roofs in the Chihuahuan Desert.

5.3.2 New Mexico Court of Appeals, University of New Mexico Campus, Albuquerque, NM

Located on the campus of the University of New Mexico, a new court of appeals building was planned and designed as a low-energy use building, and one of the first green roofs in New Mexico on a public building (Fig. 5.13). The accessible roof garden was designed with all of its vegetation native to the Rio Grande River Valley near Albuquerque. Originally designed to be a roof garden with sculpture, the accessible green roof was built without sculpture, but with all native vegetation. Water for the green roof vegetation for this LEED-certified building is sourced from runoff from the white reflected adjacent rooftop. Water is stored in a below-grade cistern.

Fig. 5.13 The light brown gravel paths are constructed from local stone to allow for maintenance access. The dark grey rounded river stone sits above perforated drainpipes and allows for drainage of the roof during monsoon events. The substrate rises gently to allow deeper substrate for taller grasses. (Photo: Bruce Dvorak, July 2018)

5.3.2.1 Project Team

Building Owner/Client: University of New Mexico
Green Roof Design Team Lead: George Radnovich
Architect: NCA Architects
Landscape Architect: SITES Southwest
Maintenance Contractor: Contracted out
Project completion: 2008
Green roof area: 240 m^2 (2600 ft^2)

5.3.2.2 Overview and Objectives

The green roof located on the New Mexico Court of Appeals Building is part of a LEED-certified Gold site and building project. The green roof is a semi-intensive system with a 30–40 cm-deep substrate (12–16 in). The main goal for the design of the vegetative system was to make use of desert succulents and perennial grasses native to the Albuquerque area. Most of the plants on the roof are found growing in the landscapes around Albuquerque and were once ubiquitous across the valley floor. The building and roofscape were designed to capture runoff from the uppermost roof of the building, the green roof, and landscapes. The roof deck has a 2% slope to achieve positive drainage for the sub-grade cistern.

5.3.2.3 Plant Establishment

All vegetation was established from pre-grown plants in containers.

Grasses

Blue grama (*Bouteloua gracilis*), Indian rice grass (*Oryzopsis hymenoides*), James' galleta (*Pleuraphis jamesii*), deer grass (*Muhlenbergia rigens*), sideoats grama (*Bouteloua curtipendula*).

Herbaceous Perennials

Broom dalea (*Psorothamnus scoparius*), feather dalea (*Dalea formosa*), Mormon tea (*Ephedra viridis*), scarlet globemallow (*Sphaeralcea coccinea*).

Succulents

Prickly pear (*Opuntia santa-rita*), Parry's Agave (*Agave parryi*), soapweed yucca (*Yucca glauca*).

Shrubs

Dwarf chamisa (*Ericameria nauseosa*), sand sage (Artimisia filifolia).

5.3.2.4 Irrigation

The irrigation system makes use of collected rainwater stored in a below-grade cistern that retains up to 30,283 liters (8000 gal) of precipitation. Each of these plant forms has different watering needs, watering rates and durations. To address these conditions, a drip irrigation system was designed to deliver water to six hydro zones about two to three times per week.

5.3.2.5 Maintenance

Maintenance crews visit the site about once a week during times when vegetation is actively growing. Activities include removing invasive plants (spurge, rye-grass, groundsel), deadheading grasses once a year, and replacing plants or irrigation parts as needed.

5.3.2.6 Observed Wildlife

Bees, birds, blue tail lizards, and butterflies have been observed during spring blooms.

5.3.2.7 Best Performing Native Vegetation

Deer grass (*Muhlenbergia rigens*), false yucca (*Hesperaloe sp.*), sideoats grama (*Bouteloua curtipendula*), prickly pear (*Opunta*), blue grama (*Bouteloua gracilis*), and *Agave.* The agave bloom on the roof, and there are bulbils present on rooftop plants. Bulbils are derived from where blooms detach and are the primary method of reproduction for agave (Fig. 5.14).

5.3.2.8 Post-occupancy Observations

Authors' Reflections

- The brown-colored south-facing facade radiates heat onto the green roof. This condition creates a micro-climate where reflected heat stunts and prohibits plant growth near the base of the facade. The wall is slightly angled towards the sky and directly takes on heat gain. Perhaps if the wall was sloped toward the roof, to self-shade and avoid direct reflection, the reflected light, and heat energy would be less damaging to vegetation.
- Some new plants, not native to the region, were added to the roof after the roof was established. If replacements or new plantings are added, the original planting plan or design team should be consulted to keep the green roof true to the original design intentions.

5.3.3 Riverfront Residence, Tucson, Arizona

This private residence located outside of Tucson is a custom-designed home with a green roof incorporated as a strategy to explore the potential of green roofs in Tucson to reduce stormwater runoff from summer monsoon storms (Fig. 5.15). A water-quality-improvement grant from the Arizona Department of Environmental

Fig. 5.14 Agave and grasses grow together on the New Mexico Court of Appeals roof, which mimics natural plant associations frequently found in the regional landscape. This aesthetic may not be familiar to those growing green roofs in milder climates; however, these local climate-adapted plants commonly grow in the Rio Valley near Albuquerque. Agave reproduce here through bulbils, as its seed rarely produces new plants. (Photo: Bruce Dvorak, July 2018)

Quality and the Environmental Protection Agency was awarded to assist with the added expenses of the green roofs. As one of the first green roofs in Tucson, there was no pilot project to learn about which kind of vegetation would grow in the shallow 12.7 cm (5 in) substrate.

5.3.3.1 Project Team

Building Owner/Client: Private residence
Green Roof Design Team Lead: Bil Taylor and Phred Bartholomaei
Architect: Bil Taylor Associates Inc.
Landscape Architect: Blue Mesa Studios
Installation Contractor: Desert Glen Landscaping
Maintenance Contractor: owner
Project completion: June 2003
Green roof area: 167 m^2 (1800 ft^2)

Fig. 5.15 Desert wildflowers and grasses are shown here growing and blooming on the residence above the first-floor living space during the initial years when irrigation was functioning. A bedroom and bathroom look out onto the green roof. Although this green roof (lower roof) has a south-facing exposure, the setback of the window and self-shading façade minimize any heat gain from the façade to the green roof vegetation. This climate-adapted architectural detail is fundamental to the success of this green roof ecosystem. (Courtesy of Bil Taylor)

5.3.3.2 Overview and Objectives

Twenty varieties of desert succulents, wildflowers, and grasses were initially installed and established on the two roofs (upper/lower). Ownership has changed several times since 2003 and the rooftop vegetation has seen a variety of levels of maintenance from frequent to none. In the initial years, the owners maintained most of the plants originally installed. Over time, some of the annuals growing in the valley below self-seeded onto the roof adding greater connectivity of the roof to the regional landscape.

5.3.3.3 Plant Establishment

Plants were selected to reflect desert landscapes local to Tucson. Vegetation was established from local nursery stock in containers of various sizes from quart-sized to gallon-sized plants. Forms of plants include grasses, herbaceous perennials (Table 5.2), and annuals.

Table 5.2 Herbaceous perennials grown on the Riverfront Residence lower green roof

Common Name	Botanical Name
Desert marigold	*Baileya multiradiata*
Prairie clover	*Dalea aurea*
Gregg dalea	*Dalea greggi*
Texas primrose	*Hymenoxys calylophus*
Blackfoot daisy	*Melampodium leucanthum*
White evening primrose	*Oenothera pallida*
Golden dalea	*Penstemon species*
Whitestem paperflower	*Psilostrophe cooperi*
Golden dyssodia	*Thymophylla pentachaeta*
Verbena	*Verbena bracteata*
Plains yucca	*Yucca glauca*
Desert zinnia	*Zinnia acerosa*
Prairie zinnia	*Zinnia grandiflora*

Grasses

Fluff grass (*Erioneuron pulchellum*) and blue grama (*Bouteloua gracilis*).

Annual Vegetation

Sand verbena (*Abronia villosa*), Arizona poppy (*Kallstroemia grandiflora*), and lupine (*Lupinus arizonicus*).

5.3.3.4 Irrigation

Initially, the roof was watered several times a week to establish vegetation. Watering was more frequent during times of drought and heat stress. The vegetation was successfully maintained on the roof in its planted form for the first 4–5 years. Changes in ownership resulted in less attention to watering. Over time, only hardy succulents were maintained on the roof. Annuals can still be observed on the roof; however, their dominance and persistence are much reduced without irrigation. The current succulent plant spacing is sparse and reflects a similar sparse plant spacing for plants in the valley below the residence (Fig. 5.16).

5.3.3.5 Maintenance

Maintenance includes periodic weeding, watering, and planting new vegetation on the roof. Annuals such as golden dyssodia (*Thymophylla pentachaeta*) are found growing on the landscape at ground level. It is thought that this plant was

Fig. 5.16 *Agave* (front center) and *Opuntia* (back right) now dominate the green roof on the Riverfront Residence roof. The irrigation system needs repair; thus, the herbaceous vegetation is no longer active on the roof, and succulent vegetation dominates. Some common annual wildflowers find their way onto the roof, and regardless of no irrigation, make for periodic blooms. (Photo: Bruce Dvorak, October 2018)

established on the green roof first, prior to its appearance below at ground level. Both the owners and neighbors welcome the golden dyssodia as it brings color to the landscape during the monsoons. Some thinning of vegetation is needed to prevent it from taking over the green roof.

5.3.3.6 Observed Wildlife

Bees, butterflies, and birds have been observed by the owners. Hummingbirds once populated the roof, when penstemon and wildflowers were maintained.

5.3.3.7 Best Performing Native Vegetation

Agave sp. and *Opuntia* spp., golden dyssodia (*Thymophylla pentachaeta*).

Owner

- Irrigation is required to maintain the wildflowers long-term. Over the past few years, when the irrigation was permanently shut off (previous owner), the succulent vegetation remained viable; however, the presence of perennial and annual wildflowers diminished over time.
- There are only a few planted rooftops in Tucson; however, on the appropriate structure, they may be viable considering that they are designed and maintained as intended.
- The green roof retains much of the precipitation during monsoons. There is no formal research to follow up the initial intent to demonstrate how green roofs can perform vital ecosystem functions such as stormwater retention.

Authors' Reflections

- This green roof demonstrates the willingness of the architect, the design team, and owner to take on a new technology. For this residence, the placement of the green roof is important for views inside the second-story of the building. The green roof also contributes to the benefit of keeping the living space below much cooler during the long summers.
- The forethought of setting a south-facing window back from the façade to reduce solar reflection was important. During times when the irrigation was running, the herbaceous vegetation thrived, even near the shaded window.

5.3.4 Museum of Northern Arizona, Easton Collection Center, Flagstaff, Arizona

The award-winning Museum of Northern Arizona (MNA) at the Easton Collection Center is a LEED Platinum-certified building that is located in Flagstaff, Arizona. The building is named after its major donors Betsy and Harry Easton. The mission of the museum is to collect artifacts regarding the native cultures of Arizona. The building and site design demonstrate respect for the tribal and cultural objects curated by MNA and the museum's commitment to sustainability. A primary goal regarding the landscape was to retain and build upon the character of a pine savanna.

The building and site appear to the visitor as an integrated landscape and building. The building has local stone used on its façade, has native vegetation surrounding the building, and a south-facing sloped green roof (Fig. 5.17). The concept for the site was to retain some of the native Ponderosa pine trees and restore a pine savanna vegetation on the ground plain. The paved portions of the site make use of porous pavement in the parking lot. The entry driveway features a circular stone wall with a rain garden in the center. The rain garden collects surface water from the green roof, landscape, and walking paths.

Fig. 5.17 Museum of Northern Arizona green roof, with a passing summer monsoon rain shower. The Ponderosa pine ecosystem (background) is the largest of its kind in the world and maintains grasses and forbs as groundcover vegetation. This south-facing montane meadow green roof is shown here emerging out of its summer dormancy, thus its transitioning from straw colors to green growth is sustained by intermittent brief rain showers the previous week. Drainage channels divert rainfall, and skylights (clear domes) channel sunlight into the building below. Although the clouds cooled the roof surface soils and vegetation during the visit to this roof, the nearby monsoon shower did not rain on the green roof on that day. Thus, precipitation events in the desert southwest can be highly localized events. (Photo: Bruce Dvorak, July 2018)

5.3.4.1 Project Team

Green Roof Design Team Lead: Rana Creek Living Architecture, Roberts/Jones Associates
Architect: Roberts/Jones Associates
Landscape Architect: WLB Group
Ecologists: Keri Stiverson, Connie Cowan, Kirstin Phillips, Jan Busco
Installation Contractor: Kinney Construction Services
Maintenance Contractor: Warner's Nursery and Landscape, Morning Dew Landscaping
Project completion: 2009
Green roof area: 929 m^2 (10,000 ft^2)

5.3.4.2 Overview and Objectives

At 2134 m (7000 ft) above sea level, the museum is located in the montane pine savanna, woodland, and meadow ecoregions. The green roof was designed to mimic the herbaceous vegetation of the grass-based pine savanna ecosystem. Because of frequent drought and dwindling monsoons over the past several years, replanting of portions of the roof is being transitioned to more drought-tolerant species and fewer Ponderosa zone species. Ponderosa pine groundcovers typically experience shade during the daytime. The rooftop is exposed to direct sunlight much of the day, and the roof has a south-facing aspect. The long-term goal is to transition the roof vegetation to species that grow in full sun and tolerate heat and drought (Fig. 5.18). Both cool season and warm season vegetation are established on the green roof. Cool-season grasses (Indian ricegrass, muttongrass) are dormant from November through March and green-up with the spring thaw. Warm-season grasses on the roof (little bluestem, blue grama) transition to dormancy during the summer prior to monsoons. Once the intense and short duration rainstorms move through the region, the grasses will re-sprout with fresh growth and transform the roof from straw-colored to bright green (Fig. 5.18). Perennial vegetation such as alliums, yarrow, and penstemons are also successful by avoiding drought with dormancy. Lupines survive through annual reseeding on the roof.

Fig. 5.18 Of the many plants that grow on the grass-based rooftop ecosystem, species were selected that fill seasonal roles. Hedgehog cactus (**a**) blooms in the late spring, Indian paintbrush (**b**) blooms during late spring, (**c**) Lambert's locoweed, *Oxytropis lambertii*, and sunflowers (**d**) bloom during the summer monsoon season. (Courtesy of Jan Busco and Lance Diskan)

5.3.4.3 Plant Establishment

Vegetation was established through seeding and live plugs. Forms of plants include annuals, herbaceous perennials (Table 5.3), grasses (Table 5.4), and succulents.

Annuals

Indian paintbrush (*Castilleja integra*), king's lupine (*Lupinus kingie*)

Table 5.3 Herbaceous perennials growing on the Museum of Northern Arizona green roof

Common Name	Botanical Name
Western yarrow	Achillea millefolium
Pussytoes	*Antennaria parvifolia*
Fendler's sandwort	*Arenaria fendleri*
Fleabane	*Erigeron divergens*
Trailing fleabane	*Erigeron flagellaris*
Sulfer flower buckwheat	*Eriogonum umbellatum*
Blanket flower	*Gaillardia aristata*
Prairie smoke	*Geum triflorum*
Lewis flax	*Linum lewisii*
Wright's deervetch	*Lotus wrightii*
Hoary tansyaster	*Machaeranthera canescens*
Purple locoweed	*Oxytropis lambertii*
Scarlet bugler	*Penstemon barbatus*
Sunset crater penstemon	*Penstemon clutei*
Eaton's firecracker	*Penstemon eatonii*
Mat penstemon	*Penstemon linarioides*
Rocky mtn penstemon	*Penstemon strictus*
Blue beardtongue	*Penstemon virgatus*
Fendler's globemallow	*Sphaeralcea fendleri*

Table 5.4 Grasses growing on the Museum of Northern Arizona green roof

Common Name	Botanical Name
Indian ricegrass	*Achnatherum hymenoides*
Pine dropseed	*Blepharoneuron tricholepis*
Side oats grama	*Bouteloua curtipendula*
Blue grama	*Bouteloua gracilis*
Arizona fescue	*Festuca arizonica*
Mountain muhly	*Muhlenbergia montana*
Switchgrass	*Panicum virgatum*
Muttongrass	*Poa fendleriana*
Little blue stem	*Schizachyrium scoparium*

Succulents

Hedgehog cactus (*Echinocereus fendleri*), desert prickly pear (*Opuntia phaeacantha).*

5.3.4.4 Irrigation

The green roof is irrigated with water captured in an above-ground cistern located behind the main structure. Maintaining the cistern and irrigation system has been a challenge for the owners, as it has required frequent maintenance to keep rooftop vegetation irrigated. The green roof is typically watered 3 days/week for 45 min, April – June, with captured rainwater in an 83,279-liter (22,000 gal) cistern. The automated KISSS [advanced subsurface irrigation] watering system delivers water below the substrate (soil) and spreads water laterally.

5.3.4.5 Maintenance

Weekly monitoring of phenology, weeding of non-native species, irrigation troubleshooting are common tasks. Regular irrigation issues include leaking pipes, running out of water in the tank, broken pipes, broken solenoids. Information about the waterproofing system installed is not known.

5.3.4.6 Observed Wildlife

Owners and staff have seen ground-nesting birds building nests in the grasses. Hummingbirds love the penstemons. Bees and butterflies frequent the roof. Some birds observed on the roof include lesser goldfinches, northern flickers, American robins, yellow-rumped warblers and pine siskins.

5.3.4.7 Best Performing Native Vegetation

Penstemon clutei, Allium cernuum, Achillea millefolium, Amaranthus powellii, Bahia dissecta, Bidens tenuisecta, Helianthus annuus, Epilobium brachycarpum, Opuntia phaeacantha.

5.3.4.8 Post-occupancy Observations

Museum Research Staff

- South-facing green roofs in Arizona may need special attention to substrate depth, watering practices, and plant selection.
- The sloped design presents its own challenges. The lowest elevations of the green roof retain more moisture while the highest elevations dry out the quickest. The substrate must be composed of stable material, otherwise, erosion could be expected.
- Subsurface irrigation requires accurate mapping to identify the location of irrigation lines if maintenance is required.
- Lack of snow (soil moisture) severely impacts the survivability of some plants through the winter.
- The number of non-native species has not risen very much even though the number of native volunteers has increased every year.
- Plant species found on the grounds may eventually be observed on the roof including shrubs and trees!

Authors' Reflections

- This amazing green roof demonstrates the willingness of the architect, the design team, and the owner to take risks regarding the establishment of a grass-based green roof in the Desert Southwest. At the high altitude of Flagstaff, this green roof does not typically experience the heat and aridity of the Sonoran Desert located below the Colorado Plateau. Conservation sites such as the Hart Prairie Preserve could be a useful resource to inform any future planting of the green roof.

5.3.5 *Tempe Transportation Center Research Roofs, Tempe, Arizona*

Located at an elevation of 358 m (1175 ft), the Tempe Transportation Center (Fig. 5.19) is located on the south side of Hayden Butte (highest elevation of 456 m or 1496 ft), near downtown Tempe. Many years ago, the site was a Hohokam village, and its remnants have been preserved in the adjacent Hayden Butte Preserve. Due to its historical significance, the Transportation Center, and its design team included an archaeologist. The designers sited the building to preserve important viewsheds of downtown Tempe and the Butte (Fig. 5.20). The adjacent Preserve provides easy access to remnant desert scrub dominated by creosote bush (*Larrea tridentata*), triangle-leaf bursage (*Ambrosia deltoidea*), brittlebush (*Encelia farinosa*), cholla (*Cylindropuntia sp.*), barrel cactus (*Ferocactus sp.*), fairy duster (*Calliandra eriophylla*), palo verde (*Parkinsonia sp.*), and prickly pear (*Opuntia sp.*) species.

Fig. 5.19 Tempe Transportation Center green roof with densely planted stands of lady's slipper (front) and yellow yucca (middle) and (background). (Photo: Bruce Dvorak, July 2018)

5.3.5.1 Project Team

Building Owner/Client: Tempe Transportation Center
Green Roof Design Team Lead: Bonnie Richardson, Project Manager for the City of Tempe
Architect: Architekton and OTAC
Structural Engineer: BDA Design
Landscape Architect: A Dye Design
Installation Contractor: Adolfson & Peterson
Maintenance Contractor: City of Tempe
Project completion: 2008
Green roof area: 546 m^2 (5880 ft^2)

5.3.5.2 Overview and Objectives

The City of Tempe's goals for the project was to have the project be environmentally and culturally responsive to its immediate site within the Sonoran Desert. The facility was designed as a demonstration of sustainability and an educational

Fig. 5.20 Tempe Transportation Center green roof is adjacent to a remnant desert habitat (**a**). The green roof features native vegetation such as lady's slippers (*Pedilanthus macrocarpus*) (**b**) which attracts a hummingbird (**c**). The green roof is adjacent to a white reflective roof (**d**) which is a standard roof in Tempe. To maximize their potential to reduce heat gain, white reflective roofs require periodic maintenance to remove layers of dust. This green roof is low maintenance and maintains a cooler rooftop surface (than the cool roof), retains stormwater from monsoons and provides habitat for local and migrating wildlife. (Photos: Bruce Dvorak, July 2018)

example for users. In particular, the green roof was intended to provide: (1) thermal buffer by insulating the roof structure and interior spaces for reduced solar gain and reductions of cooling loads; (2) stormwater harvesting and management by collecting, filtering, and storing rainfall in an underground cistern; and (3) visual connections to Hayden Butte by creating a 'fifth elevation' (in addition to the four walls of the building) to enhance the aesthetics of the site for the hikers on the Butte.

The 546 m^2 (5880 ft^2), 31 cm (12 in) deep extensive green roof system is located on the third-floor roof. The roof helped the project achieve LEED Platinum-certification, capturing 53 of the possible 69 credits. The built-up green roof system was provided by American Hydrotech. In preparation for designing the green roof system, Lerum and Thakare (2005) helped the city of Tempe tested five alternative green roof designs before deciding on the final system. The research team tested different soil depths, types of plants, and water schedules. The growing medium was manufactured by Western Organics (now GRO-WELL) and is a blend of 40% Moisturelite (pumice), 40% Concrete Sand, 10% Omni Mulch, and 10% Phoenix Black Bottom Mulch. Lerum and Thakare (2005) found that the plants did not perform well in substrate test plots in depths of 10 cm (4 in), therefore substrate depths of 20–31 cm (8–12 in) were necessary. The green roof has 20 cm (8 in) substrate depths on the eastern upslope side of the roof with 31 cm (12 in) depths on the western downslope side. Over time, plants have adapted to their preferred moisture and substrate depths. Since water drains towards the west and has thicker soils, the plants were larger and healthier on the western side of the green roof.

5.3.5.3 Plant Establishment

The vegetation on the green roof was established from one-gallon containers. Some native and a few non-native species have been trialed on the roof. The native plants are listed below.

Grass/Lily

Beargrass (*Nolina microcarpa*) above 914 m (3000 ft).

Succulents

Red aloe (*Hesperaloe parvoflora*), yellow yucca (*Hesperaloe parviflora 'Yellow'*), giant hesperaloe (*Hesperaloe funifera*), which are native to the Chihuahuan Desert; and lady's slipper (*Pedilanthus macrocarpus*) native to the Sonoran Desert of northwestern Mexico and central Baja California.

5.3.5.4 Irrigation

Irrigation frequency is set for two times a week in winter and three-four times per week in summer. All the plants are on the same system/ zone drip irrigation system. The watering schedule varies by season. The system is manually adjusted to adapt to seasonal needs. The system is getting an update to add grid management software. The City had original plans to use greywater, condensate, and stormwater, however, the cistern did not work as planned. Currently, they collect stormwater, but it's not used on the roof.

5.3.5.5 Maintenance

The budget for long term maintenance comes out of the building maintenance budget. The project manager reported that they have not had many weeds and when left on its own, it needs very little weeding. Only Mexican fan palm (*Washingtonia robusta*) seedlings need to be removed periodically. The leaves of *Alephora lutea* can burn during late summer, but it reseeds and re-sprouts each year. The spent organic debris is left in place. *Myoporum parvifolium* grows in dense stands, is short-lived, and doesn't perform well long-term.

5.3.5.6 Observed Wildlife

Bees and hummingbirds frequent the roof.

5.3.5.7 Best Performing Native Vegetation

The regionally native lady's slipper (*Pedilanthus macrocarpus*) performs very well, as well as *Hesperaloe sp.*

5.3.5.8 Post-occupancy Observations

Project Manager

- Original plans to use greywater, condensate, and stormwater cistern did not work as planned. Stormwater runoff from the white reflected roof is collected, but it's not used on the roof. Integrated water collection/ irrigation systems may be desirable on many projects in the Southwest, but it's harder to operationalize.
- It's important to the City to share results from research, methods, and trials to help everyone learn how to design and manage green roofs in the Southwest.
- Working closely with universities is useful for providing guidance in the design phase and understanding the efficacy of design after installation.

Authors' Reflections

- Future plant research could include some of the vegetation that is native to the Hayden Butte conservation site.
- This green roof has much potential to serve as a research site and a pilot project for green roofs in the region.

5.3.6 Optima Camelview Village, Scottsdale, Arizona

The Optima Camelview Village is a LEED Silver-certified mixed-use condominium project located in Scottsdale, Arizona. The 5.3-hectare (13-acre) site has 6.9 hectares (17 acres) of green roofs through a multi-dimensional landscape terrace system that was integrated into all levels of the 11, seven-story buildings (Fig. 5.21) (OCVV 2020). The building complex is supported by a 9 m × 9 m (30 ft × 30 ft) structural

Fig. 5.21 Layers of elevated roof terraces and trunks of palm trees frame a view of a native blue palo verde tree (*Parkinsonia florida*) in bloom at the Optima Camelview Village. (Courtesy of Chris Brown, Floor Associates)

bay post-tension deck system that provides support for the landscaped terraces. Expansion joints in the concrete roof deck system allow for high-temperature differentials between day and night which allows some of the landscaped terraces to cantilever up to 6 m (20 ft) out beyond support columns.

Before bringing the landscape architects onboard, the roof deck structural loads were pre-determined, which limited the green roof to a 12.7 cm (5 in) substrate depth. With a limited structural capacity, the landscape architects proposed varying the depths with a minimum of 15.2 cm (6 in) substrate depth where vegetation exists and is offset by placing no substrate or vegetation in other areas (Schuler 2018). The green roof system was provided by American Hydotech, and substrate depths range from 15 to 20 cm (6–8 in) deep on balconies and 61 cm (24 in) for trees. This vertical and cantilevered approach creates a diversity of microclimates, from deep shade to full sun, and makes this one of the largest and most unique green roofs in Arizona.

5.3.6.1 Project Team

Building Owner/Client: Optima Camelview Village
Green Roof Design Team Lead: Floor Associates, Phoenix
Architect: David Hovey and Associates Architect, Inc. Chicago
Structural Engineer: Kimley-Horn and Associates, Inc., Phoenix

Landscape Architect: Floor Associates, Phoenix
Irrigation Design: Coates Irrigation Consultants, Gilbert, AZ.
Horticulturalist: Chris Martin, Arizona State University
Installation and Maintenance Contractor: Grounds Control, Phoenix
Project completion: 2011
Green roof area: 6.9 hectares (17 acres) on a 5.3-hectare (13-acre) site

5.3.6.2 Overview and Objectives

A key challenge for the project was to understand what combination of plants, irrigation schedule, soil mix, and top treatment might work with the diversity of microclimates of the project. To better understand the complexities and what might perform better in the different microclimates, the design of the green roof system was developed through a research partnership with Dr. Chris Martin, Professor of Sustainable Horticulture, at Arizona State University and Mountain State Nursery. They created 3.7 m × 7.3 m × 12.7 cm (12 ft × 24 ft × 5 in) plots (Fig. 5.22) with American Hydrotech® systems to test irrigation methods, plants, soils, and top treatment (e.g. bark, organic mulch, and decomposed granite) to see how the different combinations impacted substrate temperature, water retention, and plant performance (Schuler 2018). Due to the complex orientations and exposures of the project,

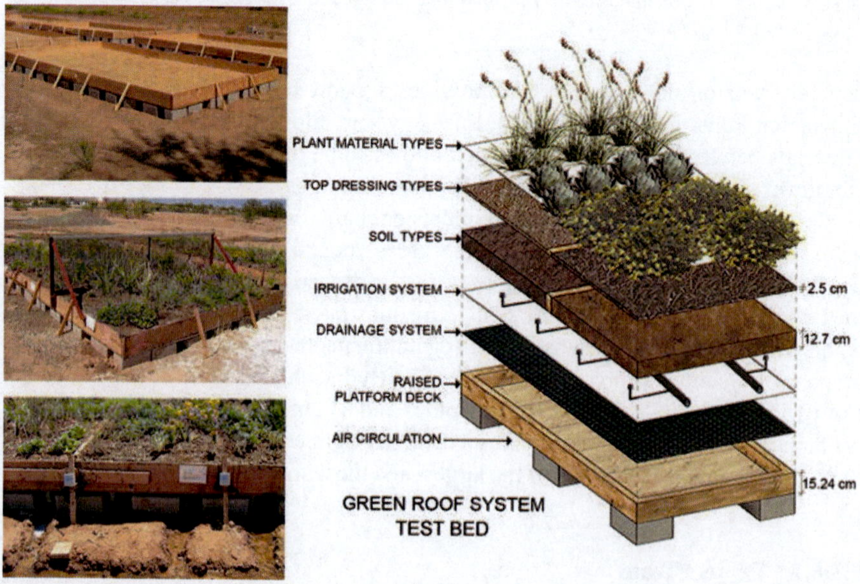

Fig. 5.22 Green roof system and plant test bed plots 3.7 m × 7.3 m × 12.7 cm (12 ft × 24 ft × 5 in) were developed in partnership with Arizona State University and Mountain State Nursery to help select plant material, and make design decisions for the many complex microclimates of the project. (Courtesy of Chris Brown, Floor Associates)

the scientist helped test different scenarios such as degrees of shade and reflected radiation (Schuler 2018).

The design concept was to follow "nature's design" of a desert canyon ecological system. In this way, vegetation was integrated into the buildings' architecture at every level of the development as a series of landscaped terraces. As in desert canyons, the hottest, arid, and sun-soaked areas occur at the canyon top. While the depth of shade and availability of water concentrate toward the canyon floor with more riparian areas with perennial water occurring on the floor of canyons. They incorporated water features to represent desert springs and ciénagas (wetlands) that would be found in canyons (Schuler 2018). The designers mimicked a slot canyon ecology with the what they called the main "Riparian Area" oriented north-south to maximize shade. This Riparian Area also serves as a pedestrian right-of-way through the project with access to the retail spaces. Other areas include a Mesquite Wash, Palo Verde Bosque, and Ironwood Bosque. The Riparian area contained higher water requirement plant species such as lady's slippers (*Pedilantus microcarpus*) native to Baja in Mexico. While at the rooftops, plants such as brittlebush (*Encelia farinose*) were selected because they perform well in drier, sunnier conditions. Ecologically minded design goals for the green roof terrace system were to provide wildlife habitat, cool the canyons through shade and evapotranspiration, provide more oxygen to the air while reducing air pollution such as dust and smog, buffer noises, and manage stormwater (Optima n.d.).

5.3.6.3 Plant Establishment

Over 60,000 plants were used representing 70 different species including six tree species, 12 shrub species, and 30 ground cover species (Schuler 2018). Due to the amount of plant material, these were grown onsite by the contractor (Schuler 2018). For locally native plant material, Floor Associates regarded the larger Sonoran and Chihuahuan desert region as regionally appropriate for plant selection. The ground level was similar to the floor of a slot canyon and the plant material reflected this shadier, moister environment (Fig. 5.23). For the ground level, this included the use of plants native to the Southwest or northern Mexico such as Tabardillo or Baja Fairduster (*Calliandra californica*), Beloperone (*Justicia californica*), toothless stool (*Dasylirion longissimum*), Texas sage (*Leucophyllum frutescens*), and red yucca (*Hesperaloe parviflora*).

The original planting design for rooftop areas took into account the microclimate matching conditions of the valley floors of the lower Sonoran Desert. The designers wanted the illusion that the building had risen from the valley floor and taken the landscape with it. For the rooftop level, the intent was to use local lower valley Sonoran native species such as creosote bush (*Larrea tridentata*), cholla (*Cylindropuntia cholla*), brittlebush (*Encelia farinosa*), and blue palo verde (*Cercidium floridum* 'AZT'). However, when implemented, all of these species except for the blue palo verde (*Cercidium floridum* 'AZT') were substituted out for other species. Today, you can see the blue palo verde on many of the upper balconies of the development.

Fig. 5.23 (**a**) The large mixed-use project has many complex spaces with many levels of roof gardens, which include ornamental water features. Some of the side spaces with views from private residences look out onto low-growing desert vegetation (**b**) and (**c**). While the project contains a number of desert plants, many are not native to Phoenix, including vegetation in the many private residential patios (**d**). (Photos: Bruce Dvorak, July 2018)

5.3.6.4 Maintenance

Most of the vegetation is thriving. However, due to some of the deeply shaded microclimates, some of the vegetation growing in the shadiest locations are struggling. The residents and Home Owners Association (HOA) maintain the balconies. The use of xeric or desert-adapted plants resulted in an estimate by the U.S. Green Building Council that the project uses about half the irrigation of a similar project that did not use xeric species and a drip irrigation system (Schuler 2018).

5.3.6.5 Irrigation

Sun plants are on one irrigation system and shade plants on another system, both with fertilizer injected into the irrigation system.

5.3.6.6 Observed Wildlife

The designers reported butterflies, birds, and lizards on terraces as high as the 6th floor (Schuler 2018).

5.3.6.7 Best Performing Native Vegetation

Due to the diversity of microclimates created, it's difficult to identify one best performing species. Many of the green roof terraces are performing well. In particular, red yucca (*Hesperaloe parviflora*) and lady's slippers (*Pedilantus microcarpus*) and blue palo verde tree (*Parkinsonia florida*) all have performed very well.

5.3.6.8 Post-occupancy Observations

Design Team

- It might be best for the HOA to take on the maintenance for consistency in this type of complex project with such extensive green roof terraces. This is especially the case because many of the terraces are located on private balconies.
- Water harvesting would have been a good addition to the project. This is not only for sustainability considerations but also because the municipal water is coming from the Central Arizona Project which has a higher salt content and the collected rainwater would have been better for plant health.
- When working on a commercial development like this, there are always tradeoffs with developers in terms of plant material, aesthetics, cost, and maintenance. The designers ended up using less locally native Lower Colorado River Valley Desert plant species and more Southwestern, regionally native plants because of expectations of developers for lusher looking vegetation.
- The project received a President's Award and an Award of Excellence from the Arizona chapter of the American Society of Landscape Architects.

Authors' Reflections

- The project stands as an exemplary example of how climate-adapted vegetation can be integrated into a large mixed-use residential project.
- The landscape architects found an inventive solution (varied soil depths) to make opportunities for a wide variety of plants to grow on a roof deck with structural limitations.
- Although there are many exotic plants on this project, the intention from the beginning was to make a larger proportion of plants from the local ecoregions. Perhaps the successes of this project demonstrate that future similar projects could use more native vegetation.

5.3.7 The Nature Center at the Estrella Mountains Regional Park, Goodyear, Arizona

The Nature Center at the Estrella Mountain Regional Park is the gateway to the 8020-hectare (19,820-acre) desert park, located west of Phoenix. The Center sits on an alluvial fan at elevation 279 m (916 ft) nestled between the Gila River and the northern base of the Sierra Estrella Mountain range with the Hayes Peak the highest elevation of 1375 m (4512 ft). The landscape surrounding the Center is composed of Lower Sonoran Desert scrub dominated by creosote bush (*Larrea tridentata*) and bursage (*Ambrosia sp.*). Stormwater basins adjacent to the Nature Center provide conditions that support higher water requirement plant species. The Center is one of three identical buildings in the County system that is a Platinum LEED-certified building that was installed 2010 and includes a green roof (Fig. 5.24).

Fig. 5.24 The green roof on the Nature Center at Estrella Mountains Regional Park is shown here in November 2018. October 2018, was the wettest month on record for nearby Phoenix Sky Harbor Airport, and resulted in a bloom of native wildflowers, such as this golden eye (*Viguiera parishii*), along with non-native exotic species, such as Sahara mustard (*Brassica tournefortii*). The vegetation has naturally adapted to drier and higher elevations (left) to lower and wetter (right) due to the accumulation of moisture at the bottom of the gently sloping roof. The wide spacing of plants with bare soil between plants matches the plant spacing seen in the desert landscape in the distance. Thus, a healthy green roof ecosystem in the Desert Southwest need not look like those typical to regions with ample rainfall. (Photo: Paul Coseo)

5.3.7.1 Project Team

Building Owner/Client: Maricopa County Parks, Don Harris, Park Supervisor
Architect: TRK Architecture & Facilities Management
General Contractor: D. L. Norton General Contracting
Landscape Architect: The Groundskeeper, a division of Environmental Earthscapes
Installation Contractor: Bret Hill Custom Roofing
Maintenance Contractor: Maricopa County Parks
Project completion: 2008
Green roof area: 160 m^2 (1722 ft^2)

5.3.7.2 Overview and Objectives

The green roof helped the Nature Center achieve LEED credits and for the County to achieve its sustainability goals more broadly. The green roof helped lower the building's impact on water and electric use by providing insulation and more natural cooling from evapotranspiration. A key goal was to minimize the building's heat island footprint. In addition, the building was designed to reduce the impact of the Nature Center on visitors' visual experience while hiking. Thus, they decided to reproduce the local desert plant communities on the green roof. The overall concept for the green roof system was to mimic the desert scrub floor surrounding the Nature Center including the dynamic nature of the landscape where plants die, seed-in, and the plant community naturalizes over time.

The Stevens GardenTop Roof system, manufactured by Stevens Roofing Systems, is 31 cm (12 in) thick continuous semi-intensive green roof system. The system includes a 60 mil Stevens EP membrane, Stevens ISO 2000 insulation, and Stevens GardenTop Drain 50RS. The system was placed on a ½:12 slope, which orients the aspect and drainage to the west. This orientation concentrates water on the western side of the green roof system. This is reflected in the concentration of plants on the western end of the green roof.

5.3.7.3 Plant Establishment

The original plant list is not available, but according to the Park Director and remaining plant evidence, plants were planted from 1 to 5-gallon containers. Evidence of original planting pallet included hedgehog cacti, currently dead, (*Echinocereus engelmannii*), brittlebush, originals are dead, but reseeding (*Encelia farinose*), and the *Agave sp.* are still thriving.

The original plantings were cacti, succulents, and perennials. Many volunteer species have since colonized the green roof. In October 2018, the Phoenix Metro area recorded its wettest October on record. This triggered many wildflower seeds to germinate and the green roof experienced a bloom of native and non-native seedlings. During the visit in fall 2018, brittlebush (*Encelia farinose*); blue palo verde

(*Parkinsonia florida*), sorrel buckwheat (*Eriogonum polycladon*); damianita daisy (*Chrysactinia Mexicana*); golden eye (*Viguiera parishii*); and desert globemallow (*Sphaeralcea ambigua*) were observed. Non-native species on the green roof included: purslane (*Portulaca oleracea*) and Sahara mustard (*Brassica tournefortii*). The non-native purslane and Sahara mustard made up roughly 50% of the vegetative mass on the roof.

5.3.7.4 Irrigation

The roof is irrigated by a six-station Intelli-Sense Controller® with WeatherTRAK® smart irrigation system. Irrigation system components were manufactured by Ewing® using 1 cm (½ in) tubing at the eastern (highest) elevation of the green roof system and 50 mm (¼ in) tubing for the drip irrigation system.

5.3.7.5 Maintenance

The Nature Center's approach to green roof maintenance over the last 10 years was to let it naturalize. According to the Park Director, they do not see the need to weed as the volunteer plants provide benefits. From their perspective, letting the roof naturalize is more sustainable and it contributes to the desired cooling and ecological benefits. The volunteer plants (weeds/wildflowers) provide effective shading and help water retention. They acknowledge that it may need some replanting and they may weed in spots to replant. Due to lack of funds and resources, the Center uses its general grounds landscape maintenance staff to periodically (every quarter) maintain the green roof. The biggest issue related to maintenance is the need to redo the irrigation system's dip lines. It is due to normal deterioration from the sun. They do not have any leaking from the green roof.

5.3.7.6 Observed Wildlife

According to the Park Director, the following wildlife makes use of the green roof over the past 10 years: say's phoebe (*Sayornis saya*), curve-billed thrasher (*Toxostoma curvirostre*), american kestrel (*Falco sparverius*), gila woodpecker (*Melanerpes uropygiali*), gilded flicker, (*Colaptes chrysoides*), purple finch (*Haemorhous purpureus*), gambel's quail (*Callipepla gambelii*), cloudless sulphur (*Phoebis sennae*), and a variety of bee species. The birds used the roof primarily to forage for seeds and insects. An American Kestrel has been seen hunting for animals.

5.3.7.7 Best Performing Native Vegetation

Agave sp. is the best performing native genera from the original design.

5.3.7.8 Post-occupancy Observations

Nature Center Staff

- It is beneficial to have stronger, longer maintenance support/schedule to maintain driplines due to the deterioration from sun and aridity.
- When green roofs are not visible, the maintenance approach can be "out of sight, out of mind". This invisibility can be good because maintenance activities can emphasize ecological performance over aesthetic desires. While on the downside, it can be bad for people to see, enjoy, and learn about green roofs. Future green roofs should be visible.
- The green roof was successful because the designers mimicked the ground desert landscape with some comparison in terms of aesthetics and plant communities.
- From a building performance perspective, the green roof system has held up well in terms of the waterproofing, and with its thick waterproofing layer, that has not been exposed to sunlight, diurnal oscillations of heat and cold (such as white reflective roof membranes), and it has not leaked.

Authors' Reflections

- This ecoregional green roof is important because it establishes a baseline comparison for ecosystem services and aesthetic expectations for green roofs in the Phoenix area. This green roof has wide plant spacing and allowances the naturalized additions of volunteer vegetation to demonstrate what a low-maintenance green roof can look like. Such green roofs used widely across the Phoenix metro area could mitigate for lost habitat, reduce energy use in buildings, reduce monsoon flooding, and mitigate urban heat islands.

5.3.8 Anza-Borrego Visitor Center, Borrego Springs, California

The Anza-Borrego State Park is the United State's largest desert state park. It is a landscape that varies greatly in elevation, vegetation, and cultural history. Thought to be populated by native Americans as long as 6000 years ago, the most recent native cultures left the site about 1600–2000 years ago. Traveled by Spanish settlers in the late 1700s, the park is named after the famed leader of migrants Juan Bautista de Anza, and the Spanish word for sheep (Borrego). The Anza-Borrego property became dedicated as a state park in 1952.

In the late 1970s, architect Robert Ferris was commissioned to design a visitor's center for the Anza-Borrego State Park. Ferris' vision was to design a building that would minimize disturbance to the desert landscape. His vision was focused on the idea that the center should become compatible with the desert landscape. Robert's understanding of the survival techniques of some desert animals led to the idea of burrowing. Many desert animals burrow during the heat of the day to avoid heat stress. This activity inspired Robert to make thc visitor's center an earth-covered structure (Fig. 5.25). Ferris said, "The use of any kind of conventionally roofed and walled structure would sound a discordant note in the natural harmony of the previously undisturbed site" (ABNC 1994).

5.3.8.1 Project Team

Building Owner/Client: State of California
Architect: Robert Ferris, FAIA
Maintenance Contractor: Visitor Center staff
Project completion: March 16, 1979
Green roof area: approximately 650 m^2 (7000 ft^2)

Fig. 5.25 This east facing zone of the Anza-Borrego Visitor Center green roof features *Encelia farinosa* (foreground) and *Agave sp.* (middle ground). (Photo: Bruce Dvorak, October 2018)

5.3.8.2 Overview and Objectives

The 650 m^2 (7000 ft^2) building is surrounded by a gently sloping and seamless desert landscape that surrounds three sides of the structure and covers the roof. The natural vegetation was planted into 61 cm (24 in) of the desert soil that was taken from the site and replaced on top of the visitor's center. Today, the vegetated roof stands as one of California's oldest vegetated roofs and was the first of the kind for the Sonoran and Colorado desert ecoregions.

5.3.8.3 Plant Establishment

Several desert cacti species grow on the roof and produce a showy bloom including several cholla species (*Cylindropuntia sp.*), beavertail cactus (*Opuntia basilaris*), which both grow on the Visitor Center roof, as well as some perennial species such as brittlebush (*Encelia farinosa*), orcutt's or borrego aster (*Xylorhiza orcuttii*), and desert lavender (*Hyptis emoryi*) (Fig. 5.26).

Fig. 5.26 Agave clusters growing on the west-facing zone of the roof. Cholla is visible growing near the roof vent (center right) and the periphery of the top of the slope. Desert willow is growing on the lower portions of the ground-level landscape not part of the roof. The wide spacing of plants and exposed soil on the roof here, match a similar spacing at the ground level. (Photo: Bruce Dvorak, October 2018)

There is no formal record of annual plant species that bloom on the visitor center roof. However, many online resources make a record of species that bloom each year. These species are frequent in the lower elevations of the Anza-Borrego desert and include hairy sand verbena (*Abronia villosa*); brown-eyed primrose (*Camissonia claviformis*); spectacle-pod (*Dithyrea californica*); hairy desert-sunflower, aka desert gold (*Geraea canescens*); Arizona lupine (*Lupinus arizonicus*); desert dandelion (*Malacothrix glabrata*); and dune primrose (*Oenothera deltoids*).

5.3.8.4 Irrigation

The rooftop vegetation is watered once a week in the summer months by staff, and less frequently during winter (once a month or less).

5.3.8.5 Maintenance

There is no fertilizer applied to the roof as animal interactions and decaying vegetation provide enough nutrients for plants to continue to survive. There are invasive exotic plant species (such as Sahara mustard) that are removed annually, as well as "undesirable" native species (palo verde seedlings in places staff don't want a tree to grow). Staff also remove plants that have died on or near the roof as needed.

5.3.8.6 Observed Wildlife

Regarding wildlife that makes use of the earth sheltered center, this includes reptiles, bees, butterflies, and other pollinators. Coyotes have been known to walk across the roof as well as evidence of rabbits and birds making frequent use of the vegetation.

5.3.8.7 Best Performing Native Vegetation

Agave sp., Echinocereus engelmannii, Opunta basilaris, Cylindropuntia echinocarpa, Cylindropuntia ganderii, Cylindropuntia X fosbergii.

5.3.8.8 Post-occupancy Observations

Nature Center Staff

- The vegetation, being native desert plants, can survive without supplemental water, but it is watered infrequently during the summer so that plants continue to grow.

- The Sahara mustard grows like crazy in the spring after the winter rains, so they try to stay on top of its removal, so that—all going well—there will be a smaller and smaller seed bank in future years. Otherwise, it outcompetes native desert wildflowers.
- There are not many poppies growing in the area around the Visitor Center. During a year with many wildflowers in bloom, there may be some wildflowers blooming on the roof.

Authors' Reflections

- This pioneering ecoregional green roof is far removed from large population centers. However, its endurance over many decades demonstrates that with an intensive substrate and wide plant spacing, even in some of the most arid regions of the Desert Southwest, green roofs are viable, with the right design, maintenance, and aesthetic expectations.
- This green roof achieves its goal of protecting the occupied space below from the heat of the desert sun. Energy conservation is perhaps the most important ecosystem service from green roofs in the Desert Southwest.
- This green roof ecosystem has protected the waterproofing system since 1979.

5.4 Plants for Desert Southwest Ecoregions

Across the eight ecoregional green roof case studies in this chapter, 106 taxa were used in total, 100 of which are native to the ecoregions in this chapter. Of those native to the ecoregions covered in this chapter, 15 species occur more than once across the case studies in this chapter. Of those occurring more than once, five are annuals, two are grasses, five are herbaceous perennials, and three are succulents. Only one species was common to three case studies (Table 5.5).

Given that plant trials and research is relatively new to the Desert Southwest, there are strategies for identifying indigenous vegetation to be used on green roofs in the Desert Southwest. Some of these include identification of short stature plants, plants with a vertical orientation to minimize solar and heat stress, annuals, perennials, cacti and other succulents that are less than one-meter-tall, are shallow-rooted or able to adapt to shallow low-nutrient substrates (See also Chap. 2, Sect. 2.1.2).

Table 5.5 Species occurring in three or more of the case study sites in the chapter

Plant Type	Common Name	Botanical Name	A	B	C
Grasses	Blue grama	*Bouteloua gracilis*	x	x	x

Key = A (New Mexico Court of Appeals), B (Riverfront Residence (AZ)), C (Easton Collection Center, Museum of Northern AZ)

Other strategies for selecting plants or ecosystems on green roofs in the Desert Southwest might include combining several tactics such as mixing cool or warm-season annuals with hardy perennials and or succulents. Since grazing and fire are not typically present on green roofs, a *de novo* or multi-variant ecosystem approach could be assembled and maintained to maximize ecosystem services. Cool-season annuals overlapped with hardy succulents could make a diverse and visually interesting arrangement. Ecologists, biologists, landscape architects, and other ecosystem experts should be consulted about which species to investigate and maintain on a green roof in the Desert Southwest. In particular, many native species occur at several different elevations in the southwest, thus the same plant species have small genetic variations and are acclimated to their microclimates. These variations in plant species' genetic and adaptation to specific microclimates may likely impact performance. Regardless, there are many montane meadow and valley desert plants that have adapted to green roofs (Table 5.6).

It is important to know how high-quality desert or meadow ecosystems function, and which flora and fauna play a dominant role in their success. Green roofs in the Desert Southwest may vary greatly depending upon plant composition and seasonal influences. One should understand how cool and hot season dormancy affects the aesthetic and functional characteristics of a green roof.

Green roof designers in the Desert Southwest might participate or consult habitat restoration organizations to learn about the potential ecosystems as a way to strategize

Table 5.6 Only a few of the many hundreds of low-profile vegetation that grow in the Desert Southwest ecoregions have been trialed on green roofs. The following list includes taxa demonstrated to be viable across the case studies on at least one green roof in this chapter

Montane Meadow	Valley Desert
Achillea millefolium	*Agave sp.*
Agave sp.	*Blue palo verde*
Allium cernuum	*Chollas sp.*
Amaranthus powellii	*Cylindropuntia echinocarpa*
Bahia dissecta	*Cylindropuntia ganderii*
Berlandiera lyrata	*Cylindropuntia X fosbergii*
Bidens tenuisecta	*Dasylirion sp.*
Bouteloua curtipendula	*Echinocereus engelmannii*
Bouteloua gracilis	*Encelia farinose*
Coreopsis sp.	*Hesperaloe parviflora*
Erioneuron pulchellum	*Larrea tridentata*
Helianthus annuus	*Opuntia sp.*
Hesperaloe sp.	*Opunta basilaris*
Kallstroemia grandiflora	*Parkinsonia florida*
Epilobium brachycarpum	*Pedilantus microcarpus*
Lupinus arizonicus	*Thymophylla pentachaeta*
Muhlenbergia rigens	*Yucca glauca*
Oenothera sp.	
Opunta sp.	
Opuntia phaeacantha	
Penstemon clutei	

and design ecoregional green roofs in the Desert Southwest. The conservation organizations and universities can help lead efforts to teach about the native plant communities and plants that may adapt to green roofs. Much could be learned by modeling the organized collaborations taking place in other locations such as the Green Roof Info Think Tank (GRiT) in Portland, Oregon (GRiT 2020), where representatives of green roof industry, universities, and municipal representatives come together to discuss green roofs, including the adoption of native plants (See also Sect. 9.3.1).

5.5 Summary

The natural vegetation of the ecoregion is outstanding in its richness, and conservation sites exist intact outside urban centers. However, within the boundaries of the major urban centers (Phoenix, AZ), there is little conservation taking place, aside from Tucson, Arizona (Crewe 2013). The quality of the conservation sites near urban centers provides great potential to learn from (Rosenzweig 2016) as the natural vegetation consists of forms that work on green roofs elsewhere: succulents, grasses, and forbs and annuals. The richness of the botanical reserves and National Parks near urban centers is fortuitous for those looking to make green roofs from the native plant communities (Rosenzweig 2016).

As the Desert Southwest continues to grow in population, the design and construction of its urban development potentially have very much to gain from the use of green roofs (Sailor et al. 2012). The monsoon rains and long summers with high daytime temperatures currently have negative impacts on the urban ecology in developed areas (Emmanuel and Fernando 2007; Middel et al. 2015). Since green roofs retain stormwater, reduce the urban heat island effects from rooftops, evapotranspire moisture back into the atmosphere, and can provide habitat for native flora and fauna, the application of green roofs should become a high priority for future development. As demonstrated in some of the case studies in this chapter, the potential benefits of ecoregional green roofs are great; however, the development of the technology in the region is just emerging.

The ecoregional green roofs in this chapter present evidence of the early success and viability of green roofs, with the appropriate application, monitoring, research, and dissemination. The case studies demonstrate that green roofs in the Desert Southwest were built to:

- demonstrate vegetation for green roofs, e.g., Biology Building, University of Texas, El Paso, Arizona State University research roof. Museum of Northern Arizona, Easton Collection Center;
- benefit ecosystem services of green roofs, e.g., Tempe Transportation Center Research Roofs, New Mexico Court of Appeals, Biology Building, University of Texas, El Paso, Arizona State University, Riverfront Residence, Tucson, Anza-Borrego Visitor Center;

- educate the public regarding ecosystem services of green roofs e.g., The Nature Center at the Estrella Mountains Regional Park, Museum of Northern Arizona, Easton Collection Center, Anza-Borrego Visitor Center;
- provide as a learning tool for green roofs e.g., Tempe Transportation Center Research Roofs;
- increase the aesthetic appearance of buildings and environments, e.g. New Mexico Court of Appeals, University of New Mexico Campus, Riverfront Residence, Tucson, Optima Camelview Village, Anza-Borrego Visitor Center;
- connect to the surrounding environment or landscapes e.g., The Nature Center at the Estrella Mountains Regional Park, Anza-Borrego Visitor Center, California;
- longevity of green roofs, e.g., Anza-Borrego Visitor Center, California since 1979, making it one of the oldest ecoregional green roofs in North America.

Of the many challenges, perhaps securing sustainable sources of water for irrigating green roofs and learning about reliable green roof plant communities are a high priority. With the long-term reliability of greywater on green roofs (see Chaps. 7, 8, and 10), green roofs in the Desert Southwest have much potential to recycle water from inside buildings as a way to secure water for green roofs. Constructed wetland green roofs could be a potential direction for buildings with integrated designs, as there is also precedent in the region (Nelson et al. 1999).

Other aspects of green roofs in the region include an emerging market where it may not yet be economically feasible to construct green roofs (Cutter 2019); however, there is a steady increase in the number of green roofs in the Desert Southwest, and research is taking place. A key component to growing the opportunity for green roofs in the Desert Southwest is an exploration of plant communities. Regardless, with all of the current interest and study of issues causing and associated with climate change, ecoregional green roofs could have the most potential to improve the quality of urban development than in any other ecoregion covered in this book.

Acknowledgments We would like to thank the following for sharing their time, knowledge and experiences which contributed to the many details in this chapter: Bonnie Richardson, City of Tempe; George Radnovich and Deborah Blea Hradek of SITES Southwest; Kris Floor and Chris Brown of Floor Associates, researchers at Arizona State University; Stacie Beute at the Desert Botanic Garden; Dr. Vanessa Lougheed and Dr. Kevin Floyd at the University of Texas-El Paso; Robert Thiele, AIA; Jan Busco, Northern Arizona Museum; Bil Taylor, AIA; Steve Kinback, The Nature Conservancy; Sally Theriault, Anza-Borrego Desert State Park; Don Harris Parks Supervisor, Maricopa County; and Don Harris at the Estrella Mountain Regional Park's Nature Center. We would also like to thank the owners of a private residence in Tuscon who allowed us access to their green roof and were willing to share knowledge about their green roof.

References

ABNC (1994) Anza-Borrego Visitor Center Fact Sheet. Pamphlet. California Department of Parks and Recreation, California

Bailey RG (1997) Ecoregions of North America. U.S. Department of Agriculture, Forest Service, Washington, DC

Baxter V (1982) CHACO/Pueblo Bonito: A Computer Analysis Applied to an Ancient Solar Dwelling. Landsc J 1(2):85–91. https://doi.org/10.3368/lj.1.2.85

Buckley S (2011) Common plants of Saguaro National Park. National Park Service, Tucson

CLIMAS (2018) Southwestern Monsoon. University of Arizona. https://www.climas.arizona.edu/sw-climate/monsoon. Accessed 8/15/19 2019

Crewe K (2013) Arizona native plants and the urban challenge. Landsc J 32(2):215–229

Crown PL (1990) The Hohokam of the American Southwest. J World Prehist 4(2):223–255

Cutter S (2019) Adapting a green roof in Tucson, Arizona. The University of Arizona, Tempe

Davis OK, Shafer DS (1992) A Holocene climate record for the Sonoran Desert from pollen analysis on Montezuma well, Arizona, USA. Palaeogeogr Palaeoclimatol Palaeoecol 92(1–2):107–199

DBG (2020) DBG research and conservation. Desert Botanical Garden. https://dbg.org/research-conservation/research-conservation-staff/. Accessed 6 Jan 2020

DeWald LE, Springer AE (2006) Incorporating ecological and nonecological concerns in the restoration of a rare, high-elevation Bebb willow riparian community. Lumber Processing in Selected Sawmills in Durango and Oaxaca, Mexico 167, pp 134

Dimmitt MA (2000) Biomes and communities of the Sonoran Desert region. In: A natural history of the Sonoran Desert. Arizona-Sonoran Desert Museum Press, Tucson, pp 3–18

Dimmitt MA, Comus PW, Phillips SJ, Brewer LM (2015) A natural history of the Sonoran Desert. Univ of California Press, Oakland

Emmanuel R, Fernando HJ (2007) Urban heat islands in humid and arid climates: role of urban form and thermal properties in Colombo, Sri Lanka and Phoenix, USA. Clim Res 34(3):241–251

Favorite J (2002) Plant guide BEBB WILLOW bebbiana Sarg, Salix. USDA

Flood Warning Branch FCDoMC (2018) WY 2018 annual precipitation Maricopa County, Arizona. Maricopa County Flood Control District, Phoenix

Gregoire BG, Clausen JC (2011) Effect of a modular extensive green roof on stormwater runoff and water quality. Ecol Eng 37(6):963–969

GRiT (2020) Green Roof Info Think Tank. GRiT. https://www.greenroofthinktank.org/what-we-do. Accessed 24 Feb 2020

Hammond E (1964) Classes of land-surface form in the United States. US Geological Survey.

Hart Prairie and the Fern Mountain Area Plant Checklist (2010) The nature conservancy. https://www.nature.org/content/dam/tnc/nature/en/documents/HPP-Plant-List.pdf

Lerum V, Thakare H (2005) Green roof in the desert: Comparing four alternative roof systems to a standard roof. In 22nd International Conference on Passive and Low Energy Architecture, PLEA 2005 (pp. 323–327)

Middel A, Chhetri N, Quay R (2015) Urban forestry and cool roofs: assessment of heat mitigation strategies in Phoenix residential neighborhoods. Urban For Urban Green 14(1):178–186

Milburn LAS, Fernández-González A, Jones T, Solano F, Martínez-Wong E (2011) Wasted space: altering building temperatures by greening barren rooftops in the Desert Southwest. In: CELA T (ed) Conference: landscape legacy: landscape architecture and planning between art and science, Maastricht, Netherlands, May 12–14, 2014. CELA, p 1

Nelson M, Finn M, Wilson C, Zabel B, van Thillo M, Hawes P, Fernandez R (1999) Bioregenerative recycling of wastewater in biosphere 2 using a constructed wetland: 2-year results. Ecol Eng 13(1–4):189–197

OCVV (2020) Optima camel view village design. Optima Web. Accessed 1-06-20 2020

Optima (n.d.) Design. Retrieved n.d. from https://optimacamelviewvillage.com/design/

Rosenzweig ML (2016) Green roofs: new ecosystems to defend species diversity. Isr J Ecol Evol 62(1–2):7–14

Sailor DJ, Elley TB, Gibson M (2012) Exploring the building energy impacts of green roof design decisions–a modeling study of buildings in four distinct climates. J Build Phys 35(4):372–391

Schuler TA (2018) Made to Disappear. Landscape Architecture Magazine vol August. American Society of Landscape Architects, Washington, D.C.

Schuler TA (2018, May) Vertical Oasis: In Scottsdale, Arizona Floor Associates Coaxes Green Waves in an Arid Climate. Landscape Architecture Magazine, pp. 84–101

Scott Nabers M (2009) UTEP installs Southwest's first commercial 'green roof'. Texas Government Insider, vol 7. Strategic Partnerships, Austin

Shreve F, Wiggins IL (1964) Vegetation and flora of the Sonoran Desert, vol 591. Stanford University Press, Stanford

Stanley LW, Stanley L (2010) UTEP's green roof thrives in desert; modular system monitored for data. Texas Architect, vol Jan/Feb. Texas Architect, Austin

Van Devender TR (1987) Holocene vegetation and climate in the Puerto Blanco Mountains, Southwestern Arizona. Quat Res 27(1):51–72

Waser NM (1978) Competition for hummingbird pollination and sequential flowering in two Colorado wildflowers. Ecology 59(5):934–944

Western Regional Climate Center (n.d.) Arizona Climate Summaries. Retrieved n.d. from http://www.wrcc.dri.edu/cgi-bin/cliMAIN.pl?azphoc

Wienhold ML (2013) Prehistoric land use and hydrology: a multi-scalar spatial analysis in Central Arizona. J Archaeol Sci 40(2):850–859

Chapter 6
Green Roofs in Intermontane Semi-Arid Grassland Ecoregions

Bruce Dvorak and Tom Woodfin

Abstract This chapter covers ecoregional case studies located within the Intermontane Plateaus. Ecoregions include semi-arid grasslands located in the valleys of the Rocky Mountains, the Columbia Plateau, the Salt Lake Valley, and Basin and Range geographic regions. This is a large geographic area (multi-state) and is ecologically complex as topographic elevation, slope, and aspect, and climatic extremes influence the development of a variety of ecoregions. Precipitation is generally limited to less than 280–500 mm annually in the lower elevations of valleys, and much of the precipitation takes place outside of the growing season. Thus, in designated natural areas, grasses and wildflowers compete with shrubs for dominance. Less than 1% of the native grass and meadow habitats remain intact. Three conservation site case studies explore a range of native plants for the biologically rich ecoregions, and ten green roof case studies demonstrate how 113 taxa of native plants were trialed on ecoregional green roofs.

Keywords High altitude · Montane · Growing season · Grass · Scrub · Drought · Winter dormancy · Desert

6.1 Ecoregion Characteristics

The Intermontane West is a vast region bounded on the east by the Rocky Mountains, the Pacific mountain ranges to the west, and the desert southwest canyons to the south. Although the region is sparsely populated, a number of counties located in

B. Dvorak (✉)
Department of Landscape Architecture and Urban Planning, 305A Langford Architecture Center, Texas A&M University, College Station, TX, USA
e-mail: bdvorak@tamu.edu

T. Woodfin
Department of Landscape Architecture and Urban Planning, Texas A&M University, College Station, TX, USA
e-mail: woodfin@tamu.edu

B. Dvorak (ed.), *Ecoregional Green Roofs*, Cities and Nature,
https://doi.org/10.1007/978-3-030-58395-8_6

Utah, Montana, Idaho, eastern Oregon, and eastern Washington state are some of the fastest-growing counties in the United States (Kiersz 2018).

The vegetation that naturally grows on the valley floors, basins and plateaus include semi-arid grasslands and desert scrub vegetation. The prevailing climate is dominated by scant rainfall during the growing season, the potential for high day-time temperatures, and heavy snowfall and cold temperatures are common during the winter. This interstitial zone is a landscape of extremes, and its vegetation is resilient and tolerant of ground fires. The Intermontane West from the Great Basin region to the northern ranges of the Columbia Plateau and northern Rockies once had vast and diverse grasslands and meadows on the plateaus, basins, and valleys of high mountains, and coniferous forests in the higher elevations. In Idaho for example, some prairies and forests sustain over 3000 vascular plants, 2200 non-vascular plants, 1000 lichens, and 1200 mosses in eight ecoregions (BLM 2019). In contrast, there are some ecosystems of the Basin and Range physiological region where a single shrub species may dominate a large area, or there may be little or no vegetation as sand or exposed rock lies exposed to full sun and receive only scant or trace levels of precipitation (Andersen 1996).

Fire, from natural causes such as lightning strikes, and humans have influenced the intermountain west vegetation for at least the past 11,000 years (Vale 2002). The grassland habitats burned with variation in fire intensity and frequency and helped replenish and sustain the ecosystems (Griffin 2002). Today about 1% of the historic prairies remain in this region and have been replaced with agricultural rangeland, horse ranches, crop agriculture, and low-density development dominate (Noss and Peters 1995). The historic prairies of the region are a critically endangered ecosystem and are conserved primarily in national and state parks and a scattering of private preserves (Ricketts 1999).

Before the nineteenth century, the Native American tribes maintained vast areas of grasslands especially along rivers and other water sources as a contiguous grassland for hunting wild game and living off selected vegetation of the land. They managed selected areas to produce foods that they ate from over 200 species of plants including grains harvested from the native grasses. They also dug up the bulbs of lilies, wild celery roots, and many other plants (Griffin 2002).

Following the Lewis and Clark expedition in the early 1800s, the magnificent landscapes of this region became the subject of a number of plein-air artists that lived in the eastern United States, or abroad. They traveled to the West and helped to make Americans living on the east coast curious about these magnificent landscapes. Artists made famous, through their paintings the largeness of the West, and its vegetation was often an important part of the backdrop of the more common main subjects such as buffalo, Indians, cowboys, or majestic mountain landscapes (Fig. 6.1). By the mid to late 1800s, many of these artists were aware that these landscapes of the Intermountain West were already disappearing and they sought to capture their essence before it was changed forever. For example, "… he (Charles Russell) was intent on capturing this place (Western Montana) and subjects that were disappearing forever. He was here in the 1880s; it was before the railroad, and he could tell everything he was experiencing was about to go away forever. So in his lifetime, his desire was to capture this place before it was gone (Nemerov 1994)."

Fig. 6.1 *The Three Tetons* (Thomas Moran 1881). Near the headwaters of the Snake River where semi-arid grasslands, wildflowers, and sagebrush dominate the valley floor of the Grand Tetons National Park. The valley here shows a contiguous uninterrupted layer of grassland vegetation across the floor and up onto the hills, before the development of the land as a National Park. (Courtesy of Gilcrease Museum, Tulsa, OK)

Other artists, such as Thomas Moran (Fig. 6.1), used their work to inspire and influence the preservation of some of the best of these landscapes, in their natural condition. Moran's work was largely influential in the development of Yellowstone National Park, and with others helped influence the development of the National Park System (Denzin 2008). Along with the majestic features of the parks of the Intermontane West, they preserved the meadows, and grasslands of the parks and prevented them from becoming overrun with exotic cattle and rangeland.

The green roofs in this chapter are all in locations that receive less than 25 mm (1 in) of rainfall during the primary growing season. In the shadows of the Sierra Nevada Mountains, precipitation is low especially in the valleys compared to higher elevations. Salt Lake City, the largest metropolitan area in the region, receives 419 mm (16.5 in) of annual precipitation, whereas the lower bench elevations of the mountains receive 508 mm (20 in) annually. Distribution of precipitation is fairly even across the months except for July and August, which are the driest and warmest (Fig. 6.2). Boise, Idaho is the second-largest urban area in the region and receives 294 mm (11.58 in) annually with even rainfall distribution except for July and August when temperatures are highest. Precipitation often occurs as intermittent and spontaneous thunderstorms, which can be intense but brief rainfall events. However, such events may deliver rain to one part of a valley or metro area, while other parts remain dry. It is this intermittent rain that grows vegetation in a land of contrasts.

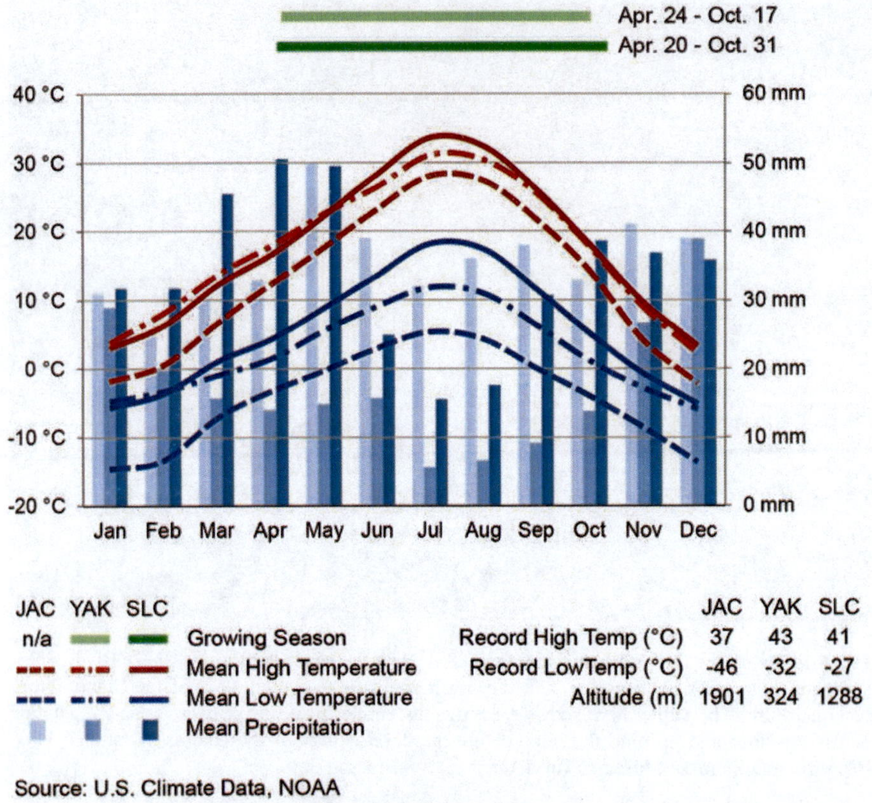

Fig. 6.2 Precipitation and air temperature monthly values for Jackson, Wyoming (JAC), Yakima, Washington (YAK), and Salt Lake City, Utah (SLC). Each of the cities can experience extreme heat and cold, which means vegetation is less diverse than ecosystems in more mild climates. (Graphic: Tess Menotti & Bruce Dvorak)

6.1.1 Vegetation of the Intermontane West

The most common cross-section of the semi-arid grassland vegetation includes perennial grasses intermixed with wildflowers, sagebrush (*Artemisia arbuscula* Nutt.), saltbush (*Atriplex*) greasewood (*Sarcobatus)* and rabbitbrush (*Ericameria nauseosa*) which formed a once common association that was widespread across the Intermontane West (Fig. 6.3). Compared to the Desert Southwest, there are only occasional yuccas and very few cacti grow in this region. Bulbs grow in the region including camas (*Camassia quamash*), which was an important food source for the native peoples. Considering the spectrum of ecoregions from grassland to forest, some parts of the Intermontane West can have high plant diversity. For example, Idaho has 3000 vascular plants, 2200 non-vascular plants, 1000 lichens, and 1200 mosses in eight ecoregions (BLM 2019).

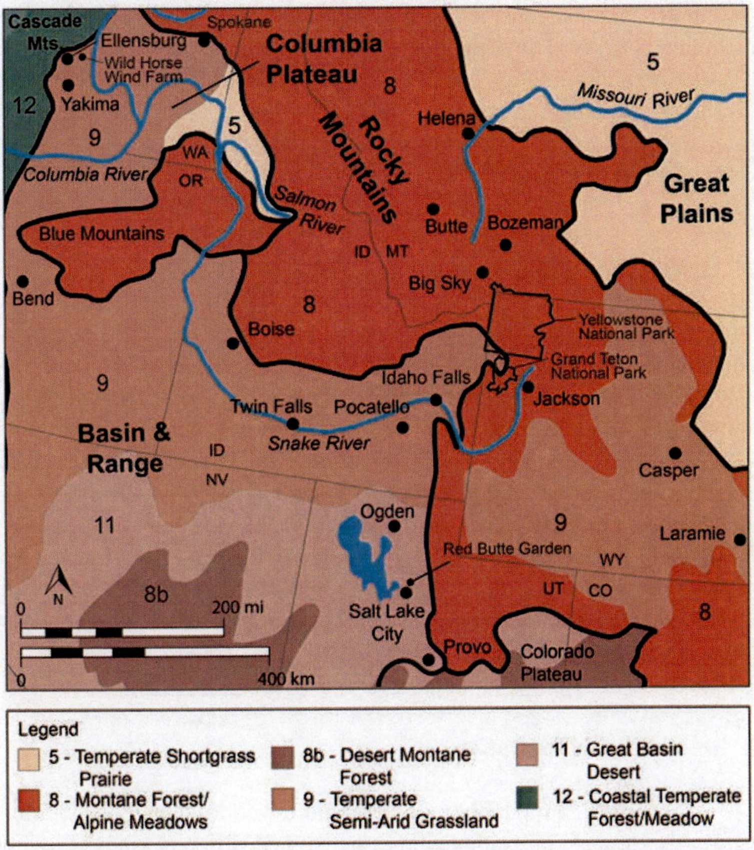

Fig. 6.3 The Intermontane Semi-Arid Grassland ecoregions (Bailey 1997). Because of the nature of the basin and range topography and plateaus, (Hammond 1964) there are many ecoregions and ecotones in the Intermontane West. Of interest to ecoregional green roofs are the Temperate Semi-Arid Grasslands (9) and the Great Basin Desert (11) ecosystems. The natural vegetation of these regions is often defined by extreme temperatures. (Graphic: Trevor Maciejewski & Bruce Dvorak)

6.1.1.1 Grasses

Grassland plant communities vary widely across the region as plant diversity and species composition are strongly influenced by topographic elevation, soils, slope, and proximity to other ecoregions. Familiar members such as grasses of the Palouse Prairie were common across the prairies of the intermontane region and included: Western wheatgrass (*Agropyron smithii*), Idaho bentgrass (*Agrostis idahoensis*), Arizona wheatgrass (*Agropyron spicatum*), squirreltail (*Elymus elymoides*), Idaho fescue (*Festuca idahoensis*), needle and thread (*Hesperostipa comate*), prairie Junegrass (*Koeleria cristata*), Sandberg bluegrass (*Poa scabrella*), and bluebunch wheatgrass (*Pseudoroegneria spicata*) (Archibold 1995; BLM 2019).

In addition to national and state parks, The Nature Conservancy owns many private landholdings to preserve prairies and scrublands across the region. The Palouse Prairie has several conservation sites, and restoration efforts work towards sustaining this critically endangered ecosystem (Stoddart 1941). Exotic grass invasion has been studied extensively as efforts to preserve grazing lands attempt to maintain land cover under grazing pressure (Davies et al. 2020).

6.1.1.2 Herbaceous Wildflowers

Some of the common herbaceous wildflowers native to prairies in the intermontane region include fragrant white sand-verbena (*Abronia fragrans*), white sand verbena (*Abronia mellifera*), Boise sand-verbena (*Abronia mellifera* var. *pahoveorum*), common yarrow (*Achillea millefolium*), many species of *Allium*, common camassia (*Camassia quamash*), Oregon sunshine (*Eriophyllum lanatum*), many species of beardtongue (*Penstemons*), upright prairie coneflower (*Ratibida columnifera*), blackeyed Susan (*Rudbeckia hirta*), and many more.

Several plants are named after Lewis and Clark in the region and include beautiful clarkia (*Clarkia pulchella*), Lewis' monkeyflower (*Erythranthe lewisii),* Sacajawea's bitterroot (*Lewisia sacajaweana*), blue flax (*Linum lewisii*), and Lewis' mock-orange (*Philadelphus lewisii*) (BLM 2019).

6.1.1.3 Desert Scrub Vegetation

Desert scrub consists of a mixture of grasses, wildflowers, and many species of low-growing woody shrubs. For example, there are over 36 different kinds of sagebrush (*Artemisia*) that grow in the region. Sagebrush steppe restoration for improved habitat competes with both introduced species and exotic grass invasion (Miller et al. 2014). Other scrub vegetation includes bitterbrush (*Purshia tridentata*) rabbitbrush (*Ericameria nauseosa*), antelope bitterbrush (*Purshia tridentate*), and in transitional areas squawbush (*Rhus trilobata*), squaw currant (*Ribes cereum*), chokecherry (*Prunus virginiana*), mountain snowberry (*Symphoricarpos oreophilus*).

6.1.1.4 Succulent Vegetation

Succulents are prevalent across the region in the deserts, talus slopes, and rocky outcrops. Several species of sedum are native to the region (Sect. 6.3) as well as several forms of cacti including barrel cactus (*Ferocactus* and *Echinocactus*), nipple cactus (*Coryphantha*), several forms of prickly-pear (*Opuntia*), and Simpson's hedgehog cactus (*Pediocactus*) (Andersen 1996; IDFW 2020).

6.1.2 Conservation Case Studies

6.1.2.1 The Greater Yellowstone Region (Rocky Mountains)

Together, Yellowstone and Grand Teton National Parks form the Greater Yellowstone Region which encompasses some of the largest contiguous semi-arid grasslands in the Intermontane West (Fig. 6.4). The parks are contiguous and wildlife are free to traverse across boundaries, and livestock grazing is not permitted. Although the altitude of the entire Yellowstone Greater region is hundreds of meters (several thousand ft) above the elevations of the largest populated cities of Salt Lake City and Boise, Idaho, the contiguous grasslands and scrub vegetation allow one to study and understand what the grasslands of the basin and range and plateau regions were like before the settlement of the west. The semi-arid grasslands of the mountainous west are perhaps best represented by large land areas where grazing by indigenous mammals and the effects of ground fires are still in operation.

Fig. 6.4 Image of a hillside meadow located inside of a fenced private ranch near Moran, Wyoming (outside of the Grand Tetons National Park). Fencing and hunting allow for the control of the population of native grazers which allows for a broader abundance and distribution of the native flora and fauna. Alliums (wild onion), yarrow, artemisia (sagebrush), and many associative plants grow at this site as this diversity is not as abundant in the open lower valley near Jackson, Wyoming, where the effect of grazing can reduce plant diversity. The high mountain meadows that extend from the Grand Teton National Park to Yellowstone National Park remain as some of the highest quality semi-arid grasslands in North America. (Photo: Bruce Dvorak, July 2018)

However, overgrazing by indigenous wildlife is evident where concentrations of elk and bison reduce the native flora. Multiple online and printed resources document the native vegetation and associative ecosystems and their functions. Hundreds of species of grasses, forbs, and annuals could be worthy of investigation for their application to green roofs in this region. Some of the first vegetated roofs in the western United States were located within the borders of these parks. See the case study Sect. 6.3.1 which covers sod roofs at Grand Teton National Park.

Grand Tetons National Park has over 1000 species of vascular plants. There are over 100 different species of grasses that grow in the region including alpine timothy (*Phleum alpinum*), bearded wheatgrass (*Elymos trachycaulus*), Idaho fescue (*Festuca idahoensis*), pinegrass (*Calamagrostis rubescens*), Sanberg bluegrass (*Poa secunda*), spike trisetum (*Trisetum spicatum*), ticklegrass (*Agrostis scabra*), timber oatgrass (*Danthonia intermedia*), tufted hairgrass (*Deschampsia cespitosabracken*).

Exotic grasses have naturalized in the park including some very invasive grasses such as: cheatgrass (*Bromus tectorum*), common timothy (*Phleum pretense*), and orchard grass (*Dactylis glomerata*). Ferns grow in open and shady sites. In open sites grows bracken fern, rockbreak or parsley fern found in more rocky soils.

With a short growing season, wildflowers grow in the valley floor, in the forest understory and alpine meadows. Common to the valley vegetation are species of *Achillea*, *Allium*, *Delphinium*, *Balsamorhiza*, *Castilleja*, and *Lupine* grow as these will tolerate the warmer temperatures during the summer (GTNP 2017).

6.1.2.2 Red Butte Garden and Canyon, Salt Lake City, Utah (Basin and Range)

The Red Butte Garden is a botanical research garden located at the upper elevations of the University of Utah campus and is located at the outlet of the Red Butte Canyon Nature Preserve (Fig. 6.5). The garden features over 250 species of penstemon which is the largest collection of penstemon in the United States. Over one hundred of them are native to the western U.S. Many of the other collections in the garden include species native to the semi-arid grasslands including blue gramma, artemisia, and many other grasses and forbs. The mission of the garden includes the protection of native biodiversity and habitat restoration.

The Red Butte Canyon Nature Preserve is located at elevation 1530 m (5020 ft) and is uphill of the garden and has one of the last remaining native lower valley grassland habitats preserved in the Salt Lake City region. The grassland species typically grow below 1829 m (6000 ft) and are common members of the mountain prairies, that were once widespread across eastern Washington State, eastern Oregon, Idaho, parts of Montana, and southern and northern Utah (Stoddart 1941). Hiking trails abound and cover many habitat types. The lower elevations of the canyon vegetation have been under the influence of aggressive naturalized (introduced) grasses such as redtop bentgrass (*Agrostis stolonifera*), cheat grass (*Bromus tectorum*) and Kentucky bluegrass (*Poa pratensis*). And although may appear to be

Fig. 6.5 A wide range of low-growing native and some exotic plants that thrive in semi-arid climates are on display at the Red Buttes Gardens. The horizontal seeds of the native bunchgrass side-oats grama can be seen in the foreground here and also adapts well to green roofs. (Photo: Bruce Dvorak, July 2018)

natural to the untrained observer, are not. Regardless, typical residents of the prairie include bluebunch wheatgrass (*Elymus spicatus*) and is found within in Red Butte Canyon. The Salt Lake basin was once a large open grassland inhabited by grasses, forbs, and big sagebrush (*Artemisia tridentata*), squawbush (*Rhus trilobata*), and bitterbrush (*Purshia tridentata*). Early Mormon settlers to the region captured images of the valley prior and during settlement and include these plants in their depictions of the valley landscapes (Kleiner and Harper 1966; RBC 2019).

6.1.2.3 Wild Horse Wind Farm, Ellensburg, Washington (Columbia Plateau)

Located in the lower elevations of the Columbia Plateau, the Wild Horse Wind Farm is located outside of Ellensburg, Washington. The previous use of the site sustained native semi-arid grassland and scrub vegetation (Fig. 6.6). The development of the wind farm preserved the natural vegetation to remain below the wind turbines. Before its development as a wind farm, environmental reports identified that there were few trees located on the property, and the vegetation was dominated by native

Fig. 6.6 Grasses and scrub vegetation growing at the Wild Horse Wind Farm outside of Ellensburg, Washington. The rocky, sloped, shallow, low-nutrient, and well-drained soils found across the wind farm are populated with drought and heat tolerant vegetation natural to the Intermountain West. Rabbitbrush is seen in yellow blooms growing between grasses and rocky openings and on a green roof in Sect. 6.3.10 of this chapter. (Photo: Bruce Dvorak, August 2018)

and naturalized grasses such as bluebunch wheatgrass (*Pseudoroegneria spicatum*), Sanberg's bluegrass (*Poa secunda*), needleand-thread grass (*Hesperostipa comata*), and Idaho fescue (*Festuca idahoensis*) which is native to the region, and exotic cheatgrass (*Bromus tectorum*) a naturalized non-native grass originally from Europe and Asia. Shrubs were also present such as sagebrush (*Artemisia rigida*), rabbitbrush (*Ericameria nauseosa*), antelope bitterbrush (*Purshia tridentate*), squaw currant (*Ribes cereum*), chokecherry (*Prunus virginiana*), mountain snowberry (*Symphoricarpos oreophilus*,) and some succulents such as hedgehog cactus (*Echinocereus engelmannii*).

Fauna surveys were conducted and they found that 59 avian species were identified including bald eagles and many songbird species. Regardless of the projected or real impacts of the wind farm operations, the semi-arid grassland scrub habitat remains intact and is a good example of this type of ecosystem and is accessible with pre-arranged visits and tours available through the Wild Horse Wind Farm visitor's center located outside of Ellensburg, Washington.

The plants and their associations found at the wind farm have a high degree of potential for use on green roofs in the region. There are several green roofs in this chapter that make use of the steppe-scrub vegetation in the case studies section.

6.2 Green Roof Research in the Intermontane Semi-Arid Grassland Ecoregions

The most extensive documentation of a prairie green roof was taken on the Temple Square green roof (Dewey et al. 2004). This study examined the competitiveness of 16 grass species and 19 wildflower species over three growing seasons. The study results and implications for green roofs in the Intermountain West are further discussed in the chapter summary.

Evaluation of the economic feasibility of a green roof installation in this region was conducted for the proposed new pharmacy college at the University of Utah (Wu and Smith 2011). Modeling the advantages of a green roof over the conventional black roof and a white reflective roof, the authors concluded that while initial capital expenditures for a green roof were greater than conventional roofing, the life-cycle economic advantages recommended that the new building include a green roof, as conventional roofing proved more expensive over the life-cycle of the building. Comparative values for different roof types relied upon previous research from other regions.

Proposed green roofs as part of a stormwater evapotranspiration study in a drainage sub-basin of the University of Utah campus were part of a larger green infrastructure modeling application (Feng et al. 2016). This study sought to establish whether pre-development runoff conditions could be restored through utilizing GI stormwater management techniques available if selected buildings had green roofs and bioretention basins were implemented. Green infrastructure was estimated to restore over 82% of the predevelopment water budget for the site. This means that conventional roofing in the area contributes to excessive runoff and flash flooding, where green roofs help reduce runoff and returns rainfall moisture into the atmosphere through evapotranspiration.

A careful study establishing evapotranspiration (ET) rates for a dry-climate intensive green roof was conducted at the Marriott Library at the University of Utah (Feng et al. 2018). The researchers compared an unvegetated control test plot with a plot covered in *Sedum* species and another planted in herbaceous plant materials including shrubs for the measurable effects of climatic conditions and irrigation on evapotranspiration rates. The findings included recommendations for planting native plant species over sedum as well as shrubs and small trees to provide shade to native species during drought periods. This study is important as the only scientific investigation of irrigation requirements on a green roof in this rain-scarce region.

6.3 Ecoregional Green Roof Case Studies (Arranged East to West)

6.3.1 Bar BC Dude Ranch at Grand Teton National Park, Jackson, WY

Dude ranches brought Old-world sod roof technology into the high plains' states and Intermontane West in the late 1800s. They were some of the first ecoregional green roofs in the west. A "dude ranch" is a cattle ranch that has guest cabins for vacationers and provisions complete with riding horses and all the basic needs of living and entertainment. They were largely popular with wealthy and influential people, city dwellers, and anyone eager to visit remote wilderness areas. The Bar BC Dude Ranch hosted many famous people including John D. Rockefeller who eventually purchased much of the land that would become Grand Tetons National Park (Longfield 2011). The Bar BC Dude Ranch at the Grand Teton National Park was built in 1912 by Struthers Burt and Horace Carncross (Fig. 6.7). It consisted of

Fig. 6.7 Restored dude ranch cabins at Grand Tetons National Park. Cabin #1369 (foreground) received pre-grown and transplanted sod vegetation (with jute netting), and cabin #1370 (background) vegetation was seeded directly onto the soil after it was placed on the roof. Both green roofs established vegetation during the 3-year establishment period, and are shown here in July during a drought. (Photo: Bruce Dvorak, 2018)

40 buildings, of which visitors could sleep, eat, dance, receive and send mail, and prepare for wilderness outings on horseback. Most of the original buildings had sod roofs, as a way to keep the building interiors cool during the summer when most used. (Cantu 2012; Wonson 2013). The era of their significant use was from 1912 through 1941 and most heavily used during the 1920s (Matero 2016).

Not much is known of the original materials used to build the Bar BC cabin green roofs; however, it is known that the sod came from near the location where the buildings were constructed and were composed of native grasses and wildflowers. The buildings were constructed of lodgepole pine from the valley, and their architectural style echoed those of dude ranches in Tennessee, Texas, and Colorado. The original sod roof decks were made from wood planks, tar paper waterproofing, and the green roof was made from native soil and native sod (Cantu 2012). Over the years, the cabins and green roofs became neglected and were completely abandoned until 1986 when the NPS began discussing their restoration (Longfield 2011).

In 2011, the potential resurrection of green roofs on cabins #1369 and #1370 came from interest within the National Park Service and the University of Pennsylvania School of Design, Department of Historic Preservation. Based upon national park guidelines and standards for green roofs, a design was conceived and then was adapted for onsite conditions and feasibility. The Grand Teton National Park received grant support from the National Parks Foundation. Faculty and students from the University of Pennsylvania teamed with park staff to restore the roof deck and build green roofs on two cabins as a test for possible replication on other restored dude ranch buildings. The waterproofing was an EPDM roofing membrane, soil came from on-site locations, and vegetation was pre-grown to develop sod with jute mat for cabin 1369, and the identical seed mix was spread across the top layer of soil on cabin 1370. Both vegetation treatments were watered and healthy vegetation was established during 2012–2015 (Wonson 2013).

6.3.1.1 Project Team

Building Owner/Client: National Park Service
Green Roof Design Team Lead: Katherine Wonson, NPS
Installation Contractor: National Park Service and the University of Pennsylvania School of Design
Project completion: Original sod roof 1912, restoration complete in 2012
Green roof area: Two green roofs each 24 m^2 (255 ft^2)

6.3.1.2 Overview and Objectives

As a way to test the feasibility of restoring sod roofs on dude ranch cabins, two of the highest quality cabins received two different treatments to restore the green roofs. The project was completed in efforts to restore these historic cabins and potentially resurrect the Bar BC dude ranch as a destination.

6.3.1.3 Plant Establishment

Native vegetation included a mix of 25% prairie junegrass (*Koelaria macrantha*), 25% Idaho fescue (*Festuca idahoensis*), 20% bluebunch wheatgrass (*Pseudoroegneria spicata*), 20% sandberg wheatgrass (*Poa secunda*), and 10% slender wheatgrass (*Elymus trachycaulus*).

These species were chosen because they are native to the valley and are drought-tolerant. Cabin #1370 received the seeds directly on top of the soil placed on the cabin, where cabin #1369 received a pre-grown sod of the same seed and was transported to the cabin roof in sod strips (Wonson 2013).

6.3.1.4 Irrigation

Sod was watered three times a week during the growing season of the first 3 years. The roofs are no longer watered.

6.3.1.5 Maintenance

Watering and removal of tree seedlings took place during the 3-year establishment period. The sod roofs are left to grow without irrigation, and with an annual maintenance visit. Today the density of grasses on the roofs without irrigation is less than when the roofs received irrigation (Fig. 6.8).

6.3.1.6 Observed Wildlife

Not part of the study.

6.3.1.7 Best Performing Native Vegetation

Fescue grasses are performing best on cabin #1370, which received only seed placed onto the soil placed on the roof.

6.3.1.8 Post-occupancy Observations

Project Team Report Findings

- While the electronic environmental (moisture) monitors will certainly provide accurate and detailed data about the performance of the roofs, we would also suggest installing lower-cost alternatives to these monitors that could provide immediate moisture readings. Lower-cost soil moisture monitoring methods

Fig. 6.8 Fescues, clover, and a few early invader species grow on the roofs today. Without irrigation, the density of bunch grasses is less than with irrigation, but it is similar to some of the disturbed soil areas near the cabins. This vegetation was established through seeding directly onto the soil of cabin #1370. (Photo: Bruce Dvorak, July 2018)

include litmus strips placed in-between the boards on the roof that will indicate the presence of water and newspaper placed on the interior of the cabin that would bleed in the event of water infiltration (Noland 2012).

- While the roofs clearly benefitted from a close analysis of seed types fit for installation, we would also recommend a careful study of not only seed mixture but also of soil composition before the roofs are installed (Noland 2012).

Authors' Reflections (Bruce Dvorak)

- From my onsite meeting with Kathy Wonson, I learned that summertime temperatures have been increasing over the past few years (since hand-watering of roofs ended) in the valley along with drought, and less than normal snowfall. These factors combined with no on-site source of irrigation has left the sod roofs with much lower vegetative cover compared to the first 3 years.
- It is interesting to note that grasses persist on the roof in shallow substrates with no irrigation and little maintenance. These roofs demonstrate that if full coverage

of grasses is desired on a sloped green roof in the region, some supplemental watering may be necessary.

6.3.2 *Residences at Jackson, Wyoming*

The tradition of building with sod green roofs in Jackson, Wyoming began with Struthers Burt and Horace Carncross and their vision of nearly 30 sod roofed log cabins at the Bar BC Ranch in 1912. Sod roofs faded from Jackson after the dude ranch green roofs and cabins went into decline after the 1940s. Once a place of summer residence for at least seven Native American tribes, and later Euro-American fur traders, Jackson now has over 10,000 permanent residents and more than 30 green roofs on private and public buildings designed and built by Ward + Blake Architects and their collaborators. Many more green roofs exist in the region, but this case study focuses on some of those designed by Ward + Blake Architects (Figs. 6.9 and 6.10).

Fig. 6.9 Green roofs on the Warshawn Residence, Jackson, Wyoming. The meadow vegetation on the ground is mimicked on the roof, as many of the same species that naturally occur on the hillside were also seeded on the green roofs. The white blooms of yarrow (*Achillea*) for example, can be seen growing behind the rock (left) and were also seeded and grow on this and other green roofs located in Jackson. (Photo: Bruce Dvorak, July 2018)

Fig. 6.10 Green roofs with grasses and a few wildflowers located in the Jackson, Wyoming area: (**a**) private residence on the west side of the valley, (**b**) residence on the east side of the valley, (**c**) residence on the east side of the valley, and (**d**) a community center with a green roof. (Photos: Bruce Dvorak, July 2018)

Beginning in 1998, some of the first sod roofs within Jackson's city limits were proposed by Ward + Blake began in Jackson, Wyoming (Ward+Blake 2014). The first green roof was the hardest to pass through the city council, but after that, many more were designed and constructed over the years. This case study looks at a green roof construction method used by one of the most prolific makers of sod green roofs in Jackson, Wyoming Ward + Blake Architects.

6.3.2.1 Project Team

Building Owner/Client: Private residences
Green Roof Design Team Lead: Varies
Architect: Ward + Blake Architects
Landscape Architect: varies
Installation Contractor: varies
Project completion: 2003-current
Green roof area: Varies

6.3.2.2 Overview and Objectives

Each of the grass roofs is intended to reflect a meadow. The plant composition includes grasses that are supplemented with wildflowers native to the Jackson valley. Some wildflower species are selected to support native wildlife and butterflies that live and migrate through the valley. The sod roof houses are integrated into the landscapes and include many local materials in their construction. A number of the green roofs are on private residences and they all have similar construction. One project at the Amangani Community has sod roofs on residences, a restaurant, and guest houses. The sod roof method for these green roofs includes a FiberTite® roofing membrane, a drainage mat provided by the American Wick Drain Corp®. (Amerigreen RS 50), a substrate depth of 30 cm (12 in) of soil from the private property mixed with Perlite, and provisions for irrigation (Barth 2015; Ward 2018).

6.3.2.3 Plant Establishment

The sod roofs are established from a seed mix provided by the Wind River Seed Company located in Manderson, Wyoming. Forms of vegetation include grasses and herbaceous perennials (Table 6.1).

Table 6.1 Herbaceous perennials used on residential green roofs in Jackson, Wyoming

Common Name	Botanical Name
Western yarrow	*Achillea millefolium occidentalis*
Arrowleaf balsamroot	*Balsamorhiza sagittata*
Rocky Mountain beeplant	*Cleome serrulata*
Sulfurflower	*Eriogonum umbellatum*
Annual sunflower	*Helianthus annuus*
Showy goldeneye	*Heliomeris multiflora*
Blue flax	*Linum perenne lewisii*
Silky lupine	*Lupinus sericeus*
Dusky penstemon	*Penstemon commarrhenus*
Blue penstemon	*Penstemon cyaneus*
Firecracker penstemon	*Penstemon eatonii*
Rocky Mountain penstemon	*Penstemon strictus*
Wilcox penstemon	*Penstemon wilcoxii*
Gooseberryleaf globemallow	*Sphaeralcea grossulariifolia*
Mules ears	*Wyethia amplexicaulis*

Grasses

20% western wheatgrass (*Pascopyrum smithii*), Idaho fescue (*Festuca Idahoensis*), bluebunch wheatgrass (*Pseudoroegneria spicata*), 15% Indian ricegrass (*Achnatherum hymenoides*), 10% slender wheatgrass (*Elymus trachycaulus*), 5% thickspike wheatgrass (*Elymus lanceolatus*), 4% galleta (*Pleuraphis jamesii*), 2% bottlebrush squirreltail (*Elymus elymoides*), 2% prairie junegrass (*Koeleria macrantha*), Sandberg's bluegrass (*Poa secunda*).

6.3.2.4 Irrigation

All of the green roofs receive irrigation daily during the growing season, but watering times differ between homeowners.

6.3.2.5 Maintenance

The roofs receive annual maintenance visits in the spring to remove tree seedlings and any unwanted vegetation. As long as the grasses are watered, weeds are not able to become established. The owners of a private residence (Fig. 6.10) mowed the grass at the end of the first growing season. He now uses a string line trimmer to cut back grasses in the fall, to present overgrowth of taller grasses and to maintain a more manicured appearance.

6.3.2.6 Observed Wildlife

One private residence was built in the path of a butterfly migration route in the valley. The green roof in this case contributed to mitigation for the loss of habitat on the site. The meadow green roof was integral to the approval of the residence. Since most of the sod roofs are located on private residences, there has been no formal records of visits by wildlife.

6.3.2.7 Best Performing Native Vegetation

The specified grasses and wildflowers establish well on all of the roofs. The mix of species and populations varies by roof location, elevation, solar aspect, and watering practices. The irrigation keeps the vegetation green during the dry periods of the growing season. The first grassed roof in the Jackson area was installed in about 1998.

6.3.2.8 Post-occupancy Observations

Authors' Reflections

- The sod roofs have formed a dense stand of grasses and sustain near full coverage of grass. The roofs on private and public structures are all well-maintained. Some owners maintain the roofs themselves and some hire local maintenance companies. Irrigation is maintained at all of the sites.
- The magnitude of grassed roofs in the region and the durability of the technology demonstrates that montane grasslands and meadows are a viable form of green roofs in the ecoregion.

6.3.3 Residence at Big Sky, Montana

This private residence is located in a high mountain community where open grasslands are dominant. With well over 50% open space, the property supports abundant grasslands and meadows across the hillsides. The community allows roof gardens, roof meadows, and sod roofs in its ordinances, and more than a few of the private homes have sod roofs (Fig. 6.11). This installation included pre-grown native sod that was established on LiveRoof® modules. With irrigation and low maintenance requirements, this residence highlights the viability of a simple rooftop planting of native grasses.

6.3.3.1 Project Team

Building Owner/Client: Private Residence
Green Roof Design Team Lead: Charisa Wagner
Architect: Miller-Roodell Architects
Installation Contractor: Intermountain Roofscape Supply
Project completion: 2017
Green roof area: 93 m^2 (1000 ft^2) on two roofs.

6.3.3.2 Overview and Objectives

The owner wanted views from inside the main residence that connects to the adjacent mountain meadows. Two simple grassed green roofs were included to create a native grass foreground across which to view the mountainside in the distance. During winter, the grasses turn a golden-brown color, and snow covers the dormant grasses until spring. During the summer, grasses blend into the distant hillside (Fig. 6.12).

Fig. 6.11 This hilltop residence blends the native grasses of the surrounding landscape onto the rooftops of the main house (upper) and garage (lower). (Photo: Bruce Dvorak, July 2018)

6.3.3.3 Plant Establishment

The grasses were pre-grown by Intermountain Roofscape Supply as sod, then established on LiveRoof® Intermountain Roofscape Supplies' Native Sod modules.

Grasses

Idaho fescue (*Festuca idahoensis*), streambank wheatgrass (*Elymus lanceolatus*), sandberg bluegrass (*Poa secunda* J. Presl), and the exotic sheep fescue (*Festuca ovina*).

6.3.3.4 Irrigation

This native grass roof requires irrigation every other day during the growing season.

Fig. 6.12 A dense stand of drought-tolerant native grasses provides a simple habitat, prevents unwanted vegetation from establishing, and blends into the mountain landscape. (Photo: Bruce Dvorak, July 2018)

6.3.3.5 Maintenance

Based on the provider's specifications, they recommend that the grasses are treated with a 14-14-14 fertilizer every spring to ensure plant health. In the spring and fall, a trimming (mower/weed trimmer) is recommended. Because the modules are vegetated with a dense cover, there is little encroachment by invasive plants, however, periodic weeding is necessary.

6.3.3.6 Observed Wildlife

Crickets, grasshoppers, birds, small insects.

6.3.3.7 Best Performing Native Vegetation

Festuca idahoensis, *Elymus lanceolatus*, *Poa secunda.*

6.3.3.8 Post-occupancy Observations

Design Team

- Every green roof needs a "champion". Every green roof needs someone to watch over it to make sure that it is performing as intended. There is no such thing as a "no maintenance" green roof. Even low maintenance green roofs such as this sod roof need someone to be sure that the irrigation is functioning, that the watering rates are appropriate, and that the plants remain healthy and that it has a "champion", someone to periodically watch over it.

Authors' Reflections

- This sod roof is easy to care for because of its simple planting design and time-testing substrate and drainage system.
- The fully established vegetation upon installation provided instant green and prevents invasive plants from becoming established.

6.3.4 Salt Lake City, UT: Conference Center of the Church of Jesus Christ of Latter-Day Saints (LDS)

When it was constructed in 2000, the semi-intensive green roof meadow on top of the LDS Conference Center at Temple Square was the largest green roof in North America. Designed to last 150 years, nearly 20 years later, the green roof stands as a testament to the viability, richness, and ecological value of rooftop meadows in the ecoregion (Fig. 6.13). The project has been well-documented in its construction and performance. The vegetated roof has as at least 14 species of native grasses and 18 species of native wildflowers that persist on the roof with consistent care, irrigation, and maintenance of the green roof (Dewey et al. 2004; Weiler and Scholz-Barth 2009). The openness of the Church of Jesus Christ of Latter-day Saints to share information about the roof meadow's design, construction, performance, and lessons learned is commendable and sets an example for other private organizations and institutions.

6.3.4.1 Project Team

Building Owner/Client: Church of Jesus Christ of Latter-day Saints
Green Roof Design Team Lead: Olin Partnership
Architect: Zimmer Gunsul Frasca Architects LLP
Structural Engineer: KPFF

Fig. 6.13 The two-hectare (5-acre) of rooftop garden features fountains and plaza includes a 0.8-hectare (2-acre) meadow. It is the largest and most diverse patch of grassland in downtown Salt Lake City. Views from surrounding buildings make visible the seasonal changes of the roof, such as bright green during the summer. (Photo: Bruce Dvorak, July 2018)

Landscape Architect: Susan K. Weiler, Olin Partnership
Stormwater consultant: (Jeffery L. Bruce Associates)
Installation Contractor: American Hydrotech, volunteers (planting)
Project completion: 2000
Green roof area: 20,234 m^2 (217,800 ft^2)

6.3.4.2 Overview and Objectives

The main concept for the 21,000-seat column-free conference center was to merge the exterior views of the building with the distant landscape while respecting the presence of other buildings in the vicinity. With the Wasatch Mountains in the background, the blocky forms of the building edges echo the forms of the distant mountains. The rooftop meadow and trees connect the site to the grasslands and forests of the distant mountains. The meadow roof was designed to be a garden without a gardener (Fig. 6.14). Thus, it has some formal plant groupings near the edges and more natural formations in the center of the meadow. The roof has a continuous 3% slope, which provides consistent drainage, a source of influence on habitat microclimates, and replicates the gentle slope of a meadow in the low-lying landscape. (Weiler and Scholz-Barth 2009; Roth 2018).

Fig. 6.14 Various habitats display native meadow flora including (**a**) Maximillian sunflower near a sunny edge (foreground), (**b**) little bluestem and short stature grasses dominate the central meadow, (**c**) purple coneflower drift (center), and (**d**) yellow blooms of goldenrod (left front) and Ashy sunflower (right). (Photos: Bruce Dvorak, July 2018)

6.3.4.3 Plant Establishment

The 0.8-hectare (2-acre) meadow consists of 75 species of trees, shrubs, grasses, bulbs, and annual and perennial wildflowers (Tables 6.2 and 6.3). The composition of wildflowers was selected for their spring, summer, and fall blooms. Some of the plants are native to Salt Lake City region, some are native to other parts of Utah, many are native to central states prairies, and some are exotic (not included here). The blended substrate is 15 cm (6 in) deep in the expanse of the roof meadow and up to 1–1.2-m (3–4 ft) in the isolated tree wells.

6.3.4.4 Irrigation

Water draining from the rooftop plaza areas and other locations on the site are harvested in a large cistern located below grade. The harvested water is applied to the green roof through a weather tracking system of moisture sensors to preserve water and prevent overwatering. During the growing season, the overhead irrigation runs

Table 6.2 Native grasses on the LDS Conference Center green roof

Common Name	Botanical Name
Mountain brome	*Bromus carinatus*
Tufted hairgrass	*Deschampsia cespitosa*
Thickspike wheatgrass	*Elymus lanceolatus*
Slender wheatgrass	*Elymus trachycaulus*
Idaho fescue	*Festuca idahoensis* Elmer
Ovina fescue	*Festuca ovina var.* ovina L.
Prairie junegrass	*Koeleria* spp. Pers.
Ricegrass	*Oryzopsis* spp. Michx.
Western wheatgrass	*Pascopyrum smithii*
Alpine bluegrass	*Poa alpina* L.
Mutton bluegrass	*Poa fendleriana*
Big bluegrass	*Poa secunda* J. Presl
Bluebunch wheatgrass	*Pseudoroegneria spicata*
Sand dropseed	*Sporobolus cryptandrus*

Table 6.3 Native herbaceous wildflowers on the LDS Conference Center green roof

Common Name	Botanical Name
Columbine	*Aquilegia* spp. L.
White sage	*Artemisia ludoviciana*
Aster	*Aster* spp. L.
Blue wild indigo	*Baptisia australis*
Bluebell bellflower	*Campanula rotundifolia* L.
Tickseed	*Coreopsis* spp. L.
Purple coneflower	*Echinacea purpurea*
Oregon daisy	*Erigeron speciosus*
Spotted joe-pye-weed	*Eupatorium maculatum*
Queen of the prairie	*Filipendula rubra*
Geranium	*Geranium* spp. L.
Rocky Mountain iris	*Iris missouriensis* Nutt.
Dense blazing star	*Liatris spicata* L. Wild.
Lupine	*Lupinus* spp. L.
Penstemon	*Penstemon* spp. Schmidel
Goldenrod	*Solidago* spp. L.
Purple meadow-rue	*Thalictrum dasycarpum*
Culver's root	*Veronicastrum virginicum*

every other day, 13–20 min. During a wet year, the irrigation ran about 5 min per irrigation event.

6.3.4.5 Maintenance

As a roof garden, the meadow receives weekly visits during the growing season to remove unwanted plants. Because the selected plant material and irrigation are compatible, infestations by unwanted plants have been minimal. Some of the plants that are frequently removed from the roof meadow include clovers, mullen, toadflax, and wild carrot. At the end of the growing season, the meadow is cut down with string trimmers to a height of about 15 cm (6 in). Plants of interest such as rabbitbrush and blue false indigo may be left in place. Other activities include collecting seed of desirable species to redistribute seeds to other locations of the meadow. This can be a very effective way to maintain a green roof, providing that the staff managing the roof has the appropriate knowledge of green roofs and local ecosystems. Thus, a full-time maintenance supervisor has been hired to facilitate the management of the green roof ecosystem.

6.3.4.6 Observed Wildlife

The roof meadow maintenance staff have observed: quail, squirrels, raccoons, migrating birds, sparrow, doves, dragonflies, monarch butterflies, bees. Butterflies have been observed hovering over blooms of the plant gaura.

6.3.4.7 Best Performing Native Vegetation

Four years after planting, Dewey et al. published a detailed report of plant performance during the early establishment of the meadow (Dewey et al. 2004). This applied research is invaluable in building the knowledge of plant selection and performance on green roofs. An onsite visit with the green roof maintenance staff (July 2018) confirmed that replanting and reseeding takes place as budget is set-aside each year to monitor and edit the meadow plant community over time. The following plants were noted to perform well on the roof meadow; (grasses) little bluestem, fescues, and crested wheat, (wildflowers) larkspur (re-seeds), Maximilian sunflower, gaura, false indigo, aster, joe pye weed, chocolate flower, yarrow, penstemon (re-seeds), pitcher sage, butterfly milkweed, Indian paintbrush, and pasque flower.

6.3.4.8 Post-occupancy Observations

Peer-Reviewed Publication (Dewey et al. 2004)

- Canada bluegrass and white sage were aggressive plants when adjacent to the other grasses and wildflowers.
- Low growing, drought-tolerant plants like ashy sunflower, aster, blue wild indigo, dense blazing star, the fescues, geranium, goldenrod, prairie junegrass, lupine, mountain brome, Oregon daisy, penstemon, purple coneflower, queen of the prairie, ricegrass, Rocky Mountain iris, sand dropseed, spotted joe pye weed, tufted hairgrass, the wheatgrasses, and wildrye were present, but not dominate.
- Because the soil was sterile upon planting, researchers found that weed competition was not an issue as unwanted plants are removed by hand each week during the growing season.

Authors' Reflections

- An on-site visit and interview with maintenance staff revealed that the visible edges of the meadow receive regular attention, so to maintain wildflowers near the plaza and perimeter walkways. The central portion of the meadow is dominated by grasses and a few wildflowers. The central meadow has shorter species grasses and appears more natural compared to the more maintained edges.
- The gentle southwest-facing slope affords many moisture microclimates in the meadow for different species. The trees along the edges create small shade pockets and habitat for birds and other species around the meadow.

6.3.5 J. Willard Marriott Library Rooftop Garden, University of Utah, Salt Lake City

As part of a major renovation of the main library on campus, an accessible roof garden was included for students, faculty, staff, and small-groups (Fig. 6.15). The meadow roof sits on top of a third floor and is viewable from student sitting areas inside the building. Designed to be a low water use garden, drought-tolerant native and some non-native plants were selected and installed. The roof garden has been a popular location for students and researchers. Recent publications about the performance of plants have highlighted the estimation of evapotranspiration of the green roof and the use of water. These are the first research studies of water use for green roofs in semi-arid environments (Feng et al. 2016, 2018).

Fig. 6.15 Native perennial wildflowers and grasses bring color to the edges of this south-facing semi-intensive green roof. Several climate-adapted exotic species were added into the garden to create a hybrid garden. (Photo: Bruce Dvorak, July 2018)

6.3.5.1 Project Team

Building Owner/Client: University of Utah
Green Roof Design Team Lead: Gary Brown
Architect: MJSA
Landscape Architect: Gary Brown
Installation Contractor: American Hydrotech
Project completion: 2008
Green roof area: 761 m^2 (82,000 ft^2)

6.3.5.2 Overview and Objectives

The roof garden was designed for students to view onto from inside the main gathering study space in the library (Fig. 6.16). A drought-tolerant roof meadow garden was the original design to protect the structure underneath and, conserve energy and provide space for research. The appearance of a natural meadow guided plant selection. There is a mixture of climate-adapted native, cultivated forms of native plants, and some exotic plants selected for ornamental effect (exotics not listed below).

Fig. 6.16 A plaza on the roof deck is surrounded by the meadow garden on both sides. Here, the wildflowers are maintained on the edges, and grasses dominate the outer side of the garden. (Photo: Bruce Dvorak, July 2018)

6.3.5.3 Plant Establishment

All of the original vegetation was installed as pre-grown vegetation in containers. Native grasses and wildflowers are presented in Table 6.4.

After the initial plants were established, other plants such as the native Butterfly milkweed (*Asclepis syriaca*) were planted to attract Monarch butterflies. Non-native Russian Sage (*Perovskia atriplicifolia*) was also installed after the initial planting in highly visible locations on the roof where there was dieback. Over time, the planting has taken on the appearance of a meadow, but replanting has not been restricted to native plants (Fig. 6.16).

6.3.5.4 Irrigation

The roof vegetation is watered with a municipal water supply and subgrade drip irrigation system. After vegetation was established, during a drought an overhead spray system was added to cover areas where the subgrade system was not functioning.

Table 6.4 Native vegetation (and cultivars) included on the library roof garden

Plant Type	Common Name	Botanical Name
Grasses	Blue grama	*Bouteloua gracilis*
Grasses	Buffalo grass	*Buchloe dactyloides* 'Legacy'
Grasses	Switch grass	*Panicum virgatum* 'Rotstrahlbusch'
Wildflowers	Moonshine yarrow	*Achillea* 'Moonshine'
Wildflowers	Hyssop	*Agastache cana*
Wildflowers	Giant hyssop	*Agastache rupestris*
Wildflowers	Purple coneflower	*Echinacea purpurea*
Wildflowers	Sulphur-flower buckwheat	*Eriogonum umbellatum*
Wildflowers	Wasatch beardtongue	*Penstemon cyananthus*
Wildflowers	Firecracker penstemon	*Penstemon eatonii*
Wildflowers	Pine-leaved penstemon	*Penstemon pinifolius*
Wildflowers	Rocky Mountain penstemon	*Penstemon strictus* 'Rocky Mountain'

6.3.5.5 Maintenance

The roof garden is maintained by crews at the University of Utah. Visits to the roof are limited to once monthly. Each visit has a focus or target for removing unwanted aggressive plants. The University has an annual budget for periodic plant replacement on the roof garden.

6.3.5.6 Observed Wildlife

A beehive program was established under the direction of the sustainable Campus Initiative Fund, under Thomas Bench. The purpose of the program is to teach the college community how to keep bees productive and produce honey for the local farmer's market, and research projects (JWM 2018).

6.3.5.7 Best Performing Native Vegetation

Butterfly milkweed, buffalo grass, rudbeckia, sideoats grama, penstemons, purple coneflower, yarrow, and native grasses are all thriving.

6.3.5.8 Post-occupancy Observations

Authors' Reflections

- During a site visit with maintenance staff, it was said "How hard it is to get the garden back into the original design once it is let go. Competition for maintaining other sites on campus does not allow more frequent visits".

- The garden has a large drift of purple coneflower. This plant is valuable to pollinators, and has not been proven on many green roofs, but is thriving on the Willard Marriott Library.

6.3.6 Salt Lake City Public Library, Salt Lake City, Utah

In 1998, citizens of Salt Lake City passed an $84 million-dollar bond to relocate and build a new main library in downtown Salt Lake City. Architect Moshe Safdie was awarded the contract to design a mixed-use building. The multi-purpose first floor injects new functions not typically associated with libraries. The building includes small shops for retail and service vendors such as a gift shop, art gallery, café, hair salon, coffee shop, and a large atrium. This open space is an "urban room" that provides public seating and transition space between the active multi-use shops and the library. The library has a rooftop reading garden that has one of the best views of downtown and the Wasatch Mountains (Fig. 6.17).

6.3.6.1 Project Team

Building Owner/Client: City of Salt Lake City
Green Roof Design Team Lead: Civitas
Architect: Safdie Architects
Landscape Architect: Civitas
Project completion: 2003
Green roof area: 1319 m^2 (14,200 ft^2)

6.3.6.2 Overview and Objectives

Designed by Civitas, this southeast facing terraced and sloped roof garden was designed to provide privacy, generous views, and a respite from the activities taking place on the mixed-use space below. Native and exotic drought-tolerant vegetation was used to create a formal garden setting, by single species planting beds (Fig. 6.18). Only native vegetation used on the roof garden is provided here.

6.3.6.3 Plant Establishment

Plants were pre-grown in potted containers and installed onsite. Forms of plants include annuals, grasses, herbaceous perennials, shrubs, and trees.

Fig. 6.17 With access from ramps and stairs, the reading garden is planted with a mix of native and exotic drought-tolerant vegetation. The native Washington hawthorn trees are planted with an understory of exotic fountain grass (shown here) and also native northern sea oats elsewhere. (Photo: Bruce Dvorak, July 2017)

Annuals

Spider flower (*Cleome hassleriana* 'cherry queen'; native to the eastern U.S.), sunflower pacino (*Helianthus annuus* 'pacino').

Grasses

Northern sea oats (*Chasmanthium latifolium*; native to the eastern U.S.).

Herbaceous Perennials

Orange coneflower (*Rudbeckia fulgida* 'goldsturm'; eastern U.S. native), cappuccino coral bells (*Heuchera* 'cappucino'), gayfeather (*Liatris spicata* 'kobold'; eastern U.S. native), michaelmas daisy (*Aster novi-belgii* 'snow flurry'), sea pink (*Armeria maritima*).

Fig. 6.18 View of native trees (*Crataegus phaenopyrum*, eastern U.S.) and shrubs on one of the rooftop terraces. Each terrace allows a private space for reading or enjoying the view. The non-native *Festuca glauca* 'Elijah Blue' is seen above, where the native northern sea oats grass can be seen in the upper far right. (Photo: Bruce Dvorak, July 2018)

Shrubs (Native to North and Eastern U.S. and Canada)

Bar Harbor juniper (*Juniperus horizontalis* 'Bar Harbor'), compact spring green cranberrybush (*Viburnum trilobum* 'compact spring green'), nannyberry viburnum (*Viburnum lentago*).

Trees

Red oak (*Quercus rubra*), skyline honeylocust (*Gleditsia triacanthos* var. inermis 'Skyline'; eastern U.S. native), Washington hawthorn (*Crataegus phaenopyrum*; eastern U.S. native).

6.3.6.4 Irrigation

A subsurface drip irrigation system provides water to plants.

6.3.6.5 Maintenance

As each planting bed consists of single species, any volunteer plants are removed. As a roof garden, the green roof receives monthly maintenance.

6.3.6.6 Observed Wildlife

Honey bees are the Utah State insect, and in 2009 the city amended a city ordinance to allow beehives to be kept on public and private properties in the city. Several beehives were added to the rooftop of the Salt Lake City Public Library in 2010. The hives produce between 34 and 113 kg (75–250 lbs) of honey each year and honey is given away at honey-tasting events at the library (Whitby 2018).

6.3.6.7 Best Performing Native Vegetation

Native grasses and trees.

6.3.6.8 Post-occupancy Observations

Authors' Reflections

- During an onsite visit in July 2018, the garden vegetation was holding well and was established in the sloped terraced garden. The approach garden has steep slopes. Some of the slopes have experienced erosion.
- The beehives are located near the roof, behind windows for viewing. The library makes use of the honey at the end of the summer during annual honey tasking events.

6.3.7 Ogden Nature Center, Ogden, Utah

Located near the foothills of the Wasatch Mountains, the Ogden Nature Center is a 61.5-hectare (152-acre) nature preserve and education center. The site conserves woodlands, meadows, ponds, and wetland habitats. Its mission is to educate youth about natural systems, to conserve habitat, and serve as a gathering space for events. Architect Robert Herman designed the nature center as a wood post and beam structure. A sod roof was included as part of the nature center structure from the beginning (Fig. 6.19). Two of the low-slope roofs support green roofs over the offices and gift shop located below.

Fig. 6.19 Ogden Nature Center Visitor's Center shown with grasses in the foreground in their dormant phase and a newly replanted sod roof. (Photo: Bruce Dvorak, July 2018)

6.3.7.1 Project Team

Building Owner/Client: Ogden Nature Center
Green Roof Design Team Lead: Barney and Della
Barnett of Willard Bay Gardens, and Jared Flynn
Architect: Robert Herman, EDA Architects
Installation Contractor: ONC, Volunteers, Jared Flynn, Willard Bay Gardens
Project completion: 2003
Green roof area: 560 m^2 (6000 ft^2)

6.3.7.2 Overview and Objectives

The original intent for the green roof was to integrate the site and building and to provide food and shelter for birds and butterflies. Secondary motivations include reducing the heat gain through the roof deck and retaining precipitation during rainstorms. As a nature center, the staff wanted the building to fit into the context of the site.

Fig. 6.20 Ogden Nature Center sod and wildflower green roof as shown here during the early summer (June) of 2012. (Photo: Courtesy of Jeanne Hunt)

6.3.7.3 Plant Establishment

The original design was completed in 1995 as a sod grassed roof. In 2003, the roof meadow was replanted with sod and wildflowers to increase diversity (Fig. 6.20). However, because of challenges with the irrigation system, the front part of the building was replanted again in 2018 with grass sod. There are no records of plant species installed.

6.3.7.4 Irrigation

The green roof receives overhead spray irrigation. Keeping the irrigation in working order has been a challenge, and has been replaced several times.

6.3.7.5 Maintenance

The green roof is a simple planting that requires the annual removal of seedling trees and any other invasive plants.

6.3.7.6 Observed Wildlife

Birds, bees, and other pollinators make use of the roof habitat.

6.3.7.7 Best Performing Native Vegetation

The grass sod is the best vegetative cover for this roof, as long as the irrigation system is functional (nature center staff).

6.3.7.8 Post-occupancy Observations

Authors' Reflections

- This project has much potential as a research project. The green roof has been replanted, but there is little documentation of the current vegetation installed. This green roof would make an excellent research project as it would be worthwhile to study vegetation, temperatures, and or stormwater retention, and use as habitat. The simple design, ease of access could make it a worthwhile green roof to learn from.

6.3.8 Mulvaney Medical Office Building, Boise, Idaho

The Mulvaney Medical Office building is a class "A" medical building located in the Saint Alphonsus Medical Center campus, in Boise, Idaho. The developer and first tenants of the private practice offices wanted a design for the new building that would provide a restorative natural environment. This LEED Gold-certified building includes a healing garden for visitors, staff, and patients to rest and view from inside offices and waiting rooms. Designed with trees, grasses, and wildflowers, the healing garden makes use of extensive (shallow) and intensive (deep) substrates (Fig. 6.21). The vertical planting (grasses and woody vegetation) provides a simple setting that uses native vegetation, where the more complex groundcovers (extensive) includes native and climate-adapted none-native vegetation.

As a facility where out-patient and minor surgeries take place daily, the entire building environment was designed to reduce stress. Research supports that even quick views to green spaces (40 s) and green roofs, use, or views to a healing garden can improve recovery times and reduced levels of stress (Ulrich 1984; Lee et al. 2015). This healing garden is a small space that achieves important functions, and together with the energy-efficient design and materials, also costs less to operate than their previous facility (Parker 2010).

Fig. 6.21 The native desert willow tree is planted in a deep planter box, within an extensive green roof planting bed that supports wildflowers, grasses, and succulents. (Photo: Bruce Dvorak, July 2018)

6.3.8.1 Project Team

Building Owner/Client: Cameron Investments
Green Roof Design Team Lead: CSHQA
Architect: CSHQA
Landscape Architect: CSHQA
Project completion: 2010
Green roof area: 232 m^2 (2500 ft^2)

6.3.8.2 Overview and Objectives

This roof garden was provided to be a restful and restorative environment for visitors, staff, and patients. Taller (woody) vegetation was included to provide visual screening of the parking lot below.

6.3.8.3 Plant Establishment

This roof garden utilizes primarily native vegetation in the vertical habitat, and a mixture of native and drought-tolerant vegetation was used as a groundcover (Fig. 6.22). The west-facing roof garden lies in shade until later in the afternoon.

Fig. 6.22 A drift of native Idaho fescue (*Festuca idahoensis*) grass stands vertical from the mix of exotic and native sedum groundcovers. (Photo: Bruce Dvorak, July 2018)

Thus, the trees located on the west side of the roof garden provide some shade during the late afternoon.

Native Vegetation

apache plume (*Fallugia paradoxaBlue*) is native to bordering states (NV, UT), flax (*Linum lewisii*), Cape Blanco stonecrop, (*Sedum spathulifolium* 'Cape Blanco') is native to bordering states (WA, OR), common woolly sunflower (*Eriophyllum lanatum*), desert willow (Chilopsis linearis), gooseberryleaf globemallow, (*Sphaeralcea grossulariifolia*).

Climate-Adapted (Exotic) Vegetation

angelina stonecrop (*Sedum repestre* 'Angelina'), *Bronzita carex* (*Carex flagellifera* 'Bronzita'), coral carpet sedum, (*Sedum album* 'Coral Carpet'), October daphne stonecrop (*Sedum sieboldii* 'October daphne'), red carpet sedum (*Sedum spurium* 'Red Carpet'), purple beauty sempervivum (*Sempervivum* 'Purple Beauty'), white diamond (*Sedum pachyclados* 'White Diamond').

6.3.8.4 Irrigation

The roof garden receives daily/weekly irrigation during the growing season.

6.3.8.5 Maintenance

The roof garden receives frequent (weekly/bi-weekly) maintenance during the growing season. Unwanted vegetation is removed, and doctors and staff participate in planting/replanting parts of the roof outside their offices.

6.3.8.6 Observed Wildlife

Occasional visits by pollinators and hummingbirds and songbirds have been observed by visitors and staff.

6.3.8.7 Best Performing Native Vegetation

All of the vegetation is performing as expected.

6.3.8.8 Post-occupancy Observations

Authors' Reflections

- As one of the few green roofs in Boise, this healing roof garden demonstrates that native plants are a viable choice for green roofs in Boise.
- This green roof requires minimal irrigation and care. The modular extensive green roof system (GreenGrid®) blends well with the custom-made tree wells.
- The doctors and staff have adopted space on the non-public portion of the green roof to interplant some of their own plants, such as strawberries, tomatoes, and a few herbs.

6.3.9 Washington Fruit and Produce Headquarters, Yakima, WA

Some of the best land in North America for growing fruit lies in the rain shadow of the Cascade Mountains in eastern Washington. Originally located on an upland riparian grassland near the confluence of the Naches and Yakima Rivers, the property of the Washington Fruit and Produce Company became an apple, pear, and cherry orchard over 100 years ago. Today, the buildings, pavement, and trucks of the headquarters' facilities and operations dominate the 36 hectares (90 acres) of land; but the owner, architect and landscape architect shared a vision in which the new headquarters building would become enveloped by a high grassed berm and the industrial facilities would be screened (Schuler 2018). Native prairie and bottomland vegetation were brought into an enclosed courtyard complete with an agricultural theme and planted with species native to the riparian corridors (Fig. 6.23). Some of the plant species in the landscape are considered early successional species; however, in their proper habitat such as this, they perform critical ecosystem functions and reflect the regional character. The green roof was built on top of the lunch house building and is seamlessly tucked into the berm. It is surrounded by native prairie and is part of the local ecosystem habitat (Schuler 2018).

6.3.9.1 Project Team

Building Owner/Client: Washington Fruit and Produce Company
Green Roof Design Team Lead: Berger Partnership
Architect: Graham Baba Architects
Landscape Architect: Berger Partnership
Installation Contractor: Elevation Contracting
Project completion: 2016
Green roof area: 139 m^2 (1500 ft^2)

6.3.9.2 Overview and Objectives

The green roof vegetation reintroduces grasses and wildflowers once pervasive in the valley. The 15 cm- deep (6 in) substrate connects directly to the berm.

6.3.9.3 Plant Establishment

Plants for the north-facing slope and meadow roof were seeded onto the green roof substrate and watered overhead during plant establishment. A low watering approach allows vegetation to become dormant during the summer (Fig. 6.24). These local grasses and herbaceous perennial wildflowers (Table 6.5) used on the green roof include:

Grasses

Bluebunch wheatgrass (*Pseudoroegneria spicata*).

Fig. 6.23 This all native meadow is located on a north-facing berm that extends onto the rooftop of the lunchroom of the Headquarters campus. (Photo: Bruce Dvorak, August 2018)

6.3.9.4 Irrigation

The green roof has a zoned drip irrigation system that is buried within the substrate.

6.3.9.5 Maintenance

The intended maintenance practice by the owner is to "let it go" so it blends into the adjacent north-facing meadow berm. Maintenance visits take place about once or twice a growing season. Except for an occasional tree seedling, the green roof vegetation is not weeded.

6.3.9.6 Observed Wildlife

With a direct connection to the ground, wildlife is frequent visitors to the roof and include birds, mice, snakes, bees, butterflies, dragonflies, and grasshoppers.

Fig. 6.24 Yellow blooms of common woolly sunflower (*Eriophyllum lanatum*) make for late summer color. Bright green vegetation and white blooms of yarrow are seen growing in the center-left. The thick layer of native grasses can be seen in the background. (Photo: Bruce Dvorak, August 2018)

Table 6.5 Native wildflowers

Common Name	Botanical Name
Common yarrow	*Achillea millefolium*
Threadleaf giant hyssop	*Agastache rupestris*
Western columbine	*Aquilegia formosa*
Butterfly milkweed	*Asclepias tuberosa*
Arrowleaf balsamroot	*Balsamorhiza sagittata*
Aspen fleabane	*Erigeron speciosus*
Common woolly sunflower	*Eriophyllum lanatum*
Blanketflower	*Gaillardia aristata*
Lewis flax	*Linum lewisii*
Big leaf lupine	*Lupinus polyphyllus*
Bigfruit evening primrose	*Oenothero macrocarpa*
Barrett's beardtongue	*Penstemon barrettiae*
Cardwell's beardtongue	*Penstemon cardwellii*
Firecracker penstemon	*Penstemon eatonii*
Bush penstemon	*Penstemon fruticosus*
Rocky Mountain penstemon	*Penstemon strictus*
Venus penstemon	*Penstemon venustus*
Canada goldenrod	*Solidago canadensis* var. *lepida*

6.3.9.7 Best Performing Native Vegetation

Grasses and wildflowers form a dense matt on the green roof. Bluebunch wheatgrass, blue flax, butterfly milkweed, common woolly sunflower, penstemon, and yarrow are present and performing well three years after planting.

6.3.9.8 Post-occupancy Observations

Authors' Reflections

- This seeded green roof has a dense vegetative cover that minimizes exposed media where volunteer vegetation could become established.
- The meadow has a rich diversity, including plants for endangered species, such as the inclusion of *Asclepius* to attract monarch butterflies.

6.3.10 Moda Building, Bend, OR

The original owner (and builder) of this multi-story office complex was the Moda Health Insurance Company. Their intention for this new building was to include a roof garden with native plants for use by their employees and occupants of this multi-story office building in Bend, Oregon (Fig. 6.25). Richard Martinson, co-owner of WinterCreek Nursery helped vision and implement this all native semi-arid green roof in Bend, Oregon. Lying in the shadows of the Cascade Mountains, Bend averages 280 mm (11.2 in) of precipitation annually, including 60.5 cm (23.8 in) of snowfall. The natural vegetation includes coniferous savanna, shrub/scrub plant communities, and grasslands (Fig. 6.26). This roof garden borrows vegetation from each of these plant communities and brings them together in a semi-arid shrub garden that reads more like a Japanese Garden than a natural landscape. The arrangement of plants, their juxtaposition, and use are a microcosm of the semi-arid landscapes but are positioned with the attention of a garden designer. The construction, materials, and maintenance practices used on this green roof are a model for semi-intensive green roofs in the Intermontane West.

6.3.10.1 Project Team

Building Owner/Client: Moda Health Services
Green Roof Design Team Lead: WinterCreek Restoration & Nursery
Architect: GBD, Russ Hale, AIA
Structural Engineer: W&H Pacific
Landscape Architect: W&H Pacific, Chelsea Schneider
Installation Contractor: WinterCreek Restoration & Nursery

Fig. 6.25 A semi-arid desert scrub themed roof garden thrives in Bend, Oregon. The apparent random distribution of plants is organized into the well-ordered composition of plant communities laid out in a multiplicity of micro-gardens. The plant arrangements mimic those sun/shade and soil associations found in their natural habitats. (Photo: Bruce Dvorak, September 2018)

Project completion: 2007
Green roof area: 884 m^2 (9500 ft^2)

6.3.10.2 Overview and Objectives

This assemblage of all native vegetation is ordered on natural plant communities: sagebrush/rabbitbrush, grassland, lodgepole pine. With the larger plants providing a structure to the garden, the grasses and wildflowers support the natural plant associations in both aesthetic arrangement and ecological function.

6.3.10.3 Plant Establishment

Plant choices were based on natural plant community structure, rooting depths, and tolerance to exposure to sun, wind, and reflected heat. Substrate depth ranges from 20 cm (8 in) to 40 cm (16 in) in herb/shrub areas and up to 76 cm (30 in) at tree

Fig. 6.26 The yellow blooms of rabbitbrush (center) and dark foliage of the manzanita shrub (center right) dot the landscapes surrounding central Oregon. These taller plants and their associative grasses and wildflowers become a focal point of this semi-arid roof garden. (Photo: Bruce Dvorak, September 2018)

locations. Soil composition accounted for plants that adapt to similar soil moisture conditions, soil fungal, and bacterial communities to make for a self-sustaining system. Forms of vegetation include grasses, herbaceous perennials (Table 6.6), groundcover, shrubs, and trees (Fig. 6.27).

6.3.10.4 Grasses

Idaho fescue (*Festuca idahoensis*), and prairie junegrass (*Koeleria cristata*).

Groundcovers

Kinnikinnick (*Arctostaphylos uva-ursi*), and strawberry (*Fragaria sp.*)

Table 6.6 Herbaceous perennials on the Moda roof garden

Common Name	Botanical Name
Pussytoes	*Antennaria microphylla*
Linear-leaf fleabane	*Erigeron linearis*
Naked buckwheat	*Eriogonum nudum*
Sulphur buckwheat	*Eriogonum umbellatum*
Cushion buckwheat	*Erogonum caespitosum*
California poppy	*Eschscholtzia californica*
Scarlet gilia	*Ipomopsis aggregate*
Blue flax	*Linum lewisii*
Sickle keeled lupine	*Lupinus albicaulis*
Lowly penstemon	*Penstemon humulis*
Pine mat penstemon	*Penstemon pinnifolius*
Rocky Mountain penstemon	*Penstemon strictus*

Fig. 6.27 The roof garden features low-growing drought and heat tolerant shrubs such as the silvery foliage (center) of sagebrush (*Artemisia tridentata*), rabbitbrush (*Ericameria nauseosus*) seen to the right (yellow blooms), the dark green reflective foliage of the manzanita shrub (*Arctostaphylos patula) is center-right, and the red/brown blooms of* cliff spiraea (*Holodiscus microphyllus*) center-left. Lodgepole pine (*Pinus contorta*) is seen in the background to the right. (Photo: Bruce Dvorak, September 2018)

Shrubs

Big sagebrush (*Artemesia tridentata*), cliff spiraea (*Holodiscus microphyllus*), desert sweet (*Chamaebatiaria millifolium*), gray rabbitbrush (*Ericameria nauseosus*), green rabbitbrush (*Chrysothamnus viscidiflorus*), manzanita (*Arctostaphylos patula*), and prostrate Oregon grape (*Mahonia repens*).

Trees

Lodgepole pine (*Pinus contorta*) and mountain mahogany (*Cercocarpos ledifolius*).

6.3.10.5 Irrigation

A subgrade irrigation system supplies water to the roots of plants. Plants are arranged in moisture zones and irrigation zones are separated to deliver irrigation to each zone based upon the needs of the plants.

6.3.10.6 Maintenance

The roof garden receives maintenance activities about once a month. There are a limited number of qualified companies that can care for this roof since there is a learning curve for most companies to understand the design, arrangement, and plant material. No fertilizers or pesticides are used.

6.3.10.7 Observed Wildlife

Bees, insects, chipmunks, monarch butterflies, hummingbirds, chickadees, finches, blackbirds, pine siskins, robins have been observed on the green roof.

6.3.10.8 Best Performing Native Vegetation

Idaho fescue, prairie junegrass, globemallow, California poppy, desert sweet, rabbitbrush, desert ocean spray, sagebrush, manzanita, and many more.

6.3.10.9 Post-occupancy Observations

Published

- "Not all landscape designers and contractors understand how to use native plants in an ecological landscape. It is worth taking the time to find the right one to get the results you want" (Danler and Langellotto-Rhodaback 2015).

Design Team (Dr. Richard Martinson)

- Changes in ownership can bring changes in maintenance practices. Design a roof garden to anticipate changes in ownership and changes in maintenance practices. Select plants that can tolerate natural changes in climate from year to year and the sometimes-abrupt human changes to maintenance and irrigation. Irrigation was shut down for some time but is now up and functioning again.
- Overall, the roof has exceeded expectations in growth rate (biomass production), and ecological function. As one of the first completely native green roofs in a semi-arid environment, the ability of typically deep-rooted xerophytic species to acclimate to a completely artificial setting is remarkable. The resilience of the planting is expressed through continued recruitment and spread of installed species, but also through the influx of vegetation not installed but appropriate for the designed plant community.
- The long-term viability of the planting on the roof was unknown, but a risk MODA was willing to accept to complete something unique in this building. I think the roof provides an excellent example of what can be accomplished through thoughtful planning and confidence in the process of designing a project based on ecological principles.

Authors' Reflections

- This roof garden is a masterful study of the regional scrub ecosystems. The garden takes ecological associations found in the semi-arid scrub plant communities and arranges them by basic design principles such as visual contrast, texture, and repetition. The forms of plants are used to achieve a sense of harmony. Most of the plants are growing in only 20–30 cm (8–12 in) of growing media and are very healthy.
- The vernacular shrubs of this Intermontane Mountain region include sagebrush and rabbitbrush. These plants may be associated with rural or wild places, but their use in this garden highlights the delicate beauty of these wild plants.

6.4 Plants for the Intermontane Semi-Arid Grasslands Ecoregions

The variety of plants that have been specified on green roofs in this region indicates a high level of experimentation among designers and green roof specifiers. 113 taxa of native and native-adjacent plants were used on the ten green roofs in this chapter. Of those native to the ecoregions, 19 species occurred more than once. Seven species occurred in three or more of the ten case study sites in this chapter (Table 6.7).

The wide range of climatic settings, depths of green roof installations, and maintenance regimes employed suggests that no single list of recommended plants has been created for green roofs in the vast Intermontane West. From the ten ecoregional green roof case studies located respectively in the subregions of Basin and Range desert, Rocky Mountain valleys and Columbia Plateau grasslands a broad set of grasses, forbs, succulents, bulbs, shrubs, and trees can be identified. The following discussion focuses on the occurrence and repetition of green roof plants within and across the three subregions. Occurrences of species are observed in several patterns: by the number of species representing the same genera and the genera most widely distributed among the greatest number of roofs.

Table 6.7 Native plant species used on three or more ecoregional green roof case studies in this chapter

Plant Type	Common Name	Botanical Name	A	B	C	D	E	F	G	H
Herbaceous Perennials	Sulfurflower	*Eriogonum umbellatum*	x	x	x					
Herbaceous Perennials	Firecracker penstemon	*Penstemon eatonii*	x	x		x				
Grasses	Sandberg wheatgrass	*Poa secunda*					x	x	x	
Grasses	Bluebunch wheatgrass	*Pseudoroegneria spicata*				x	x	x		
Grasses	Idaho fescue	*Festuca idahoensis*			x		x	x	x	
Herbaceous Perennials	Blue flax	*Linum lewisii*	x		x	x				x
Herbaceous Perennials	Rocky Mountain penstemon	*Penstemon strictus*	x	x	x	x				

Key = A (Jackson, WY), B (J. Willard Marriott Library, SLC, UT), C (Moda Building, Bend, OR), D (Washington Fruit and Produce, Yakima, WA), E (Bar C Dude Ranch), F (LDS Center, SLC, UT), G (Big Sky, MT), H (Mulvaney Medical Office Building)

6.4.1 Grasses

Thirty-nine genera of grasses are planted on the ten green roof case studies, but only one species occurs on four roofs in all three subregions: *Fescue idahoensis*, making it the most widely-distributed grass species on the green roofs. Seven grass genera have at least two species represented. In terms of geographic distribution, only four grasses occur on roofs in all three subregions: Idaho fescue (*Fescue idahoensis*), Big bluegrass (*Poa secunda*), Blue bunch wheatgrass (*Pseudoregnaria spicata/Elymus spicatus*) and Prairie Junegrass (*Koelaria macrantha*).

Bluebunch wheat grass (*Pseudoregnaria spicata/Elymus spicatus*) is found on three roofs while another three *Elymus* species are distributed across four (some roofs have more than one *Elymus* species present) indicating four wheatgrass varieties present on five of the nine roofs across all three subregions.

In terms of grass species diversity, *Poa* is represented by five species, four of them native (*P. alpine*, *P. compresa*, *P. fendleriana* and *P. secunda*) with the most prevalent being Sandburg or big bluegrass (*P*. secunda) on three roofs in the Rocky Mountains and Basin & Range. *Poa* species are present on three roofs across those same subregions. *Festuca ovina* was the only other grass present on at least two roofs in two subregions, Rocky Mountains and Basin and Range. Another 18 grass species appear once on a single roof in one subregion although four of the single specie/single occurrence were noted as best performers in the Basin and Range subregion: crested wheatgrass, blue grama, sideoats grama, and 'Legacy' buffalo grass.

The greatest diversity of grass species in a single location is the 15 different genera present on the LDS Conference Center roof in Salt Lake City. Fourteen of the 15 grasses are native to the intermontane semi-desert grasslands – only crested wheatgrass (*Agropyron cristatum*) would be considered non-native.

6.4.2 Herbaceous Perennials, Sedums, Bulbs

A pattern of species diversity relative to roof area is observed in the planting of herbaceous perennials on the chosen case study roofs: the Salt Lake City roofs have 67 different perennials and forbs, the Columbian Plateau roofs 47 and the Rocky Mountain roofs only 15 perennials. In terms of geographic distribution, 13 species of *Penstemon* occur on four roofs with one roof in each of the subregions – the widest distribution of any forb in the case studies. The Jackson residences (Rocky Mountains), Marriott Library SLC (Basin & Range), Washington Fruit & Produce, and MODA Office Building (both Columbian Plateau) contain blue penstemon, Rocky Mountain penstemon (*P. eatonii* and *P. strictus*) and yarrow (*Achillea millefolium*).

The plantings on each roof represent the best professional knowledge in plant selection and the choice of plantings varies widely. For example, only the Salt Lake

City Public Library has flowering bulbs and the Mulvaney Office Building has the most sedum species; eight of the nine mostly non-native sedums recorded.

6.4.2.1 Species Diversity

On the ten green roof case studies, 99 perennial species were recorded across all three subregions. Of these, 56 genera are represented with 47 having two or more species. The largest genera contain the 13 varieties of *Penstemon*. As native forbs, this could be expected since Red Butte Gardens has a notable collection of over 250 species of *Penstemon*. Even more than the grasses, the perennial plant selection represented by the ten case studies contains many plant families.

The diversity of herbaceous perennials across multiple subregions is indicated with the number of plants on roofs in different subregions. Yarrow is one of the most common natives found on five roofs across all three subregions. Eight species are found on roofs in at least two different subregions (Table 6.8).

The growing conditions that encourage hyssop, milkweed, and fleabane in both the Basin & Range and Columbian Plateau need to be better understood. Balsamroot, blue flax, and globemallow are growing on green roofs in both the Rocky Mountain intermontane valleys and the Columbia Plateau grasslands. That these six species successfully grow on roofs in two subregions would indicate the need to investigate conditions that encourage their survival and perhaps encourage wider usage especially if these species can improve the ecological value of the roofs. What are the conditions that promote a broader range for species? Can ecosystem richness improve with a more consistent palette of viable plant species for this region?

Ten perennial species are found on two case study roofs which can encourage closer comparison of plant performance over time. As with the grasses, most of the roofs have singular perennial species represented; 84 different species or 65% of all the perennials recorded are planted on only one roof.

Table 6.8 Native herbaceous perennial plant diversity across physiographic regions of the Intermontane West

Common Name	Botanical Name	Rocky Mts.	Basin & Range	Columbia Plateau
Threadleaf giant hyssop	*Agastache rupestris*		x	x
Milkweed	*Asclepias tuberosa*		x	x
Oregon daisy/aspen fleabane	*Erigeron speciosus*		x	x
Blue penstemon	*Penstemon cyananthus*	x	x	
Balsamroot	*Balsamorhiza sagittate*	x		x
Sulphurflower buckwheat	*Eriogonum umbellatum*	x		x
Blue flax	*Linum perenne lewisii*	x		x
Globemallow	*Sphaeralcea glossularifolia*			x

6.4.3 *Shrubs and Trees*

Shrubs and small trees are recorded on only the Great Basin and Columbia Plateau green roof examples. While sage is a dominant a dominant plant in the ecoregion, it is only found on roofs that are roof gardens with deeper substrates. The same is noted for tree planting: deeper planting pits or raised planting boxes permit successful tree growth. The Salt Lake City Public Library and the Moda Office Building have 23 of the 25 shrub and tree species planted on them.

6.5 Summary

Since the construction of the first dude ranches with sod roofs in Grand Teton National Park, and roof gardens on hotels and clubs in Salt Lake City in the early twentieth century, many modern green roofs have been constructed across the Intermontane Semi-Arid Grassland ecoregions. The case studies highlighted in this chapter demonstrate that green roofs were built to:

- connect visually with the surrounding environment, e.g., LDS Conference Center, Washington Fruit and Produce Headquarters, Salt Lake City Library, private residences;
- provide a learning tool for green roofs, e.g., J. Willard Marriott Library, Ogden Nature Center, Bar C Dude Ranch, LDS Conference Center;
- provide an outdoor laboratory for observing nature, especially pollinators and other visitors to the green roofs, e.g., LDS Conference Center, University of Utah, Moda Building, Salt Lake City Library;
- demonstrate benefits generally associated with green roofs such as energy conservation, runoff amelioration, temperature, evapotranspiration, e.g., University of Utah, LDS Conference Center;
- offset the effects of habitat loss, not via a mitigation process per se, but rather through the intentional introduction of habitat that increases the number of plants of a particular habitat once common to the site, e.g., Moda building, Washington Fruit and Produce Headquarters, LDS Conference Center;
- increase the aesthetic appeal of buildings especially where the function of the building itself may be associated with increased stress for visitors, (hospitals) or employees in private offices, e.g. Mulvaney Medical Office Building, Washington Fruit and Produce, and the Moda building;
- provide an environment where people can rest or take a break, e.g. Mulvaney Medical Office Building, University of Utah, Moda Building, Salt Lake City Library.

Analysis of the ecoregional green roof case studies in the three subregions reveals a pattern differentiating the types of roofs observed. The Great Basin roofs in Salt Lake City are all large installations on public and institutional buildings. The variety of grasses, forbs, and shrubs planted on these roofs is the greatest with more than 27 different grasses, 67 different forbs, and 11 different shrubs and trees. The Columbia Plateau case study roofs are generally second in square footage, are all private businesses, and were planted with four grass species, 47 perennial species, and nine different shrubs and trees. The Rocky Mountain case study roofs are primarily at a residential rooftop scale and exhibit the narrowest range of species: nine different grasses, 15 perennials, and no shrubs or trees. The available square footage of green roof area exhibited in the three sets of case studies differentiated by decreasing roof area display a narrowing variety of all types of plant species. The greater species variety on the larger scale roofs would not necessarily be considered a common pattern in green roof plant specification protocols and may only be a pattern visible in this limited sampling of representative roofs.

Further analysis of the plants to identify the best pollinator attractors during different life stages is warranted. Specific varieties of the species currently living the case study roofs are identified through pollinator organizations as attractive to insects at different life stages. With public interest in the Intermontane West growing for protecting pollinators, the green roofs in this region represent both convenient habitats as well as public education opportunities.

With the variety of plants specified and the majority of species being planted on only one roof, it will be difficult to draw comparisons in terms of survival and hardiness. The relatively small sample of plants occurring on roofs across at least two subregions points the way towards regional plant studies for resilience and survivability.

Acknowledgments We would like to thank the following individuals for making time and resources available to explore the green roofs in this chapter including Charisa Wagner with Intermountain Roofscape Supply; Thomas Ward of Ward+Blake Architects; Kathy Wonson at the National Park Service; Kyle Hemly with CSHQA; Dr. Paul Johnson at Utah State University; Jay Warnick, Jen Udy and staff at the Church of Jesus Christ of Latter-day Saints; Sue Pope at the University of Utah; Mark Johnson of Civitas; Jonathan Morley at Berger Partnership; staff at the Washington Produce and Fruit Growers Headquarters; and Dr. Rick Martinson owner WinterCreek Environmental. We would also like to thank the owners of several private residences who allowed us access to their green roofs and were willing to share knowledge about their green roofs.

References

Andersen BA (1996) Desert plants of Utah. Utah State University, Logan

Archibold O (1995) Temperate grasslands. In: Ecology of world vegetation. Springer, Cham, pp 204–237

Barth B (2015) Green building, Teton style. gb&d Magazine, vol 6. gb&d Magazine, Chicago

Bailey RG (1997) Ecoregions of North America. U.S. Department of Agriculture, Forest Service, Washington, DC

BLM (2019) Bureau of Land Management Idaho native plants program. BLM. https://www.blm.gov/programs/natural-resources/native-plant-communities/native-plant-and-seed-material-development/collection/idaho. Accessed 17 July 2019

Cantu RJ (2012) Green roofs for historic buildings: case study of the Bar BC dude ranch at Grand Teton National Park. University of Pennsylvania, Philadelphia

Danler S, Langellotto-Rhodaback GA (2015) Ecological design of urban landscapes: economic, social, and ecological benefits, vol 17. Oregon State University, Extension Service, Corvallis

Davies KW, Boyd CS, Bates JD, Hamerlynck EP, Copeland SM (2020) Restoration of sagebrush in crested wheatgrass communities: longer-term evaluation in Northern Great Basin. Rangel Ecol Manag 73(1):1–8

Denzin NK (2008) Drawn to Yellowstone. Qual Res 8(4):451–472

Dewey DW, Johnson PG, Kjelgren RK (2004) Species composition changes in a rooftop grass and wildflower meadow: implications for designing successful mixtures. Native Plants J 5(1):56–65

Feng Y, Burian S, Pomeroy C (2016) Potential of green infrastructure to restore predevelopment water budget of a semi-arid urban catchment. J Hydrol 542:744–755

Feng Y, Burian S, Pardyjak E (2018) Observation and estimation of evapotranspiration from an irrigated green roof in a rain-scarce environment. Water 10(3):262

Griffin D (2002) Prehistoric human impacts on fire regimes and vegetation in the northern Intermountain West. In: Vale TR (ed) Fire, native peoples, and the natural landscape. Island Press, Washington, DC, pp 77–100

Hammond E (1964) Classes of land-surface form in the United States. US Geological Survey

GTNP (2017) Grand Teton/learn about the park/nature/plants. National Park Service. https://www.nps.gov/grte/learn/nature/plants.htm. Accessed 16 July 2019

IDFW (2020) Idaho Department of Fish and Game: Idaho Species. Idaho Department of Fish and Game. https://idfg.idaho.gov/species/taxa/family/Plantae/Anthophyta/Dicotyledoneae/Caryophyllales/Cactaceae?idaho=1. Accessed 5 Feb 2020

JWM (2018) J. Willard Marriott Libray/Rooftop Garden. University of Utah. https://lib.utah.edu/info/green/rooftop-garden.php. Accessed 19 July 2019

Kiersz A (2018) Here are the fastest growing and shrinking counties in the US. Business Insider. Insider, Inc., New York City

Kleiner EF, Harper K (1966) An investi-gation of association patterns of prevalent grass-land species in Red Butte Canyon, Salt Lake County, Utah. Proceedings of Utah Academy of Science. Arts Lett 43:29–36

Lee KE, Williams KJ, Sargent LD, Williams NS, Johnson KA (2015) 40-second green roof views sustain attention: the role of micro-breaks in attention restoration. J Environ Psychol 42:182–189

Longfield K (2011) Conservation and management plan for the Bar BC Dude Ranch Grand Teton National Park (trans: Preservation H). National Park Service, Washington, DC

Matero F (2016) Saving the Bar BC Dude ranch: a new method for setting preservation priorities. George Wright Forum 33(1):47–58

Miller RF, Chambers JC, Pellant M (2014) A field guide for selecting the most appropriate treatment in sagebrush and piñon-juniper ecosystems in the Great Basin: evaluating resilience to disturbance and resistance to invasive annual grasses, and predicting vegetation response. General Technical Report RMRS-GTR-322-rev Fort Collins, CO: US Department of Agriculture, Forest Service, Rocky Mountain Research Station 66 p 322:76

Nemerov A (1994) Projecting the future: film and race in the art of Charles Russell. Am Artist 8(1):71–89

Noland S (2012) Bar BC cabins #1369 and #1370 sod roofs project completion report. National Park Service, Grand Teton National Park

Noss RF, Peters RL (1995) Endangered ecosystems of the United States: a status report and plan for action. Washington, DC

Parker L (2010) Video: grow your practice clean, profitable and green. Idaho

RBC (2019) Red Butte Canyon/vascular flora. U.S. Forest Service. https://redbuttecanyon.net/flora.html. Accessed 2 Aug 2019

Ricketts TH (1999) Terrestrial ecoregions of North America: a conservation assessment. Island Press, Washington, DC

Roth D (2018) Church of Jesus Christ of Latter-Day Saints (LDS) Conference Center. Greenroofs.com. Accessed 19 July 2019

Schuler TA (2018) Made to disappear. Landscape Architecture Magazine, vol August. American Society of Landscape Architects, Washington, DC

Stoddart LA (1941) The Palouse Grassland Association in Northern Utah. Ecology 22(2):158–163. https://doi.org/10.2307/1932210

Ulrich R (1984) View through a window may influence recovery. Science 224(4647):224–225

Vale T (2002) Fire, native peoples, and the natural landscape. Island Press, Washington, DC

Ward T (2018) Jackson sod roofs. Ward+Blake, Jackson

Ward+Blake (2014) In the shadows of the Tetons. Gorden Goff, China

Weiler SK, Scholz-Barth K (2009) Green roof systems: a guide to the planning, design, and construction of landscapes over structure. John Wiley & Sons, Inc., Hoboken

Whitby F (2018) Honeybee info book. The City Library and Frank Whitby, Salt Lake City

Wonson K (2013) Bar BC cabins 1369 and 1370 sod roofs project. Penn State University, State College

Wu T, Smith R (2011) Economic benefits for green roofs: a case study of the Skaggs pharmacy building, University of Utah. Int J Des Nat Ecodyn 6(2):122–138

Chapter 7
Green Roofs in California Coastal Ecoregions

Bruce Dvorak and Philippa Drennan

Abstract This chapter presents case studies of four conservation sites and 11 green roofs located in the broadly coastal ecoregions of western California. Geographically complex, California's many ecoregions are differentiated by topography and microclimate along the coast. This chapter focuses on the chaparral/shrub/meadow/and dune ecoregions from San Francisco to San Diego. Precipitation varies greatly along the coastal region from 900 mm annually in the north to 350 mm in the south, most of which falls as winter rain. Thus, grasses, herbaceous perennials, and annual forbs compete with shrubs for dominance. Intermixed are a wide range of larger shrubs, oaks, and conifers. Less than 30% of the native habitat remains intact in these ecoregions; introduced grasses are widespread and have nearly replaced native grasses. Eleven green roof case studies demonstrate how 191 taxa native to California coastal ecoregions can be employed on green roofs in highly built-up areas.

Keywords Dunes · Coastal meadow · Rainwater harvesting · Native · Endangered · Restoration · Microbiota · Environmental education · Living roof · Policy

7.1 Ecoregion Characteristics

The range of ecosystems and the biological diversity of California's coastal ecoregions are some of the most complex ecosystems in North America. Further, they have been influenced by human activity since the beginning of the Holocene (Munz 1970). Before the Spanish exploration and later settlement of the California coast in

B. Dvorak (✉)
Department of Landscape Architecture and Urban Planning, 305A Langford Architecture Center, Texas A&M University, College Station, TX, USA
e-mail: bdvorak@tamu.edu

P. Drennan
Department of Biology, Loyola Marymount University, Los Angeles, CA, USA
e-mail: pdrennan@lmu.edu

B. Dvorak (ed.), *Ecoregional Green Roofs*, Cities and Nature,
https://doi.org/10.1007/978-3-030-58395-8_7

the late eighteenth century, many Native American tribes lived widespread throughout the California coast and the Central Valley. Since the early Holocene, humans occupied the coastal landscapes and made widespread modifications to the vegetation in terms of anthropogenic fire, and the distribution of plants used for food and medicine (Keeley 2002; King 1988). There is evidence of extensive intentional burning of the land by Native Americans in coastal California, which increased the proportion of grasslands and meadows offering greater utility to those living there (King 1988; Huenneke 1989; Keeley 2002). With the influx of people following the Gold Rush of 1849, the vegetation became altered in dramatic ways. Habitat loss in the coastal ecoregions continues as population and urban development have increased.

California has over 7000 native plant species, including subspecies, and varieties (Calscape 2019) up to 30% of which are endemic occurring only in California. Some plants, such as lupine (*Lupinus albifrons* and *Lupinus formosus*), support endangered wildlife such as the Mission blue butterfly which is endemic to the meadow ecosystems once common across the San Francisco Bay region (Fig. 7.1)

Fig. 7.1 *Blue Lupines of the Sand Dunes of San Francisco*, by Theodore Wores (1912). Prior to its urbanization, the low-elevation ecosystems of the San Francisco area included sand dunes, meadows, dune scrub, mudflats, and tidal marshes. The bedrock hills were covered with grasses, clusters of conifer trees, oaks and scrub vegetation. In his many paintings of these habitats, Theodore Wores captured the occurrence of native plants including lupines, yarrow, penstemon, cactus, poppies, grasses, shrubs, and trees. These and the red hummingbird sage (lower right) shown in this painting are used in several case studies in this chapter. (Triton Museum, photo courtesy of Jan Looper Smith)

(Green 2019). The coastal vegetative communities have experienced significant change as ground fires were suppressed, exotic grasses invaded coastal meadows and valley prairies, and landscapes were plowed, and planted as orchards, vineyards, or with other crops. The great diversity of native plants and plant communities remain, but their extent and richness are stressed and altered by human development and activities (Huenneke 1989; Mattoni and Longcore 1997; Cox and Allen 2007; Stromberg et al. 2007). Thus, preservation and restoration of ecosystems are critical for maintaining biodiversity as urbanization expands (Munz 1970; CDFW 2019; Green 2019).

Conservation and restoration of coastal meadows, sagebrush communities, prairies, and chaparral takes place throughout the state, but invasive species, drought, and disagreement over fire management practices make restoration difficult to achieve. Green roofs can be used to increase biodiversity and reintroduce native plants in urban areas, although they should not be used as a form of mitigation or replacement for a quality native habitat (Lundholm and Walker 2018). Parts of California, such as urban Los Angeles, could greatly benefit from the use of ecoregional green roofs to increase the connectivity between remaining native landscapes and provide opportunities for research and the exploration of native plants on green roofs. The remaining and protected California coastal plant communities are composed of plants that tolerate and avoid drought (Commission 1975; Keeley 2002; Alexander and D'Antonio 2003), characteristics that potentially make them well suited for use in green roofs.

The climates of the California coast are typical of a Mediterranean climate (Fig. 7.2). Generally, there is more precipitation along the northern California coast and less in Southern California. San Francisco and San Diego experience mild temperatures, where Los Angeles is warmer during the summer. Precipitation generally falls in gentle low-intensity events, but high-intensity storms can occur during winter. Thus, vegetation for green roofs of the ecoregions may not be interchangeable across the north-to-south gradient. The climate of the interior valley and Sierra Mountains is quite different from the coastal climates, and are not included here.

7.1.1 Vegetation in the California Coastal Ecoregions

Dominant ecosystems located in the ecoregions of this chapter include grasslands, shrub habitats, chaparral, and dunes (Fig. 7.3). Although little of the pre-Columbian 8,000,000 ha. (30,888 mi^2) of grassland native to California remain, there is much interest in restoring grassland ecosystems, including those that were once thought impossible to restore (Huenneke 1989; Stromberg et al. 2007). Some of the most common kinds of grasslands found growing in California include valley and hill grasslands, north coastal prairies, and in higher altitudes, there are montane meadows and alpine tundra habitats (Baker 1989). Plants common to grasslands include a wide range of genera and species and different compositions that depend on location. It is not clear that there was a typical grassland community, but there were

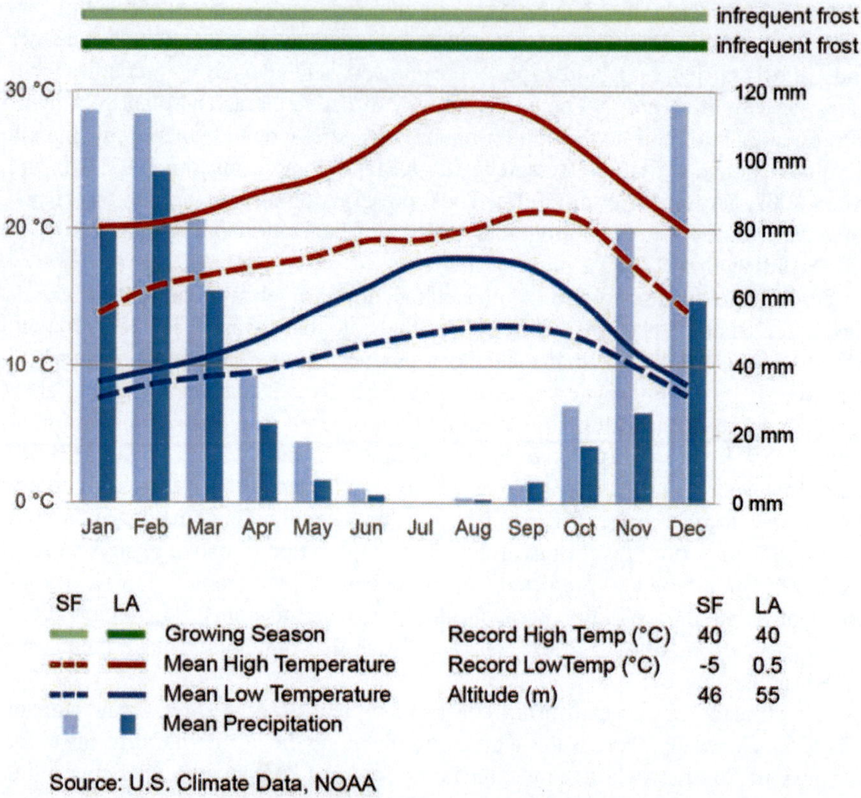

Fig. 7.2 Precipitation and temperature characteristics for San Francisco (SF) and Los Angeles (LA), California. Generally, more precipitation falls near San Francisco annually 584 mm (23 in) with a gradual decline towards San Diego (300 mm/12 in). Locations north of San Francisco such as Sonoma and Santa Rosa, still near the vicinity of San Francisco, receive about 780–965 mm (31–38 in) of precipitation annually, as they are higher in elevation than San Francisco. Much of the precipitation falls during the winter, thus a significant dry/dormant season is present along the coast. The climate in Los Angeles and Southern California is much warmer than San Francisco, where its daily average high and low temperatures are significantly warmer, and with less annual precipitation 384 mm (15 in)

species that were common to many locations, and anthropogenic burning by native cultures was common across the coastal regions for thousands of years (Huenneke and Mooney 1989; Cox and Allen 2007; Vale 2013). Current grassland restoration is taking place in many parts of California, and fire is being re-introduced to remove exotic invasives and regenerate species once common to California (D'Antonio et al. 2002; Keeley 2002).

Because the flora of coastal California is so diverse, we present some of the more common native vegetation of plant communities that grow along the coast near urban regions. A few invasive species are listed, to make them known and avoid their use on green roofs. The native vegetation includes different forms of plants

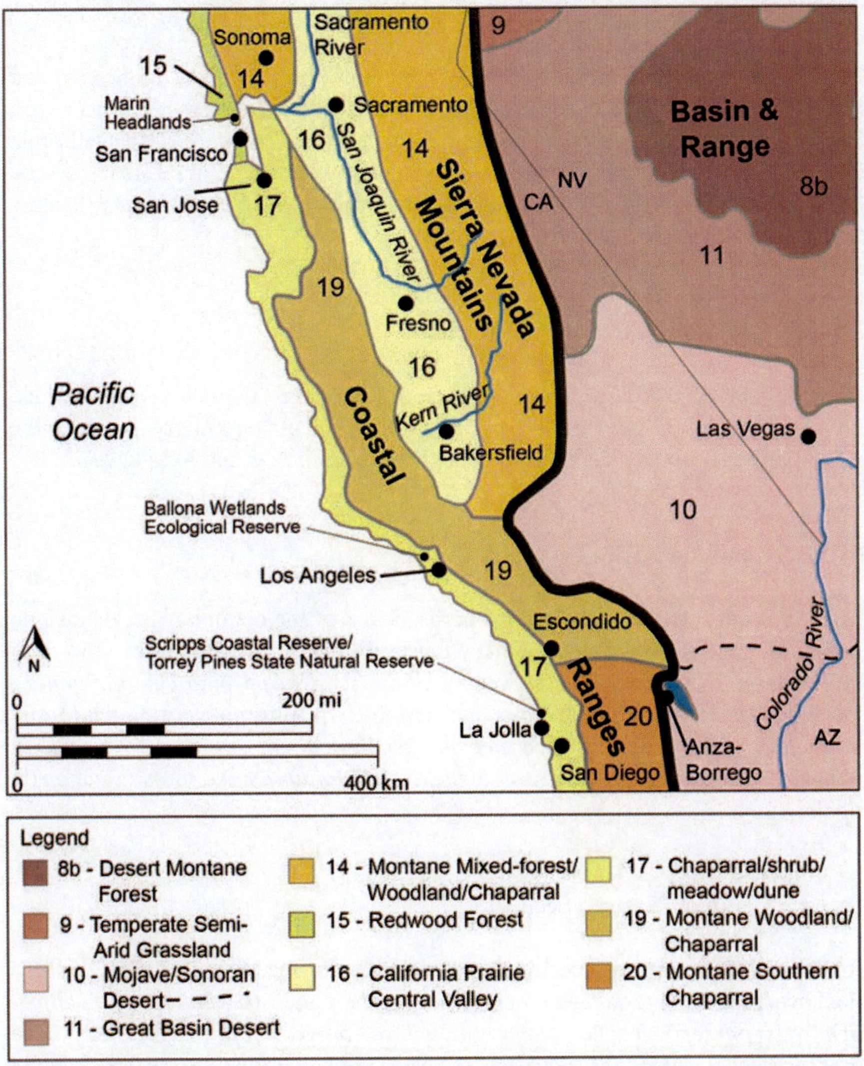

Fig. 7.3 The western California ecoregions (Bailey 1997) covered in this chapter include Chaparral (17) also intermixed in this region are shrub, meadow, and dune ecosystems. Succulents grow in the ecoregion and are more frequent south of Los Angles. The Montane Woodland and Montane Chaparral (19) ecoregions dominate the coastal mountains. Urban sprawl associated with Los Angeles, San Diego, and Escondido border and encroach these ecoregions. (Graphic: Trevor Maciejewski and Bruce Dvorak)

such as annual and perennial grasses and herbs, trees and shrubs belonging to scrub, chaparral, grassland, and desert communities (Biswell 1956; Munz 1970; Alexander and D'Antonio 2003; Keeley et al. 2006; Koteen et al. 2011).

7.1.1.1 Common Native Perennial Grasses/Sedges

There are over 600 native grass species growing in California. Some of the common species include: purple needlegrass (*Stipa pulchra*), pine bluegrass (*Poa scabrella*), creeping wildrye (*Elymus triticoides*), melic grass (*Melica californica*) California oatgrass (*Dassthoniia californica*), squirrel grass (*Sitanion hystrix*), Idaho fescue (*Festuca idahoensis*), carex (*Carex*), deschampsia (*Deschampsia* spp.), Pacific reedgrass (*Calamagrostis nutkaensis*).

7.1.1.2 Common Native Annual Herbs

There are over 2100 native annual species. Some of the common annuals include clovers, (*Trifolium*), tarweed, (*Hemizonia*), Spanish clover (*Lotus americanus*), ground lupine (*Lupinus bicolor*), and poppies (*Eschscholzia* and *Argemone*).

7.1.1.3 Common Native Perennial Herbs

There are over 3300 native perennial herbs. Some of the common species include those with large genera such as monkeyflower (*Diplacus* syn. *Mimulus* is known to have at least 77 species), beach evening primrose (*Cammisonia chieranthifolia),* coastal golden yarrow (*Eriophyllum staechadifolium),* California aster (*Corethrogyne filaginifolia),* blue springs (*Penstemon heterophyllus),* woodland strawberry (*Frageria vesca),* common yarrow (*Achillea millefolium)* spike moss (*Selaginella bigelovii*).

7.1.1.4 Common Native Succulents

There are 179 species of succulents native to California. Some of those include blochman's dudleya (*Dudleya blochmaniae*), cane cholla (*Cylindropuntia californica* var. *parkeri*), chalk dudleya (*Dudleya pulverulenta*), chaparral yucca (*Hesperoyucca whipplei*), coastal agave (*Agave shawii*), coastal cholla (*Cylindropuntia prolifera*), coastal prickly pear (*Opuntia littoralis*), San Diego fingertips (*Dudleya edulis*), fish hook cactus (*Mammillaria dioica*), red maids (*Calandrinia ciliata*), and San Diego barrelcactus (*Ferocactus viridescens*).

There are also several species of native sedum including cascade stonecrop (*Sedum divergens*), canyon creek stonecrop (*Sedum obtusatum* ssp. *paradisum,* rose flowered stonecrop (*Sedum laxum*), oblongleaf stonecrop (*Sedum oblanceolatum*), Sierra stonecrop (*Sedum obtusatum*), Davidson's stonecrop (*Sedum niveum*), yellow stonecrop (*Sedum spathulifolium*), Oregon stonecrop (*Sedum oreganum*), spearleaf stonecrop (*Sedum lanceolatum*), and cream stonecrop (*Sedum oregonense*).

7.1.1.5 Common Native Shrubs

There are over 1054 native shrub species. Some of the common species that may translate to some green roofs include coyote brush (*Baccharis pilularis*), wedgeleaf ceanothus (*Ceanothus cuneatus*), ceanothus (*Ceanothus* spp.) (some as trees), hollyleaf buckthorn (*Rhamnus crocea*) manzanita (*Aretostaphylos* spp.), yucca (*Yucca* spp.), California brittlebush (*Encelia californica*), artemisia (*Artemisia* spp.) eriogonum (*Eriogonum* spp.), current or gooseberry (*Ribes* spp.), purple sage (*Salvia leucophylla*), and other salvias (*Salvia* spp.).

7.1.1.6 Common Native Trees

There are 201 native tree species including 80 species of oak. Some of the common oaks include blue oak (*Quercus douglasii*), interior liveoak (*Quereus wislizeni*), (*Quercus* spp.), buckeye (*Aesculus californica*), foothill pine (*Pinus sabiniana*), and hollyleaf buckthorn (*Rhamnus crocea*).

7.1.1.7 Common Exotic and Invasive Annual Herbs, Grasses, and Shrubs

Broadleaf filaree (*Erodium botrys*), bur clover (*Medicago hispida*), common foxtail (*Hordeum hystrix*), common ryegrass (*Lolum multiflorum*), foxtail fescue (*Festuca megalura*), french broom (*Genista monspessulana*), napa thistle (*Centaurea melitenmis*), scotch broom (*Cytisus scoparius*), soft chess (*Bromus mollis*), red brome (*Bromus rubens*), red-stem filaree (*Erodium cicutarium*), ripgut grass (*Bromus rigidus*) slender oat (*Avena barbata*), and wild oat (*Avena fatua*).

7.1.1.8 Common Exotic and Invasive Succulents

Iceplant or highway iceplant (*Carpobrotus edulis*) is found all along the coast. This plant has been introduced to California from South Africa, is highly invasive in the coastal ranges of California, is on California's invasive species list, and should not be planted on green roofs. This plant can outcompete and suppress native ecosystems and is a major hindrance to conservation and restoration efforts (Fig. 7.4). Another invasive succulent, common ice plant (*Mesembryanthemum crystallinum*) is native to southern Europe. Its unique foliage, flowers and bladder cells, and adaptability make this ornamental exotic difficult to eradicate.

Fig. 7.4 The upright dark green leaves of iceplant (*Carpobrotus edulis*) spread with lateral underground stems (rhizomes) on a foredune in Playa del Rey. Icecplant forms large mats that encroach on native species such as beach bur (*Ambrosia chamissonis)* shown above. (Photo: Bruce Dvorak, October 2018)

7.1.2 Ecoregional Conservation Case Studies (Arranged North to South)

7.1.2.1 Marin Headlands, San Francisco, California

As part of the Golden Gate National Recreation Area, the Marin Headlands is a hilly peninsula located north of San Francisco that preserves native scrub vegetation (Fig. 7.5). With a long history that reaches back to Native American cultures, the preserved land is centrally located in the California Floristic Province, a biodiversity hotspot that extends all along the California coast. It is a unique community of grasses, wildflowers, trees, and shrubs, and groundcovers. Of interest in identifying vegetation that may adapt to green roofs are the grasslands, meadows, and scrublands (Fiedler and Leidy 1987; Steers and Spaulding 2013; NPS 2019).

Native perennial grasses include purple needlegrass, tufted hair grass, blue wild rye, and California oat grass. Wildflowers include Indian paintbrush, California poppies, mission bells, lizard tail, lupines, yarrow, columbine, monkeyflower, daisies, buttercups, blue-eyed grass, and many more (NPS 2006).

Fig. 7.5 Many perennials, low-growing shrubs, and grasses grow along the rocky, open, and wind-prone slopes at Marin Headlands preserve. The yellow blooms of lizard tail (*Eriophyllum staechadifolium*) grows here with a community of other plants. It also grows on the California Academy of Sciences living roof, also in a dynamic plant community (Sect. 7.3.4). (Courtesy of Tess Menotti, June 2013)

Shrub vegetation that may adapt to semi-intensive or intensive green roofs includes California sagebrush, coyote brush, lizard tail, cream bush, osoberry, coffeeberry, and twinberry. Other hardy plants include chamise, manzanita, yerba santa, and ceanothus (NPS 2019).

7.1.2.2 Ballona Wetlands Ecological Reserve, Los Angeles, California

The Ballona Wetlands is the last remaining major coastal wetlands in Los Angeles County (Fig. 7.6). It is situated to the south of Marina del Rey and approximately 3.2 km (2 mi) north of the Los Angeles International airport (Dorsey 2007; Read 2015). The Native Americans that lived in the region for at least the past 8000–10,000 years, originally survived as hunter-gatherers with a shift about 1000 years ago to a greater reliance on resources from the lagoon and open ocean (Boytner and Dorsey 2007). Anthropogenic burning was known to occur along the coast potentially modifying grasslands and prairies (King 1988). Increasing human impact occurred following the arrival of the Spaniards. In addition to cattle grazing

Fig. 7.6 Western goldenrod (*Euthamia occidentalis*) blooms along the fringes of the Ballona wetlands. (Photo: Bruce Dvorak, October 2018)

and agriculture, the ecosystem became fragmented by rail and roads. In the 1930s the Ballona Creek was channelized limiting tidal flow to the wetlands. Together with the construction of the largest man-made marina, Marina del Rey to the north, and increasing urbanization, the future existence of the wetlands was jeopardized. Today the land is owned by the State of California and large-scale restoration planning is going through the permitting process. Restoration provides the opportunity to reintroduce historical flora into the area including rare and endangered species (Read 2015).

Community-based restoration by-hand is undertaken currently to remove exotic invasive species and to encourage native vegetation to grow. An example of restoration efforts is the Ballona Dunes Restoration Program (initiated in 1994 by Friends of Ballona Wetlands) in which non-native vegetation consisting largely of iceplant (*Carpobrotus edulis)*, is removed from a coastal dune remnant along the western boundary of the Ballona Wetlands, and replaced with native species. The native dune lupine (*Lupinus chamissonis)*, branching phacelia (*Phacelia ramosissima)*, deerweed (*Acmispon glaber* syn. *Lotus scoparius)* and seacliff wild buckwheat (*Eriogonum parvifolium*) are now common in the dunes. El Segundo blue butterflies are federally endangered endemic butterflies that have codependency with seacliff wild buckwheat, a low-growing evergreen shrubby perennial that was once common to the coastal areas near Los Angeles. Previously extirpated, the El Segundo blue butterfly was discovered in the restored dune area in 2011. Green roofs could be explored as a mechanism for increasing and reintroducing habitat available for species in such tight ecological relationships and to increase the abundance of rare and endangered plant species of the surrounding ecosystem (Dixon 1999; Drennan et al. 2011).

7.1.2.3 Scripps Coastal Reserve/Torrey Pines, San Diego, California

Pre-historic use of the Scripps property dates back to 5000 years ago when the Kumeyaay Dieguero tribe inhabited the knoll. From 1822 onward, various uses by settlers took place on the site including farming, ranching, and training grounds for the U.S. Army. In 1967, the University of California purchased the 65-hectare (162-acre) property as a place for research. As a coastal scrub community, the vegetation is well-adapted to limited precipitation. Many succulents hold onto water, and many low-growing woody shrubs complete their growing season during the brief periods of rain during late fall through spring. Bush sunflower, for example, is a common plant on the knoll and blooms January to June with dark-centered flowers. The remainder of the summer, the plant enters a summer dormancy to avoid the summer heat and drought (Fig. 7.7). Over 200 plant species are growing on the preserve that makes habitat for at least 88 bird species. Other common plants include bladder pod (*Peritoma arborea*), California sagebrush (*Artimesia californica*), cane cholla (*Cylindropuntia californica* var. *parkeri*), coast prickly pear (*Opuntia littoralis*) California box thorn (*Lycium californicum*), flat top buckwheat (*Eriogonum fasciculatum*), fingertips (*Dudleya edulis*), lemonadeberry (*Rhus integrifolia*), and San Diego barrelcactus (*Ferocactus viridescens*) which is endangered and endemic to San Diego.

7.1.2.4 Torrey Pines State Natural Reserve®

Torrey Pines is a 607-hectare (1500-acre) conservation site known for its preservation of an endemic and rarest pine tree in the United States, *Pinus torreyana*. As a coastal sage scrub plant community, the reserve presents visitors with an authentic landscape for San Diego (Fig. 7.8). As irrigation from the Colorado River keeps San Diego green throughout the year, this landscape holds plant communities that have endured here for thousands of years through drought tolerance and avoidance and only natural rainfall. The Torrey pine is grown in horticultural production and used on the landscape. Little is known about its use on roof gardens.

The reserve recognizes five dominant plant communities based upon elevation, exposure to wind, solar radiation, soil composition, salinity, moisture, and interactions with other organisms: chaparral, coastal sage scrub, coastal strand, salt marsh, and Torrey pine woodland (Brothers and Halsey 2020). At least 425 taxa of plants

Fig. 7.7 Coastal prickly pear (foreground) is a native cactus that established in a community of low-growing drought-adapted California sagebrush at the Scripps Coastal Reserve. The grey twigs of sagebrush are near the end of their summer dormancy cycle shown here in late October. This site includes many species that were planted at the NOAA Southwest Fisheries Science Center green roof (Sect. 7.3.10) in La Jolla, located just a few kilometers away below the development shown here. (Photo: Bruce Dvorak, October 2018)

are native to the reserve. Some of the plants that grow at the reserve and on green roofs in the case studies include Shaw's agave (*Agave shawii*), chalk dudleya (*Dudleya pulverulenta*), California aster (*Corethrogyne filaginifolia*), coyote bush (*Baccharis pilularis*), California buckwheat (*Eriogonum fasciculatum*), fingertips (*Dudleya edulis*), California poppy (*Eschscholzia californica*), purple three awn (*Aristida purpurea*), and many more.

7.2 Research in the Ecoregion

Much of the original green roof research in California with native vegetation has been conducted *in situ* through green roof projects, beginning with the Gap headquarters living roof, in San Bruno. This is one of the first large-scale green roof installations in North America that pioneered the use of native species (EarthPledge 2005; Edgerton 2011). There, a green roof of native grasses and annuals provides the benefits associated with green roofs in general, of which noise reduction was

Fig. 7.8 Shaw's agave is shown in the foreground and is native to the San Diego Bay, about 20 km (15 mi) away. It is not native to this site but was introduced here for the preservation of the species. It is growing along with summer dormant native woody vegetation and Torrey pine in the background. The Shaw agave grows successfully on the NOAA Southwest Fisheries Science Center (Sect. 7.3.10). (Photo: Bruce Dvorak, September 2009)

especially important given the location. The roof also increased habitat. Native species selection took into account environmental factors such as aspect and wind exposure. In developing such roofs, Kephart emphasized a holistic approach in design in which green roofs are not just offsets for detrimental impacts, but are part of positive solutions that improve the quality of the environment by, for example, also providing habitat for pollinators (Kephart 2005).

In comparing species native to coastal southern California versus non-native species in green roof mesocosms established in Los Angles, it was shown that native succulents *(Dudleya* spp.) outperformed nonnative sedums which typically succumbed to heat stress especially if irrigation was withheld (Drennan et al. 2011). The difference in heat tolerance between sedums and the native live-forevers is also evident at the cellular level where greater cell death in sedums occurred at lower temperatures than for *Dudleyas* (McDonald and Drennan 2013). In the same research, succulents generally outperformed other growth forms. However, the substrate depth used in the experiment was limited to 10 cm (4 in) as is typical for extensive roofs emphasizing mat-forming succulents. Research establishing minimum growth depth for some of the species in the drier parts of California would be useful as well as the best methods for establishing plants. In the research conducted above to determine a plant palette for use in coastal southern California it was observed that for herbaceous species, plant establishment was the most critical phase. While the planting of seedlings may be aesthetically pleasing, seeding some of the matrix species possibly gives a more consistent cover.

7.3 Green Roof Case Studies (Arranged North to South)

7.3.1 Sonoma Academy High School, Janet Durgin Guild & Commons, Sonoma, California

Sonoma County lies at the juncture of coastal and montane ecoregions. It is a biodiversity hotspot where at least 1210 taxa of plants live in remnant grasslands, oak woodlands, chaparral, and redwood forests (Klein et al. 2015). Located adjacent to prairies and oak woodlands, the Sonoma Academy is a place of learning where "nature is an asset" (WBDG 2019). In this light, the Sonoma Academy High School administration and the design team set out to achieve aggressive sustainability goals and benchmarks for a new multi-purpose building on campus (WBDG 2019). They approached the development with three industry sustainability challenges forming the decision-making process: LEED Platinum, the Living Building Challenge, and aspirations to achieve zero net energy of −4.85 kBTU/sf/year.

The multi-purpose guild and commons buildings are set up to teach and learn about "farm to table" programs, issues, and functions that take place on campus. Classrooms, kitchens, studios, and indoor/outdoor dining space makes this a one of a kind educational setting for young adults. With environmental education as one of the academy's learning objectives for students, the buildings were designed to help teach students to engage their environments, to be critical thinkers and able to participate in the practical observations from data collected by the building systems, including a green roof (Fig. 7.9), solar panels and a cistern (WBDG 2019). This is the kind of education that can help inform future generations about climate-adapted and integrated-designs (building/environment).

7.3.1.1 Project Team

Building Owner/Client: Sonoma Academy
Green Roof Design Team: Rana Creek (lead), SYMBIOS
Architect: WRNS Studio
Structural Engineer: Mar Structural Design
Landscape Architect: RHAA Landscape Architects
Installation Contractor: SYMBIOS
Project completion: September 2017
Green roof area: 564 m^2 (6074 ft^2)

Fig. 7.9 The guild and commons building at the Sonoma Academy makes multiple uses of rooftop space. The roof is programmed with a biodiverse green roof (on top of conditioned space), solar panels, and a white reflective stone ballasted roof to capture water in a cistern for later on-site use. (Courtesy of SYMBIOS, July 2019)

7.3.1.2 Overview and Objectives

The main portion of the green roof was inspired by groundcover vegetation associated with Garry oak ecosystems. The vegetation grows in substrates 20 cm (8 in) to 30 cm (12 in) deep. Substrates are not uniform, but varied in depth to aid in the development of a more diverse ecosystem, and anticipated to support local insects and birds, or those migrating through the region.

7.3.1.3 Plant Establishment

Plants were pre-grown in containers of various sizes.

Grasses

Deer grass (*Muhlenbergia rigens*), foothill sedge (*Carex tumulicola*), nodding needlegrass (*Nassella cernua*), purple needlegrass (*Nassella pulchra*).

Herbaceous Perennials

Blue-eyed grass (*Sisyrinchium bellum*), California fuchsia (*Epilobium canum* 'Everett's choice'), foothill penstemon (*Penstemon heterophyllus* 'Blue Springs'), hummingbird sage (*Salvia spathacea*), scarlet bugler (*Penstemon centranthifolius*), self-heal (*Prunella vulgaris*), yarrow (*Achillea millefolium*), yerba buena (*Satureja douglasii*).

Shrubs

Cleveland sage (*Salvia clevelandii* 'Pozo Blue'), dwarf coyote brush (*Baccharis pilularis*).

Succulents

Stonecrop (*Sedum spathulifolium*).

Photo-Voltaic Understory

These plants are used to keep photo-voltaic panels cool, and are planted beneath solar panels foothill sedge (*Carex tumulicola*), self-heal (*Prunella vulgaris*), stonecrop (*Sedum spathulifolium*) yerba buena (*Clinopodium douglasii* syn. *Satureja douglasii*). In spaces between photo-voltaic panels nodding needlegrass (*Nassella cernua*), and purple needlegrass (*Nassella pulchra*) are added (Fig. 7.10).

7.3.1.4 Irrigation

The irrigation system uses water captured from onsite in a 19 m^3 (5000-gal) cistern that is connected to a system of cisterns on campus. Together cisterns supply 88% of the building's non-potable water demand used for flushing toilets and irrigating plants. The irrigation system for the green roof has eight hydration zones. Each zone is based upon the water needs of the plants and is guided by a weather tracking system that assesses the moisture needs of the plants.

7.3.1.5 Maintenance

The roof is maintained by outside expertise (SYMBIOS) through periodic visits determined by seasonal changes taking place with the weather and on the landscape. This approach is possible with a budget set-aside for the care of the green roof. The

Fig. 7.10 (**a**) *Nassella cernua* and *Nassella pulchra* grow upright between the solar panels. Both grasses are native to Sonoma County and are native to chaparral ecosystems such as the one seen adjacent to the property. (**b**) Lightweight scoria volcanic rock and native *Sedum spathulifolium* cover the roof at the dripline of the solar panels. (**c**) Golden yarrow is in bloom with grasses that are beginning to resume growth from summer dormancy. (**d**) Rain gardens at the ground-level manage runoff from a central gathering space. (Photos: Bruce Dvorak, September 2018)

green roof is integral to the performance of the building systems, and education of the students, staff, and administration, thus infrequent but persistent maintenance visits take place. Activities include the removal of unwanted vegetation, redistribution of seeds from desirable plants, removal of aggressive vegetation, and the review of irrigation and monitoring systems. Nutrient levels of the substrate are measured annually in the spring. Maintenance visits to the roof were weekly during the first week of planting, followed by bi-weekly visits the first month. Monthly visits are typical to keep the roof maintained.

7.3.1.6 Observed Wildlife

Hummingbirds visit salvia, dragonflies, and native bees have been observed on the roof.

7.3.1.7 Best Performing Native Vegetation

All of the vegetation has adapted well (Fig. 7.11).

7.3.1.8 Post-occupancy Observations

SYMBIOS

- Our approach to the maintenance of the green roof is to be aware of the seasonal cycles of the regional landscape. We visit the roof at strategic times (cool season, warm season) to remove unwanted vegetation to prevent potential outbreaks of invasive plants on the rooftop ecosystem. It is much healthier for the green roof to prevent than to restore the green roof ecosystem after invasive plants take hold.
- Matching the irrigation to the water needs of the plants is critical to maintaining the ecosystem. Too much water can cause as much or more damage to the ecosystem as too little water. Through the zoning of irrigation, we aim to maintain the healthy growth of the plants we want. Sometimes only minimal amounts of water are needed to keep the system going. Other times more water is needed.

Fig. 7.11 Sonoma Academy living roof shown here during the second growing season. *Salvia clevelandii* 'Pozo Blue' is in bloom in front (left) along with common yarrow (right). (Courtesy of SYMBIOS, July 2019)

- With "Nature as an asset, and a source of learning" the Academy administration and staff understand that the green roof vegetation may go dormant during the summer. Water is managed to sustain plants, not as a garden setting, but as a climate-adapted ecosystem.

Authors' Reflections

- This integrated design provides educators and students a new model for the design of teaching environments that include nature, and green building practices. Faculty can engage students with green roofs, rain gardens, food crops, and native vegetation.
- The inclusion of monitoring devices on the green roof allows for regular assessment of soil moisture, evapotranspiration, current weather conditions, and water demand for green roof vegetation.

7.3.2 *Slide Ranch Nature Center, Muir Beach, California*

Before the mid-nineteenth century, the Miwok tribes walked and lived along the coastal meadows and cliffs now known as the Golden Gate National Recreation Area. They fished and gathered food. They taught their children how to do these things, and so they lived off the land generation after generation for thousands of years (Anderson 1997). During the settlement of San Francisco, the coastlands property where the Slide Ranch lies today was undeveloped. In 1970, the 54-hectare (134-acre) Slide Ranch property was purchased by a non-profit group to preserve the property from commercial development. The group conceived of the concept of the Slide Ranch as a place to teach children about nature, food, and connections between the two.

The mission of the Slide Ranch is to "Connect children to nature" (SR 2019). In 2017, the Slide Ranch opened a new Farm-to-Table Teaching Center to help connect with the 10,000 school children and families that visit the ranch each year. The farm has goats, sheep, chickens, ducks, deer, and produce that is grown in an organic farm on site. The food feeds the 18 staff and family members who live on the Ranch year-round. A green roof was built on the food center to help keep energy use to a minimum inside the center, reduce runoff on the steep slopes of the property, and connect the built environment with the living environment (Fig. 7.12).

7.3.2.1 Project Team

Building Owner/Client: Slide Ranch Nature Center
Green Roof Design Team Lead: SYMBIOS
Architect: Mark Cavagnero

Fig. 7.12 The living roof on the Farm-to-Table Center is highly visible from the arrival sequence above. Signage along a walking path from the parking area down to the farm describes the purpose and elements of the living roof. (Photo: Bruce Dvorak, October 2018)

Landscape Architect: Rana Creek (landscape master plan)
Installation Contractor: SYMBIOS
Project completion: 2017
Green roof area: 33.6 m^2 (740 ft^2)

7.3.2.2 Overview and Objectives

The green roof was designed to showcase the native plants that grow along the coastal meadows located on the property, retain stormwater, and provide habitat for wildlife. The substrate is 15.24 cm (6 in) deep and is composed of local materials. All of the plants were established from seed that was collected on the property by staff (Fig. 7.13).

7.3.2.3 Plant Establishment

Grasses

Small flowered needlegrass (*Stipa lepida*).

Fig. 7.13 The roof meadow is dominated by herbs and a few patches of the native grass. The ranch is seen below, and the native coastal meadows and pockets of cypress trees are seen in the distance. (Photo: Bruce Dvorak, October 2018)

Herbaceous Perennials

California aster (*Symphyotrichum chilense*), coast buckwheat (*Eriogonum latifolium*), seaside daisy (*Erigeron glaucus*), wild strawberry (*Fragaria vesca* ssp. *californica*), yarrow (*Achillea millefolium*), yerba buena (*Clinopodium douglasii*).

Succulents

Bluff lettuce (*Dudleya farinosa*).

7.3.2.4 Irrigation

An overhead spray irrigation system waters the roof. The roof is watered as needed.

7.3.2.5 Maintenance

The roof is maintained about once a month. Staff at the ranch have maintained the roof for its first two years. Maintenance is now being outsourced since the staff has recognized that maintenance of this green roof is a specialized activity.

7.3.2.6 Observed Wildlife

Pollinators such as butterflies, bees, and birds frequent the living roof.

7.3.2.7 Best Performing Native Vegetation

All of the vegetation has established.

7.3.2.8 Post-occupancy Observations

SYMBIOS

1. The frequency of maintenance is not as important as timing. Preventing an outbreak of aggressive invasive plants such as clover (from the sheep pasture) is more efficient and effective than recovering the roof after an outbreak occurs.
2. Once when the power source to the irrigation lines went out, the staff had to water the roof by hand. Provisions for these type of unplanned events should be discussed and addressed during the design phase of a green roof.

Authors' Reflections

- The design of this green roof makes a clear connection to the concept of an ecoregional roof. The on-site seed sourcing and composition of selected species reinvests the life of the landscape onto the rooftop.

7.3.3 One South Van Ness Avenue, San Francisco, California

The first green roof pilot project for the City of San Francisco and San Francisco County is located above a mixed-use building that includes city and county offices at One South Van Ness Avenue. The Bay Area uses the term "living roof" to discuss and regulate green roofs in San Francisco. Inherent to "living roofs" is their intended replacement of habitat that was once on the ground and displaced by building construction. This traditional view of green roofs is expressed in their inaugural pilot project. The living roof was placed on an existing building in need of a new waterproofing system. The city and county worked with Rana Creek to establish a green roof test plot for staff to test, learn, and teach (Palmer 2015). Although the green roof (Fig. 7.14) is not an equal replacement of the original habitats of the Bay Area (Fig. 7.1), some of the vegetation, insects, birds that once belonged to those plant communities make use of this living roof.

7.3.3.1 Project Team

Building Owner/Client: City and County of San Francisco
Green Roof Design Team Lead: Paul Kephart, Rana Creek
Architect: Douglas Ullman and Glenn Hunt, San Francisco Public Works
Landscape Architect: Rana Creek

Fig. 7.14 The green roof at One South Van Ness Avenue sits on top of the eight-floor in the core administrative district and is viewable from the city offices across the street. The vegetation is partially selected from species growing in local coastal meadows. Since 2010, this pilot project has helped launch a green roof program for the city that educates the public about native plants on green roofs. The green roof guidelines published by the City includes a list of native plants, and examples of how to support local biodiversity (SFLRM 2015). (Photo: Bruce Dvorak, October 2018)

Installation Contractor: Rana Creek
Project completion: 2010
Green roof area: 890 m^2 (9579 ft^2)

7.3.3.2 Overview and Objectives

This green roof is the first installation by the city or county as a pilot project to learn from and as a showcase for those interested in learning about green roofs. Project goals were set to make the green roof resilient in its planting, self-propagating, low maintenance, low chemical input, habitat for urban wildlife, and no use of potable water.

The green roof includes two systems, a biodegradable Biotray® system, and a monolithic (layered) system. The substrate depth is 15 cm (6 in) which includes 7.6 (3 in) of the growing medium inside the Biotrays® and 7.6 cm (3 in) underneath the trays. Biotrays® were made from coconut coir, organic latex, and wood chips. The substrate is a mixture derived from scoria (volcanic rock), sand, compost, and fir bark, and includes macronutrients of nitrogen, phosphorus, potassium, calcium, magnesium, and sulfur; and micronutrients of boron, chlorine, copper, iron, manganese, molybdenum, cobalt, and zinc. The substrate was inoculated with mycorrhizal fungi, to jumpstart a healthy growing environment for plant establishment (Garcia and Gaberšek 2010).

7.3.3.3 Plant Establishment

Plants native to the coastal ecoregions of California (Fig. 7.15) and a few non-native and climate-adapted plants (not included here) were selected for the pilot project (Garcia and Gaberšek 2010).

Annuals

Coast poppy (*Eschscholzia californica* ssp. *maritima*).

Ferns

California polypody fern (*Polypodium californicum*), western sword fern (*Polystichum munitum*).

Grasses

Tufted hairgrass (*Deschampsia caespitosa*).

Fig. 7.15 (**a**) Seacliff buckwheat hybrid (white blooms), and manzanita in the background. (**b**) California fushia in bloom late September. (**c**) Blooms of the California daisy (*Erigeron*) were a later planting addition to the roof. (**d**) California poppy shows its golden blooms. (Photos: Bruce Dvorak, October 2018)

Herbaceous Perennials

Blue-eyed grass (*Sisyrinchium bellum*), California fuchsia (*Zauschneria californica*), common yarrow (*Achillea millefolium*), wild ginger (*Asarum caudatum*).

Shrubs

Little Sur manzanita (*Arctostaphylos edmunsii*), seacliff buckwheat (*Eriogonum parvifolium*).

7.3.3.4 Irrigation

Irrigation for the green roof is provided by a 25.6 m^3 (6500 gals.) cistern and pump system that delivers water to a drip irrigation system that is buried in the growing medium below the trays. Rana Creek provided an irrigation schedule that outlines different watering needs during different seasons. During the rainy season, irrigation is not typically run. During the summer and fall (dry season) irrigation should run once weekly, with supplemental watering at edges if needed. The city, however, runs irrigation daily during the dry season.

7.3.3.5 Maintenance

The vegetation is maintained by the City, and no pesticides or herbicides are used. Rana Creek educated managers of the roof about the rooftop ecosystem, and its required routine maintenance activities and schedule. Unwanted vegetation is removed during the growing season about twice monthly. Plants and plant parts are removed by hand and with tools. Activities include removal of some annual invasive plants by removing the plant or their seeds, and some seasonal pruning of California Fuchia (Nov-Jan), and deadheading of poppy, hairgrass, and fescue as needed in October (Garcia and Gaberšek 2010). This kind of attention to detail regarding educating the owner of a green roof is worthwhile and should be included for all green roofs.

Some of the invasive plants removed each year include bur clover (*Medicago polymorpha)* common groundsel (*Senecio vulgaris)* common sow thistle (*Sonchus oleraceus)*, petty spurge (*Euphorbia peplus*), spring vetch (*Vicia sativa)*. Other maintenance activities include occasional replanting where vegetation has died back.

7.3.3.6 Observed Wildlife

Hummingbirds, bees, dragonflies, butterflies, flies (in wet areas).

7.3.3.7 Best Performing Native Vegetation

All the vegetation specified grows on the roof. Over the years, some species have adapted to their preferred microclimates.

7.3.3.8 Post-occupancy Observations

Authors' Reflections

- This pilot project has been an invaluable resource for the City. It has been used extensively for education, promotion, and experimentation.
- The roof visually reads as a flat meadow, with grasses inter-mixed with local wildflowers. It is far from a standard sedum roof, and it is the City's project to teach and demonstrate how living roofs work in San Francisco. The design, installation and maintenance of a pilot project with native vegetation was intentional, as a way to demonstrate the concept of ecoregional green roofs.

7.3.4 *California Academy of Sciences, San Francisco, California*

The California Academy of Sciences living roof (Fig. 7.16), is a one-of-a-kind visitor experience where the coastal meadow vegetation that is native to the hills of San Francisco (Fig. 7.17) is on display and interpreted in form, material, signage, docent tours and through online media for a dynamic learning experience about the multiple ecosystem services of green roofs. Before its renovation, the academy building was damaged from an earthquake, and its form was a standard concrete box structure. Renzo Piano, Paul Kephart, and a diverse team envisioned and designed this LEED double Platinum-certified building with a living roof as an exhibit, and to perform important building functions such as moderating building temperature and capturing 98% of rainfall to the rooftop. This ecoregional green roof was designed with a designated roof deck that is accessible to the public and includes signage to interpret ecosystem services of green roofs, the native coastal meadows, species of plants on the roof, and some of the many kinds of birds, butterflies, and insects that visit the roof.

Fig. 7.16 Accessible roof deck with night lighting and interpretive signage informs visitors about green roofs, native plants, and native birds, butterflies, and raptors that visit the roof. Online data found on the iNaturalist project page for the CAS living roof expand the visitor experience. Observers can contribute to the hundreds of observations of plants, insects, birds, and other critters found on the roof. (Photo: Bruce Dvorak, October 2018)

Fig. 7.17 (**a**) An ad-hoc maintenance shed (hidden behind one of the rooftop domes) stores tools and equipment frequently used to maintain the roof. (**b**) Logs are strategically placed on the roof, and researchers observe how the addition of woody debris may help green roof substrates develop in their microflora and microbiota. (**c**) Native seaside daisy grows near and between bones that have been scattered on the roof for solarization. The bones are from archeological digs and need solarization before being brought into exhibits or storage. The rooftop makes a safe and secure location for this process to take place. (**d**) This south-facing top of meadow functions as a xeric habitat for plants that adapt to drier and warmer habitats. (Photos: Bruce Dvorak, October 2018)

7.3.4.1 Project Team

Building Owner/Client: California Academy of Sciences
Green Roof Design Team Lead: Rana Creek (Paul Kephart)
Architect: Renzo Piano Building Workshop, Stantec Architecture
Landscape Architect: SWA Group, Rana Creek
Installation Contractor: Rana Creek
Project completion: 2008
Green roof area: 1 hectare (2.5 acres)

7.3.4.2 Overview and Objectives

With resistance and resilience as a driving theme for the living roof vegetation, a 15 cm-deep (6 in) custom-blended substrated was designed for structural stability, made from local materials, and included microbes to help establish vegetation

(Kephart 2005). The initial aesthetic conceived by the architect was a single species for a uniform appearance. The design team, however, convinced the architect and their client to go with nine pre-tested species that were initially installed. Very soon after occupation, the CAS leadership began to expand, monitor, and make use of the green roof as a place to explore a rich interpretation of California coastal meadow species on the green roof, and quickly expanded the plant palette to 70 taxa (Hauser 2013).

7.3.4.3 Plant Establishment

The California Academy of Sciences keeps a periodically updated list of plants trialed on the green roof on their website for the living roof. Near one-hundred taxa have been trialed on the roof including annuals, ferns, grasses (Table 7.1), herbaceous perennials (Table 7.2), shrubs (Table 7.3) and succulents. The following 93 taxa have been installed on the green roof, including the initial nine species (*), and many others added since 2014 (Kephart 2005; CAS 2014; Greenroofs.com and Rana Creek 2020):

Annuals

Baby blue eyes (*Nemophila menziesii*), California phacelia (*Phacelia californica*), *California plaintain (*Plantago erecta*), *California poppy (*Eschscholzia californica*), California poppy (*Eschscholzia californica* 'Maritima'), *goldfields (*Lasthenia californica*), hedge nettle (*Stachys bullata* California), seep monkeyflower (*Mimulus guttatus*), sky lupine (*Lupinus nanus*), *tidy tips (*Layia platyglossa*).

Table 7.1 Native grasses on the California Academy of Sciences green roof

Common Name	Botanical Name
California sweet grass	*Anthoxanthum occidentale* (syn. *Hierochloe occidentalis*)
Sand dune sedge	*Carex pansa*
Clustered field sedge	*Carex praegracilis*
Point Joe fescue	*Festuca brachyphylla* ssp. *breviculmis* 'Pt. Joe'
California fescue	*Festuca californica*
Red fescue	*Festuca rubra* 'Patricks Point'
Baltic rush	*Juncus balticus*
Spreading rush	*Juncus patens*
Irisleaf rush	*Juncus xiphioides*
American dune grass	*Leymus mollis*
Deer grass	*Muhlenbergia rigens*
Golden-eyed grass	*Sisyrinchium californicum*
Purple needle grass	*Stipa pulchra* (syn. *Nassella pulchra*)

Table 7.2 Native Herbaceous Perennials on the California Academy of Sciences green roof
(*) = initial species planted

Common Name	Botanical Name
Common yarrow	*Achillea millefolium*
Deerweed	*Acmispon glaber*
Beach bur	*Ambrosia chamissonis*
Pearly everlasting	*Anaphalis margaritacea*
Western colombine	*Aquilegia formosa*
*Sea pink	*Armeria maritima* ssp. *californica*
Mugwort	*Artemisia douglasiana*
Beach sagewort	*Artemisia pycnocephala*
Wild ginger	*Asarum caudatum*
Nuttall's milkvetch	*Astragalus nuttallii*
Yerba buena	*Clinopodium douglasii*
Common sandaster	*Corethrogyne filaginifolia*
California fuchsia	*Epilobium canum* ssp. *canum*
California fuchsia	*Epilobium canum* ssp. *canum* 'Everetts choice'
Seaside daisy	*Erigeron glaucus*
Redflower buckwheat	*Eriogonum grande* var. *rubescens*
Coast buckwheat	*Eriogonum latifolium*
Sulphur flower buckwheat	*Eriogonum umbellatum*
Lizard tail	*Eriophyllum staechadifolium*
*Beach strawberry	*Fragaria chiloensis*
Hairy gumweed	*Grindelia hirsutula*
Gumweed	*Grindelia stricta*
Island alumroot	*Heuchera maxima*
Alumroot	*Heuchera micrantha*
Douglas iris	*Iris douglasiana*
*Miniature lupine	*Lupinus bicolor*
Scarlet monkeyflower	*Mimulus cardinalis*
Coyote mint	*Monardella villosa*
Yellow evening primrose	*Oenothera elata* ssp. *hookeri*
Foothill penstemon	*Penstemon heterophyllus*
*Selfheal	*Prunella vulgaris* ssp. *lanceolata*
California buttercup	*Ranunculus californicus*
Hummingbird sage	*Salvia spathacea*
Bee plant	*Scrophularia californica*
Checker mallow	*Sidalcea malvaeflora*
Blue-eyed grass	*Sisyrinchium bellum*
Blue-eyed grass	*Sisyrinchium bellum* 'rocky point'
California goldenrod	*Solidago velutina* ssp. *californica*
Pacific aster	*Symphyotrichum chilense*

Table 7.3 Native shrubs on the California Academy of Sciences green roof

Common Name	Botanical Name
Big Sur manzanita	*Arctostaphylos edmundsii* 'Big Sur'
Kinnikinick	*Arctostaphylos uva-ursi*
California sagebrush	*Artemisia californica*
Coyote bush	*Baccharis pilularis*
Coyote bush	*Baccharis pilularis* 'Twin Peaks'
Maritime ceanothus	*Ceanothus maritimus*
Yankee Point lilac	*Ceanothus thyrsiflorus* var. *griseus*
Seacliff buckwheat	*Eriogonum parvifolium*
California coffeeberry	*Frangula californica* 'Eve case'
Pink honeysuckle	*Lonicera hispidula*
Yellow bush lupine	*Lupinus arboreus*
Manycolored lupine	*Lupinus variicolor*
Sticky monkeyflower	*Mimulus aurantiacus*
California wax myrtle	*Morella californica* (syn. *Myrica californica*)
Fuchsia-flowered gooseberry	*Ribes speciosum*
California wild rose	*Rosa californica*
California blackberry	*Rubus ursinus*
Black sage	*Salvia mellifera*
Sonoma sage	*Salvia sonomensis*
Common snowberry	*Symphoricarpos albus* 'Tilden Park'

Ferns

Bracken fern (*Pteridium aquilinum* var. *pubescens*), giant chain fern (*Woodwardia fimbriata*), scouringrush horsetail (*Equisetum hyemale*), western sword fern (*Polystichum munitum*).

Succulents

Chalk dudleya (*Dudleya pulverulenta*), checker mallow (*Sidalcea malvaeflora*), coast dudleya (*Dudleya caespitosa*), lanceleaf liveforever (*Dudleya lanceolata*), powdery liveforever (*Dudleya farinosa*), *stonecrop (*Sedum spathulifolium*).

7.3.4.4 Irrigation

Harvested rainwater supplies irrigation water to subsurface drip lines beneath the biodegradable green roof trays. Moisture sensors regulate irrigation from May through September. Driplines are spaced 30 cm (12 in) apart below the biodegradable modules. Watering rates were about 757 cm^3 (0.20 gals) per 930 m^2 per week. Irrigation is typically not used during the winter.

7.3.4.5 Maintenance

The California Academy of Science staff maintains the green roof along with trained volunteers. The academy has an annual maintenance plan that is guided by the vision and mission of the academy. The living roof is directed by a small group of highly trained leaders that accomplish their goals with the aid of volunteer citizens. Volunteers also make science observations or perform testing as part of the upkeep and management of the roof. The green roof has gone through several different management approaches over the years, thus the aesthetic characteristics and floristic composition changes over time (Fig. 7.17).

7.3.4.6 Observed Wildlife

In addition to common visitors to green roofs such as bees, birds, hawks, snails, butterflies, the online resource iNaturalist (acquired by the California Academy of Sciences) hosts a project website for the rooftop where citizen scientists have logged and mapped biodiversity observed on the rooftop and landscape below. Insect observations include ants, mealy bugs, spiders, bumble bees, hornets, yellowjackets, beetles, moth flies, sandflies, midges, lady beetles, moths, grasshoppers, crickets, eight-spotted skimmers, and more. Butterflies include American lady, California pipevine, checkerspot, fiery skipper, red admiral, swallowtail, umbra skipper. Bird observations include brewer's blackbird, dark-eyed junco, American robin, red-tailed hawk, Steller's jay, California scrub-jay, Anna's hummingbird, brown-headed cowbird, European starling, song sparrow, hooded oriole, red-winged blackbird, black phoebe, wild turkey, and many more (IN 2019). Other volunteer databases such as e-bird identify 99 species of birds that have visited the green roof from February 2011 up to March 2019 (CLO 2019).

7.3.4.7 Best Performing Native Vegetation

Pacific Horticulture published an excellent summary of successful vegetation on the roof. It was written by staff at the academy that frequent the roof. The excerpt highlights the adaptation of plants to rooftop microclimates:

> Plants on the roof grow in six inches of substrate and, in addition to topography, their success is influenced by salt spray, summer fog, and frequent high winds. On the most arid regions of the domes, Powdery Liveforever (*Dudleya farinosa*) and Sea Lettuce (*D. caespitosa*) have flourished. 'Pigeon Point' Coyote Brush (*Baccharis pilularis* 'Pigeon Point'), Redflower Buckwheat (*Eriogonum grande* var. *rubescens*), and Coast Buckwheat (*E. latifolium*) also perform well on the steep slopes. On the flat areas, which have a higher water-holding capacity than the domes, Yarrow (*Achillea millefolium*), Coast Strawberry (*Fragaria chiloensis*), Manycolored Lupine (*Lupinus versicolor*), Seaside Daisy (*Erigeron glaucus*), California Fescue (*Festuca californica*), and Yellow Evening Primrose (*Oenothera elata*)

have proliferated. In the valleys between the domes, we have experimented with species that tolerate more soil moisture and shade including Yellow-eyed Grass (*Sisyrinchium californicum*) and Seep Monkeyflower (*Mimulus guttatus*), which thrive in these areas (Hauser 2013).

7.3.4.8 Post-occupancy Observations

In Publication

- The California Academy Building and site is the world's first Double Platinum LEED-certified museum, and the largest Double Platinum building on the planet (CAS 2011).
- "One of the most interesting lessons we have learned is how the varying topography of the roof's seven domes creates unique microclimates that, in turn, influence Plant Establishment and promote diversity. On the tops and steep sides of the domes, the substrate loses moisture more quickly than other locations on the roof. Conversely, the valleys between the domes create pockets of the continually moist substrate. These areas are blocked from the wind, shaded by the domes, and accumulate moisture from coastal San Francisco fog. Even the smallest topographical change creates unique microclimates (Hauser 2013)."

Authors' Reflections

- The dynamic effect of topography on the substrate is intentionally applied in this project to influence biodiversity. A lesson learned is that observations at the bottoms and tops of slopes indicated that even minor topographic variations may be valuable for any green roof (Fig. 7.18). Since flat roof decks and uniform substrate depths produce a single kind of microclimate, any topographic relief on green roofs may achieve more ecologically dynamic green roofs.
- The intentional use of annuals on this living roof suggests an approach not used by many designers of green roofs but should be considered much more frequently. Since annuals can sustain critical functions in climates with bi-modal distributions, or those in regions with frequent drought, annuals may add a necessary function in restoring and sustaining vegetative cover on green roofs. The successful mixing of multiple forms of plants is important to note and should be explored as a part of ecoregional green roofs where annuals also thrive in the natural environment.
- The public display and interpretation deck for this green roof is outstanding. In addition, its online connections expand the interpretation (i.e. CAS website, iNaturalist, e-bird Hotspot) and likely makes this one of the most popular green roofs in North America.

Fig. 7.18 The mounded design creates much diversity across the roof meadow. This view between mounds shows a gradation in microclimates created by the various slopes, moisture regimes, and exposures to sunlight. (Photo: Bruce Dvorak, October 2018)

7.3.5 *Sutter Hospital at Van Ness and Geary, San Francisco, California*

Research has now long demonstrated that patients that have multi-day stays to hospitals and have views of vegetation outside their recovery rooms have lower levels of stress and quicker recovery times (Ulrich 1984; Marcus and Barnes 1999; Suppakittpaisarn et al. 2017). The Sutter Health network considers these environmental benefits as essential components of the hospital built environment. Aside from the monetary benefits of green roofs from their conservation of energy on buildings, and preservation of roof membranes, reduction of urban flooding and habitat benefits, the psychological and emotional benefits of green roof vegetation were a driving factor when Sutter included five green roofs on a new hospital. The Sutter Health Hospital is located in downtown San Francisco, has 11 floors, 274 beds, 35 NICU beds and five green roofs that are visible from inside locations of visitor areas, recovery rooms and other strategic locations (Fig. 7.19).

Fig. 7.19 A central visitor waiting area (inside, behind the glass wall) extends the visitor space to include an outdoor seating area. Drifts of native grasses and herbaceous plants make for an informal and colorful space. The plants were selected because they are native to San Francisco and attract wildlife such as hummingbirds, butterflies, and other pollinators. (Photo: Bruce Dvorak, October 2018)

7.3.5.1 Project Team

Building Owner/Client: Sutter Health
Green Roof Design Team Lead: Rana Creek
Architect: SmithGroup
Structural Engineer: Degenkolb Engineers
Landscape Architect: Rana Creek
Installation Contractor: Rana Creek
Project completion: March 2019
Green roof area: 2489 m^2 (26,800 ft^2.)

7.3.5.2 Overview and Objectives

Rana Creek designed the green roofs so that all of the vegetation was native from the region and that the beds were attractive and simple to maintain. They wanted patients to have physical and visual access to nature, but they did not want formal gardens nor a random environment. The team designed multiple green roofs with large drifts of plants that were located in beds with informal edges. Several species make up each bed, however, it is okay if some plants migrate across boundaries. Informed maintenance keeps the designed assemblages intact, but allows the removal of invasive or exotic species.

7.3.5.3 Plant Establishment

The main goal of this planting design was to achieve diversity, stability, and longevity. With a stable, long-living, and diverse plant community infestations of plants or animals can be reduced. Native flora favors fibrous-rooted herbaceous vegetation and local pollinators (Fig. 7.20). Compared to a monoculture of succulents, these plants will help retain and cleanse stormwater and attract local wildlife. Forms of plants include annuals, ferns, grasses, and herbaceous perennials (Table 7.4).

Fig. 7.20 (**a**) Western sword fern occupies a shady portion of a green roof; (**b**) views from inside the hospital waiting area out to a green roof; (**c**) native penstemon and sage attract local pollinators; (**d**) an open and sunny location is planted with vegetation adapted to a warmer and drier green roof. (Photos: Bruce Dvorak, October 2018)

Table 7.4 Native herbaceous perennials on the Sutter Hospital green roofs

Common Name	Botanical Name
Red yarrow	*Achillea millefolium* 'strawberry seduction'
Narrowleaf milkweed	*Asclepias fasicularis*
Dwarf California aster	*Aster chilensis* 'Point St. George'
California fuchsia	*Epilobium canum*
Pearly everlasting	*Gnaphalium californicum*
Giant coral bells	*Heuchera maxima*
Seaside alumroot	*Heuchera pilosissima*
White douglas iris	*Iris douglasii* 'canyon snow'
Miniature lupine	*Lupinus bicolor* (innoculated)
Scarlet bugler	*Penstemon centranthifolius*
Royal penstemon	*Penstemon spectabilis*
Hummingbird sage	*Salvia spathacea*

Annuals

Hedge nettle (*Stachys bullata*).

Ferns

Deer fern (*Blechnum spicant*), western sword fern (*Polystichum munitum*).

Grasses

California fescue (*Festuca californica*), Idaho blue fescue (*Festuca idahoensis* 'siskyou blue'), purple needle grass (*Stipa pulchra*).

7.3.5.4 Irrigation

Approximately 1400 m^3 (370,000 gals) of precipitation is collected annually in a cistern for reuse on the green roofs. This water is stored in a 340 m^3 (90,000 gals) cistern in the parking level of the hospital (Davis 2016).

7.3.5.5 Maintenance

Maintenance crews visit the green roofs once a month.

7.3.5.6 Observed Wildlife

The green roofs were new and had not yet been observed.

7.3.5.7 Best Performing Native Vegetation

All of the vegetation installed is established and healthy.

7.3.5.8 Post-occupancy Observations

Authors' Reflections

- The hospital takes advantage of a multi-story and multi-building complex to strategically distribute green roofs throughout. This approach ensures that views to green are dispersed to many locations inside.
- Although the green roof was installed nearly a year before the hospital was ready to open, Rana Creek maintained the green roofs to ensure that the vegetation was well-established.
- A maintenance manual was published for the owner to hand-off maintenance of the green roof to other contractors. The manual includes laminated pages with photos of plants installed, common invasive plants with common and botanical names.

7.3.6 Salesforce Transit Center, San Francisco, California

Located at a major hub of activity, the Salesforce Transit Center (formerly Transbay Transit Center) forms one of the largest and most diverse green roofs in San Francisco and California. Nearly a decade in the making, the 2 billion-dollar (USD) center, include the public park and botanic garden which sit on top of the reconstructed transit building. The design solution merges multiple objectives to make one of the most dynamic spaces in the city. The 4.5-block long roof garden has several sky bridges that link to adjacent buildings for ease of access, has an aerial tram that provides access from a plaza below, and has several escalators and elevator access points for universal access. The project has stimulated new mixed-use, office, and residential development nearby (Fig. 7.21).

7.3.6.1 Project Team

Building Owner/Client: Transbay Joint Powers
Green Roof Design Team Lead: PWP/Rana Creek

Fig. 7.21 The Salesforce Transit Center rooftop park and botanic gardens shown from Salesforce Tower. (Creative Commons, ust 2018, Fullmetal2887, https://commons.wikimedia.org/wiki/File:Salesforce_Park_and_bus_bridge,_seen_from_Salesforce_Tower.jpg)

Architect: Pelli Clarke Pelli
Structural Engineer: Arup
Landscape Architect: Peter Walker Partnership (PWP)
Installation Contractor: Brightview
Project completion: 2018
Green roof area: 1.82 hectares (5.4-acres)

7.3.6.2 Overview and Objectives

In the fall of 2015, the City of San Francisco published a Living Roof Manual that includes much detail regarding the selection of plants based upon improving biodiversity in the city, and the use of native vegetation (SFLRM 2015). In line with the intent of the manual, this massive rooftop park makes a living laboratory that was designed with an ecological mission to interpret and teach visitors about plants from multiple ecoregions. The vegetation was selected to make the park didactic, as a garden to inform visitors about California's native ecoregions, and ecoregions around the world that also have Mediterranean-like climates. Themed from wet to dry in an east to west direction, the gardens hold collections ranging from wetlands to shrub and forest communities, to desert plants. Plant communities cover plants from coastal meadows, chaparral, oak woodlands, desert scrub communities, and succulent plant communities (PWP 2019). California coastal collections include native plants selected to attract endangered species, such as the Bay checkerspot butterfly. The roof garden is loaded with pollinator plants to respond to the local needs of ecosystem members such as the Bay checkerspot.

7.3.6.3 Plant Establishment

The collections include over 55 species of trees and 208 species of understory plants, and 23 species of groundcovers (Fig. 7.22). California related garden themes include a California Garden, Desert Garden, Fog Garden, Redwood Forest Garden, and a Wetland Garden. Other themed gardens include Prehistoric Garden, Palm Garden, South African Garden, Mediterranean Basin Garden, and a Chilean Garden. The following vegetation is native to California and is included in the garden collections. The many exotic species planted in the botanic gardens are not covered in this case study. Forms of plants include ferns, grasses, herbaceous perennials, shrubs (Table 7.5), trees (Table 7.6), and succulents (Table 7.7).

Fig. 7.22 (**a**) Street view of the transit center building; (**b**) view of a succulent collection in the Dessert Garden with native and non-native plants; (**c**) the constructed wetland has vegetation native to marsh habitats in California and other Mediterranean climates. These plants clean recycled rooftop water that is used for flushing toilets; (**d**) the oval walking path that meanders through the Fog Garden visible to the right of the railing. (Photos: Bruce Dvorak, October 2018)

Table 7.5 Native shrubs on the Salesforce Transit Center green roofs

Common Name	Botanical Name
Howard McMinn manzanita	*Arctostaphylos* 'Howard McMinn'
John Dourley manzanita	*Arctostaphylos* 'John Dourley'
Paradise manzanita	*Arctostaphylos pajaroensis* 'paradise'
Sentinel manzanita	*Arctostaphylos* 'sentinel'
Concha ceanothus	*Ceanothus* 'concha'
Frosty blue ceanothus	*Ceanothus* 'frosty blue'
Ray Hartman ceanothus	*Ceanothus* 'Ray Hartman'
Yankee Point ceanothus	*Ceanothus* 'Yankee Point'
Giant sea dahlia	*Coreopsis gigantea*
Red-twig dogwood	*Cornus sericea* 'isanti'
St. Catherine's lace	*Eriogonum giganteum*
Coast silktassel	*Garrya elliptica*
Toyon	*Heteromeles arbutifolia*
Dwarf silver bush lupine	*Lupinus albifrons* var. *collinus*
Bush lupine	*Lupinus arboreus*
Pacific wax myrtle	*Myrica californica*
Eve case coffeeberry	*Rhamnus californica* 'Eve case'
Lemonade berry	*Rhus integrifolia*
Pink-flowering currant	*Ribes s. glutinosum* 'Claremont'
White sage	*Salvia apiana*
Winnifred Gilman blue sage	*Salvia clevelandii* 'Winnifred Gilman'
Common snowberry	*Symphoricarpos albus*

Table 7.6 Native trees on the Salesforce Transit Center green roofs

Common Name	Botanical Name
California buckeye	*Aesculus californica*
Marina strawberry tree	*Arbutus* 'marina'
California incense-cedar	*Calocedrus decurrens*
Monterey cypress	*Cupressus macrocarpa*
Santa Cruz Island ironwood	*Lyonothamnus floribundus* ssp. *asplenifolius*
Torrey pine	*Pinus torreyana*
Coast live oak	*Quercus agrifolia*
Engelmann oak	*Quercus engelmannii*
Netleaf oak	*Quercus rugosa*
Island oak	*Quercus tomentella*
Southern live oak	*Quercus virginiana*
'Aptos blue' coastal redwood	*Sequoia sempervirens* 'aptos blue'
Weeping sequoia	*Sequoiadendron giganteum* 'pendulum'
Hybrid California fan palm	*Washingtonia robusta x filifera*

Table 7.7 Succulents on the Salesforce Transit Center green roofs native to adjacent ecoregions

Common Name	Botanical Name	Native Region
Foxtail agave	*Agave attenuata*	Central Mexico
Blue glow agave	*Agave* 'blue glow'	Mexico
White agave	*Agave celsii albicans*	Central Mexico
Thread agave	*Agave filifera*	Central Mexico
whale's tongue agave	*Agave ovatifolia*	Mexico
Hedgehog agave	*Agave stricta*	Southern Mexico
Queen Victoria agave	*Agave victoriae-reginae*	Chihuahuan Desert in Mexico
Mexican lily	*Beschorneria yuccoides*	Mexico
Giant chalk dudleya	*Dudleya brittonii*	Baja California, Mexico
Afterglow echeveria	*Echeveria* 'afterglow'	Texas, Mexico, Central and South America
Cante echeveria	*Echeveria cante*	Texas, Mexico, Central and South America
Hens and chicks	*Echeveria elegans*	Mexico
Red edge echeveria	*Echeveria* 'red edge'	Mexico
Macdougall's furcraea	*Furcraea macdougallii*	Mexico

Ferns

Giant chain fern (*Woodwardia fimbriata*), western sword fern (*Polystichum munitum*).

Grasses

California fescue (*Festuca californica*), deer grass (*Muhlenbergia rigens*), elk blue California gray rush (*Juncus patens* 'elk blue'), fiber optic plant (*Isolepis cernua*), Mendocino reed grass (*Calamagrostis foliosus*), quartz creek soft rush (*Juncus effusus* 'quartz creek').

Herbaceous Perennials

Coast buckwheat (*Eriogonum latifolium*), cow parsnip (*Heracleum lanatum*), Douglas iris hybrids (*Iris douglasiana* hybrids), elk clover (*Aralia californica*), fringecups (*Tellima grandiflora*), Matilija poppy (*Romneya coulteri*), red-flowered buckwheat (*Eriogonum grande* var. *rubescens*), rosada hybrid alum root (*Heuchera* 'rosada'), seaside daisy (*Erigeron glaucus*), white inside-out flower (*Vancouveria hexandra*).

Succulents

Frank Reinelt dudleya (*Dudleya* 'Frank Reinelt').

7.3.6.4 Native to Adjacent Ecoregions

Grasses

Dwarf horsetail (*Equisetum scirpoides*) (native to Canada and the northern U.S.).

Shrubs

Mexican grass tree (*Dasylirion longissimum*) (native to northeastern Mexico).

7.3.6.5 Irrigation

Rainwater from the green roof and water from the sinks inside the transit center is collected and cleansed in the rooftop Wetland Garden. Water that is not used by the plants is stored for use in restrooms throughout the transit center (PWP 2019). Multiple irrigation zones and moisture sensors regulate the flows of subsurface irrigation based upon current weather, moisture levels in the substrate, and generalized needs of plants in the moisture-themed plant collections.

7.3.6.6 Maintenance

As a high-profile public park, the gardens are designed in beds with drifts of plants, and traditional garden appearance is expected (Fig. 7.22). The gardens receive frequent maintenance from daily to weekly based upon seasons and current conditions.

7.3.6.7 Observed Wildlife

Thus far, citizen scientists at the online iNaturalist.org have observed many birds including hummingbirds, and many insects including bees, on the green roof.

7.3.6.8 Best Performing Native Vegetation

Most of the vegetation planted on the green roof is established and thriving.

7.3.6.9 Post-occupancy Observations

Authors' Reflections

- This massive rooftop botanic garden and a public park is one of the largest, most diverse, and accessible rooftop gardens constructed in California, if not North America (Fig. 7.23).
- The interpretive program for the gardens and the transit center is accessible and informative about the ecoregions of California, and elsewhere.
- The garden layout follows a moisture gradient from wet to dry. This living laboratory should become an outstanding place for future green roof monitoring and research.

Fig. 7.23 Photograph of the massive rooftop public park and botanic garden under construction. Constructed in phases from 2011 to 2018, the image shows a partially planted roof (upper right), orange waterproofing layer (front left), and walking paths are in various levels of completion. (Creative Commons, Pi.1415926535, https://upload.wikimedia.org/wikipedia/commons/6/60/Transbay_Transit_Center_construction_from_TJPA_office%2C_August_2017.jpg)

7.3.7 *EcoCenter at Heron's Head Park, San Francisco, California*

Located on the southeastern edge of San Francisco on the bay, the Heron's Head Park was constructed on infill from an incomplete and unbuilt pier project begun in the 1970s. Today it is an 8.9-hectare (22-acre) peninsula park made for passive recreation, education, and wildlife. The wetland habitats attract over 100 species of local and migratory birds each year.

The EcoCenter (education) building at Heron's Head Park is the City of San Francisco's first building designed to be nearly "off-the-grid". The LEED Platinum – Zero Net Energy Building attains electricity and treats sewage on-site. Potable water is sourced from the City. All of the building's internal and external functions are derived from renewable on-site resources such as solar panels, rainwater harvesting tanks, and on-site wastewater treatment with a constructed wetland (FA 2019). The center is designed to educate the public about green roofs, renewable energy, pollution, greenhouse gas reduction, wastewater treatment, sustainable building materials, rainwater harvesting, and the green economy (Fig. 7.24).

7.3.7.1 Project Team

Building Owner/Client: Port of San Francisco
Green Roof Design Team Lead: Habitat Gardens with Eco Catalyst (Lisa Lee Benjamin)
Architect: Feldman Architecture
Installation Contractor: Habitat Gardens
Project completion: 2010
Green roof area: 102 m^2 (1100 ft^2)

7.3.7.2 Overview and Objectives

Two small green roofs share rooftop space with solar panels and skylights to provide energy, light, and energy conservation for the building systems. The green roof functions to reduce heat gain, provide habitat for wildlife, and retain stormwater (Fig. 7.25). The green roof was made from recycled materials wherever possible, including materials found on the site or nearby. The substrate is made from local materials including 15% well-aged compost, 15% forest humus, 50% mineral aggregate, and 20% course sand. The substrate depth is not uniform as it varies from 5 cm–20 cm (2 in–8 in). One isolated 30 cm (12 in) depth of media is placed for the manzanita shrub. The shallow pond is made with an underlying and isolated membrane that rises to make a shallow depth pond (5 cm/2 in) that is filled with gravel.

Fig. 7.24 The north-facing habitat roof includes native plants (yarrow front/center) common to upland habitats of the bay area (manzanita shrub is on the left, grasses & succulents rear), driftwood branches (lower right) and a small shallow-elevation water pond (lower right) is designed to provide a source of pond water to attract birds and wildlife. A black plastic irrigation head is seen in the center-right. Irrigation water is sourced from harvested rainwater. A gravel quarry lies adjacent to the property (upper right) and is the source of the stone in the pond. The City of San Francisco lies in the background. (Photo: Bruce Dvorak, October 2018)

Only partially visible from the ground, the green roof was designed to be low-maintenance and was designed more for wildlife than as a garden for people. Local materials include gravel that was placed in wire baskets as edge containment of the substrate was purchased from a local quarry just off the property (Fig. 7.13), and rocks and driftwood were collected from Heron's Head Park (Lyon 2019). The roof is used as an educational tool for the youth showcasing that plants that grow on the property can also grow on the roof (Lyon 2019).

7.3.7.3 Plant Establishment

Pre-grown plants were installed during initial planting, and successive plants have been re-seeded over time. Both initial and successive plantings are provided below. Several non-native species of plants have been installed on the living roof and are not reported here. Forms of plants include ferns, grasses, herbaceous perennials (Table 7.8) shrubs, and succulents.

Fig. 7.25 (**a**) Skylight and native vegetation on the south-facing green roof. (**b**) Driftwood branch for biodiversity and the gravel edge allows for draining the substrate. (**c**) Indoor wastewater treatment system (constructed wetland). (**d**) Outdoor classroom and one of two cisterns that are sized for landscape and rooftop irrigation. (Photos: Bruce Dvorak, October 2018)

Ferns

Deer fern (*Blechnum spicant*), lady fern (*Athyrium filix-femina*), western sword fern (*Polystichum munitum*).

Grasses and Rushes

Irisleaf rush (*Juncus xiphioides*), junegrass (*Koeleria macrantha*), saltgrass (*Distichlis spicata*), spreading rush (*Juncus patens*), tufted hairgrass (*Deschampsia caespitosa*).

Shrubs/Groundcovers

Hairy manzanita (*Arctostaphylos columbiana*), woodland strawberry (*Fragaria vesca*).

Table 7.8 Native herbaceous perennials on the EcoCenter at Heron's Head Park green roofs

Common Name	Botanical Name
Yarrow	*Achillea millefolium*
Sea thrift	*Armeria maritima*
Thrift seapink	*Armeria maritima* 'Splendens'
Mexican whorled milkweed	*Asclepias fascicularis*
Harvest brodiaea	*Brodiaea elegans*
Beach coreopsis	*Coreopsis maritima*
California aster	*Corethrogyne filaginifolia*
California fuchsia	*Epilobium canum*
Seaside fleabane	*Erigeron glaucus*
Coast buckwheat	*Eriogonum latifolium*
California perennial poppy	*Eschscholzia californica* var. *maritima*
Golden aster	*Heterotheca sess. bolanderi* 'San Bruno Mtn.'
Island alum root	*Heuchera maxima*
Bush monkey flower	*Mimulus aurantiacus*
Sticky monkey-flower	*Mimulus aurantiacus*
Monkeyflower	*Mimulus* 'Georgie Red'
Hummingbird sage	*Salvia spathacea*
Blue eyed grass	*Sisyrinchium bellum*

Succulents

Broadleaf stonecrop (*Sedum spathulifolium*), Catalina Island live-forever (*Dudlyea hassei*), sea lettuce (*Dudleya caespitosa*).

7.3.7.4 Irrigation

The green roof is watered with an overhead irrigation system. Harvested rainwater is provided from two above-ground 19 m^3 (5000-gallon) galvanized tanks. Irrigation runs during the summer about 15 min in each of the roofs, three times a week.

7.3.7.5 Maintenance

Citizen volunteers assist with roof maintenance about once a month under the direction of a trained leader. The main tasks include the removal of unwanted vegetation and the re-seeding of desirable vegetation from the seeds of existing plants. Common plants removed off the roof include an annual clover and annual grasses. Because irrigation is provided through a rainwater harvesting system, maintenance includes checking irrigation levels of the cisterns, and that there are no blockages or leaks. Driftwood branches placed on the roof (for habitat) are inspected after major storms to make sure they did not blow off the roof.

7.3.7.6 Observed Wildlife

Birds, butterflies, bees, insects.

7.3.7.7 Best Performing Native Vegetation

Yarrow, woodland strawberry, monkeyflower, coast buckwheat, California poppy, dudleya, stonecrops, ferns (in shady areas), grasses, seaside fleabane.

7.3.7.8 Post-occupancy Observations

Authors' Reflections

- As one of the smaller green roofs on the west coast, it is also a very important green roof, in terms of demonstrating connections between the use of vegetation from local plant communities, habitat structure, and provisions for biodiversity. For example, the small and shallow water pond on the roof invites wildlife to visit the roof. Driftwood and rocks from the site connect to local materials and habitat complexity. Strategically placed larger plants make micro-climates for smaller plants.
- This green roof has been managed through a vision informed by knowledge of local ecosystems, aesthetics, and the intuition of a garden designer. After the green roof was initially planted, over time through selective thinning, replanting, and re-seeding, a series of eco-microcosms were intentionally developed on the roof. The lead maintenance staff is highly qualified and knowledgeable of plants native to the region, and their ecological and aesthetic qualities.
- Since the green roof is only visible from a ladder, a future accessible observation area should be designed at the EcoCenter so that visitors can see the green roof.

7.3.8 Los Angeles Museum of the Holocaust, Los Angeles, California

Located at the northern edge of Pan Pacific Park in Los Angeles, the Museum of the Holocaust (LAMOTH) is the oldest Holocaust museum in the United States and is the first with a green roof. From the outside, the museum is partially submerged below ground and is hidden behind layers of formed concrete and drifts of grasses that ascend the sloping roof (Fig. 7.26). These materials appear to intertwine to rise above the park and evoke an undulating landscape that metaphorically represents

Fig. 7.26 A view of the ascending ramps that rise above the museum below, and above include views back to Pan Pacific Park. The simple planting design blends three species of grasses that are similar in height, water use requirements, and texture. (Photo: Bruce Dvorak, October 2018)

the virtues of hope, love, and beauty. These qualities are present in the characteristics of the three grasses that were selected in that they have fine blades and feathery seed heads that are enhanced by the effects of wind and light. The grasses and their light shadows wave in the wind above the descending ramp to the entrance into the museum below. Since 1961, the museum has moved four times and to this current location in 2010. The LEED Gold-certified building has won numerous awards, including its green roof (Dawkin 2013).

7.3.8.1 Project Team

Building Owner/Client: Los Angeles Museum of the Holocaust
Architect: Belzberg Architects
Structural Engineer: William Koh & Associates
Green roof design: Lisa Lee Benjamin of Evo Catalyst with Karla Dakin
Landscape architect: Karla Dakin of K. Dakin Design
Plant consultant: John Greenlee
Installation Contractor: Roofmeadow

Project completion: 2010
Green roof area: 1180 m^2 (12,700 ft^2)

7.3.8.2 Overview and Objectives

As an extension of the park, the roof garden provides two functions. First, the roof garden is a terminus to the north end of Pan Pacific Park with its requirements to endure year-round access 24-hours a day by park users. This means that the vegetation and its containing vessel must be durable and resilient to sustain many unsupervised visits by park users. Second, the roof garden needs to serve the function of a roof garden of the museum, and respond to the purposes, needs, and expectations of museum visitors. Thus, the selection of plants must respond to both contexts.

The plants were selected to withstand wide temperature fluctuations and sustained high daytime temperatures. They were also selected to create a visual effect that is appropriate as a sloped-façade of the museum (Hauser 2013). The substrate varies in depth based upon the root depth of the plants. Substrates were laid in depths at 10 cm (4 in), 15 cm (6 in) or 30 cm (12 in) deep. Grasses and some non-native succulents hold the steep grades (45-degree pitches) in place, along with a reinforced geocell blanket. Prickly pear is planted at strategic locations to define edges of the museum. This ecoregional green roof introduces a view of the kind of vegetation that was once widespread in Los Angeles plain: grasses. Historically, the Los Angeles plain was covered with sagebrush, prairie grasses, chaparral, and wetlands and estuaries in the low-lying landscapes where streams and rivers meet the Pacific Ocean (Cooper 2008; Griffith et al. 2016).

7.3.8.3 Plant Establishment

The planting plan includes a simple palate of grasses and a few succulents. The grasses were selected to be tall enough to be seen from a distance, but not too tall so that safety becomes an issue (Fig. 7.27). Other factors influencing selection included the availability of plants in the market, and their seasonal color and texture (Benjamin et al. 2013).

Grasses

The three grasses selected for the roof garden include blue grama grass (*Chondrosum gracile* syn. *Bouteloua gracilis*) which is native to southern California; pine muhly grass (*Muhlenbergia dubia*) which is native to Arizona, New Mexico, and Texas; and esparto grass (*Lygeum spartum*) which is native to the Mediterranean region.

Fig. 7.27 Pine Muhly Grass is planted in drifts to make a mass effect. The east to the west alignment of plants allows sunlight and shade to reinforce the aesthetic qualities of the grasses and echo themes of dark and light. (Photo: Bruce Dvorak, October 2018)

Succulents

Coastal prickly pear (*Opuntia littoralis*) is native to the coastal ecoregions of California and is planted on the green roof parapet, where it is visible in a few select locations (Fig. 7.28). A few non-native succulents were also included in the project, but are not reported here.

7.3.8.4 Irrigation

The green roof is irrigated from rainwater that is collected and stored in subgrade cisterns. Runoff from the green roofs is captured and recycled to water the roof during the summer. A surface drip and capillary irrigation system deliver water to the roots of plants (Benjamin et al. 2013).

7.3.8.5 Maintenance

A maintenance manual was produced for the project to maintain the simple planting. Activities include weeding, watering, fertilizing (organic only), trimming,

Fig. 7.28 Coastal prickly pear (*Opuntia littoralis*) grows as a focal point, in contrast to fine-textured grasses. This prickly pear is native to the Los Angeles basin and is well-adapted to the xeric soils in the valley

thinning, and pruning. When planted, the vegetation was to be established up to 80% cover of the roof by the end of the second growing season (Benjamin et al. 2013). There has been some plant loss at locations where steep slopes occur.

7.3.8.6 Observed Wildlife

Bees, sparrows, doves, and pigeons have been observed.

7.3.8.7 Best Performing Native Vegetation

The three grasses have performed well since 2010. Some dieback has occurred at the steepest locations of the sloped roofs.

7.3.8.8 Post-occupancy Observations

Authors' Reflections

- The simple and refined concept is striking and very effective. The grassed roof makes a bold statement and connects to the prairie ecosystems once common in the valley.
- The rainwater harvesting system that was built into the project from the beginning makes this ecoregional green roof work, as there is no stress on potable water and demonstrates how grassed green roofs are possible in Southern California.
- Some of the steep sections of the green roofs have lost vegetation and are bare. Budget and maintenance should be secured to retain at least 80% cover as is outlined in the original design and maintenance manual. The original planting plan should be consulted.

7.3.9 Palomar Medical Center, Escondido, California

The Palomar Medical Center leadership and staff believe that contact with nature helps in the healing and recovery process (Ulrich 1984; Marcus and Barnes 1999; Suppakittpaisarn et al. 2017), and have demonstrated this belief in the design of a new facility. From day one of the design process, the project team worked together to integrate the exterior and interior environments around a central theme of "empathy" (Timm 2017). This concept is physically made visible in the quantity and quality of healing gardens that lead visitors from the heavily planted parking lot, along the planted walkway, the entrance into the building, and up the 11 floors with interior gardens. The front of the hospital has vertical greening along the facades leading up to the front entrance. Interior green space accompanies visitors up a central elevator spine that has interior garden terraces on each floor. As such, patients, doctors, nurses, staff, visitors, and those that maintain the property benefit from continuous contact with nature. The 360-bed acute-care hospital also incorporates natural daylighting, views of vegetation from recovery rooms, multiple accessible roof gardens (Fig. 7.29), natural materials, and many other sustainable practices (Spurlock 2015; Timm 2017).

7.3.9.1 Project Team

Building Owner/Client: Palomar Medical Center
Green Roof Design Team Lead: Spurlock with Rana Creek as consultant
Architect: CO Architects with Stantec
Landscape Architect: Spurlock Poirier Landscape Architects

Fig. 7.29 This view from the helicopter pad looks down onto the third-floor roof meadow which is visible from patient recovery rooms (center right) and a second-floor roof accessible roof garden for visitors and staff (center left), and the terminus of the garden walk spine that extends into the parking lot is seen in the background. (Photo: Bruce Dvorak, October 2018)

Installation Contractor: ValleyCrest Landscape Companies
Project completion: 2012
Green roof area: 5574 m^2 (60,000 ft^2)

7.3.9.2 Overview and Objectives

With healing and sustainability as driving goals for the development of the built environments, the entire project was designed to be water-efficient, energy-efficient, and built with materials that are from recycled or renewable sources with priority given to local materials. A coastal meadow inspired the target plant community for the extensive green roof. Escondido lies at the ecotone of montane woodlands, chaparral, and coastal meadows (Fig. 7.30). The green roof meadow was designed to address the ecosystem needs of local biodiversity and to connect to nearby conservation sites. The green roof was intended to replace lost open space, attract wildlife, and provide views to greenspace.

Fig. 7.30 Native to the southwestern regions of the U.S., pink muhly grass (in bloom) was added to the otherwise locally native mix. Pink muhly grass was added to provide visual interest for patients with views out of the recovery rooms. (Photo: Bruce Dvorak, October 2018)

7.3.9.3 Plant Establishment

This plant list is for the green roof over the third floor (Fig. 7.31). The vegetation was established through hydroseeding. There are multiple roof gardens around the hospital, but most don't have a native palette as a primary goal. The 10 cm (4 in) substrate was provided by American Hydrotech.

Annuals

Golden yarrow (*Eriophyllum confertiforum*).

Grasses

California fescue (*Festuca californica*), canyon prince wild rye (*Leymus condensatus* 'canyon prince'), Idaho blue fescue (*Festuca idahoensis*), purple three awn (*Aristida purpurea*).

Fig. 7.31 (**a**) roof garden on the second floor, (**b**) one of the many light wells for natural daylighting, (**c**) view onto the extensive green roof from a patient recovery room, (**d**) visual garden in an interior courtyard on the first floor. (Photos: Bruce Dvorak, October 2018)

Herbaceous Perennials

California fuchsia (*Epilobium canum*), common yarrow (*Achillea millefolium*), foothill penstemon (*Penstemon heterophyllus*), hummingbird sage (*Salvia spathacea*), pink yarrow (*Achillea* 'rosea'), red California fuchsia (*Zauschneria californica*), royal beard tongue (*Penstemon spectabilis*), woodland strawberry (*Fragaria vesca*).

Shrubs

California buckwheat (*Erigonum fasciculatum*), Cleveland sage (*Salvia clevelandii*), coastal sagewood (*Artemisia pycnocephala* 'David's choice'), Eastwood's Manzanita (*Arctostaphylos glandulosa*), southern bush monkeyflower (*Mimulus longiflorus*), white sage (*Salvia apiana*).

Succulents

Our Lord's candle (*Yucca whipplei*).

7.3.9.4 Irrigation

The green roof is irrigated with harvested rainwater through a subsurface drip irrigation system.

7.3.9.5 Maintenance

The green roof is visited by maintenance crews about once a month, or on-demand. Vegetation removed from the roof is hauled off in bags and taken down an exterior stairwell or exterior elevator. This pre-planned route was designed to prevent maintenance crews working on the green roof from walking inside the hospital with plants being hauled off the roof. Voluntary vegetation such as coyote bush (*Baccharis pilularis*) is native to the region and has self-seeded onto the green roof. The plant is left to grow on the roof as it is very attractive to insects including pollinators and the native Gabb's checkerspot and Fatal metalmark butterflies (CNPS 2020).

7.3.9.6 Observed Wildlife

Butterflies, bees, hummingbirds, migratory birds.

7.3.9.7 Best Performing Native Vegetation

All of the vegetation planted has adapted to the roof.

7.3.9.8 Post-occupancy Observations

In Publications

"Palomar Pomerado Health spokesman Andy Hoang said the green roof is more expensive than a traditional roof, but "based on our analysis we will have recovered the cost through energy and water savings in a seven-year time period." (Breier 2010).

Authors' Reflections

- This project is a model for the future of health care in North America. Clearly, patient health and positive environmental experiences are a high priority at this medical center from the entrance drive to the front door and into the building.
- The level of integrated and systems thinking is outstanding. The collaboration and integration of landscape, building, and interiors make for a healthy and inspiring setting that has a lasting return on the investments. This type of result can only take place when all of the representatives are at the table (owners, architect, landscape architect, facilities, staff) and are free to contribute from the first conception of a project.
- This project demonstrates how green roofs can be incorporated into a building program, and repay its cost in a short time. This knowledge should inspire others to investigate how green roofs can become integrated into future projects. However, since specialized knowledge was required to make this project work, the positive outcomes of the interdisciplinary collaborations should be emulated and respected as a critical part of the success. A project like this may not be possible if a collaborative environment is not present.

7.3.10 NOAA Southwest Fisheries Science Center, La Jolla, California

Located in La Jolla, north of San Diego, the National Oceanic and Atmospheric Administration (NOAA) Southwest Fisheries Science Center is a 1.3-hectare (3.3-acre) site located near a hub of research institutes and conservations sites. The San Diego-La Jolla Underwater Park and Ecological Reserves, and the Scripps Coastal Reserve and Institution of Oceanography, and the University of California at San Diego are all located within the same neighborhood. The Scripps Coastal Reserve (see Sect. 7.1.2) is a 51-hectare (126-acre) conservation site set aside to preserve native plant communities. It is also the source of inspiration for some of the vegetation in this case study (Bruce 2018). Each of the properties and institutions shares a similar goal: conservation. The previously undeveloped property now has a LEED Gold-certified building for researcher staff working at NOAA. The property has a 30 m (100 ft) change in elevation and has remnants of coastal scrub and meadow communities. The 7-story building was designed to step up the slope, and incorporate 9 green roofs to make an integrated building and site for its 275 employees (Fig. 7.32). The project team designed the building to emulate the form of the offshore subgrade topography and site. The building has a nestled form where building layers and vegetation are intertwined into the site. Interior spaces were designed so that everywhere, there are views to the

Fig. 7.32 A main outdoor rooftop greenspace shares views from offices, the main conference room (left), and outdoor seat walls that have views to the Pacific Ocean. Part of the intended aesthetic of the green roofs includes open space between drifts of plants. The rock mulched open spaces replicate vegetation patterns in natural systems and benefit biodiversity. (Photo: Bruce Dvorak, October 2018)

ocean and rooftop vegetation. All of the green roofs were designed to complement the coastal ecosystems on and near the site, and characterize some of the common and rare plants or plant communities near La Jolla and San Diego (Bruce 2018; NSWF 2019).

7.3.10.1 Project Team

Building Owner/Client: National Oceanic and Atmospheric Administration (NOAA)
Green Roof Design Team Lead: Jeffrey L. Bruce & Company
Architect: Gould Evans
Landscape Architect: Jeffrey L. Bruce & Company, Wimmer, Yamada, and Caughey
Installation Contractor: Recon Native Plants
Project completion: 2013
Green roof area: 2508 m^2 (27,000 ft^2)

7.3.10.2 Overview and Objectives

The green roofs cover about 30% of all roof space, and the remainder of the roof space houses photo-voltaic panels which supply 7% of the energy used in the building. The green roofs were designed to manage stormwater, to conserve energy, enhance biodiversity, and provide views for the resident researchers and visitors. The undulating green roof growing media averages about 23 cm (9 in) in-depth, and consists of an engineered mixture of locally available materials. The porosity was designed to manage the functional needs of native plants growing in a semi-arid climate. Primary components of the growing media include locally procured sand, organic material, and a local rock mulch (Bruce 2018; NSWF 2019).

7.3.10.3 Plant Establishment

Vegetation was selected from coastal chaparral that is native to San Diego County (Fig. 7.33). Species were chosen in part to showcase their shape, color, and

Fig. 7.33 (**a**) Shaw's agave is a rare and endangered slow-growing plant and was planted here on the roof to supplement conservation sites on the ground, and not as mitigation for lost habitat. Shaw's agave is not native to this site but is native to the San Diego area (but became rare due to lost habitat) and is more frequent in northern Baja California. (**b**) Waves of native grass and succulents make visual connections to the preserved off-site habitats. (**c**) Native grasses grow on an upper roof deck and (**d**) native succulents grow near an overlook to the Pacific Ocean. (Photo: Bruce Dvorak, October 2018)

diversity, and were grouped in mosaic drift patterns to enhance the corresponding viewing angles. Featured plants on Level 3 (main viewing deck) include velvet cactus (*Bergerocactus emoryi*), coast prickly pear (*Opuntia littoralis*), and Shaw agave (*Agave shawii*). On Level 4, (above the main deck) featured plants include woolly blue curls (*Trichostema lanatum*), rock rose (*Helianthenum scoparium*) and deer grass (*Muhlenbergia rugen*) (Bruce 2018). Forms of vegetation used on the green roofs include annuals, ferns, grasses, herbaceous perennials (Table 7.9), shrubs, and succulents.

Annuals

Canchalagua (*Centaurium venustum*), golden yarrow (*Eriophyllum confertiflorum*).

Ferns

Coastal wood fem (*Dryopteris arguta*).

Grasses

Barbara sedge (*Carex barbarae*), California brome (*Bromus carinatus* var. *c.*), cave bluestem (*Botriochloa barbinodis*), deer grass (*Muhlenbergia rigens*), Pacific rush (*Juncus effusus* ssp. *austrocaliforinicus*), spike sedge (*Eleocharis macrostachya*).

Table 7.9 Native herbaceous perennials on the NOAA Southwest Fisheries Science Center green roofs

Common Name	Botanical Name
Santa Barbara milkvetch	*Astragalus trichopodus* var.
Beach evening primrose	*Camissonia cheiranthifolia*
Woolly paintbrush	*Castilleja affinis* ssp. *a.; C. foliolosa*
Datura	*Datura wrightii*
Buckwheat	*Eriogonum grande* var. *rubescens*
Wood strawberry	*Fragaria vesco*
Alkali heath	*Frankenia salina*
Felt-leaf monardella	*Monardella hypoleuca* ssp. *lanata*
California peony	*Paeonia californica*
Blue springs	*Penstemon heterophyllus*
Hummingbird sage	*Salivia spathacea*
Checkerbloom	*Sidalcea malviflora*
Cattail	*Typha domingensis*

Shrubs

Rock rose (*Helianthenum scoparium*), woolly blue curls (*Trichostema lanatum*).

Succulents

Canyon liveforever (*Dudleya cymosa*), coast prickly pear (*Opuntia littoralis*), dudleya (*Dudleya adolphia*), fingertips dudleya (*Dudleya edulis*), golden club (*Bergerocactus emoryi*), pincushion cactus (*Mammillaria dioica*), shaw agave (*Agave shawii*).

Trees

Narrow leaf willow (*Salix exigua*).

7.3.10.4 Irrigation

The green roof irrigation system includes multiple irrigation zones that make use of harvested rainwater. The rooftop irrigation is separated from the landscape irrigation system in order to maximize water efficiency. Excess rainwater is captured by drains and is conveyed to storage basins for later use. Irrigation runs about 5–7 min at 4 a.m., twice a week during the dry season, and is controlled with moisture sensors.

7.3.10.5 Maintenance

The green roof ecosystems are maintained by a crew of three trained personnel about once a week when vegetation is actively growing. Frequent activities include removing unwanted vegetation. Gloves are necessary when working around cacti. The vegetation is fertilized up to two times a year, based upon soil samples. Once a year there is an annual meeting to discuss the direction of the management of the green roofs and decisions are made through discussion between the owners, the head of maintenance, and a consultant.

7.3.10.6 Observed Wildlife

The vegetation provides habitat for bees and hummingbirds and many insects.

7.3.10.7 Best Performing Native Vegetation

All of the vegetation has adapted. Some of the grasses that were initially planted adjacent to cacti have been removed for easier maintenance.

7.3.10.8 Post-occupancy Observations

In Publications

- "With the little bit of rain that we get every so often this winter, the sage smells utterly fabulous and fills the walkways with the scent of the chaparral. The hummers are everywhere and the first blossoms of spring are filling the courtyard with yellows and purples." ~ Sarah Mesnick, SWFSC Science Liaison (Bruce 2018).

Authors' Reflections

- This is an amazing green roof that serves as a role model for the integrated design of buildings and their environments, and the beauty and aesthetics (sights, sounds, aromas) of ecoregional green roofs.
- The vision for maintaining the roof is exemplary. The idea that there is an annual meeting between the owners, the head of maintenance and a green roof consultant is integral to its success. The little time and money invested engage those with knowledge and those active on the site with a common vision.

7.3.11 Good Earth Plants Building #1 San Diego, California

In 1977, Jim Mumford established his business career selling plants, in the founding of Good Earth Plants. The aim of Jim's work follows a belief in the adage, "We don't inherit the earth from our ancestors; we borrow it from our children". This belief led Jim's passion for plants to connect to vegetation in urban places. His experiences with growing plants turned skyward to green roofs and living walls beginning around 2006. Mumford was working with plants in indoor environments but became inspired by the idea of growing plants vertically when he attended a local landscape conference. He began exploring with green roofs on his own buildings and instead of using exotic sedums, he looked to the native plants of the region (Fig. 7.34). There were no other green roofs in the Southern California region at the time, so his work was a pioneering effort in San Diego as the green roof on Building #1 was the first permitted green roof in San Diego (Mumford 2017).

Fig. 7.34 Located on Building #1, this green roof has several themed collections of native plants. Later, several non-native plants have been trialed on the roof. The intended maintenance regime for this roof is to let the vegetation compete, with minimal upkeep. As such, this ecoregional green roof demonstrates a biologically diverse roof with an ecological aesthetic. (Photo: Bruce Dvorak, October 2018)

7.3.11.1 Project Team

Building Owner/Client: Good Earth Plants Company, Inc.
Green Roof Design Team Lead: Jim Mumford
Architect: Robert Thiele, AIA
Landscape Architect: David Solmes, MLA
Project completion: 2007
Green roof area: 158 m^2 (1700 ft^2)

7.3.11.2 Overview and Objectives

This extensive green roof was designed to feature California and southwest native plants including, but not limited to: cacti, succulents, grasses, sun rose, desert marigold, and beach primrose. The substrate has two designs. One design has a 10–12 cm deep (4–5 in) built-up (layered) green roof system and the other consists of 24 modular units provided by Tournesol Siteworks®. The northeast side of the building features a staircase, a small viewing platform, a boardwalk, and a maintenance path (Mumford 2018).

The project and research goals for the green roof include:

1. Conduct a performance evaluation of specific plant species for the rate of establishment, environmental tolerances, plant competition, ability to exclude invasive weeds, survival, and persistence.
2. Evaluate mixed plant communities and succession over time.
3. Evaluate growth media mix for drainage, water retention, compaction, organic break down, viability to chosen plant material, and resistance to wind erosion.
4. Determine irrigation rates, volumes, seasonality, and systems and methods.
5. Examine the pros and cons of using a built-up system versus a modular system.
6. Monitor stormwater run-off for volume reduction and filtration of particulates.
7. Record rooftop temperature reduction in comparison to a control roof.
8. Cost analysis
9. Community outreach
10. Evaluate biodiversity." (Mumford 2018)

7.3.11.3 Plant Establishment

Planting of the living roof was achieved in phases (Fig. 7.35). The initial planting was followed with later installations to infill and trial new species. The following plants have been trialed on the green roof.

Native Grasses

Nodding needle grass (*Nassella cernua*), purple three awn (*Aristida purpurea*).

Native Herbaceous Perennials

Beach primrose (*Camissoniopsis cheiranthifolia*), blue eyed grass (*Sisyrinchium bellum*), California aster (*Lessingia filaginifolia*), California sunflower (*Encelia californica*), desert mallow (*Sphaeralcea ambigua*), desert marigold (*Baileya multiraoiata*), evening primrose (*Oenothera speciosa* ·rosea'), golden yarrow (*Eriophyllum confertiflorum*), red-skinned onion (*Allium haematochiton*), silver beach burr (*Ambrosia chamissonis*), silver wormwood (*Artemisia ludoviciana*), sun rose (*Helianthemum scoparium*), yarrow (*Achillea millefolium*).

Fig. 7.35 Prickly pear (*Opuntia littoralis*) was planted and is spreading, flowering, and fruiting on the roof. It grows with grasses, as it does at conservation sites along the California coast. This aesthetic is consistent with a low-maintenance biodiverse approach and benefits wildlife. (Photo: Bruce Dvorak, October 2018)

Native Succulents

Agave (species), chalk dudleya (*Dudleya pulverulenta*), silver dollar plant (*Dudleya britton*), stonecrop (*Sedum* (species various)), ladies' fingers (*Dudleya edulis*), prickly pear (*Opuntia littoralis*).

Native Shrubs/Groundcovers

Cliff spurge (*Euphorbia misera*).

7.3.11.4 Irrigation

Irrigation is a subsurface drip system set to operate for 20 min, three times per week during the summer.

7.3.11.5 Maintenance

Maintenance activities have evolved and vary with seasonal needs. Some of the invasive (non-native) species removed from the green roof include wild pea (pea family), and pampas grass (*Cortaderia selloana*).

7.3.11.6 Observed Wildlife

Crows, ravens, doves, spiders, bees, Monarch butterflies, aphids, ladybugs.

7.3.11.7 Best Performing Native Vegetation

Some of the top-performing plants include agave, grasses (native), milkweeds (*Asclepius*), onion grass, prickly pear, yarrow. Coyote brush (*Baccharis pilularis*), is a California native woody shrub that is easy to establish. It is part of the coastal chaparral ecoregion. Coyote brush has established itself on the roof, and a small population is retained on the roof for its benefits to pollinator insects and wildlife.

7.3.11.8 Post-occupancy Observations

Owner's Reflections (Jim Mumford)

- The living roof has evolved over time. The initial plantings have been edited over time, and new species have been added where some plants have not established. There was never any intention to make this a formal roof garden, but more of an exploration of a native green roof palette for the San Diego area.
- I have had an interest in following up with my research agenda. However, I have had difficulty locating research partners.

Authors' Reflections

- Although the green roof is not large, the complexity and diversity of native plants on this roof is amazing and could have a wide impact. This living roof demonstrates that a variety of plants native to the coastal ecoregion can grow on extensive green roofs in Southern California with minimal watering and maintenance. This kind of low-maintenance extensive roof, with some supplemental watering, can provide for biodiversity in urban centers, conservation of energy in buildings and retention of stormwater.

7.4 Plants for the California Coastal Ecoregions

Following the richness of the coastal ecoregions, the green roof case study sites in Chap. 7 explore 207 taxa in total (native and native-adjacent). Of the 207 taxa, 191 are native to the ecoregions in this chapter, and 49 species occur more than once across the case studies in this chapter. Of those occurring more than once, three are annuals, three are ferns, eight are grasses, one is a groundcover, 24 herbaceous perennials, seven are shrubs, and three are succulents. Eighteen species occur in three or more of the case study sites in the chapter, see Table 7.10.

Table 7.10 Plant species that occur on three or more ecoregional green roofs in this chapter

Plant Type	Common Name	Botanical Name	A	B	C	D	E	F	G	H	I
Annual	California poppy	*Eschscholzia californica*	x	x		x					
Fern	Western sword fern	*Polystichum munitum*	x	x		x		x			x
Grass	California fescue	*Festuca californica*	x				x	x			x
Grass	Spreading rush	*Juncus patens*	x	x				x			
Grass	Deer grass	*Muhlenbergia rigens*	x		x			x		x	
Groundcover	Beach strawberry	*Fragaria vesca*		x	x		x		x		
Herbaceous perennial	Yarrow	*Achillea millefolium*	x	x		x	x		x	x	x
Herbaceous perennial	California fuchsia	*Epilobium canum*	x	x			x			x	x
Herbaceous perennial	Seaside fleabane	*Erigeron glaucus*	x	x				x	x		
Herbaceous perennial	Red-flowered buckwheat	*Eriogonum grande* var. *rubescens*	x		x			x			
Herbaceous perennial	Coast buckwheat	*Eriogonum latifolium*	x	x				x	x		
Herbaceous perennial	Island alum root	*Heuchera maxima*	x	x							x
Herbaceous perennial	Douglas iris	*Iris douglasiana*	x					x			x
Herbaceous perennial	Foothill penstemon	*Penstemon heterophyllus*	x		x		x			x	
Herbaceous perennial	Hummingbird sage	*Salvia spathacea*	x	x			x			x	x
Herbaceous perennial	Blue-eyed grass	*Sisyrinchium bellum*	x	x		x				x	
Shrub	Cleveland sage	*Salvia clevelandii*					x	x		x	
Succulent	Broadleaf stonecrop	*Sedum spathulifolium*	x	x						x	

Key = A (California Academy of Sciences), B (Heron's Head), C (NOAA SW Fisheries), D (One Van Ness), E (Palomar Medical Center), F (Salesforce Transit Center), G (Slide Ranch), H (Sonoma Academy), I (Sutter Hospital)

7.5 Summary

Since the construction of the Gap Headquarters living roofs in 1997 (EarthPledge 2005), there has been a slow but growing interest in implementing green roofs in California using native vegetation for various purposes. The case studies here demonstrate that green roofs were built to:

- connect visually with the surrounding environment, e.g., Sonoma Academy, California Academy of Sciences, Slide Ranch, NOAA Southwest Fisheries Science Center, LAMOTH;
- integrate buildings into the surrounding topography, e.g., Holocaust Museum, NOAA Southwest Fisheries Science Center, LAMOTH;
- provide a learning tool for green roofs, e.g., Salesforce Transit Center, Good Earth Plants Building, Sonoma Academy, California Academy of Sciences, Heron's Head.
- provide an outdoor laboratory for observing nature, especially pollinators and other visitors to the green roofs, e.g., California Academy of Sciences, Salesforce Transit Center;
- benefit from the ecosystem services generally associated with green roofs, viz., energy conservation, runoff amelioration, temperature, and noise abatement (and this would include roofs using non-native species as well), e.g. Sonoma Academy, California Academy of Sciences, NOAA;
- offset the effects of habitat loss, not via a mitigation process *per se*, but rather through the intentional introduction of habitat that either increases the number of plants of a rare species in an area, e.g., Shaw's agave NOAA SW Fisheries green roof; or counters the effect of habitat degradation and destruction for animal species reliant on particular plant species, e.g., Bay checkerspot butterfly, Salesforce Transit Center;
- increase local habitat diversity, e.g., Heron's Head, California Academy of Sciences, Slide Ranch;
- increase the aesthetic appeal of buildings especially where the function of the building itself may be associated with increased stress for visitors, e.g., hospitals (Sutter Health and Palomar Medical Center) and Holocaust Museum;
- serve as a pilot project to educate policymakers and developers about green roofs with natives, e.g. One South Van Ness Avenue.

The environment of the broadly coastal region of California is challenging for green roof development because of the dry, hot summers. Nonetheless, as can be seen from the examples in the chapter, there are many successful roofs, and some of these are quite large. More roofs have been developed in northern California than southern California, possibly associated with the slightly wetter and cooler conditions in the north; however, these areas still are subject to difficult periods in summer. It should be noted that while the southernmost and driest region of coastal California does have several green roofs that reflect the vegetation of native habitats, there is a paucity of green roofs in the Los Angeles area, especially green roofs with native plants. The few examples of green roofs with native vegetation in southern California have not been studied nor made known such as is the case in northern California. However, ecoregional green roofs are viable in southern California with an integrated-design approach where harvested rainwater or greywater is used to supplement natural rainfall. Indeed, there are a few green roofs in the greater Los Angeles region that are well-developed, but do not emphasize native plants, e.g., Premier Automotive North America HQ/Taco Bell HQ; or are designed as rooftop urban gardens, e.g., Judy Skalsky Memorial Rooftop Garden, Art Center College of Design. These private organizations value green roofs but were designed as gardens.

Little formal research into native-planted green roofs in California is publicly available, nor is there much-concerted effort in this area. However, many of the green roofs in this chapter were designed by or partnered with the same design team (Rana Creek) that now has the most empirical experience with ecoregional green roofs, especially in northern California. Research partnerships with design firms that do this work are one possible venture to advance knowledge. Research funding through universities is highly competitive and limited, so the role of corporate social responsibility could prove extremely valuable if private corporations include in their corporate mission statements sharing of knowledge from their green infrastructure projects, especially corporations where millions of daily users support the wealth of the company.

There are several elements common to the green roofs we observed. Most green roofs have a minimum substrate depth of 15 cm (6 in) allowing semi-extensive mixes of grasses, annuals, and herbaceous perennials. However, the Palomar Medical Center roof meadow has an extensive 10 cm-deep (4 in) substrate and the vegetation is thriving with proper care and watering. Thus, native plants on extensive substrates may be possible with an appropriate design. Deeper areas or pockets accommodate larger shrubs and trees increasing diversity of growth form and species. Direct planting, biodegradable trays, or hydroseeding are used to establish many of the roofs. Modular trays, more suitable for extensive roofs comprised of mat-forming succulents (of which there are few native to coastal California) are seldom used. Deeper substrates also buffer more against environmental extremes, particularly water stress. The accommodations for semi-intensive systems was not a barrier to construction costs since most of the green roofs featured here are new construction. All of the roofs have irrigation systems: various systems such as overhead spray or capillary systems are used. Most green roofs are irrigated during the drier summer through fall months only, and most water only several times a week. Consistent with the environmental goals established for many of the roofs/buildings, irrigation often uses rainfall collected in the wetter months. Species for the most part reflect the particular ecosystems of the area, although horticultural cultivars and a few locally non-native species have been used. Common to most of the roofs is a matrix of grasses and/or herbaceous species that is then interspersed with annuals, larger shrubs, and succulents. Succulents native to the region can be quite large plants that are upright in stature and are not mat-forming succulents. Thus for the green roofs of California, succulents seldom form a continuous matrix, rather they add to the visual interest and attract pollinators and are members of a plant community.

Many of the green roofs are still relatively new. However, visually, functionally, and environmentally they seem to be successful, despite the challenging extremes of summer. Underlying their success is detailed planning, not only for their creation and installation but also for ongoing maintenance. The maintenance is either carried out by contracted specialists or transitioned to occupants of the building who are trained. Vigilance with respect to potentially invasive species is critical to maintaining the ecological integrity of these green roofs. For several of the roofs, it is noted that the frequency of maintenance is not as important as the timing of maintenance. Ecoregional green roofs reflect both the species and timing of their local environment.

Acknowledgments We would like to thank the following individuals for the time, sharing of knowledge, access to green roofs and their dedication to ecoregional green roofs: Paul Kephart and Matt Yurus with Rana Creek Habitat Restoration; Kevin & Kerrie Lee with SYMBIOS; Jeffery L. Bruce with Jeffery L. Bruce & Company; Andrew Spurlock and Emily Dowgiallo with Spurlock Landscape Architects; Ron Alameida and Christine Falvey with the City of San Franciso; Annemarie Dompe of the EcoCenter at Heron's Head Park and Tai Trang at Port of San Francisco; Jim Mumford owner Good Earth Plants; Gerald Sui with the City of San Francisco; Elizabeth Bagley and William Silver with the California Academy of Sciences; Matthew Vogel with NOAA; and the staff at the Slide Ranch, Sonoma Academy, Palomar Medical Center, Sutter Hospital, LAMOTH, Ballona Wetlands Ecological Reserve, and Torry Pines Reserve® for allowing access to their property and sharing knowledge about their work.

References

Alexander JM, D'Antonio CM (2003) Seed bank dynamics of French broom in coastal California grasslands: effects of stand age and prescribed burning on control and restoration. Restor Ecol 11(2):185–197

Anderson MK (1997) From tillage to table: the indigenous cultivation of geophytes for food in California. J Ethnobiol 17(2):149–169

Bailey RG (1997) Ecoregions of North America. U.S. Department of Agriculture, Forest Service, Washington, DC

Baker HG (1989) Sources of the naturalized grasses and herbs in California grasslands. In: Grassland structure and function. Springer, New York, pp 29–38

Benjamin LL, Dakin K, Pantiel M (2013) The professional design guide to green roofs. Timber Press, Portland

Biswell HH (1956) Ecology of California grasslands. J Range Manag 9(1):19–24

Boytner R, Dorsey JH (2007) A baseline survey of the Ballona Outdoor Learning and Discovery (BOLD) area. Report submitted to the California Coastal Conservancy and Santa Monica Bay Restoration Commission by the Ballona Wetlands Foundation. California State Coastal Conservancy, Los Angeles

Breier M (2010) Green roof at new hospital is more than a cover. The San Diego Union-Tribune, San Diego

Brothers W, Halsey R (2020) Plants at Torrey Pines State Natural Reserve. Plants at Torrey Pines State Natural Reserve®. https://torreypine.org/nature-center/plants/. Accessed 27 Jan 2020

Bruce JL (2018) NOAA La Jolla Southwest Fisheries Science Center Laboratory. greenroofs.org. https://www.greenroofs.com/projects/noaa-la-jolla-southwest-fisheries-science-center-laboratory-replacement-project-uc-san-diego. Accessed 18 Sept 2019

wildlife.ca.gov/Conservation/Plants (2019) California Department of Fish and Wildlife. https://wildlife.ca.gov/Conservation/Plants. Accessed July 16

Calscape (2019) Calscape: California Native Plant Society. California Native Plant Society. https://calscape.org/. Accessed 30 Aug 2019

CAS (2011) The California Academy of Sciences receives second LEED platinum rating from U.S. Green Building Council. California Academy of Sciences, San Francisco

CAS (2014) Native plants species on the living roof. CAS, San Francisco

CLO (2019) California Academy of Science--Living Roof Cornell Lab of Ornithology: eBird. https://ebird.org/hotspot/L2721356. Accessed 10 Sept 2019

CNPS (2020) Coyote brush: Baccharis pilularis ssp. consanguinea. California Native Plants Society. https://calscape.org/Baccharis-pilularis-ssp.-consanguinea-(). Accessed 23 June 2020

Commission CCZC (1975) California coastal plan. California Coastal Zone Conservation Commission

Cooper DS (2008) The use of historical data in the restoration of the avifauna of the Ballona Wetlands, Los Angeles County, California. Nat Areas J 28(1):83–91

Cox RD, Allen EB (2007) Composition of soil seed banks in southern California coastal sage scrub and adjacent exotic grassland. Plant Ecol 198(1):37. https://doi.org/10.1007/s11258-007-9383-9

D'Antonio C, Bainbridge S, Kennedy C, Bartolome J, Reynolds S (2002) Ecology and restoration of California grasslands with special emphasis on the influence of fire and grazing on native grassland species. Report to the Packard Foundation

Davis S (2016) Project manager's report out: the Hospital's Green Roofs. Sutter Health Newsletter. Sutter Health CPMC, San Francisco

Dawkin K (2013) Wind and light. Pac Hortic 74:04

Dixon D (1999) Dunes alive: the endangered Smith's blue and marina blue butterflies. A closer look at coastal dune wildlife of south Monterey Bay. Tideline 19(3):1–3

Dorsey J (2007) Introduction. In: Dorsey JH, Bergquist S (eds) A baseline survey of the Ballona Outdoor Learning and Discovery (BOLD) area. Report submitted to the California Coastal Conservancy and Santa Monica Bay Restoration Commission by the Ballona Wetlands Foundation. Ballona Wetlands Foundation, Santa Monica

Drennan PM, Leon A, Tyrrell C (2011) Green roofs: developing a native plant palette for Southern California. Proc Water Environ Fed 2011(10):5972–5978

EarthPledge (2005) Green roofs ecological design and construction. Schiffer Books, Atglen

Edgerton MF (2011) Green roofs and regenerative design. Fabric Architecture. IFAI, Roseville

FA (2019) Heron's Head EcoCenter. Feldman Architecture. https://feldmanarchitecture.com/herons-head-ecocenter/. Accessed 19 Aug 2019

Fiedler PL, Leidy RA (1987) Plant communities of Ring Mountain Preserve, Marin County, California. Madrono 34(3):173–192

Garcia R, Gaberšek O (2010) One South Van Ness Green roof maintenance. Rana Creek, San Francisco

Green J (2019) Restoring the remnant ecosystems of San Franciso. The American Society of Landscape Architects, Washington, DC

Greenroofs.com, Rana Creek (2020) California Academy of Sciences (CAS) Living Roof. greenroofs.com. https://www.greenroofs.com/projects/california-academy-of-sciences-cas-living-roof/. Accessed Oct 28 2020

Griffith GE, Omernik JM, Smith DW, Cook TD, Tallyn E, Moseley K, Johnson CB (2016) Ecoregions of California. US Geological Survey, Reston

Hauser K (2013) An evolving landscape: the living roof at California Academy of Sciences. Pacific Horticulture Journal. Pacific Horticulture Society, Berkeley

Huenneke LF (1989) Distribution and regional patterns of Californian grasslands. In: Huenneke LF, Mooney HA (eds) Grassland structure and function: California annual grassland. Springer Netherlands, Dordrecht, pp 1–12. https://doi.org/10.1007/978-94-009-3113-8_1

Huenneke L, Mooney H (1989) The California annual grassland: an overview. In: Grassland structure and function. Springer, Kluwer, pp 213–218

IN (2019) iNaturalist. https://www.inaturalist.org. https://www.inaturalist.org. Accessed 31 Aug 2019

Keeley JE (2002) Native American impacts on fire regimes of the California coastal ranges. J Biogeogr 29(3):303–320

Keeley JE, Fotheringham C, Baer-Keeley M (2006) Demographic patterns of postfire regeneration in Mediterranean-climate shrublands of California. Ecol Monogr 76(2):235–255

Kephart P (2005) Living architecture – an ecological approach. In: Paper presented at the Third Annual Greening Rooftops for Sustainable Communities Conference, Washington, DC, May 4–6

King C (1988) Prehistoric resource use and settlement in the Santa Ynez River Basin. In: Draft Report prepared by Science Applications International Corporation, Santa Barbara, California, vol I. Submitted to Unocal Corporation, Santa Barbara, p 62

Klein A, Keeler-Wolf T, Evens J (2015) Classification of the vegetation alliances and associations of Sonoma County, California volume 1 of 2–introduction, methods, and results. California Department of Fish and Wildlife, Byron

Koteen LE, Baldocchi DD, Harte J (2011) Invasion of non-native grasses causes a drop in soil carbon storage in California grasslands. Environ Res Lett 6(4):044001. https://doi.org/10.1088/1748-9326/6/4/044001

Lundholm JT, Walker EA (2018) Evaluating the habitat template approach applied to green roofs. Urban Natural S 1:39–51

Lyon C (2019) A Green roof that does the environment justice. Habitat Gardens. Accessed 19 Aug 2019

Marcus CC, Barnes M (1999) Healing gardens: therapeutic benefits and design recommendations. Wiley, New York

Mattoni R, Longcore TR (1997) The Los Angeles coastal prairie, a vanished community. Crossosoma 23(2):71–102

McDonald J, Drennan P (2013) A comparison of heat tolerance of Dudleyas and Sedums used for greenroofs in southern California. In: Poster presented at the 5th Undergraduate Research Symposium, Los Angeles, CA, March 2013. Loyola Marymount University, p 1

Mumford J (2017) Building integrated vegetation art, commerce and design. In: Green roof conference. Good Earth Plants, San Diego, p 8

Mumford J (2018) Good Earth Plants building #1. greenroofs.com. https://www.greenroofs.com/projects/good-earth-plants-building-1. Accessed 24 Sept 2019

Munz PA (1970) A California flora. University of California Press, Berkeley

NPS (2006) Wildflowers of Golden Gate, vol 2019. National Park Service: Department of the Interior, Washington, DC

NPS (2019) NPS.govPark HomeLearn About the ParkNaturePlantsWildflowers. NPSS. https://www.nps.gov/goga/learn/nature/wildflowers.htm. Accessed 9 Aug 2019

NSWF (2019) NOAA Southwest Fisheries Science Center La Jolla, California. Gould Evans. Accessed 19 Sept 2019

Palmer M (2015) Living roof at one South Van Ness Avenue. San Francisco Department of Environment, San Francisco

PWP (2019) Salesforce Transit Center Park (Transbay Transit Center). PWP. http://www.pwpla.com/salesforce-transit-center-park

Read EA (2015) A flora of the Ballona wetlands and environs. Bull, Southern California Acad Sci 114(3):149–164

SFLRM (2015) San Francisco living roof manual. City of San Francisco. SFLRM, San Francisco

Spurlock A (2015) Awards submittal package: Palomar Medical Center, Spurlock Poirier. Spurlock Landscape Architecture, San Diego

SR (2019) Slide Ranch Mission and history. Slide Ranch. http://www.slideranch.org/company. Accessed 3 Sept 2019

Steers RJ, Spaulding HL (2013) Native Component Grasslands of the Marin Headlands. National Park Service, Fort Collins

Stromberg MR, D'Antonio CM, Young TP, Wirka J, Kephart PR (2007) California grassland restoration. California grasslands: ecology and management. University of California Press, Berkeley, pp 254–280

Suppakittpaisarn P, Jiang X, Sullivan WC (2017) Green infrastructure, green stormwater infrastructure, and human health: a review. Curr Landsc Ecol Rep 2(4):96–110

Timm N (2017) Palomar Medical Center. CO Architecs, San Diego

Ulrich R (1984) View through a window may influence recovery. Science 224(4647):224–225

Vale T (2013) Fire, native peoples, and the natural landscape. Island Press, Washington, DC

WBDG (2019) Sonoma Academy's Janet Durgin Guild & Commons. Whole Building Design Guide 2019 (May 28, 2019)

Chapter 8
Green Roofs in Puget Lowland Ecoregions

Bruce Dvorak and Nancy D. Rottle

Abstract This chapter presents case studies of five conservation sites and nine green roofs mainly located in the Puget Lowland ecoregions of western Washington state, and foothills of the Cascade Mountains. The landscape is geographically complex with many diverse ecoregions such as coniferous forests, mixed forests, prairie-oak ecosystems, rocky balds, wetlands, and riparian habitats. Precipitation averages 800–950 mm annually in the lowlands near the Seattle and Tacoma metro areas, and over 2540 mm in the western Cascades. Much of the precipitation takes place during the fall, winter, and early spring. Historically, prairies, oak/coniferous savanna and forest vegetation dominated at elevations from sea level to the foothills of the Cascade Mountains. Less than 3% of prairie-oak habitat remains and about 5% of the native forested habitat remains intact. Nine ecoregional green roof case studies demonstrate how 60 taxa of plants native to the coastal ecoregions can be employed on green roofs.

Keywords Summer drought · Greywater · Brownfields · Integrated design · Maintenance training · Leak detection · Corporate social responsibility · Wetlands · Living roof

8.1 Ecoregion Characteristics

The Willamette Valley-Puget Trough-Georgia Basin physiographic regions lie positioned between coastal mountain ranges to the west and the Cascade Mountains to the east. These form a linear landscape of inland and coastal valleys that extends

B. Dvorak (✉)
Department of Landscape Architecture and Urban Planning, 305A Langford Architecture Center, Texas A&M University, College Station, TX, USA
e-mail: bdvorak@tamu.edu

N. D. Rottle
Department of Landscape Architecture, University of Washington, Seattle, WA, USA
e-mail: nrottle@u.washington.edu

B. Dvorak (ed.), *Ecoregional Green Roofs*, Cities and Nature,
https://doi.org/10.1007/978-3-030-58395-8_8

611 kilometers (380 miles) from the southern edges of the Willamette Valley to Vancouver British Columbia at the north. This chapter focuses on the mid-way point in the Puget Trough, the Puget Lowlands, as a way to introduce vegetation of the ecoregions and ecoregional green roofs of the Pacific Northwest.

About 17,000–20,000 years ago, the Puget lobe of the Cordilleran Ice Sheet was spread across the land that would become the Puget Lowlands, where Olympia, Tacoma, and Seattle lie today. About 7000 years ago, the glaciers were retreating and the landscape vegetation was changing from tundra to a mosaic of different ecosystems, such as grasslands and open forested landscapes. As the climate was warming, sea levels were rising and intermittent layers of volcanic were deposited upon the glacial till, glacial outwash, and lacustrine deposited soils formed across the lowlands (Franklin and Dyrness 1988; Whitlock 1992).

About 4000 years ago, the climate became wetter and more closely represented the climate of the Pacific Northwest today, with a shift from drier vegetation such as prairies with pines, oaks and Douglas fir towards a vegetative cover that included a smaller proportion of prairies and oak ecosystems intermixed with western redcedar and mixed-species forests. (Barnosky 1985; Chappell and Crawford 1997; Leopold and Boyd 1999). The Puget Lowlands evolved into a complex mosaic of forests, oak and conifer savannas, prairies (Fig. 8.1), wetlands, and estuaries. (Hebda and Mathewes 1984; Franklin and Dyrness 1988; Thysell and Carey 2001).

The coastal lowlands had become one of the first regions of North America to be settled by humans at least 13,500 years ago (Haynes 2002). The dominant Native

Fig. 8.1 *Morning, Mt. Tacoma, 10 Miles South of Tacoma City, Washington*, 1891, by James Everett Stuart. This painting shows an open prairie in the foreground and gives way to savanna and forest beyond. This historical painting shows the edges of the extent of open prairie vegetation that once stretched from Commencement Bay near present-day downtown Tacoma, and southward near the Columbia River. (Courtesy: Altermann Galleries & Auctioneers)

American tribes living in the region at the time of European settlement included the Coast Salish tribes of the Puyallup, Snohomish, Muckleshoot, Snoqualmie, Nisqually, Skagit, Suquamish, Squaxin, Swinomish, Stillaguamish, and Sauk-Suiattle. These tribes developed important cultural relationships with the land that had a significant influence on the formation and composition of the plant communities including forests, woodlands, and fire-resistant plant communities such as prairies and savannas (Storm and Shebitz 2006). These were maintained by the intentional burning of the ground vegetation to sustain habitat for hunting wild game, and the growing and gathering of plants and plant parts (Turner et al. 2013). Important to their culture were many species of forbs, bulbs, shrubs, and trees, including western redcedar (Turner et al. 2013).

Early European explorers of the Pacific Northwest describe an almost Eden-like landscape that was unparalleled to anywhere they had experienced (Sage 1854): "No country in the world affords a better soil, or a more romantic scenery" (p. 215). As settlers from the eastern United States settled the region in the nineteenth century, burning of the ground vegetation as a land management practice was suppressed, and much of the remaining prairie and oak savanna became enclosed and forested due to the absence of fire (Crawford and Hall 1997; Leopold and Boyd 1999; Schultz et al. 2011). Today, in the Puget Lowlands, about 5% of the original historic forested areas remain, and about 3% of the prairie-oak ecosystems remain on managed conservation sites (DellaSala et al. 2019).

In modern times, the lowland vegetation has been very heavily influenced by logging, agriculture, urbanization, fire suppression, invasive species, and changes to watershed management (DellaSala et al. 2019). A serious threat to native plant communities, especially prairie-oak woodlands, is invasion by Scotch broom (*Cytisus scoparius*) and other exotics (DellaSala et al. 2019). Few conservation sites exist where one can learn about the historic vegetation native to the lowlands (Chappell and Crawford 1997; Crawford and Hall 1997).

The climate of the region is defined by summers that are typically dry and warm followed by a fall, winter, and early spring that is characterized by persistent precipitation and cool temperatures (Fig. 8.2). Precipitation is greatest in the lowlands along the west-facing slopes of the Cascade Mountains. A rain shadow from the Olympic Mountains exists at the western edge of the Puget Sound near Sequim (410 mm/16 in), where east and across the sound near Seattle and Everett annual precipitation is about 940 mm (37 in). These rainfall patterns favor the presence of prairie, savanna, and evergreen vegetation. During the summer dry season, some herbaceous species adapt through a summer dormancy period.

8.1.1 Vegetation Within the Ecoregions

Pre-Columbian vegetation types and ecoregions of the Puget Lowlands includes prairies, oak savannas, mixed-forests, coastal temperate forests, and other kinds of plant associations (Fig. 8.3). These include a wide variety of forms of vegetation

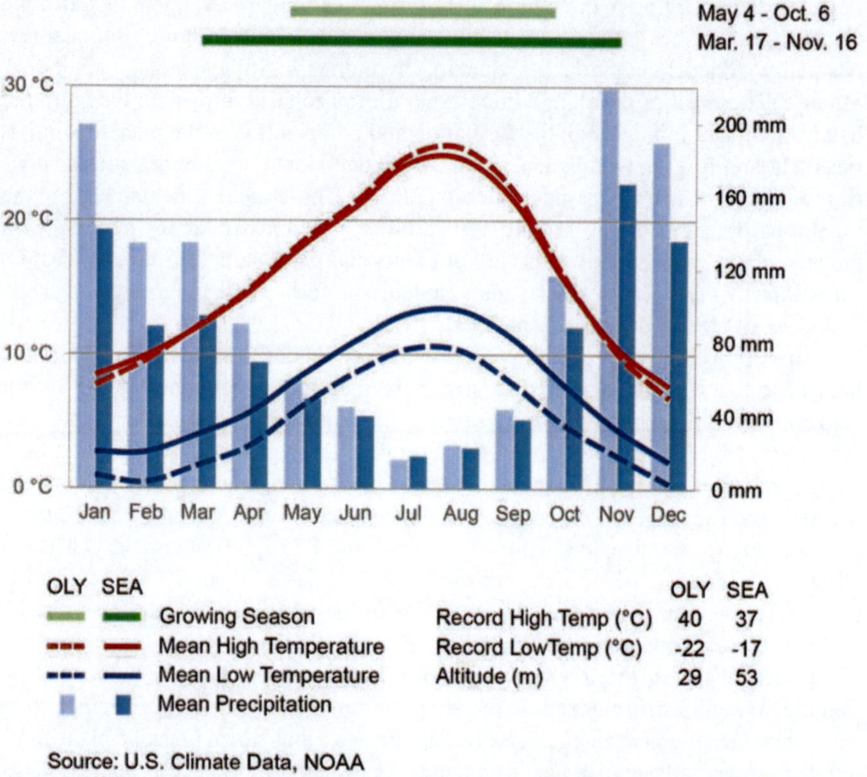

Fig. 8.2 Climate characteristics of the Puget Lowlands for Seattle and Olympia, Washington. The Puget Lowlands climate is modified by the effects of cool ocean waters that surround and permeate the lowlands. Annual precipitation for Olympia is 1270 mm (50 in), where Seattle is drier at 940 mm (37 in). Olympia, being further inland than, Seattle experiences cooler temperatures during winter, and slightly warmer temperatures during summer. Thus, the natural vegetation of the region favored prairie-oak ecosystems in the south end of the lowlands near Olympia, because of its drier and warmer summers. (Graphic: Tess Menotti & Bruce Dvorak)

including trees, shrubs, grasses, herbaceous perennials, annuals, bulbs, ferns, and mosses (Figs. 8.4 and 8.5).

8.1.1.1 Coastal Temperate Forests

In western Washington, coastal temperate forests include a range of coniferous trees, groundcovers, shrubs, grasses, and herbaceous plants. Major conifer tree species in the Puget Lowlands (sea level to about 1800 m/ 5–6000 ft.), include western redcedar (*Thuja plicata*), Douglas-fir (*Pseudotsuga menziesii*) western hemlock (*Tsuga heterophylla*), grand fir (*Abies grandis*), spruces (*Picea sp.),* and ponderosa pine (*Pinus ponderosa*) in drier microclimates. Other native conifers found in the

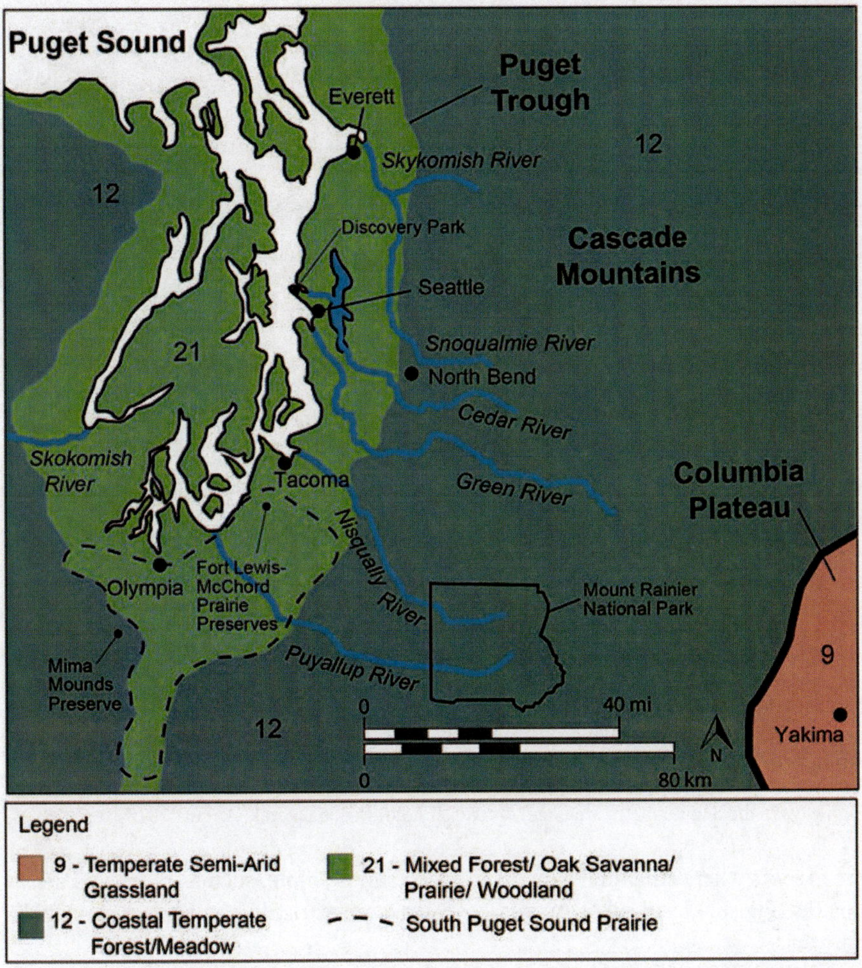

Fig. 8.3 Ecoregions of the Puget Lowlands and the Cascade Mountains, and vicinity (Omernik 1995; Baily 1997). Located up out of the lowland plains, the biologically diverse Coastal Temperate Forests are widespread across the Cascade Mountains and foothills (12). Mixed forest, oak savanna, and prairie existed in the lowlands from sea level to elevations of about 305 m (1000 ft) and closely follows the footprint of the Puget lobe ice sheet (21). The dashed black line shows the approximate boundary of the South Sound Prairies where large open prairies existed. Semi-Arid Grassland ecoregions (9) preside in the rain shadow east of the Cascade Mountains. Complex ecotones exist between ecoregion boundaries. (Graphic: Trevor Maciejewski & Bruce Dvorak)

ecoregion include western larch (*Larix occidentalis*) at high elevations, yew (*Taxus*), and juniper (*Juniperus*) in dry areas. Western redcedar was very important to the Native cultures, and there is evidence that they propagated the expanse of western redcedar because of their extensive use of the tree species to build canoes, house planks, clothing, and more (Hebda and Mathewes 1984). Some of the trees, shrubs, and groundcovers of the Coastal Temperate Forests may be adaptable to intensive

Fig. 8.4 Freeway Park in Seattle, Washington is a roof deck park covering over Interstate-5 and is planted with many native groundcovers, shrubs, and trees such western hemlock. Freeway Park reconnects neighborhood blocks that were bisected from the construction of I-5. Built in 1976, this iconic roof garden has inspired numerous roof deck plantings, trails, and parks over suppressed highways in the Seattle metro area, and abroad. (Photo: Bruce Dvorak, August 2018)

green roofs where structural support for deep soil is ample, such as Freeway Park in Seattle (Fig. 8.4) and other overpass on-structure parks in the Seattle metropolitan area.

Not a coniferous tree, but evergreen and native to fire-tolerant communities such as oak and ponderosa pine, the Pacific madrone (*Arbutus menziesii*) grows on well-drained coastal and drier sites (Pojar and MacKinnon 2004). It is losing habitat range due to being overtaken by taller coniferous trees, which expand into madrone habitats that previously experienced ground fires.

8.1.1.2 Deciduous Trees

Dominant deciduous trees that are native to the region include Oregon white oak (*Quercus garryana*), vine maple (*Acer circinatum*), Oregon ash (*Fraxinus latifolia*), Pacific dogwood (*Cornus nutallii*) and bigleaf maple (*Acer macrophyllum*).

Fig. 8.5 Native mosses planted and growing on a shady unirrigated green roof at Evergreen State College in Olympia, Washington. (Photo: Bruce Dvorak, September 2018)

8.1.1.3 Shrubs

Native shrubs common to the ecoregion include snowberry (*Symphoricarpos albus* var. *laevigatus*), Oregon grape (*Mahonia/Berberis aquifolium*), elderberry (*Sambucus*), Douglas spiraea (*Spiraea douglasii*), beaked hazelnut (*Corylus cornuta* var. *californica*), ocean spray (*Holodiscus discolor*), ninebark (*Physocarpus capitatus*), salal (*Gaultheria shallon*), red huckleberry (*Vaccinium parvifolium*), evergreen huckleberry (*Vaccinium ovatum*) and rhododendron (*Rhododendron macrophyllum*) which is Washington's State Flower. Many of the small trees and shrubs may be adaptable to intensive green roofs in western Washington.

8.1.1.4 Ferns and Groundcovers

Native ferns and groundcovers such as western sword fern (*Polystichum munitum*), kinnikinnick (*Arctostaphylos uva-ursi*), and Oregon oxalis (*Oxalis oregano*) grow throughout the forested and open forest habitats and may adapt to deep-extensive or semi-intensive green roofs (Franklin and Dyrness 1988). Other herbaceous plants native to sunny or part-shade sites include several species of strawberry such as wild strawberry (*Fragaria virginiana*), coastal strawberry (*Fragaria chiloensis*), and wood strawberry (*Fragaria vesca*). Hundreds of species of mosses (Fig. 8.5) are native to western Washington and many can be found in the Puget Lowlands and the Cascade Mountains (Pojar and MacKinnon 2004).

8.1.1.5 Bulbs and Tubers

Some of the native bulbs and tubers that grow throughout the region include the culturally important camas bulbs (*Camassia quamash* and *Camassia leichtlinii*) and species of wild onion (*Allium* sp.). Native species of bulbs were used as food and were cultivated by Native American tribes and include brodiaea (*Brodiaea* spp.), tritelia (*Triteleia* spp.), checker lily (*Fritillaria affinis*), roots of biscuitroot (*Lomatium* spp.), yampah plants (*Perideridia* spp.), seeds of tarweed (*Madia* spp.) and most parts of the balsamroot (*Balsamorhiza* spp.) (Christy and Alverson 2011).

8.1.1.6 Prairies and Garry Oak Ecosystems

The Garry oak ecosystem is a broad assemblage of plants that are based on grassland, rocky outcrop, and savanna habitats. Oaks grow in gravelly and open soils in the lowlands, especially in the glacial outwash soils and paths where ancient volcanic flows from Mt. Rainer spilled into the valleys (Dunn 1998; Chappell 2004; Easterly et al. 2005). There are four major oak woodland plant community assemblages including Oregon white oak/long-stolon sedge-camas, Oregon white oak/common snowberry/long-stolon sedge, Oregon white oak-Douglas fir/snowberry/sword fern, and Oregon white oak/snowberry/moist forb communities (Chappell and Crawford 1997).

Prairies of the lowlands once covered 60,702–72,843 hectares (150,000–180,000 acres) in western Washington. These included prairies that established on dry soil profiles, on upland slopes or south-facing slopes, and an extensive wet prairie matrix in the southern range of the prairie ecoregion (Easterly et al. 2005; Mima Mounds: A Special Prairie 2018). Other prairie regimes include mesic and wet plant communities. There are many associative relationships between plants, insects, birds, and other biological lifeforms indigenous to the Puget Trough prairies and oak woodlands and should be considered part of the ecosystem (Schemske et al. 1994; Wilson et al. 1997; Schultz et al. 2011). Today, less than 3% of the pre-settlement area of prairie-oak ecosystems remain (Crawford and Hall 1997).

Dominant overstory vegetation includes *Quercus garryana* (Oregon white oak), *Populus tremuloides* (cottonwood), *Fraxinus latifolia* (Oregon ash), and *Pinus ponderosa* (ponderosa pine). Shrubs of the prairie-oak habitats include *Rosa nutkana* (nootka rose), *Symphoricarpos albus* (common snowberry), and *Holodiscus discolor* (ocean spray). “Salal (*Gaultheria shallon*) prairies” were noted by early Euro explorers in the Snoqualmie Valley. Dominant herbs include *Camassia quamash* (common camas) and *C. leichtlinii* (great camas).

Regarding upland or dry prairies, those on the glacial outwash soils were represented by the Idaho fescue-white-topped aster community type (Chappell and Crawford 1997). There are over 200 species of grasses that grow in the Pacific Northwest (most are native); however, there has been a major shift from perennial native grasses to invasive annual grasses near urban and agricultural sites (Alaback 1994; Chappell and Crawford 1997). Main native grasses include Roemer's fescue (*Festuca idahoensis* ssp. *roemeri*) which was a dominant species (Dunn 1998), long-stolon sedge (*Carex pensylvanica*), California oatgrass (*Danthonia californica*), and prairie junegrass (*Koeleria cristata*) (Dunn 1998).

Hundreds of herbaceous perennials are native to the prairies, including asters, penstemons, buttercups, figworts, larkspurs, and some drought tolerant grass-like herbs such as western blue-eyed grass (*Sisyrinchium bellum*). Some early invader perennial species such as fireweed (*Chamerion angustifolium*) grow on disturbed sites and is commonly found on sites after fire-based disturbances. It is beneficial for browsers, and a variety of bees and insects. It also has medicinal value for humans (Fleenor 2016). This annual makes appearances on green roofs in the Pacific Northwest including the case studies described in Sects. 8.3.2 and 8.3.7 where it is a non-intrusive and welcome addition to the green roof.

Regarding wet prairies, much of the wet prairie habitat was associated with clay soils and floodplains. Dominant species included tufted hairgrass (*Deschampsia cespitosa*), rushes (*Juncus* spp.), and sedges (*Carex* spp.). There was also much diversity of perennial and annual forbs and bulbs growing in upland and wet prairies. Camas (*Camassia quamash* ssp. *maxima* and *Camassia leichtlinii* ssp. *suksdorfii*) grew in great numbers, especially where Native cultures cultivated camas for food (Christy and Alverson 2011). One study at Scatter Creek Preserve found these native grasses and herbaceous species of the wet/moist prairie: *Polygonum bistortoides, Plagiobothrys figuratus, Carex arcta, C. arthrostachya, C. unilateralis, Deschampsia cespitosa, Hordeum brachyantherum, Ranunculus orthorhynchus,* and *Equisetum hyemale* (Chappell 2004).

The study of wet prairies may become important for green roofs as wetland green roofs are being designed to internally function for building systems as well as stormwater management. See case studies 8.3.3 and 8.3.4 for examples of wetland green roofs in this chapter. Identifying plants and plant communities that serve multiple ecosystem services, such as wetlands, may become vital as green roofs are integrated into more sustainable and resilient urban solutions (MacIvor et al. 2011; Song et al. 2013).

Volunteering and participating in habitat restoration is a good way to learn about ecosystem conservation. Many private organizations are working to restore prairies in the Puget Trough: The Nature Conservancy, the Cascadia Prairie-Oak Partnership, South Sound Prairies, South Sound Prairie Grazing Project, the National Audubon Society, and others. A serious threat to native plant communities, especially prairie and prairie-oak woodlands, is invasion by Scotch broom (*Cytisus scoparius*) and Himalayan blackberry (*Rubus armeniacus*) (DellaSala et al. 2019). A major activity of prairie and savanna restoration is the removal of these species.

8.1.1.7 Succulents and Mountain Meadows

In western Washington, a wide range of vegetation is native to mountain environments including herbs, grasses and succulents. Succulents grow where shallow soils exist along the foothills of the Cascade Range, where rocky balds, rocky outcrops, and disturbed sites occur. Of particular interest to ecoregional green roofs are rocky bald habitats and open slopes with vegetation growing in exposed areas. *Sedum spathulifolim* (grows at low elevations), *Sedum oreganum, Sedum divergens,* and *Sedum lanceolatum* are all native to Washington and all along the Pacific Northwest, each growing in specific elevation ranges.

One other succulent native to the Puget Lowlands region includes the prickly pear cactus (*Opuntia fragilis*). This succulent grows on the San Juan Islands in the rain shadow of the Olympic Mountains in shallow soils or rocky outcrops, and grows on green roofs along the west coast (See Chap. 9 Sect. 9.3.3) (Pojar and MacKinnon 2004; Barclay 2007).

Mountain meadow habitats can be found on the slopes of the Cascade Mountains and its associated habitats where soils are shallow and winter snowpack is common. Some of the plants common to mountain meadows include purple milkweed (*Asclepias cordifolia*), monkeyflower (*Mimulus guttatus*), paintbrush (*Castilleja hispida*), yellow Oregon sunshine (*Eriophyllum lanatum*), creamy white northern buckwheat (*Eriogonum compositum*), pink farewell-to spring (*Clarkia amoena*), fleabane (*Erigeron inornatus*), Davidson's penstemon (*Penstemon davidsonii*), and Cardwell's penstemon (*Penstemon cardwellii*). These are members of meadow communities in the mid-to-lower elevations of the Cascades and grow in association with grasses and other herbaceous plants in rocky or dry sloped habitats, or wet habitats. Evergreen groundcovers such as juniper (*Juniperus communis*) and kinnikinnick (*Arctostaphylos uva-ursi*) are adapted to dry habitats and grow on well-drained habitats in the Cascade Mountains (Pojar and MacKinnon 2004; Harvey 2019).

8.1.1.8 Plant Identification Resources

Several outstanding online resources allow citizen scientists to learn about plants by family, genera, and species in the field. The *Washington Native Plant Society* has an online resource that allows multiple functions for identifying and learning about plants, and the location of sites where native plants can be found. The *Northwest Native Plant Guide* is an online resource that offers plant lists and locations where to find native plants. *Mountain Plants of the Western Cascades* is another online resource with locations of mountain meadows in the Cascade Range. This site provides plant lists and blogs about native vegetation of these habitats (Harvey 2019). One of the classic plant identification texts is *Plants of the Pacific Northwest Coast* by Pojar & Mackinnon. This photographic guide references plants according to family, with drawings that include root structure and general location. Only a few of the plants native to the Cascade Mountains have been trialed in green roof research or public demonstration projects.

8.1.2 Conservation Site Case Studies

8.1.2.1 Cascade Mountains, National Forests and Parks, Washington

Within 96 km (60 miles) from downtown Seattle, Tacoma, and Olympia, the Cascade Mountains are in close vicinity to the Puget Lowlands and include national forests, state parks, and national parks. Various habitats such as forests, conifer savannas, meadows, rocky balds, and wetland habitats are common. These habitats are formed upon differences in altitude, slope, aspect, and soil moisture gradients. Plant forms growing in these habitats include trees, shrubs, groundcovers, succulents, bulbs, grasses, annual and perennial wildflowers, horsetails, mosses, and ferns. Very few plants native to these habitats have been trialed on green roofs.

Mount Rainier National Park

Established in 1899, Mount Rainier National Park has one of the largest, oldest, and most contiguous preserves of forests and mountain meadows along the Cascade Range. The park is 956 km^2 (369.3 mi^2) in land area and includes old-growth forests, subalpine meadows, alpine meadows, and tundra habitats. Mountain meadow habitats can be found at elevations of 1500–2100 m (5000–7000 ft) above sea level (Fig. 8.6).

Regarding forest composition, old-growth forests at Mt. Rainier National Park exemplify how many of the coastal conifer communities formed varied-age stands of trees with complex sun/shade microcosms that included pockets of open sunny habitats where large trees have fallen. Most contemporary even-aged forests that are maintained for harvesting may not represent the community makeup and structure of old-growth forests (Kuiper 1988). Thus, conservation sites such as these at Mt. Rainier preserve habitats that have been intact for millennia.

About 23% of the park is vegetated with subalpine meadow. Five plant communities of subalpine meadows include heather/bell-heather/huckleberry communities *(Phyllodoce/Cassiope/Vaccinium);* Sitka valerian/showy sedge communities (*Valeriana sitchensis/Carex spectabilis*); black alpine sedge communities (*Carex nigricans*); low herbaceous communities, dominated by mosses; and green fescue (*Festuca viridula*) grassy meadows (Henderson 1973). The soils are generally well-drained and vary in degree of slope and aspect. As the climate has been trending warmer since the Little Ice Age, the distribution and composition of these meadow communities have changed (bloom earlier and shorter) and will likely continue to change as the climate is in a warming trend (Rochefort and Peterson 1996; Breckheimer et al. 2020).

Some of the annual or perennial species of herbs, grasses, and groundcovers of the alpine meadows may grow at lower elevations, but some are particular to the habitats in which they evolved. Thus, some plants growing in these communities

Fig. 8.6 As a massive volcanic dome, Mt. Rainier has many microclimates that form distinct vegetative zones, including a range of mountain meadows. South facing meadows in the east-side rain shadow of the mountain favor drier assemblages of plants while west and north-facing meadows in the path of Pacific moisture favor wetter assemblages. This sub-alpine meadow is at about 1950 m (6400 ft) above sea level near Sunrise Visitor Center and has a variety of vegetation that is adapted to the well-drained soils. (Photo: Bruce Dvorak, August 2018)

may translate to green roofs at lower elevations, but others may be restricted to green roofs at their adapted elevations.

North Cascades National Park

Located above the foothills of the Cascade Mountain range, situated between Seattle and Vancouver, B.C., the North Cascades National Park is thought to host 1630 vascular plant species, more than any other U.S. national park, in eight distinctive life zones. Some of the plants that grow in the park include over 150 species of grasses, and many ferns including sword, deer, licorice, lace, parsley, maidenhair, bracken, lady, oak and wood fern. The park's website has a plant database listing all known species in the park's diverse greater ecosystems (NCNP 2019).

8.1.2.2 Mima Mounds Preserve, Littlerock, Washington/Fort Lewis-McChord Prairie Preserves

Two remnant prairies remain in the south end of the Puget Lowlands: Mima Mounds Preserve and the Garry oak ecosystems and prairies at the Fort Lewis-McChord military bases. Mima Mounds is open to the public; the Fort Lewis-McChord military base is available only by appointment.

Mima Mounds Preserve

Mima Mounds is a unique landscape that was formed thousands of years ago. The mounds and prairies formed sometime after the retreat of the Vashon glacier about 12,000 years ago. Although it is not known exactly how the mounds were formed, there are several theories. Some of these include their being formed by massive wind deposits, or earthquakes, or deposits left from the retreat of the glaciers. Regardless of origin, the thousands of low mounds are evenly distributed upon an outwash plain, have gravelly well-drained soils, and are covered with prairie vegetation. Over the years, Garry oak and Douglas-fir encroach and retreat upon the open landscape depending on the extent of fire. Historically, the forces of fire helped maintain the land as open prairie, and the porous soils also contribute to the absence of forest on the mounds (Del Moral and Deardorff 1976).

Each mound has a unique microclimate as slope, aspect, height, and soil composition influence the arrangement of plants covering the mounds. The nature of the mounds and ground plain soils are dry and open, similar to engineered substrates often used on green roofs. Thus, the Mima Mounds are a prime location to study plant communities naturally growing on a variety of soil types and slope aspects (Fig. 8.7).

Forms of vegetation include grasses, woody groundcovers, mosses, bulbs, and herbaceous perennials and annuals. Mosses on the drier portions of the mounds include: *Hacomitrium canescens* var. *ericoides* and *Polytricum juniperinum*; mosses growing between the mounds in depressions and wetter locations include *Dicranum scoparium* and *Eurhynchium oreganum*. Some of the herbaceous species include *Achillea millefolium*, *Brodiaea coronaria*, and *Chrysanthemum leucanthemum*. The grass *Festuca idahoensis* is widespread across the mounds and grows with *Agrostis diegoensis* and *Agrostis tenuis*. The bulb *Camassia quamash* grows extensively across the prairie and blooms each spring (Del Moral and Deardorff 1976). The drought-tolerant evergreen groundcover kinnikinnick (*Arctostaphylos uva-ursi*) grows from the forested edges outwards onto the mounds (See Sect. 8.3.1).

Fort Lewis-McChord Prairie Preserves

At just over 2428 hectares (6000 acres) spread across five properties, the prairies and oak/pine savanna at Fort Lewis-McChord are some of the highest quality and largest collection of native prairies in the South Puget Sound region. Major

Fig. 8.7 Mima Mounds preserve. The small, thick, and light-reflective leaves of kinnikinnick (foreground) allow this drought-resistant plant to grow in the well-drained soil and full sun conditions. Mounds are also covered with fescues and mosses and herbaceous wildflowers. (Photo: Bruce Dvorak, September 2018)

restoration efforts have made this a conservation site where prescribed burns and invasive species removal are active. Douglas-fir and Scotch broom are primary invaders of the prairie habitats. Prescribed burns and physical removal of the plants make control of these invasive plants slow, and difficult work. Douglas-fir is a native conifer tree adapted to disturbed sites. Scotch broom is an invasive exotic species that is difficult to control as it has established widespread across the Puget Trough along transportation routes, disturbed sites, and open fields. A number of rare native species exist in the prairie preserves including the Columbian whitetop aster (*Sericocarpus rigidus*). Research is taking place on the preserve, and knowledge about the plant communities and restoration efforts are accessible through published studies and pre-arranged visits.

A diversity of wet and dry prairie habitats exist on the military's preserve. Prairie vegetation is dominated by Idaho fescue (*Festuca idahoensis*) bunchgrass, at up to 70% cover. Other native grasses and sedges are present but are not significant in the percentage of cover. Associative species include long-stolon sedge (*Carex pensylvanica*), California oatgrass (*Danthonia californica*), and prairie junegrass (*Koeleria cristata*). Forbs on the prairie include bluebells-of-Scotland, bracken fern, broadpetal strawberry, chocolate lily, common camas, common vetch, cutleaf microseris, early blue violet, Henderson's shooting star, houndstongue hawkweed, meadow death-camus, Nuttall's peavine, pomo-celery lomatium, prairie lupine, Puget balsamroot, slender cinquefoil, spikelike goldenrod, western buttercup, white-top aster, woods strawberry, and wooly sunflower (Dunn 1998).

8.1.2.3 Discovery Park, Seattle, Washington

Located along the shores of the Puget Sound, the 216-hectare (534-acre) park is Seattle's largest public park which includes many native plant communities. The park has a long history of use before Euro-American settlement, but most of the changes have taken place since the nineteenth century. When first surveyed back in the mid-1800s, the land was covered by a mature climax coniferous forest. It was soon logged and made into a military base, then later gifted to the City of Seattle as parkland. As property owners changed so did the plant communities. Regardless, a mixture of historic and new influences on the vegetation present a variety of vegetation natural to the region including prairie, rocky cliffs, shrub meadows and conifer forests with all of the usually associated plant community members such as ferns, salal, understory trees and shrubs (Williams 2015).

Today, the park is a resource for learning about vegetation typical to a lowland mixed-forest community such as bigleaf maple, red alder, bitter cherry, Douglas-fir, western red cedar, and western hemlock. These consist of a large percentage of the tree cover in the forested areas of the park. The south end of the park includes a more open and drier habitat where Pacific madrone trees grow, along with a large meadow with small trees and shrubs such as salal, snowberry, and exotic hawthorns. The rocky bluffs and beach dunes retain some of their original vegetation. A number of invasive species such as Scotch broom and Himalayan blackberry persist throughout the park, thus knowledge of exotic and native vegetation is helpful to discern plants and their natural assemblages (Williams 2015).

Several restoration efforts have taken place over the years, on sections of the open meadow and in several wooded areas. Some of the native vegetation planted during restoration efforts include Douglas-fir, Pacific ninebark, black hawthorn, black twinberry, mock orange, red-flowering currant, tall Oregon grape, quaking aspen, Indian plum, and western sword fern (Williams 2015).

8.2 Green Roof Research in the Puget Lowlands

King County published a report on some of the first green roofs built in the Seattle metro area. The purpose of the study was to determine how plants and green roofs systems had fared on publicly owned green roofs. Overall, most plants on green roofs were early in the process of adapting to their rooftop conditions, and a couple of the green roofs needed to be replanted. Recommendations included having a design team with at least one person with prior green roof experience, provisions for adequate slope, and providing watering during plant establishment and as needed for periods of summer drought. Building owners should be made aware of the dormant cycling of vegetation, provide adequate levels of maintenance for desired aesthetics, and be provided with a detailed maintenance guide (KCP 2006).

One of the few peer-reviewed publications regarding green roof plant trials took place at the Woodland Park Zoo at the Zoomazium building. Plant survival, light levels, substrate temperature, and moisture levels were monitored for the summer and fall. Substrate temperatures ranged from 50 °C to 60 °C in open sunny portions of the roof during the summer, where temperatures in the shaded sections were about 20 °C on a day when maximum air temperatures were 22–25 °C. Plants were installed from 10 cm (4 in) pot containers onto a geometric design. The roof was irrigated by zoo staff about two times a week. Native vegetation was used on the roof with kinnikinnick, allium, and lupines performing well (Martin and Hinckley 2007). See case study 8.3.1 for more information.

The City of Seattle has been a longtime supporter of green roofs and has a green roof policy with a focus on detention and cleaning of stormwater (Taylor 2008). In 2012, the City of Seattle and partners produced a report of the hydrological performance of green roofs in Seattle. Five green roofs were included in the study. The volume of precipitation retained on the green roofs ranged from 60% to 71.7% across the five monitored green roofs. Individual rain event performance varied greatly from 7% to 100% depending upon the duration of the rain event, antecedent moisture in the substrate before a rain event, duration, and intensity of rainfall (SPU 2012). One study compared several green roof test plots for their stormwater detention capacity compared to other results from the region. The experimental green roofs at the University of Washington detained or evapotranspired 30–56% of all precipitation, and held up to 99% during dry periods. The results were similar to other results from the region (Yocom and Spencer 2013).

8.3 Ecoregional Green Roof Case Studies

8.3.1 Woodland Park Zoo, Seattle, Washington

The Zoomazium building at the Woodland Park Zoo is a nature-inspired indoor play space designed for children 8 years old or younger. In addition to green strategies inside the building, a 743 m^2 (8000 ft^2) green roof was installed to showcase primarily native plants mixed with a few drought-tolerant non-native plants (Fig. 8.8). The large interior space below the green roof is kept cool during the summer from the inclusion of the green roof. The sustainable building received Gold certification from LEED.

Fig. 8.8 Annual lupines (purple) and perennial nodding onions (pink/white in the background) are in bloom on the roof during the early summer of the first growing season (2006). All of the original plants survived the establishment period. Over time, invasive species have populated parts of the roof, and nuisance birds such as crows have contributed to plucking some plants up and out of the substrate. In 2018, most of the native species still remain on the roof, however, in reduced numbers. (Courtesy of Woodland Park Zoo)

8.3.1.1 Project Team

Building Owner/Client: Woodland Park Zoo.
Green Roof Design Team Lead: Mithun.
Architect: Mithun.
Landscape Architect: Mithun.
Installation Contractor: Kirtley/Cole Associates.
Maintenance Contractor: Woodland Park Zoo.
Project completion: 2006.
Green roof area: 743 m^2 (8000 ft^2).

8.3.1.2 Overview and Objectives

The green roof was originally designed to mimic a lowland meadow habitat, with groupings or drifts of plant assemblages placed together. The green roof is partially visible from below, but the arrangement of plants is not discernable when viewed from below (Fig. 8.9). Over time, the zoo staff has maintained the roof more like a natural habitat, as the original assemblages or drifts of plants are not maintained, nor are they discernable on the green roof.

Fig. 8.9 (**a**) Installation of vegetation on the Zoomazium green roof (November 2005). (**b**) View of the building and blooming green roof as seen from below (June 2006). (**c**) Image of Alliums growing near the lower and moist edge of the green roof in August 2018. (**d**) During the summer, grasses on the high sections of the green roof go dormant. (Photos: (**a**) & (**b**) Courtesy of Woodland Park Zoo, (**c**) & (**d**) Bruce Dvorak, August 2018)

8.3.1.3 Plant Establishment

Several microclimates exist on the roof, due to its east-facing roof deck and variable slope. The deck slope ranges from 0:12 pitch at the upper, flatter section to 3:12 pitch on the lower edge. Trees on the north side of the roof cast shade onto the upper flat portion of the northeast corner of the roof. The substrate depth was set at 15 cm (6 in) across all sections of the roof. Over time, the substrate has settled at 10 cm (4 in) in some sections of the roof. The custom-designed substrate included 65% mineral (pumice), 10% coarse sand, 25% aged organics, and 5% fertilizers & amendments. Substrate samples before installment reported organic matter content at a 6% dry weight basis (Martin and Hinckley 2007).

Plant selection was closely tied to the design of the substrate. The design team led efforts to collaborate with soil experts and the zoo horticulture staff to make selections reflect that of a glacial outwash meadow (Martin and Hinckley 2007). Thus, the appearance of the green roof goes through seasonal dynamics with bright green and blooms in the spring and tans and browns during summer dormancy.

Herbaceous Perennials

beach strawberry (*Fragaria chiloensis*), bigleaf lupine (*Lupinus polyphyllus*), narrowleaf blue-eyed grass (*Sisyrinchium angustifolium*, native to the eastern U.S. and Canada), nodding onion (*Allium cernuum*), western sword fern (*Polystichum munitum*).

Shrubs

kinnikinnick (*Arctostaphylos uva-ursi*), salal (*Gaultheria shallon*).

8.3.1.4 Irrigation

The roof vegetation is watered with overhead sprinklers on a timer set to run about 15 min every other day. The overhead spray reaches some parts of the green roof but not all. The coverage is not head-to-head and is run only as needed. Since the source of irrigation water is municipal, during droughts irrigation is not allowed.

8.3.1.5 Maintenance

The plant composition has shifted towards plants that require shade only persisting in the shady sections of the roof and grasses that self-seeded are widespread. Alliums and lupines have survived the roof; however, their numbers are reduced. The Zoomazium staff had not added any new species to the roof. The goal is to maintain a vegetative cover with a natural meadow appearance. No fertilizer has been added to the growing medium, however, birds do frequently visit the green roof.

8.3.1.6 Observed Wildlife

Seagulls and crows sometimes visit the roof and can exhibit aggressive behavior there, and elsewhere in the zoo. They use the green roof as a place to hang out. Bare spots can result if birds attempt to nest on the roof. Bumblebees and yellow jackets visit the roof.

8.3.1.7 Best Performing Native Vegetation

Invasive grasses dominate the roof. In open sunny areas alliums and some lupine grow along with a matrix of additive grasses. Salal and a couple of western sword fern grow in shady areas.

8.3.1.8 Post-occupancy Observations

Publications

- It is not clear if or how the substrate placed on the roof met design specifications (Martin and Hinckley 2007).
- During drought events, the substrate heats up to temperatures far above the ambient air temperature and dries out vegetation near exposed media. The substrate may benefit from amendments to retain more moisture (Martin and Hinckley 2007).

Zoo Maintenance Staff

- Green roof vegetation has adapted to their preferred microclimates on the roof.
- The zoo grounds are maintained to appear natural. There are few formal gardens in the zoo, as the animal exhibits are designed to look natural. Therefore, the green roof is intentionally left to take on a natural appearance.
- Alder and birch seedlings are removed each year.

Authors' Reflections

- This green roof has endured the test of time. Regardless of the loss of the initial planting patterns, potentially reduced levels of amendments in the growing media, and an inconsistent irrigation system, the original and volunteer vegetation has adapted and maintained cover since 2006.
- As one of the first native green roofs in the region, it would be worthwhile for the zoo or a volunteer group to fund a small grant to restore the irrigation system, revisit the original planting concept, and conduct a follow-up study. Additional vegetation native to glacial outwash meadows might be worth seeding or planting on the green roof.

8.3.2 Ballard Library, Ballard, Washington

Located in an urban neighborhood northwest of downtown Seattle, the Ballard Branch Carnegie Library was a new addition to its library system. It was conceived of and designed to become a "community front porch" (Vinh 2005). This innovative library was programmed to incorporate public space for lounging, organized events, public engagement, and to demonstrate and educate about sustainable building practices. As such, the library has a periscope on the first floor that looks out onto its green roof. The library has a tower for viewers to climb up the stairs to see the green roof directly from within the tower (Fig. 8.10) (Vinh 2005). A kiosk is located in the tower that informs about the green roof's design, its vegetation, and the general benefits of green roofs.

Fig. 8.10 The parabolic sloped roof forms a variety of dry and wet microclimates. Most of the original species planted on the roof remain, however, in much less proportion to the self-seeded invasive exotic grasses. Skylights and an observation tower allow for natural light and engagement of views with the green roof. (Photo: Bruce Dvorak, August 2018)

8.3.2.1 Project Team

Building Owner/Client: Ballard branch of the Seattle Public Library.
Green Roof Consultant: American Hydrotech.
Architect: Bohlin Cywinski Jackson.
Structural Engineer: PCS Structural Solutions.
Landscape Architect: Swift & Co.
Installation Contractor: PCL Construction Services Inc.
Project completion: March 2005.
Green roof area: 1858 m^2 (20,000 ft^2).

8.3.2.2 Overview and Objectives

The green roof was included on the new library roof to address common environmental effects of traditional rooftops, and as a gesture to connect with the local community. Covering 75% of the urban site, the green roof was designed to reduce runoff loads on the undersized and old stormwater infrastructure, and to reduce the city's heat island effect. The design team proposed meadow-like vegetation to connect to the Scandinavian heritage of the surrounding community, thus recalling traditional sod roof construction made popular in Scandinavia (Swift 2019).

The green roof on this LEED Gold-certified building functions to capture all of the rooftop stormwater runoff and helps to keep the building cool. The saddle-shaped roof has variable slopes from 2.5:12 to 0.5:12 on the north and from 1:12 to 0.5:12 on the south concave profile. The substrate is 10–12 cm (4–5 in) deep and includes 45% mineral, 15% coarse sand, and 40% approved aged organic fertilizers and amendments (SPU 2012).

8.3.2.3 Plant Establishment

Although a meadow with native vegetation was the intended plant community, there was no particular reference ecoregion or plant community. Plants were selected on their potential to establish on the green roof. The vegetation was established through seeding and some plugs.

Annuals

common woolly sunflower (*Eriophyllum lanatum*).

Grasses

Idaho fescue (*Festuca idahoensis*), long-stolon sedge (*Carex inops*), red fescue (*Festuca rubra*).

Herbaceous Perennials

Idaho blue-eyed grass (*Sisyrinchium idahoense*), moss phlox (*Phlox subulate*), thrift seapink (*Armeria maritima*), western yarrow (*Achillea tomentosa*), white brodiaea (*Triteleia hyacinthine*).

Succulents

Oregon stonecrop (*Sedum oreganum*).

8.3.2.4 Irrigation

The original irrigation installed was a subsurface drip irrigation system. However, the water did not effectively reach the vegetation during its run cycles and many plants died. An overhead irrigation system was later added. The green roof has received a variety of irrigation schedules over the years. During establishment, the roof received only rainfall; and fortuitously, it was a rainy summer. However, the following summer was drier than normal and 5–10% of the 18,000 plants died, primarily the perennial wildflowers. An overhead irrigation system was subsequently installed and the bare spots of the roof replanted with grasses, mostly the native Idaho fescue (*Festuca idahoensis*) and sedums, both of which have re-seeded or spread. The roof now receives irrigation for 40 min, three times a week. Since the library roof is visible to the public, it is watered regularly to keep plants active and not dormant during the summer (Currie 2018).

8.3.2.5 Maintenance

The green roof has had different management schedules over the years. Currently, the roof has unwanted plants removed several times in the spring and fall. This rooftop meadow is not designed to be maintained like a garden. Thus, vegetation is free to adapt to its most advantageous micro-climate on the roof (Fig. 8.11).

Fig. 8.11 A few volunteer wildflowers persist in the saddle of the roof slope, where moisture lingers. Fireweed (*Chamaenerion angstifolium*), shown here, is an early invading native perennial of the Pacific Northwest, and it found its way onto the roof. Its bright pink blooms bring color to this section of the green roof near the observation tower and attract pollinators. (Photo: Bruce Dvorak, August 2018)

8.3.2.6 Observed Wildlife

In the spring, many birds use the roof. Hawks have been observed on the roof, feeding on pigeons.

8.3.2.7 Best Performing Native Vegetation

Original species performed well during their establishment period. Maintenance practices had to change when watering restrictions were enforced, and as a result, much of the native vegetation died. After reestablishing irrigation, new self-sown grasses have reoccupied the former space of the native vegetation.

8.3.2.8 Post-occupancy Observations

Maintenance Staff

- Irrigation must be provided to ensure plant survival through Seattle's dry summers
- A maintenance budget for roof upkeep needs to be incorporated in the planning.
- Crows uprooted some plants during the establishment period.
- In the future, we could try seeding native fescue grasses rather than planting plugs.

Authors' Reflections

- This green roof is directly observable, and open to public viewing from within the library. It has had consistent attention and has endured the test of time. However, since watering restrictions are consistently enforced during droughts, much of the original vegetation has been reduced. It would be worthwhile for the library to explore some minor re-planting of the green roof through seeding.

8.3.3 Bill and Melinda Gates Foundation Campus, and Seattle Center Parking Garage, Seattle, Washington

Located in downtown Seattle, the Bill and Melinda Gates Foundation established its world headquarters near the core of downtown adjacent to the Seattle Center at the base of the Space Needle. The building and site are the first and largest LEED-NC Platinum-certified project for a non-for-profit organization in the world. The triangular-shaped property was previously a 4.8-hectare (12-acre) parking lot for visitors of the Seattle Center, built over a former bog. The impervious surface discharged runoff into a combined sewer, with overflows during severe storm events draining polluted sewer water into the Puget Sound. The property was large enough for the placement of three six-story buildings (two currently built), and a public visitor center integrated with a parking garage (Fig. 8.12). The Foundation headquarters

Fig. 8.12 View from the Space Needle of the Gates Visitor Center and Seattle Center parking garage green roof. A 1020 space-parking garage is located below the extensive green roof. The Gates Foundation Headquarters site collects all of the precipitation that falls on the property (rooftops and pavements), stores it in underground cisterns and uses it for toilet flushing and to irrigate the landscape and green roofs. (Photo: Nancy Rottle)

buildings are arranged to form one large central enclosed open space for its hundreds of employees that use the site each day. The site is easily accessible to public transportation.

A public parking garage is located below a 6070 m^2 (1.5-acre) arched green roof to capture rainfall and reduce runoff in the combined sewer basin. The 12 cm-deep (5-in) substrate grows a variety of sedums, which include natives. This building supports the largest contiguous stretch of extensive green roofs in the city. Sections of semi-intensive green roofs were added with the construction of the headquarters complex, with rainwater collected in an underground cistern and central water feature. Harvested rainwater is used in the buildings and to irrigate the roofs and gardens.

8.3.3.1 Project Team

Building Owner/Client: Bill and Melinda Gates Foundation.
Architect: NBBJ.
Landscape Architect: GGN.

Civil Engineer: KPFF.
Consultants: Jeffrey Bruce and Co., Irrigation and Soils, Rana Creek (plants).
Maintenance Contractor: Pacific Earth Works.
Project completion: 2008 (garage) 2011 (headquarters).
Green roof area: 8094 m^2 (87,120 ft^2) (garage and site).

GGN's design of the Gates headquarters landscape is meant to bring nature close to the Foundation's employees and restore lost ecological functions. The plantings were designed as habitat vignettes that echo back to Seattle's native plant communities and the former wetland, with the central water feature providing both additional storage and evaporation as well as habitat for ducks and aquatic plant species (Fig. 8.13). Much of Seattle's native landscape was marsh and prairie, and the tribes that lived where Seattle is located today used a word that means "prairie" to describe the local area's landscape. The intensive and semi-intensive green roofs were designed to reflect those historic plant communities in the form of outdoor garden rooms. The 4.8-hectare site (12-acre) includes open wetlands, sunny grass gardens, shady understory gardens, and scores of big-leaf maple gardens (Easton 2011).

Fig. 8.13 Native gardens at the Bill and Melinda Gates Foundation Headquarters are themed after wetlands, prairies, and woodlands. Western sword fern (foreground) frames the view of a rooftop pond and wetland which lies above underground parking and cisterns. (Photo: Bruce Dvorak, August 2018)

8.3.3.2 Overview and Objectives

Primary ecological objectives for the site include a regenerative use of rainwater and the establishment and maintenance of formal gardens throughout the site. The property is designed to capture and harvest rainwater in subgrade storage tanks that hold 3785 m^3 (1 million gals) of water. During a typical year, this results in 13,248 m^3 (3.3 million gals) of rainwater harvested and used by the green roofs, buildings (toilet flushing), and wetlands. Because of this efficient use of water, municipal water use is reduced by 79%. Solar panels located on the top of the six-story buildings help reduce municipal energy consumption by 39%. As a building that is designed to last 100 years, these investments will reduce operation costs, which will allow for the building to pay for itself in fewer than 30 years (Velazquez 2017).

Garden spaces were designed for a variety of potential users in mind such as employees, staff, and international guests. Therefore, the design team used a traditional garden aesthetic (well-maintained). Particular attention was given to the ornamental and visual texture of plants. Thus, ecological goals and aesthetic goals drove plant selection towards a hybrid of native and climate-adapted non-native plants (Easton 2011).

The 1020 space-parking garage is located adjacent to the Foundation building and is intended for use by visitors to the organization's headquarters, the Foundation's visitor center, visitors to the Seattle Center and some of the Foundation's employees. Many of the Foundation's employees use public transportation or ride bicycles. The parking garage and Visitor Center have extensive green roofs with native and exotic sedums (Fig. 8.14d).

8.3.3.3 Plant Establishment

Many species of native plants were used in the gardens, as well as cultivars of native plants, plants native to other parts of North America, and plants that have become naturalized in the region or are exotic. Throughout the Gates Headquarters property there are open and sunny habitats, shady and partly shady gardens, and all are on-structure but at the landscape level. So, these gardens are not elevated above the landscape (Fig. 8.14a–c). The use of irrigation from the beginning of the design and implementation has helped the green roof plantings to thrive. The names of primarily native plants used in gardens are below.

8.3.3.4 Extensive Green Roofs

Succulents

Oregon stonecrop (*Sedum oreganum*), Pacific stonecrop (*Sedum divergens*), also two species of naturalized sedums, four species of exotic sedums.

Fig. 8.14 A variety of gardens receive maintenance each week to retain the garden themes and desired aesthetic. Green roof habitat vignettes show (**a**) a sunny grassland habitat, (**b**) a sword fern and redwood sorrel habitat on shady and sloped deck, (**c**) bulrushes (*Schoenoplectus/Juncus*) and cattails (*Typha*) grow in the sunny aquatic habitat used to filter all of the harvested water, and (**d**) native and exotic sedums grow on top of the parking garage (distant) and Foundation roofs (foreground). (Photos: Bruce Dvorak, August 2018)

8.3.3.5 Roof Gardens

Grasses/rushes

common rush (*Juncus effusus*), hardstem bulrush (*Schoenoplectus acutus*), tufted hair grass (*Deschampsia cespitosa*), also five species of exotic grasses, and one species of exotic sedges.

Groundcovers

beach strawberry (*Fragaria chiloensis*), British Columbia wild ginger (*Asarum caudatum*), inside-out flower (*Vancouveria hexandra*), Pacific trillium (*Trillium ovatum*), redwood sorrel (*Oxalis oregana*), vanilla leaf (*Achlys triphylla*), western bleeding heart (*Dicentra formosa*), also four species of exotic herbaceous perennial groundcovers.

Ferns

common lady fern (*Athyrium filix-femina*), ostrich fern (*Matteuccia struthiopteris*, native to Canada and the northeastern U.S.), Pacific oak fern (*Gymnocarpium dryopteris*), Robust male fern (*Dryopteris filix-mas* 'Robusta'), western sword fern (*Polystichum munitum*), also one species of exotic fern, one species of exotic grasses, and one species of exotic sedges.

Bulbs and Herbaceous Perennials

camas (*Camassia quamash*), lavender (*Lavandula* x *intermedia*), narrow leaf cattail (*Typha angustifolia*), Niveum bishop's hat (*Epimedium* x *younganium* 'Niveum'), also one exotic perennial species.

Shrubs

Arctic Fire red twig dogwood (*Cornus sericea* 'Farrow' Arctic Fire b.), dwarf Oregon grape (*Mahonia nervosa*), enkianthus (*Enkianthus campanulatus*), Nootka rose (*Rosa nutkana*), Northsky lowbush blueberry (*Vaccnum angustilolium* 'Northsky'), ocean spray (*Holodiscus discolor* b.), red-flowering currant (*Ribes sanguineum*), thimbleberry (*Rubus parviflorus*), Top Hat blueberry (*Vaccinium* 'Top Hat'), western serviceberry (*Amelanchier alnifolia*), also two exotic species of shrubs.

Trees

bigleaf maple (*Acer mocrophyllum*), vine maple (*Acer circinotum*), also two exotic tree species.

8.3.3.6 Irrigation

Daily watering of garden areas with overhead spray heads and water sensors. Extensive green roofs are irrigated with subgrade drip irrigation.

8.3.3.7 Maintenance

Weekly maintenance to remove unwanted vegetation, tree seedlings.

8.3.3.8 Observed Wildlife

Users of the garden spaces (Foundation green roofs) have observed butterflies, seagulls, crows, ducks, bees, and dragonflies using the green roofs.

8.3.3.9 Best Performing Native Vegetation

All of the vegetation is adapting; however, due to reflective sunlight, bigleaf maple is having difficulty growing where it is located near solid panels of reflective glass near building facades.

8.3.3.10 Post-occupancy Observations

Design Team

- The rushes are very successful and need to be cut back to keep them from overgrowing.
- A full-time maintenance crew is needed to care for the expanse of formal green roof gardens.

Authors' Reflections

- This massive integrated site and building project provide multiple examples of how urban design will look and function in the future. The investments made here are already paying back through reduced expenditures on utilities, reduced negative effects on urban flooding, reduced urban heat islands, and increased positive environmental and aesthetic environments.
- The design team achieved an amazing integration of good design form and execution on what was a left-over and odd-shaped property. The quality of its integration into the urban fabric achieves its prominence as a focal point in the community.
- The gardens surround the accessible and paved open spaces. There is little indication that a massive cistern lies below the network of gardens, ponds, and open spaces.
- The planting design had a particular focus on plant foliage color, texture, and density. The lack of blooming plants is only subtly noticeable; however, the presence of clear water and lush green which is often back-illuminated is fitting and appropriate for its diverse group of users.

8.3.4 Bullitt Center, Seattle, Washington

There are more than a few green roofs located near the Bullitt Center. What makes this green roof and building unique is that it is currently the most sustainable Class A office building in the world. The Bullitt Center, located in the Capitol Hill

Fig. 8.15 (**a**) The Bullitt Center wetland green roof treats greywater from building uses such as sinks and showers before being further filtered and infiltrated through soil and gravel of the subgrade level of a rain garden at the street level, located uphill from this view (**b**). The rooftop wetland vegetation shown here (**a**) is about 4 years after installation. (Photos: (**a**) Courtesy of Phillip Thompson, (**b**) Bruce Dvorak, August 2018)

neighborhood of Seattle, is certified by the International Living Future Institute and has met the Living Building Challenge. This project exemplifies how green roofs and rain gardens can play a vital role in green buildings (Fig. 8.15).

8.3.4.1 Project Team

Building Owner/Client: Bullitt Foundation.
Green Roof Design Team Lead: Berger Partnership.
Architect: Miller Hull.
Landscape Architect: Berger Partnership.
Project completion: Construction began 2011, completed April 22, 2013.
Green roof area: 700 m^2 (477 ft^2 wetland greywater treatment).

8.3.4.2 Overview and Objectives

The primary objective for the building was for it to become an example of how buildings can produce their own energy and water supply, generate zero waste, and be healthy, beautiful, and durable for at least the next 250 years. The Bullitt Foundation and project team worked closely with the City of Seattle staff during the design and planning phase to accomplish these goals. The wetland green roof on this project is small and simple, but its function is integral to the daily functions of the building. The green roof is a greywater constructed wetland system that is designed to remove nutrients and pollutants from wash water used inside the building. The bioretention cells with a deep gravel layer on the exterior site at the streetscape function as the last filter for the greywater before it infiltrates into the ground.

One of the six Living Building Challenge focus areas is water. There are two aspects of water use: Imperative 5: Net Zero Water and Imperative 6: Ecological Water Flow. Under Imperative 5, all of the water use (for occupants) must come from captured precipitation or closed-loop water systems. These waters must be purified without the use of chemicals. Imperative 6: Ecological Water Flow states that all of the building and site stormwater discharge must be managed on-site to either supply the project's internal water needs, or be released as surface flow (not to exceed regulated rates), as groundwater recharge, for agricultural use or to be used by other buildings (Seattle and Cascadia 2011).

What this means is that all of the water used inside (and outside) the six-story office building comes from the sky, not public utilities. Rooftop water is stored in cisterns before it is purified and pumped back into the building's distribution system for drinking, showering, washing, and for reuse in the composting toilets. Wastewater from sinks, showers, and dishwashers is collected and treated by recirculating it several times through the special soils and equisetum growing in the constructed roof wetland. Excess treated water that is not evapotranspired is discharged and further filtered and infiltrated into the ground through 3.6-m deep (12 ft) soil and gravel layers in a sidewalk-level rain garden (Mitchell and van Daalen 2019). As a notable feature of the Living Building, the first-story roof wetland is adjacent to and visible from the building's central stairway. There is no sewer connection to the city from the composting toilets, so there is no connection to the municipal waste treatment plant, and all biological waste is treated on-site, dried, and sold as fertilizer for plant growth around Seattle (Mitchell and van Daalen 2019).

8.3.4.3 Plant Establishment

The constructed wetland green roof is planted with *Equisetum arvense* (common horsetail) and *Juncus* (bulrush). Edging of the wetland includes native plants of low Oregon grape (*Mahonia/Berberis nervosa*), redwood sorrel (*Oxalis oregana*), and western trillium (*Trillium ovatum*).

8.3.4.4 Irrigation

What makes this green roof unique is that the water used to sustain wetland plants comes from inside the building, making irrigation integral to the success of the vegetation. The irrigation source is generated by rainwater channeled from the sky, through the building fixtures, through the wetland, and then back to the site. A 212-m^3 (56,000-gals) cistern located on the base floor collects 69% of the rooftop runoff. The water is filtered to remove particulates, collected and stored in the cistern, and then treated with ultra-violet (UV) filters to remove pathogens before being used inside the building. While drinking water was initially supplied through the municipal system, in 2018 the local health department allowed potable water to be sourced from the building's harvested rainwater. Greywater from sinks and showers which irrigate the green roof wetland is then returned, cleaned, into the ground outside the building for groundwater recharge. Each year, approximately 380 m^3 (100,600 gals) of greywater from inside the building is treated and either evaporated or infiltrated onsite (WBDG 2016).

8.3.4.5 Maintenance

Green roof maintenance includes the occasional removal of unwanted plants and cutting back of equisetum tops.

8.3.4.6 Observed Wildlife

No formal observations are reported.

8.3.4.7 Best Performing Native Vegetation

Equisetum arvense has performed exceptionally well, especially since the northern exposure and overhanging solar roof creates shady conditions.

8.3.4.8 Post-occupancy Observations

Researchers

- The roof areas were significantly reworked after refinement of the greywater system and replacement of the planting medium which had become clogged with fines, presumably from building construction. Once the system was functioning properly, the wetland species especially thrived with all of the added nutrients from the greywater (Phillip Thompson, Seattle University).

Authors' Reflections

- If community leaders are looking for practical ways to promote buildings that reduce energy use, prevent urban flooding, recharge groundwater, and promote healthy environments, then this project is a real and working example to learn from. This example of net-zero design should be explored, researched, and repeated elsewhere.

- The abundance of post-construction publications and online sharing of design ideas, performance, and maintenance is exemplary. Much will be learned from this cooperative project that exemplifies corporate social responsibility, through proactive partnering with designers, municipal, and university interests. The additional expenses invested by the foundation will help make similar projects more affordable in the future.

8.3.5 SRM Development, Kirkland, Washington

SRM Development specializes in transforming properties that are neglected and difficult to develop (such as many brownfield sites) and transforms them into exemplary development projects. A two-hectare (5-acre) remnant railroad brownfield site in Kirkland, Washington was purchased by SRM Development and transformed from a derelict site that had 11 toxic soil spots that needed to be removed and cleaned, into a LEED Platinum-certified project with accessible green roofs (Fig. 8.16). Today, the building and property are healthy, beautiful, safe, and

Fig. 8.16 The native black-eyed Susan (yellow bloom), wild onion (with pom-pom like blooms), and a mixture of native and naturalized sedums make for a formal rooftop garden that is accessible to employees and staff. This high-maintenance extensive green roof receives attention each week during the growing season to retain the composition of plants in zones and sustain the plant community boundaries. (Photo: Bruce Dvorak, August 2018)

inviting. Two buildings are connected with a bridge, and the property below is a linear trail that is accessible as a public greenway and park. (ENRNorthwest 2016). The company philosophy is to create a stress-free working environment for everyone, with views to green which includes an accessible roof garden. The project includes many sustainable practices such as a site that is connected to greenways, bike parking, food grown onsite and prepared in the cafeteria, natural daylighting, energy efficiency, and more.

8.3.5.1 Project Team

Building Owner/Client: SRM Development.
Green Roof Design Team Lead: Thomas Rengstorf & Associates (TRA).
Architect: DLR Group.
Structural Engineer: DCI Engineers.
Landscape Architect: Thomas Rengstorf & Associates (TRA).
Green roof systems: DIADEM USA, Inc., Etera.
Project completion: 2015.
Green roof area: 1593 m^2 (17,150 ft^2).

8.3.5.2 Overview and Objectives

Planning for the green roof began early in the design stages for the site and building integration. It was determined that the rooftop of the 16,722 m^2 (180,000 ft^2) building would be shared to provide 60 kW of solar energy (1000 lightbulbs for an hour), 1672 m^2 (18,000 ft^2) of extensive green roof and the remaining unplanted rooftop was retained to capture precipitation for a 189 m^3 (50,000 gals) rain catchment system that is used for toilet flushing and green roof irrigation.

The green roof is centrally located on the building, has access for its near 1000 employees, and is designed as a roof garden (Fig. 8.17). This extensive green roof has a 10 cm-deep (4 in) substrate, a formal appearance, and features native and exotic vegetation. As a formal garden, it was designed and budgeted to receive frequent maintenance. The combination of pre-planning, integrated design, and an experienced design team allowed a beautiful formal garden on a roof that is reliant on only the precipitation that falls from the sky.

8.3.5.3 Plant Establishment

Green roof plants were established through the use of a sedum tile mat and pre-grown plants in containers.

Fig. 8.17 The campus includes (**a**) a private roof deck access for employees; (**b**) *Allium Schoenoprasum,* a naturalized wild onion (round blooms), is featured in the gardens; (**c**) beehives are located on a rooftop, supporting ecosystem health and a food garden at the ground level; (**d**) a public trail and path connection from the site to a community trail passes under a bridge between buildings. A caboose and rail section was restored to honor the site's past use. (Photos: Bruce Dvorak, August 2018)

Native Herbaceous Perennials

common yarrow (*Achillea millefolium*), orange coneflower (*Rudbeckia fulgida* 'Goldstrom', native to the central U.S.), wild chives (*Allium schoenoprasum*) introduced/naturalized.

Native Succulents

Cape Blanco stonecrop (*Sedum spathulifolium* 'Cape Blanco'), Oregon stonecrop (*Sedum oreganum*), Pacific stonecrop (*Sedum divergens*), also nine exotic species of sedums, two exotic species of grass, and two exotic herbaceous perennials were included.

8.3.5.4 Irrigation

Subgrade drip irrigation water is sourced from a rainwater harvesting system. No municipal water is used on the rooftop.

8.3.5.5 Maintenance

The formal extensive green roof receives weekly and sometimes daily maintenance. Removal of unwanted plants is the most frequent activity, as the formal bed lines and formal planting arrangements require regular maintenance. The client considered a garden aesthetic for the employees as a high priority. Harvested irrigation water was planned from the beginning stages, to not burden local water supplies, and to provide a sustainable source of irrigation water.

8.3.5.6 Observed Wildlife

The project has two beehives maintained on one of the two buildings. In addition to bees, butterflies, birds, and crows frequent the roof.

8.3.5.7 Best Performing Native Vegetation

All of the vegetation is thriving.

8.3.5.8 Post-occupancy Observations

Authors' Reflections

- The project developer (SRM) led the motivations for many of the innovations that take place with the project. This integrated design demonstrates what is possible when invested interests come together prior to development, and developers lead with high aspirations.
- The level of detail throughout the site and project is outstanding. The cooperation and corporate leadership of Google demonstrate how private organizations can make significant contributions to sustainable planning, construction, and design.
- Corporate social responsibility is exemplified in this project both by the developer and client. In this case, the project site is accessible to the public without interfering with the activities of the private corporation. Also, transforming an unusable site into a fabulous work/play environment unloads the burden of a derelict property that likely never would have been developed.

8.3.6 Bellevue City Hall, Bellevue, Washington

The City of Bellevue, Washington needed more space for its City Hall functions and staff. Eager to set an example of how City facilities can grow efficiently and more sustainably, it launched into a study of options for future expansion. Instead of building a new structure, the city identified an unoccupied building that was constructed in 1983 that needed updating. The city purchased the building in 2002 and began the process to design and implement upgrades. The upgrades included the addition of two green roofs, one extensive and the other intensive. The extensive green roof is visible from inside the building but is not accessible. The intensive green roof is accessible to the public from within and outside the building (Fig. 8.18).

8.3.6.1 Project Team

Building Owner/Client: City of Bellevue.
Green Roof Design Team Lead: Hapa Collaborative.

Fig. 8.18 Single species drifts of mostly native plants are arranged and maintained in diagonal bands for visual effect and to help keep maintenance simple. Native and exotic drought-tolerant plants are used on this north-facing green roof over a parking garage. Half of the roof is in shade much of the day near the building facade. Plants selected can tolerate both full sun and part-shade. (Photo: Bruce Dvorak, August 2018)

Architect: SRG.
Landscape Architect: Hapa Collaborative.
Installation Contractor: Pacific Earth Works, Inc.
Project completion: 2006.
Green roof area: Approximately 2090 m^2 (22,500 ft^2)

8.3.6.2 Overview and Objectives

The goal for this semi-intensive green roof was to provide a space outdoors that would extend the usable space from the main conference hall wing of the City Hall. The city rents out this space regularly. A walkway connects access from the parking garage at the main entrance at the street level into a rooftop garden. The green roof includes vegetation native to the Pacific Northwest (Fig. 8.19) intermixed with some exotics. Some of the funds for the green roofs were provided by a $90,000 (US dollars) King County Conservation District grant. Stormwater retention, energy conservation, and native vegetation were important factors for obtaining the grant to support the green roofs. The prior roof condition was gravel-ballasted roofing (Fry 2019).

Fig. 8.19 The shrub Oregon grape (*Mahonia aquifolium, Berberis aquifolium*) is native to open disturbed sites at the fringe of forest edges and open areas in the Pacific Northwest. On this roof, the Oregon grape attracts birds to the roof deck and acts as habitat relief from an urban center that is dense with midrise buildings and freeways. (Photo: Bruce Dvorak, August 2018)

8.3.6.3 Plant Establishment

Due to the structural limitations of the pre-existing roof deck, plant selection was limited to species that would grow in 15–30 cm (6–12 in) of growth media. For this reason, trees and large shrubs were not included in the design, except for one row of shrubs in elevated planters.

Grasses/Sedges

switchgrass (*Panicum virgatum*, native to central and eastern U.S.), also one exotic species of sedge.

Herbaceous Perennials

nodding onion (*Allium cernuum*), also three exotic species of herbaceous perennials.

Ferns

western sword fern (*Polystichum munitum*).

Shrubs

Oregon grape (*Mahonia aquifolium, Berberis aquifolium*), yellow twig dogwood (*Cornus sericea* 'Flaviramea'), also three exotic species of shrubs.

8.3.6.4 Irrigation

The green roof is irrigated regularly and monitored by a centralized computer system called Maxicom. The frequency of irrigation is based on soil moisture, weather conditions, and evaporation rates. So, irrigation can run daily or intermittently based on weather temperatures and rainfall. The rooftop is irrigated between May and September (Harris 2019).

8.3.6.5 Maintenance

The formal beds receive weekly maintenance during the growing season. Since turnover can be frequent in the maintenance staff, education is needed to pass along information regarding the care for each species.

8.3.6.6 Observed Wildlife

Butterflies and birds have been observed by city staff.

8.3.6.7 Best Performing Native Vegetation

Most of the plants specified are performing well. A few replacements have been made. *Mahonia aquifolium* was added to the roof as a substitute during installation.

8.3.6.8 Post-occupancy Observations

City Staff

- The landscape is maintained by a private landscape contractor. Education and training take place to adjust how the plants should be maintained.
- The irrigation system and gardens need frequent monitoring, just to be sure that they are doing what they are supposed to.

Authors' Reflections

- This project demonstrates resourcefulness. The City reduced financing needs by re-purposing an existing building in an already built-up downtown. The inclusion of green roofs serves as a pilot project that demonstrates a successful example and educates the public about green roof benefits.
- The difficult microclimate (sun/shade) relationships were dealt with by looking to native vegetation that adapts to both conditions.
- The City has become well-versed with green roofs and has a process in place to pass along education about how to maintain them, to the various private companies which may have frequent personnel changes. This is important since maintenance staff turnover can be high, and specialized knowledge can be effectively passed along.

8.3.7 Mercer Slough Environmental Center, Bellevue, Washington

"Treading lightly" is a concept that is sometimes used to describe the way delicate and precise placement of buildings, pavements, and walking surfaces can have very little impact on a natural setting. The 129-hectare (320-acre) Mercer Slough Environmental Education Center in Bellevue, Washington exemplifies this concept. The slough was formed by the establishment of backwaters from Lake Washington over millennia, and by the artificial lowering of the lake elevation in the early twentieth century. It is located at the lake's intersection with the low-lying land at the mouth of Kelsey Creek. At this intersection, the two water bodies meet to form the largest remaining freshwater slough and wetlands in the metropolitan region. An environmental learning center was built on the upper slopes of the preserve. The project team designed green roofs on several of the environmental center buildings (Fig. 8.20).

8.3.7.1 Project Team

Building Owner/Client: City of Bellevue Parks & Recreation.
Green Roof Design Team Lead: Jones & Jones Architects and Landscape Architects.
Architect: Jones & Jones Architects and Landscape Architects.
Structural Engineer: Lund + Everton Structural Engineering.
Landscape Architect: Jones & Jones Architects and Landscape Architects.
Installation Contractor: Berschauer Phillips Construction Company.
Project completion: 2008.
Green roof area: 204 m^2 (2200 ft^2) on four buildings.

Fig. 8.20 Green roof over one of the education classrooms. This low maintenance extensive green roof is intended to appear natural and receives maintenance only a few times a year for the removal of tree seedlings or other invasive plants. (Photo: Bruce Dvorak, August 2018)

8.3.7.2 Overview and Objectives

Jones & Jones designed the buildings and site and worked closely with representatives of a partnership between the Pacific Science Center and City of Bellevue Parks Department. The center includes classrooms, wet labs where visitors can learn about urban wetland ecology, restrooms, porous pavements, boardwalks, overlooks to the slough, trail connections, a community building, and a visitor center.

The buildings were built into the steep site above the slough. Half of the buildings are perched on piers to float above the sloping hillside, which allows surface drainage to flow through the forest and lifts the visitor into the tree canopy. One of the buildings was designed as an actual treehouse. Four of the eight structures sit on foundations and feature living roofs (Fig. 8.21). The use of pervious concrete, green roofs, rock-filled porous gabion walls, and catchment ponds serve to infiltrate and recharge groundwater. These features help to preserve the natural hydrology of the site, protect vegetation on the steep slopes and in the slough, and prevent soil erosion. The green roofs were designed to be low maintenance and accepting of vegetation that naturally adapts to the green roof ecology.

Fig. 8.21 This open and sunny habitat sustains grasses and sedum on an extensive green roof on top of a restroom. Beneficial self-seeding annuals such as fireweed are allowed to populate the green roof and benefit local pollinators. (Photo: Bruce Dvorak, August 2018)

8.3.7.3 Plant Establishment

Original plantings were native sedums, ferns and ground covers. Without irrigation and little maintenance in the first years of establishment, many plants needed to be replaced. The roofs were subsequently replanted with native and some non-native sedums.

Grasses

small-flowered wood rush (*Luzula parviflora*).

Herbaceous Perennials

deer fern (*Blechnum spicant*), fringe cup (*Tellima grandiflora*), oak fern (*Gymnocarpium dryopteris*), parsley fern (*Cryptogramma crispa*), small-flowered alum root (*Heuchera micrantha*), western sword fern (*Polystichum munitum*).

Mosses

gooseneck moss (*Rhytidiadelphus loreus*).

Succulents

broadleaved stonecrop (*Sedum spathulifolium*), Oregon stonecrop (*Sedum oreganum*).

8.3.7.4 Irrigation

Water is collected from metal roofs and is held in cisterns, sometimes using potable water as a back-up. Irrigation was added post-occupancy and is used only during times of summer drought.

8.3.7.5 Maintenance

Since several roofs are located in the forest, maple and fir seedlings easily take root on the green roofs and the seedlings need to be removed annually. Scotch broom, blackberries, and other unwanted plants have sprouted in bare areas and need to be removed.

8.3.7.6 Observed Wildlife

Typical wildlife that visits the nature preserve includes birds and bees.

8.3.7.7 Best Performing Native Vegetation

Sedums and grasses.

8.3.7.8 Post-occupancy Observations

Staff (City of Bellevue)

- Green roofs that are located on the south and west areas where they receive hot sun, especially in drought years, have suffered plant losses, opening up bare ground that allows invasive plants to establish and grow. After an irrigation failure during a hot summer caused major plant loss, some of the roofs were replanted with sedums.

Authors' Reflectio+ns

- Designers and operational staff must recognize that green roofs in the dry, warm, and sometimes hot summers of the Puget Lowlands typically require regular summer irrigation unless plant dormancy is desired and planned with plants adapted to summer dormancy.
- This project demonstrates a high level of integrated design of the building to the site. As multiple low impact features were required to accomplish a "light touch" (of development), the concept was established at the beginning of the design process, before the siting or design of buildings on the site. These goals could not have been achieved if the architects and landscape architects did not take a unified approach at the beginning.

8.3.8 Pacific Plaza Office Building, Tacoma, Washington

Once a blight in downtown Tacoma, the 1960s dilapidated parking garage was targeted to receive a 30-million-dollar renovation and additional programming of seven floors of class A office space above which included a green roof (Figs. 8.22 and 8.23). The project includes a meadow-based green roof and a rainwater harvesting system that helped the project receive a LEED Platinum certification.

Fig. 8.22 *Clarkia pulchella* in bloom during the first year of establishment in October 2009. This annual native herb came back the next few growing seasons but in significantly reduced numbers. Sometimes plant competition (between species) or changing maintenance practices such as allowing invasive species to establish or changing irrigation rates can alter plant composition. In the case of this green roof, other species eventually took over and filled the space where Clarkia was one dominant. (Courtesy of Ben Ferguson)

8.3.8.1 Project Team

Building Owner/Client: Pacific Plaza Office Building.
Architect: Ben Ferguson.
Landscape Architect: Alan McWain.
Installation Contractor: Roofmeadow.
Project completion: 2009.
Green roof area: 2648 m^2 (28,500 ft^2).

8.3.8.2 Overview and Objectives

Inspired by lowland coastal meadow ecosystems, this green roof includes the microclimatic pairing of meadow and succulent habitats. The gently saddle-sloped green roof deck has succulent vegetation located on the drier upland habitats and the meadow vegetation is located on the lower saddle of the living roof (Matthews 2009).

The green roof has an engineered growing media that is 5–15 cm (3″–5.5″) deep with two-thirds of the substrate as growing media, and 1/3 as a pumice drainage

Fig. 8.23 Coreopsis (yellow bloom) and yarrow (white/tan bloom) grow intermixed with dormant grasses in the foreground and lupines (in the background). Compare this image taken near the same location as Fig. 8.22, but about 9 years later. In this condition (above) the green roof is more diverse with perennial vegetation than the earlier years where annuals were dominant. (Photo: Bruce Dvorak, August 2018)

media that is 5 cm (2 in) deep below the substrate. A capillary fabric was used to capture and retain moisture.

8.3.8.3 Plant Establishment

Succulents were established through live cuttings and 2.5 cm (1 in) root plugs. Grasses and perennials were established through seeding. A scour blanket was placed on top of the substrate after cuttings and seeding were installed to control wind erosion and deter birds from eating seed.

8.3.8.4 Succulent Roof

Four exotic species of succulents were used.

8.3.8.5 Meadow Roof

Annuals

desert sand verbena (*Abronia villosa*, native to the southwestern U.S.), pinkfairies (*Clarkia pulchella*), tufted poppy (*Eschscholzia caespitosa*, native to Oregon and California).

Grasses

blue wildrye (*Elymus glaucus*), tufted hairgrass (*Deschampsia cespitosa*), western blue-eyed grass (*Sisyrinchium bellum*, Oregon and California), also three exotic species of grasses.

Herbaceous Perennials

beach suncup (*Camissonia cheiranthifolia*, Oregon and California), common yarrow (*Achillea millefolium*), largeflower fameflower (*Talinum calycinum*, central U.S.), nodding onion (*Allium cernuum*), Pike's Peak purple penstemon (*Penstemon mexicali* 'Pike's Peak Purple', central U.S.), red columbine (*Aquilegia canadensis*, central and eastern U.S.), sea thrift (*Armeria caespitosa*, California), also two exotic species of herbaceous perennials, and one exotic bulb.

8.3.8.6 Irrigation

Base level drip irrigation system that runs twice daily during the summer. A 719 m^3 (190,000 gals) cistern is used to collect rainwater. The cistern holds a majority of the rainfall that falls on the roof.

8.3.8.7 Maintenance

The roof membrane has a leak detection system installed. This means that the entire roof is sectioned by grids with an electronic detection system to locate any moisture that may leak. The system was tested when one leak occurred and the leak was easily found. It was repaired with minimal disturbance to the vegetation of the green roof.

TAGRO is applied to the green roof annually (COT 2020). The fertilizer is made from solid waste from the City of Tacoma. Its ingredients include biosolids and sand, sawdust, bark, and other ingredients. It is made for landscape applications and is Class A-rated by the EPA.

8.3.8.8 Observed Wildlife

Bees and birds frequent the roof. Bees made their appearance during the first growing season.

8.3.8.9 Best Performing Native Vegetation

camas, yarrow, coreopsis, grasses, and lupines.

8.3.8.10 Post-occupancy Observations

Architect

- The leak detection system installed below the waterproof membrane was well worth the investment. It works, as it had identified one leak immediately.
- Wildflowers blooms were extensive during the first years. Thereafter, their numbers have decreased. The annual reseeding of wildflowers is necessary to continue their presence.

Authors' Reflections

- This project exemplifies resourcefulness, in the restoration of a parking garage and added program. The harvested rainwater allows the green roof to maintain meadow vegetation without using potable water. As a highly visible rooftop from buildings above, the roof meadow improves the view for those with overhead views of the roof.
- The strategy to use succulent vegetation at the upper edges of the saddle and meadow vegetation in the moist zones of the rooftop demonstrates awareness of the microclimate of the roof deck and slope. This approach has proved successful.

8.3.9 Cedar River Watershed Education Center, North Bend, Washington

The Cedar River Watershed Education Center is located at the gateway of Seattle's Cedar River Municipal Watershed and the Rattlesnake Lake Recreation Area, on a site that was once the location of a cross-country train depot. Since 1901, the watershed has been a source of drinking water for Seattle, today providing potable water to 70% of the greater Seattle area. The center's buildings are energy-efficient, constructed from local materials, and are designed to teach the next generation about watersheds, hydrology, ecology, and green building practices. The center hosts hundreds of school education sessions each year and is set up to teach children and the public about where Seattle's drinking water comes from, habitat conservation, forest ecosystems, and ways to sustain clean water in urban areas, green roofs (Fig. 8.24), and porous pavements.

Fig. 8.24 One of the southwest-facing, sloped green roofs at the education center in a sunny location. Native sedums and grasses have maintained cover on the roof since 2002. (Photo: Bruce Dvorak, August 2018)

8.3.9.1 Project Team

Building Owner/Client: City of Seattle.
Green Roof Design Team Lead: Jones & Jones Architects and Landscape Architects.
Architect: Jones & Jones Architects and Landscape Architects.
Landscape Architect: Jones & Jones Architects and Landscape Architects.
Installation Contractor: Berschauer Phillips Construction Co. & SH Landscape.
Project completion: 2002.
Green roof area: approximately 235 m^2 (1450 ft^2).

8.3.9.2 Overview and Objectives

The design is revelatory, in that metal roofs juxtpose the green roofs. Fast runoff from the metal roofs metaphorically represents impervious surfaces in Seattle, while the minimal outflows from the thick green roofs represent the "sponge" habitat and ecosystem functions of a healthy watershed such as stormwater retention, biodiversity and the beauty of forest floor vegetation (Viani et al. 2007).

The 36,421 hectares (90,000 acres) of forested habitats in the watershed include a cross-section of Coastal Temperate forest habitats from the Puget Sound lowlands to the central Cascade Mountains crest. The protected habitat provides a rich biological diversity of plants and animals that are integral to the ecosystem. The watershed contains old-growth and second-growth forest and a variety of non-forested terrestrial habitats.

The green roofs were designed to make use of local vegetation species and genotypes. They cover a sixth of the buildings at the education center, with many plant species selected based upon plants native to rocky balds located within the recreation area and other plants native to the forest floor. The substrate depth is 30 cm (12 in) and is composed of 30% sphagnum peat, 30% pumice, 30% compost (which was substituted for specified composted bark), and 10% clean sand.

8.3.9.3 Plant Establishment

Plants specified to grow on the education center green roofs were selected from a list of plants that a local botanist team found growing on the rocky outcrop of Rattlesnake Ledge, which is visible from the Center, and along the steep forested trail leading to the Ledge.

Several rooftops were anticipated to receive sun much of the day so a "Sunny Green Roof" plant list was developed, with criteria for plant selection including plants that naturally grow on local exposed and sunny sites (Fig. 8.24). As some of the roofs were anticipated to receive substantial shade from tall surrounding trees and overhanging vegetation, a Shady Green Roof (Fig. 8.25) plant list was also designed. All of the plants are native to the shaded and sunny areas of Rattlesnake Ledge and trail, with the mixes assigned in groupings to potentially similar rooftop conditions. (Rottle 2000). Some of these species were contract grown from seed collected from Rattlesnake Ledge.

8.3.9.4 Shady Green Roof

Grasses

small-flowered woodrush (*Luzula parviflora*).

Herbaceous Perennials

deer fern (*Blechnum spicant*), fringecup (*Tellima grandiflora*), oak fern (*Gymnocarpium dryoperteris*), parsley fern (*Cryptogramma crispa*), small-flowered alum root (*Heuchera micrantha*), western sword fern (*Polystichum munitum*), locally-collected moss.

Fig. 8.25 This portion of a shady green roof covers walkways between buildings and contains a deep-shaded habitat where the native western sword fern and other shade-loving plants persist. (Photo: Bruce Dvorak, August 2018)

8.3.9.5 Sunny Green Roof

Herbaceous Perennials

matted saxifrage (*Saxifraga bronchialis*), small camas (*Camassia quamash*), small-flowered alum root (*Heuchera micrantha*), tufted saxifrage (*Saxifraga caespitosa*), western saxifrage (*Saxifraga occidentalis*), western yarrow (*Achillea millefolium*)

Succulents

broadleaf stonecrop (*Sedum spathulifolium*), Oregon stonecrop (*Sedum oreganum*)

8.3.9.6 Irrigation

Irrigation water is sourced from non-potable sources, using excess water tapped from local hydroelectric generation. A drip irrigation system was installed on the green roofs. Over the years, there have been different watering practices ranging from original periodic watering in the establishment period, to the curtailment of irrigation during a hot, dry summer, to re-institution of regular summer watering.

8.3.9.7 Maintenance

The green roofs receive attention a couple of times in the spring to remove invasive tree seedlings.

8.3.9.8 Observed Wildlife

Insects, butterflies, bees are frequently observed.

8.3.9.9 Best Performing Native Vegetation

Vegetation has self-selected to its preferred microclimate. Initially, all of the vegetation established. During disruptions to the irrigation system, portions of the roof suffered dieback where they were replanted with grasses. There has been no formal follow up on the current species growing on the roof. Sedums and grasses grow in the sunny areas, and western sword fern and other ferns grow in shady areas. The small camas has persisted, blooming in the spring.

8.3.9.10 Post-occupancy Observations

Owner

- There is a need for consistent training for the care of the living roof. When different managers arrive at the center, institutional knowledge of how to care for the green roof (specific to each roof) is lost.
- Different aesthetic expectations have been maintained over time, including letting the invasive grasses grow long and keeping the grasses trimmed. Different techniques for maintaining the grass have been used such as using weed trimmers. The current approach is to let the vegetation grow as it is and not trim the grass.

Design Team

- This was one of the first extensive green roofs in the Seattle area. It was designed and installed before the build-up of the modern green roof industry in place today. The longevity of the green roof points to the resilience of the green roof and the use of a variety of different plant species.
- Continuity of care and knowledge of the design intent and expectations is critical to long-term plant success.

- Summer irrigation is usually needed to sustain living roofs in Pacific Northwest summers, especially during drought conditions.
- Interpretation of the living roof versus metal roof juxtaposition may benefit from interpretive signage.

Authors' Reflections

- The endurance of the vegetation on these green roofs demonstrates how native vegetation can be resilient, and adapt to green roofs in the ecoregion, even when mishaps with irrigation systems take place.
- A unique and perhaps critical step used during the planning stages was the anticipation of the shading of the green roofs. Measures were taken during the planning phases to articulate a plant list from which plants could adapt to preferred microclimates over time.
- Some plants were pre-grown from local genotypes. The attention to ecological sourcing of plants is outstanding and contributes to the persistence of the ecosystems with minimal disturbance.

8.4 Plant Species

Across all of the ecoregional green roof case study sites in Chap. 8, there are 73 species in total, 60 of which are native to the ecoregions in this chapter. Of those native to chapter ecoregions, 16 species occur more than once across the case studies. Of those occurring more than once, two are grasses/rushes, 10 are herbaceous perennials, one is a shrub, and three are succulents. Five species occurred on three or more green roofs (Table 8.1).

Table 8.1 Five species occur in three or more of the case study sites in the chapter

Plant Type	Common Name	Botanical Name	A	B	C	D	E	F	G	H	I
Herbaceous Perennials	Common yarrow	*Achillea millefolium*	x	x	x						
Herbaceous Perennials	Nodding onion	*Allium cernuum*				x	x	x			
Herbaceous Perennials	Oak fern	*Gymnocarpium dryoperteris*			x				x	x	
Herbaceous Perennials	Western swordfern	*Polystichum munitum*			x	x	x		x	x	
Succulents	Oregon stonecrop	*Sedum oreganum*	x		x				x	x	x

Key = A (SRM Development/Google), B (Pacific Plaza Office Building), C (Cedar River), D (Woodland Park Zoo), E (Belleview City Hall), F (Pacific Plaza Office Building), G (Bill and Melinda Gates Headquarters), H (Mercer Slough Environmental Center), I (Ballard Library)

8.5 Summary

Green roofs made an early appearance to the Puget lowlands in 1899 with the construction of the luxurious Lincoln Hotel roof garden in Seattle (Fig. 1.8). Public greenspace as roof gardens arrived in 1976 with the opening of Freeway Park (Fig. 8.4), and thereafter numerous roof deck plantings over suppressed highways in the metro region. Since 2002, many extensive and semi-intensive types of green roofs have been constructed in the Seattle/Tacoma and Olympia metro areas including those highlighted in this chapter. The case studies here demonstrate that green roofs were built to:

- connect visually with the surrounding environment, e.g., Bill and Melinda Gates Headquarters, Mercer Slough Environmental Center, Cedar River Watershed Education Center;
- make use of a brownfield site that was previously unbuildable but now has accessible rooftop gardens, e.g., Bill and Melinda Gates Headquarters, SRM Development/Google, Mercer Slough Environmental Center (steep slopes), or refurbish a building for new use and green roofs Belleview City Hall, Pacific Plaza;
- provide a learning tool for green roofs, e.g., Cedar River Watershed Education Center, Mercer Slough Environmental Center, Bullitt Center;
- provide an outdoor laboratory for observing nature, especially pollinators and other visitors to the green roofs, e.g., Ballard Public Library, Mercer Slough Environmental Center, Bullitt Center;
- demonstrate benefits generally associated with green roofs such as energy conservation, runoff amelioration, temperature, and noise abatement, e.g. SRM Development/Google, Belleview City Hall, the Bill and Melinda Gates Headquarters (municipal water use reduced by 79%, solar panels reduce municipal energy consumption by 39%), and the Bullitt Center which is off-grid, meaning the project is self-sufficient from public utilities;
- offset the effects of habitat loss, not via a mitigation process *per se*, but rather through the intentional introduction of habitat that increases the number of plants of a particular habitat once common to the site, e.g., Bill and Melinda Gates Headquarters (wetlands), Mercer Slough Environmental Center (preserve site hydrology), Cedar River Watershed Education Center (incorporate plants from local genotypes);
- increase local habitat diversity, e.g., Zoomazium, Ballard Library, Pacific Plaza, Belleview City Hall;
- increase the aesthetic appeal of buildings especially where the function of the building itself may reduce stress for workers or visitors, e.g., Belleview City Hall, SRM Development/Google, Ballard Public Library, Bill and Melinda Gates Headquarters.

The case studies presented here represent a cross-section of green roofs from the region that are accessible and that made use of native vegetation and represented a range of kinds of green roofs and various aesthetic intentions. Some of the vegetation used on these green roofs were native to the prairies, rocky outcrops, meadows, and forest understories of the Puget Lowlands. There is a range of aesthetic intentions of the case studies from natural to formal gardens. The more natural-looking green roofs required maintenance at least a couple of times in the spring and fall. All of the formal designs required weekly and sometimes daily maintenance, and those buildings owners have funds to support ongoing maintenance. The Cedar River Watershed Education Center was the only case study where plants local to the site were directly observed by local botanists and were included in the design and maintenance protocols.

Most of the projects made use of harvested rainwater to irrigate the roofs during the summer. One wetland green roof that treated greywater from inside the building was integral to the internal function of the wastewater treatment process. Several green roofs experienced dieback during drought periods when watering with municipal water was restricted or harvested water insufficient. If municipal water is used, then backup sources of water may be needed during drought periods when water restrictions are enforced. If harvested rainwater is used, then appropriate sizing of holding tanks is needed to sustain vegetation through periods of drought, or municipal backup supplied. The wetland green roof may have the most reliable source of water as long as the building has occupancy and people are using water inside the building.

Regarding exploration of the native plants, only 60 of potentially hundreds of plants native to the Puget Lowlands and Coastal Temperate Forests have been tested on green roofs. There are only a few publications regarding plant trials to disseminate knowledge about which plants have been trialed, about their growing and maintenance conditions. The natural sites in the lowlands have vegetation growing on flat to sloped sites, and vegetation in the Cascades is predominantly sloped or steeply sloped sites. About half of the case studies had sloped roof decks. From discussions with owners, the sloped green roofs dry out more quickly than flat roofs, and the flat roof tended to retain more water and have more invasive plants.

It is not clear how influential the availability of native plants at local nurseries was in the selection of vegetation. Several projects made use of the non-native *Festuca glauca;* however, native blue fescue is in production (*Festuca idahoensis* Elmer ssp. *Idahoensis)* and could possibly serve as a replacement, as *Festuca idahoensis* was successfully grown on the Ballard Library (8.3.2) before the droughts and no irrigation for 2 years. Native plant communities were referenced as a source for many of the grassed green roofs; however, the plant lists for some of those projects also included a number of exotic plants. The natural plant communities of the Cascades and alpine meadow plant communities were not trialed. Prairie conservation sites exist in the lowlands; however, no project attempted a Garry oak ecosystem prairie planting on a green roof. Only four of 200 grass species native to the Pacific Northwest were used, and one species of sedge (*Carex)* and four rushes (*Juncus*).

The City of Seattle has made concerted efforts to publish results of some post-occupancy studies of their major green infrastructure investments. When green roofs are designed sustainably, and with appropriate plant-to-substrate and microclimate associations, green roofs can become a reliable technology and investment for people and the wildlife that use them. The City of Seattle also includes living roofs as a pathway for developers to earn "Seattle Green Factor" points that are required to obtain building permits, and this innovative code requirement has spurred the design and installation of many living roofs within the city limits in efforts to increase green space, ecosystem services, and the attenuation of stormwater (Roehr et al. 2009; Rottle and Yocom 2010).

Some private organizations are making use of green roofs for their employees and were open to sharing knowledge about them. The provision of sharing knowledge about living roofs is being conceived of as a corporate social responsibility. Much was learned here about how some developers and private organizations have designed their working environments, maintain them, and contribute or give back to the community at large (Dudley 2013). These types of relationships are critical to the development of new technologies such as green roofs, as many publicly-funded projects have limited budgets, and opportunities to fund green roof research or access to outstanding built projects are not yet a high priority for many funding agencies (Rowe 2019). The case studies here begin to build a base from which to learn about the various goals, technologies, eco-regional plant selection possibilities, and benefits of green roofs.

Acknowledgments We would like to thank the following individuals for the time, sharing of knowledge, access to green roofs, and their dedication to ecoregional green roofs: Bernie Alonzo and Cheryl dos Remedios with GGN Seattle; Forrest Jammer Thomas Rengstorf & Associates, Inc.; Dave Tomson with SRM Development; Scott Morgan The Evergreen State College; Ben Ferguson, Tracy Taylor with Ferguson Architecture; Josh Clarke and Eric Huseby with the City of Tacoma; Lisa Lange with the Cedar River Watershed Education Center; Deborah Sigler with the Bullitt Center; Jonathan Morley with Berger Partnership; Gabe Varga with Diadem USA, Inc., Darcy Drysdale with Etera; Mike Brandvold Chris Dariotis Pacific Earth Works, Inc.; Joe Fry from Hapa Collaborative; and Pat Harris and Andrew Heider with the City of Bellevue.

References

Alaback PB (1994) Plants of the Pacific Northwest Coast: Washington, Oregon, British Columbia & Alaska. Lone Pine Pub, Vancouver

Bailey RG (1997) Ecoregions of North America. U.S. Department of Agriculture, Forest Service, Washington, DC

Barclay I (2007) A Primer on Washington Native Cacti. The Desert Northwest. Sequim, Washington, DC

Barnosky CW (1985) Late Quaternary vegetation near Battle Ground Lake, southern Puget Trough, Washington. Geol Soc Am Bull 96(2):263–271

Breckheimer IK, Theobald EJ, Cristea NC, Wilson AK, Lundquist JD, Rochefort RM, HilleRisLambers J (2020) Crowd-sourced data reveal social–ecological mismatches in phenology driven by climate. Front Ecol Environ 18(2):76–82

Chappell CB (2004) Upland Plant Associations in Washington's Puget Trough Ecoregion. Washington Natural Heritage Program. Washington Department of Natural Resources, Olympia

Chappell CB, Crawford RC (1997) Native vegetation of the South Puget Sound prairie landscape. Ecology and conservation of the South Puget Sound Prairie Landscape. The Nature Conservancy of Washington, Seattle, pp 107–122

Christy JA, Alverson ER (2011) Historical vegetation of the Willamette Valley, Oregon, circa 1850. Northwest Science 85(2):93–108

COT (2020) TAGRO, Tacoma Grow. City of Tacoma, WA. https://www.cityoftacoma.org/. Accessed 20 Feb 2020

Crawford RC, Hall H (1997) Changes in the south Puget prairie landscape. In: Ewing PDK (ed) Ecology and conservation of the South Puget Sound prairie landscape. The Nature Conservancy, Seattle, pp 11–15

Currie I (2018) Ballard library green roof maintenance, Ballard

Del Moral R, Deardorff DC (1976) Vegetation of the Mima Mounds, Washington State. Ecology 57(3):520–530

D. DellaSala, G. Orians, K. Kavanagh, Sims M (2019) Puget lowland forests. World Wildlife Fund. https://www.worldwildlife.org/ecoregions/na0524. Accessed 18 June 2019

Dudley B (2013) Google will double Kirkland campus. The Seattle Times

Dunn P (1998) Prairie habitat restoration and maintenance on Fort Lewis and within the South Puget Sound Prairie landscape. Final report and summary of findings. The Nature Conservancy of Washington, Seattle

Easterly RT, Salstrom DL, Chappell CB, Dunn P (2005) Wet prairie swales of the South Puget Sound, Washington. The Nature Conservancy of Washington, Olympia, WA, Seattle

Easton V (2011) Shannon Nichol puts her mark on Gates Foundation landscape. Seattle Times

ENRNorthwest (2016) Best Green Project – Google Kirkland Campus – Building D. ERN Northwest

Fleenor R (2016) Plant guide for fireweed (Chamerion angustifolium). USDA-Natural Resources Conservation Service, vol 99201. USDA, Spokane

Franklin JF, Dyrness CT (1988) Natural vegetation of Oregon and Washington, vol PNW-8. Oregon State University Press, Corvallis

Fry J (2019) Bellevue City Hall green roof design. e-mail communication, USA

Harris P (2019) Bellevue City Hall roof garden maintenance. e-mail communication, USA

Harvey T (2019) Mountain plants of the Western Cascades. Tanya Harvey. http://westerncascades.com/. Accessed 18 June 2019

Haynes G (2002) The early settlement of North America: the Clovis era. Cambridge University Press, Cambridge

Hebda RJ, Mathewes RW (1984) Holocene history of cedar and native Indian cultures of the North American Pacific Coast. Science 225(4663):711–713

Henderson JA (1973) Composition, distribution and succession of subalpine meadows in Mount Rainier National Park. Oregon State University, Corvallis

KCP (2006) King County green roof case study report. King County & Palandino & Company, Seattle

Kuiper LC (1988) The structure of natural Douglas-fir forests in Western Washington and Western Oregon, vol 88-5. Wageningen Agricultural University, Wageningen

Leopold EB, Boyd R (1999) An ecological history of old prairie areas in southwestern Washington. In: Indians, fire and the land in the Pacific Northwest. Oregon State University Press, Corvallis, pp 139–163

MacIvor JS, Ranalli MA, Lundholm JT (2011) Performance of dryland and wetland plant species on extensive green roofs. Ann Bot 107(4):671–679

Martin MA, Hinckley TM (2007) Native plant performance on A Seattle Green Roof. Paper presented at the Fifth Greening Rooftops for Sustainable Communities Conference, Minneapolis, Minnesota, April 29–May1, 2007

Matthews T (2009) Outside Pacific Plaza: a garden grows high above downtown Tacoma. Tacoma Daily Index

Mima Mounds: A Special Prairie (2018). National Park Service, Kiosk at Mima Mounds

Mitchell C, van Daalen C (2019) Greywater treatment and infiltration at the Bullitt Center, vol 2019. Building Innovations Database, Seattle

NCNP (2019) Plants of North Cascades National Park. National Park Service. https://www.nps.gov/noca/learn/nature/plants.htm. Accessed 3 July 2020

Omernik JM (1995) Ecoregions: a framework for managing ecosystems. In: The George Wright Forum. vol 1. JSTOR, pp 35–50

Pojar J, MacKinnon A (2004) Plants of the Pacific Northwest Coast. Partners Publishing Group, USA

Rochefort RM, Peterson DL (1996) Temporal and spatial distribution of trees in Subalpine Meadows of Mount Rainier National Park, Washington, U.S.A. Arct Alp Res 28(1):52–59. https://doi.org/10.1080/00040851.1996.12003147

Roehr D, Laurenz J, Kong Y (2009) Green envelopes: contribution of green roofs, green facades, and green streets to reducing stormwater runoff, CO2 emissions, and energy demand in cities. Paper presented at the Low Impact Development for Urban Ecosystem and Habitat Protection, Seattle, WA, November 16–19

Rottle N (2000) Water comes to life at the Cedar River Watershed Education Center. Seattle Daily Journal of Commerce, April 30, p 3

Rottle N, Yocom K (2010) Basics landscape architecture 02: ecological design. AVA Publishing, Lausanne

Rowe B (2019) Reflections on 20 years of green roof research, Living architecture monitor, vol 21. Green Roofs for Healthy Cities, Toronto

Sage RB (1854) Scenes in the Rocky Mountains, and in Oregon, California, New Mexico, Texas, and the Grand Prairies: Or, Notes by the Way, During an Excursion of Three Years, with a Description of the Countries Passed Through, Including Their Geography, Geology, Resources, Present Condition, and the Different Nations Inhabiting Them. Henry Carey Baird, Philadelphia

Schemske DW, Husband BC, Ruckelshaus MH, Goodwillie C, Parker IM, Bishop JG (1994) Evaluating approaches to the conservation of rare and endangered plants. Ecology 75(3):584–606

Schultz CB, Henry E, Carleton A, Hicks T, Thomas R, Potter A, Collins M, Linders M, Fimbel C, Black S (2011) Conservation of prairie-oak butterflies in Oregon, Washington, and British Columbia. Northwest Sci 85(2):361–389

Seattle, Cascadia (2011) Regulatory pathways to net zero waterguidance for innovative water projects in Seattle. vol Phase II Summary report. Seattle

Song U, Kim E, Bang JH, Son DJ, Waldman B, Lee EJ (2013) Wetlands are an effective green roof system. Build Environ 66:141–147

SPU (2012) FINAL green roof performance study Seattle Public Utilities. Seattle Public Utilities & Cardno TEC, Seattle, WA

Storm L, Shebitz D (2006) Evaluating the purpose, extent, and ecological restoration applications of indigenous burning practices in southwestern Washington. Ecol Restor 24(4):256–268

Swift B (2019) Ballard public library green roof design

Taylor BL (2008) The stormwater control potential of green roofs in Seattle. In: Nian S, Michael C (eds) 2008 International low impact development conference. ASCE, p 11

Thysell DR, Carey AB (2001) Quercus garryana communities in the Puget Trough, Washington. Northwest Sci 75(3):219–235

Turner NJ, Lepofsky D, Deur D (2013) Plant management systems of British Columbia's first peoples. BC Stud Br Columb Q (179):107–133

Velazquez L (2017) Project of the week: Bill & Melinda Gates Foundation & Seattle Center 5th Avenue North Parking Garage. Greenroofs.com. https://www.greenroofs.com/projects/bill-melinda-gates-foundation-seattle-center-5th-avenue-north-parking-garage/

Viani LO, Rottle ND, Pennypacker E, Puddy M (2007) The feel of a watershed: the Cedar River Watershed Education Center teaches by sensory experience-should it do more? Landsc Architect 97(8):24

Vinh T (2005) Ballard celebrates its new library. The Seattle Times, May 15, p 1

WBDG (2016) Case studies: Bullitt Center. Whole Building Design Guide, Washington, DC

Whitlock C (1992) Vegetational and climatic history of the Pacific Northwest during the last 20,000 years: implications for understanding present-day biodiversity. Northwest Environ J 8:5–5

Williams DB (2015) Discovery Park (Seattle): Natural History. (11161)

Wilson MV, Hammond PC, Schultz CB (1997) The interdependence of native plants and Fender's blue butterfly. Conservation and management of native flora and fungi. Native Plant Society of Oregon, Corvallis, pp 83–87

Yocom K, Spencer B (2013) Green roof performance study: Puget Sound Region. Paper presented at the Space, Time/Place, Duration CELA Conference, Manhattan, Kansas, March 28–30

Chapter 9
Green Roofs in Willamette Valley Ecoregions

Bruce Dvorak and Olyssa Starry

Abstract This chapter covers case studies of four conservation sites and nine green roofs located in the Willamette Valley ecoregions of western Oregon. The region is geographically complex with multiple ecoregions including prairies, rocky outcrops, oak woodlands, mixed woodlands, forests, wetlands, and riparian habitats. Historically, grasslands were widespread in the valley, and these were intermixed with oak and conifer woodlands from Portland to Eugene. Less than 1% of the native prairie habitat remains intact. Precipitation averages 800–950 mm annually in the valley, and the region experiences warm to hot and dry summers and long wet winters, with infrequent snow. Much of the precipitation takes place from October through May. Although green roofs were introduced into the valley since early 2000, much of the vegetation on the over 500 green roofs in the Portland metro region has apparently not been intended to mitigate the displacement of native ecosystems by building construction. However, the nine ecoregional green roof case studies in this chapter demonstrate how 69 species of vegetation native to the Willamette Valley ecoregions can be employed on green roofs.

Keywords Prairie · Wildlife habitat · Municipal · Native American · Healing garden · Maintenance · Think tank (green roof)

B. Dvorak (✉)
Department of Landscape Architecture and Urban Planning, 305A Langford Architecture Center, Texas A&M University, College Station, TX, USA
e-mail: bdvorak@tamu.edu

O. Starry
University Honors College, Portland State University, Portland, OR, USA
e-mail: ostarry@pdx.edu

B. Dvorak (ed.), *Ecoregional Green Roofs*, Cities and Nature,
https://doi.org/10.1007/978-3-030-58395-8_9

9.1 Ecoregion Characteristics

The Willamette Valley was first settled by Paleo-Indians after crossing the Bering land bridge at least 10–11,000 years ago, or more (Williams et al. 1985). Before the land bridge became submerged after the last glacial maximum, multiple waves of new settlers from Asia and/or Eurasia came to settle in the Willamette Valley and they formed many tribes (bands), including the Chinook, Clackamas and Kalapuya peoples of the Willamette Valley (Zenk 2008). The landscape vegetation was heavily influenced by these first human settlements such as their intentional burning of the land. They managed the land (near settlements) for plant food sources and hunting grounds (Kimmerer and Lake 2001; Deur and Turner 2005). Research regarding the landscape habitats of the Willamette Valley before European settlement confirms that the valley floor vegetation largely consisted of prairies, oak woodlands, mixed-woodlands, and wetlands for at least the previous 4000 years before settlement by Europeans (Williams 2002; Christy and Alverson 2011).

In October and November of 1805, Meriwether Lewis and William Clark and company were the first U.S. citizens to visit the Willamette Valley from an eastern land route. They were also the first to make a written record of animals, ecosystems, and 240 kinds of plants along their journey across the west including many plants native to the Pacific Northwest that bear the names Lewis or Clark. They described some of the cultural habits of the native settlements in the Willamette Valley and recorded that there was a mosaic of prairie and woodlands and evidence of the widespread effects of burning the land (Ambrose and Abell 1998; Reveal et al. 1999).

Several historic renderings of native habitats that existed before the settlers arrived were painted by artists Paul Kane and later, Albert Bierstadt. During the 1840s, Paul Kane traveled the Willamette Valley and captured the broadness of the prairie across the Willamette Valley in numerous sketches and his painting titled "The Wilhamet River from a Mountain" (Fig. 1.1). Several decades later, Albert Bierstadt visited the Columbia River Gorge. His painting titled *Mount Hood* (Fig. 9.1) was painted on-site in rough form and then later completed in his studio. It is thought that the foreground elements were painted as he saw them. The proportion of open land with prairie and open woods is significant. The exact location is not known but is thought to be several miles southwest of Portland (Carr and Bierstadt 1997).

Plant communities growing in the Pacific Northwest interior valleys including the Willamette Valley were subject to decades-long intervals of dry to wet climates. The extent of woodland, forest, and grass habitats was subject to the changes in climate as dry periods favored the expansion of prairie and wet periods favored the expansion of woodland and forest (Walsh et al. 2010). Today, less than 1% of native prairie habitats are maintained in the valley; however, many of the plant species local to the interior valley grasslands are still present in mostly private preserves throughout the valley. Public universities, State and Federal agencies, and private organizations such as The Nature Conservancy, the Greenbelt Land Trust, Cascadia Prairie Oak Partnership, and other groups are working to restore prairie habitats in the valley.

The Willamette River is a major feature of the valley. It extends from the southern boundary of Oregon near Eugene to its confluence with the Columbia River near

Fig. 9.1 Detail of *Mount Hood*, 1863. On Albert Bierstadt's second trip to Oregon, he painted Mount Hood from a point several miles southwest of town. The painting shows a prairie habitat widespread in the landscape with isolated pockets of open woods. According to the source, he painted the mountain from a study including this view painted in November of 1863. (Carr and Bierstadt 1997. Courtesy of Wikiart https://uploads2.wikiart.org/images/albert-bierstadt/mount-hood-1863.jpg)

Portland. The river has gone through extensive change and modification over time, with significant modifications made during the second haft of the nineteenth and twentieth centuries, as the Army Corps of Engineers made changes to the river channel to increase navigability. Regardless, most of the historic wet prairies, riparian woodlands, sloughs, open woodlands no longer remain. (Benner and Sedell 1997). Wetland habitats, however, may be useful to study to learn about how plants native to these hydric communities may play a role with green roofs and ecosystem services for sustainable sites and buildings (MacIvor et al. 2011; Ahiablame et al. 2012; Everett et al. 2018).

Climate remains one of the greatest challenges for green roofs in the Willamette Valley as summers are typically dry and warm, with periods of dry and hot weather followed by months of persistent precipitation and cool temperatures during the winter. Due in part to variations in the elevation of the coastal mountains, precipitation across the valley is greater near Eugene, and less in the northern section of the valley near Portland (Fig. 9.2).

Comparatively, the climate in the Willamette Valley is warmer than the marine-influenced interior valleys of where Seattle, WA and Vancouver, B.C. are located. The ecosystems native to the Willamette Valley include some of the same species and families of plants as Seattle and Vancouver, but there is a shift towards species that favor drier and warmer microclimates.

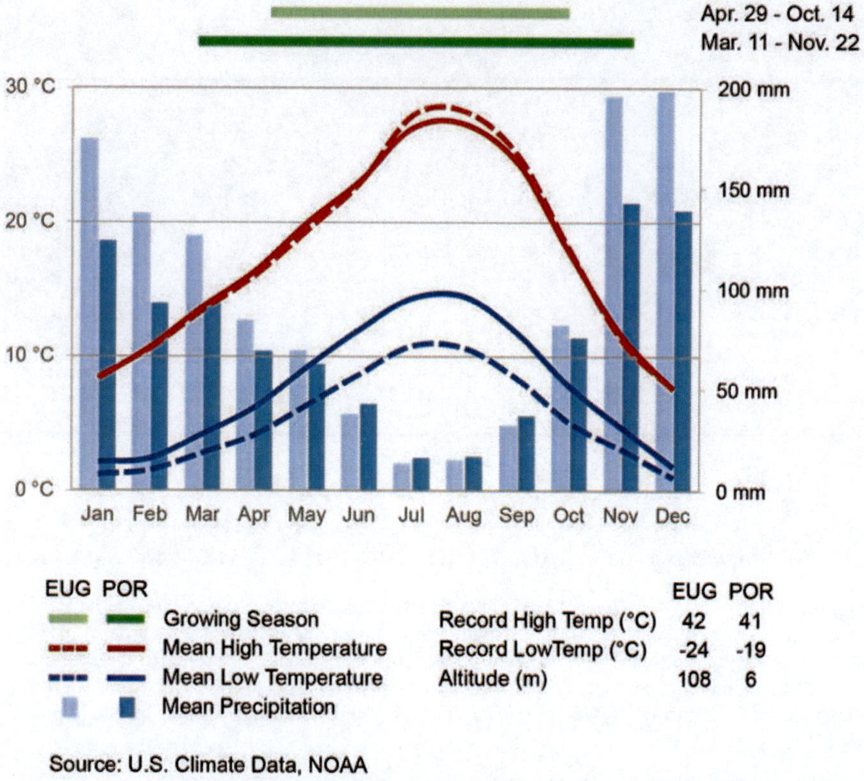

Fig. 9.2 Climate characteristics of the Willamette Valley at Portland (POR) and Eugene (EUG). Eugene receives more precipitation than Portland and has warmer summers and cooler winters. (Graphic: Trevor Maciejewski and Bruce Dvorak)

9.1.1 Vegetation Within the Ecoregions

This section covers some of the dominant forms of vegetation native to the valley.

The dominant native ecoregions of the Willamette Valley consist of prairie, oak woodlands, mixed-woodlands, coniferous forest, montane meadows, and rocky outcrops (Fig. 9.3). These include a wide variety of forms of vegetation including trees, shrubs, grasses, herbaceous perennials, annuals, bulbs, ferns, and mosses.

9.1.1.1 Temperate Coastal Forests

Common and widespread throughout western Oregon is a wide range of temperate coastal forest plant communities that are comprised of coniferous trees, groundcovers, shrubs, grasses, and herbaceous plants. Major conifer tree species

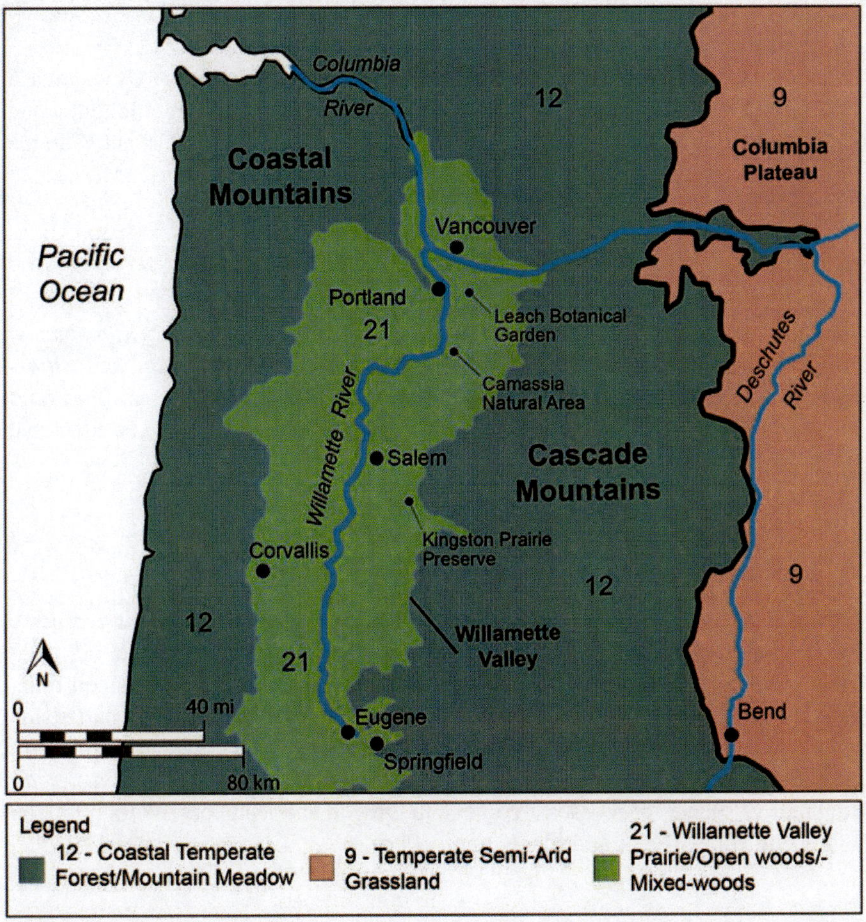

Fig. 9.3 Willamette Valley Ecoregions. The Willamette Valley natural vegetation consists of prairies, open woodlands, oak ecosystems, forests, and wetlands (21); and Coastal Temperate Forests with mountain meadows established on disturbed sites, steep-sloped environments, and windy or high elevation sites (12) (Thilenius 1968; Bailey 1997; Baker et al. 2004; Christy and Alverson 2011). East of the Cascade Mountains lies Temperate Semi-Arid grasslands (9). See Chap. 6 for temperate semi-arid grassland green roofs in central Oregon. Complex ecotones exist between boundaries, especially the transitions between forested and open habitats. (Graphic: Trevor Maciejewski and Bruce Dvorak)

in the Willamette Valley include grand fir (*Abies grandis*), Douglas-fir (*Pseudotsuga menziesii*), and western redcedar (*Thuja plicata*), ponderosa and sugar pines. Other native conifers trees found in the ecoregion include larch, yew, and juniper.

Major Deciduous Trees

Common native deciduous trees include vine maple (*Acer circinatum*), Oregon ash (*Fraxinus latifolia*), Oregon white oak (*Quercus garryana*), Pacific dogwood (*Cornus nutallii*) and bigleaf maple (*Acer macrophyllum*). Madrone is native to the region, is evergreen, and grows on drier sites (Pojar and MacKinnon 2004).

Shrubs

Common native shrubs include Oregon grape (*Mahonia*), elderberry (*Sambucus*) Douglas spiraea (*Spiraea douglasii*) beaked hazelnut (*Corylus cornuta* var. *californica*) ninebark (*Physocarpus capitatus*) and salal (*Gaultheria shallon*) and *Rhododendron macrophyllum*. Many of the small trees and shrubs may be adaptable to intensive green roofs.

Ferns and Groundcovers

western sword fern (*Polystichum munitum*), kinnikinnick (*Arctostaphylos uva-ursi*), and Oregon oxalis (*Oxalis oregano*) also grow throughout the forested and open forest habitats and may function on deep extensive or semi-intensive green roofs (Franklin and Dyrness 1988). Other herbaceous plants native to sunny or part-shade include several species of strawberry such as wild strawberry (*Fragaria virginiana*), coastal strawberry (*Fragaria chiloensis*), and wood strawberry (*Fragaria vesca*). Hundreds of species of mosses are native to Oregon and many can be found in the Willamette Valley (Pojar and MacKinnon 2004).

Bulbs

Common bulb species that grow throughout the region include the culturally important camas and species of allium. Bulbs as plant food cultivated by the Native American tribes include brodiaea (*Brodiaea* spp. a native wildflower, *Triteleia* spp.), and checker lily (*Fritillaria affinis*), roots of biscuitroot (*Lomatium* spp.), yampah (Perideridia spp.) plants, and seeds of tarweed (Madia spp.) and balsamroot (Christy and Alverson 2011).

9.1.1.2 Prairies and Oregon White (Garry) Oak Ecosystems

Of interest to this book, we will focus on vegetation native to the variety of prairie and oak habitats native to the valley. This is a broad assemblage of plants that are based on grassland and savanna-like habitats. Prairies of the valley include those that are considered dry are located on upland or south-facing slopes. Other prairie regimes include mesic and wet plant communities. There are many associative plant and insect, bird, and other biological relationships implicit to the Willamette Valley prairies and oak woodlands (Schemske et al. 1994; Wilson et al. 1997; Schultz et al. 2011).

Dominant vegetation includes Oregon white oak (*Quercus garryana*), shrubs such as nootka rose (*Rosa nutkana*), common snowberry (*Symphoricarpos albus*), ocean spray (*Holodiscus discolor*). Dominant herbs include common camas (*Camassia quamash*) and great camas (*C. leichtlinii*).

Regarding upland or dry prairies, there is little record of the number of taxa composing the Willamette Valley prairies but it was likely similar to other prairies and consists of several hundred species or more. There are over 200 species of grasses that grow in the Pacific Northwest and most of those species are native (Alaback 1994).

Dominant grasses of the upland prairies include Roemer's fescue (*Festuca idahoensis* ssp. *roemeri*), Sandberg bluegrass (*Poa secunda*), Scribner panicum (*Panicum Scribnerianum* Nash), prairie June grass (*Koeleria macrantha*), Lemmon's needlegrass (*Achnatherum lemmonii*), wheatgrass (*Elymus trachycaulus*), California oatgrass (*Danthonia californica*). There are hundreds of herbaceous perennials native to Oregon and the prairies included: asters, penstemons, buttercups, figworts, larkspurs, and some drought tolerant grass-like herbs such as western blue-eyed grass (*Sisyrinchium bellum*). Upland prairies covered about two-thirds of the prairies in the valley before 1850 (Pojar and MacKinnon 2004; Christy and Alverson 2011).

Wet prairies covered about one-third of the valley lowlands before 1850. Much of the wet prairie habitat was associated with clay soils and floodplains. Dominant species included: tufted hairgrass (*Deschampsia cespitosa*), rushes (*Juncus* spp.), and sedges (*Carex* spp.). There was also much diversity of perennial and annual forbs growing in upland and wet prairies, and camas. Camas (*Camassia quamash* ssp. *Maxima* and *Camassia leichtlinii* ssp. *suksdorfii*) grew in great numbers, especially where the native cultures cultivated camas for food (Christy and Alverson 2011). In the wet prairies remaining today, studies in the Fern Ridge research natural area, found, for example, at the Rose Prairie site, three different communities dominated by either *Deschampsia cespitosa* and *Danthonica californica*, *Vaccinium cespitosum*, or *Anthoxanthum odoratum* and *Rosa nutknaa* (Finley 1994; Connelly and Kauffman 1991).

9.1.1.3 Succulents

In western Oregon, succulents grow where shallow soils exist along the Columbia River Gorge or near edges of rocky landscapes. One study of the vegetation of McCord Creek Falls area found over 167 taxa of plants growing in a variety of forested, rocky, open, dry, and wet habitats (Harvey 2012). Of these, some are adaptable to green roofs, such as *Sedum spathulifolim* and *Sedum oreganum*. The Columbia River Gorge may have other succulent or drought-adapted species that have yet to be trialed on green roofs (Franklin and Dyrness 1988).

Sedum divergens and *Sedum lanceolatum* are native to western Oregon. Prickly pear (*Opuntia fragilis*) grows in southern, central, and northeastern Oregon and may adapt to green roofs with dry or exposed microclimates in parts of the valley (see Sect. 9.3.3) (Pojar and MacKinnon 2004).

9.1.1.4 Mountain Meadows

Meadows on the slopes of the Cascades and open areas along the Columbia River Gorge are another location where vegetation may found and may be adaptable to green roofs. Some of the plants that may be useful to roof meadows and important for conservation purposes include purple milkweed (*Asclepias cordifolia*) monkey-flower (*Mimulus guttatus*) paintbrush (*Castilleja hispida*) yellow Oregon sunshine (*Eriophyllum lanatum*), creamy white northern buckwheat (*Eriogonum compositum*), pink farewell-to spring (*Clarkia amoena*), fleabane (*Erigeron inornatus*), Davidson's Penstemon, (*Penstemon davidsonii*), and Cardwell's penstemon (*Penstemon cardwellii*). These are members of meadow communities in the mid-to-lower elevations of the Cascades and grow in association with grasses and other herbaceous plants in rocky or dry sloped habitats, or wet habitats. Evergreen groundcovers such a juniper (*Juniperus communis*) and kinnikinnick (*Arctostaphylos uva-ursi*) are adapted to dry habitats and grow on well-drained habitats in the Columbia River Gorge and exposed sites found along the mountains (Pojar and MacKinnon 2004).

Several online resources allow one to search by plant family, genera, species, listing and location of where the plant has been observed. The *Flora of Oregon* online resource allows multiple functions and is maintained by Oregon State University. *Mountain Plants of the Western Cascades* is another online resource with locations of mountain meadows in the Cascade Range and the Columbia River Gorge. This site provides plant lists and blogs about native vegetation of these habitats. The author is Tanya Harvey, an amateur botanist, and photographer that has contributed to the *Flora of Oregon* and many other publications.

Natural vegetation throughout the valley today consists of a much-reduced matrix, with prairie taking a small proportion to farms, ranches, developed land, and forest preserves. The quality of prairies suffers from a widespread infusion of invasive species. Some prairie preserves in the Willamette Valley have become dominated by non-native plants including reed canary grass (*Phalaris arundinacea*), roughstalk bluegrass (*Poa trivialis*), Himalayan blackberry (*Rubus discolor*), nipplewort (*Lapsana communis*), English ivy (*Hedera helix*) and bittersweet nightshade (*Solanum dulcamara*). These are very aggressive species and are extremely difficult to control once they have invaded an area (Titus et al. 1996).

9.1.2 *Conservation Site Case Studies (Arranged North to South)*

9.1.2.1 McCord Creek Falls, Cascade Locks, Oregon

The Columbia River Gorge is an incredibly biologically rich and diverse region. It was formed by the cutting of the Columbia River, glaciers, and uplift of the Cascade Mountains over millennia. On the western side of the Cascades lies temperate rain forests, and east of the Cascades lies semi-arid landscapes. The Columbia River Gorge is a linear landscape that bridges these two contrasting wet and dry landscapes. There are over 800 species of wildflowers growing in the Gorge; 15 are endemic to the Gorge (CRG 2020).

The McCord Creek Falls lies midway between the wet and dry habitats and is generally north-facing slopes. One study of the flora of the McCord Creek Falls area found over 167 taxa of plants growing in a variety of forested, rocky, open, dry and wet habitats in the vicinity of the upper and lower falls (Harvey 2012). Some of the plants on the list that may translate to green roofs include yarrow (*Achillea millefolium*), nodding onion (*Allium cernuum*), shining Oregon grape (*Berberis aquifolium*), Cascade Oregon grape (*Berberis nervosa*), Oregon sunshine (*Eriophyllum lanatum*), salal (*Gaultheria shallon*), Cascade penstemon (*Penstemon serrulatus*), spreading phlox (*Phlox diffusa*), western sword fern (*Polystichum munitum*), Oregon stonecrop (*Sedum oreganum*), broadleaf stonecrop (*Sedum spathulifolium*).

There are many natural areas in the Gorge and are worthwhile to investigate plants and their microhabitats they thrive. The north side of the gorge (Washington State) has many south-facing slopes, with sunny and drier habitats compared to the north-facing Oregon side. Plants that thrive in the gorge are exposed to persistent wind, wildfires, and rapid changes in temperature.

9.1.2.2 Leach Botanical Garden, Portland, Oregon

Since 1983, the Leach Botanical Garden has provided a setting for a collection of over 2000 hybrids, cultivars, native and non-native plants. Named after John and Lilla Leach, the 6.5-hectare (16-acre) garden features a large rock garden collection which includes native sedum, allium, and drought-tolerant groundcovers (Fig. 9.4). Collections include a Willamette Valley Savanna Garden that features vegetation native to the prairies and oak openings.

9.1.2.3 Camassia Natural Area, West Linn, Oregon, the Nature Conservancy

This 10-hectare (26-acre) remnant prairie barrens and oak and wooded landscape is named for the common camas bulb (*Camassia quamash*) which blooms each year in the spring (Fig. 9.5). There are over 300 plant species native to the preserve,

Fig. 9.4 Native Sedum, groundcovers, and other drought-tolerant vegetation growing on a west-facing rock outcrop inside the gardens. (Photo: Bruce Dvorak, September 2018)

Fig. 9.5 (**a**) The Nature Conservancy is working with local high school students who help maintain the conservation site. Rosy plectritis and camas lilies are in full bloom during early May 2019. (**b**) An open grassy barren at the Camassia Natural area during September. This image shows the fall colors of the native grasses and the dark-colored native basalt bedrock. (Photos: (**a**) courtesy of Jennifer L. Morse, (**b**) Bruce Dvorak, September 2018)

including several rare Willamette Valley species such as the white rock larkspur (*Delphinium leucophaeum*) which occurs fewer than six other places in the world (CNA 2019). Other species growing that are abundant in the open prairie include blue-eyed Mary (*Collinsia grandiflora*), Rosy plectritis (*Plectritis congesta*), and species of buttercups.

Because the basalt rock under the preserve is fractured, the land is not developable. The site is surrounded by urban development, but the development stops at the border of the preserve that the Nature Conservancy acquired in 1962. This prairie is one of the few remnant prairies in the Portland metropolitan area. The preserve has several open habitats with vegetation that may be relevant to a variety of kinds of green roofs: meadows, herbaceous balds, oak woodlands, and rocky bluffs.

9.1.2.4 Kingston Prairie Preserve

The Kingston Prairie Preserve is one of the last remaining remnant prairies in the Willamette Valley. Located outside Stayton, Oregon, the 62.7-hectare (155-acre) conservation site is managed by Greenbelt Land Trust (Fig. 9.6). Because of the shallow soils and the underlying basalt rock, the land is not suited to development and was initially donated to The Nature Conservancy. Today, naturalists study the

Fig. 9.6 Located in a rural landscape outside of Stayton, Oregon the 155-acre preserve is one of the largest remnant upland prairies in the state. Blooming plants include saxifrage (white), shooting star/dodecatheon (pink), and camas (purple). (Courtesy of Greenbelt Land Trust, Corvallis, Oregon)

site for the habitat of the western meadowlark and other songbirds known to be in decline in the Willamette Valley due to loss of habitat. Rare plant populations are monitored by ecologists, and teams of volunteers work to remove the invasive Scots broom and Himalayan blackberry from the preserve. Some of the rare plants include the federally endangered Bradshaw's lomatium and Willamette daisy. The preserve is part of a larger collection of preserves that are in the process of becoming integrated and protected (KPR 2019). Guided tours can be arranged.

Other prairie preserves in the valley near urban regions that can be visited include Sauvie Island-Scappoose (Portland), Dundee Oaks (Yamhill), Eloa Hills (Salem), Finley-Muddy Creek, Salem Hills-Ankeny (Salem), West Eugene Area (Eugene) (OCS 2019).

9.2 Green Roof Research in the Willamette Valley

Oregon has a long history of research regarding vegetation for extensive and semi-intensive green roofs. The City of Portland Bureau of Environment Services conducted studies of vegetation established on some of Portland's first ecoroofs (green roofs) (Hauth and Liptan 2003). Termed "ecoroofs" in Portland, their research highlighted the mixture of native and exotic succulents on several extensive green roofs. Early vegetation studies identified some promising native species for use on green roofs such as hardy iceplant, 'Cape Blanco' broadleaf stonecrop (*Sedum spathulifolium*), and the bulb small camas (*Camassia quamash*). For the other taxa, plant survival and growth generally decreased with decreasing irrigation and many species did not survive at all without irrigation, in shallow substrates.

A study about the habitat value of native vegetation grown on a variety of green roof substrate types (engineered, native soil) was compared to natural sites in the Portland area. The green roofs were producing equally, and more so than the natural sites for many research parameters including abundant blooms, insects, and species richness (Wood 2018).

Researchers at Portland State University investigated the community composition of green roofs designed for habitat as compared to ground sites and more streamlined roofs designed expressly for stormwater management. Their study used beetles as indicator organisms and encompassed six roof sites and five ground sites. Researchers found that ground sites, habitat roofs, and stormwater roofs each grouped distinctively in terms of beetle community composition and that biodiversity was positively correlated with roof age, percent plant cover, average plant height, and plant species richness (Gonsalves 2016; Starry et al. 2018).

Much of the initial stormwater research on green roofs came from Portland and Corvallis when low impact development practices were emerging (Liptan 2017). Green roofs were found to reduce the volume and initial flush of water during storm events (Hutchinson et al. 2003; Kurtz 2008; Spolek 2008). Stormwater retention rates for summer rain events on three ecoroofs ranged from 7% to 85% (average of 42%). During wintertime when many of the plants used on the ecoroof are dormant, retention ranged from 0% to 52% with an average of 12% (Spolek 2008). Corvallis is near the central valley and several studies coincide with findings in Portland. At the watershed level, a study in Corvallis looked at the role of green roofs in combination with other LID features (e.g. porous pavements) that could play a significant role in the long-term stormwater management for urban watersheds (Zhou et al. 2018).

Green roofs were monitored for their capacity to reduce energy, and cool rooftops is another common ecosystem service from green roofs. Portland State University and the City of Portland was one of the early leaders in studying microclimates on green roofs. (Sailor et al. 2008; Scherba et al. 2011).

One of the important topics regarding the installation of green roofs is their sustainment over time and their ability to meet the goals and objectives of the building owner and designer. One of the first in-depth investigations into these issues took place in Portland. With over nearly 400 green roofs in Portland by 2016, there was little knowledge of how green roofs have performed. Thurston, investigated these questions conducted a survey of 52 green roof owners and managers and found that there were some unmet expectations regarding installed green roofs on approximately 9% of mostly private installations. Common issues included a failure or partial failure regarding the owner's expectations for appearances and maintenance expectations. These unmet expectations included maintenance (lack of knowledge), poor plant establishment or performance (plant dieback), aesthetics (did not understand seasonal variations of plants) and issues with irrigation/drainage (was told irrigation was not needed/or that drains could become clogged) (Thurston 2017). This important follow-up study on how owners feel about green roofs is a needed process to grow and educate the green roof industry.

9.3 Ecoregional Green Roof Case Studies (Arranged North to South)

9.3.1 Multnomah County Building (Amy Joslin Eco Roof), Portland, Oregon

One of the characteristics of a successful green roof is one that receives persistent attention and editing over time. The Amy Joslin Eco Roof has been planted, partially replanted, and edited with new planting additions over time. Today, the roof meadow symbolizes the goals of the City of Portland has for its urban district (Fig. 9.7). In June of 2018, the City of Portland passed a mandate for green roofs to be used on new construction on green roofs in its central district (Hamrick 2018). The mandate encourages climate-adapted vegetation. The Multnomah County Building green roof has made an extensive exploration of the use of native, naturalized, and exotic vegetation.

The substrate is 15 cm (6 in) deep and is composed of pumice, perlite, organic matter, and paper pulp. Supplemental watering of the vegetation is provided during dry periods through a buried drip tube irrigation system.

Fig. 9.7 View of the roof meadow looking towards downtown Portland. (Photo: Bruce Dvorak, September 2018)

9.3.1.1 Project Team

Building Owner/Client: Multnomah County
Architect: Carlton Hart Architecture
Structural Engineer: KPFF Consulting Engineers
Landscape Architect: Macdonald Environmental Planning
Green Roof consultant: Soprema (Marie-Anne Boivin)
Project completion: June 2003
Green roof area: 1104 m^2 (11,893 ft^2) of vegetated area

9.3.1.2 Overview and Objectives

The purpose of the retrofitted green roof was to provide an accessible green roof pilot project as a venue where the county, city, and local researchers could monitor green roofs and learn about their functions, challenges, and benefits. The previous roof covering was a gravel ballasted roof, without plantings. The County and City wanted to build an accessible roof, the intention was to make the learning and outcomes available and in participation with professionals.

9.3.1.3 Plant Establishment

The green roof was designed as a meadow with a mixture of native and exotic vegetation. Forms of plants include grasses, forbs (wildflowers), bubs, and succulents. Some of the prairie species planted are native to the tallgrass prairies of the central U.S. including plains coreopsis, black-eyed Susan, lance-leaved coreopsis, and prairie coneflower. Plants native to the Portland area include Cape Blanco stonecrop, and purple stonecrop. Plant installation was accomplished with hydroseeding. Several meadow mixes were first installed and included some non-native (European) flowering perennials. Over the years, additional plantings included interplanting of vegetation native to the Willamette Valley.

A number of invasive species have become self-seeded onto the roof including a non-native clover, and native fireweed, which both sustains a presence. Other additions planted onto the roof include lupines and other annuals. Today the roof meadow is a mixture of planted species, additive species (wind, birds), and additional species added for color after the initial planting (Fig. 9.8). The result is a colorful grass and wildflower green roof that maintains one of the best views of downtown Portland.

Fig. 9.8 A cluster of western pearly everlasting (*Anaphalis margaritacea*) is native to the Willamette Valley and grows near the accessible viewing deck (white bloom) with a few tall spikes of native fireweed (*Chamerion angustifolium* or *Epilobium oreganum*) rising above to the left. The fireweed was self-seeding but is welcome as it is not an aggressive species on this roof. (Photo: Bruce Dvorak, September 2018)

Annuals

baby blue eyes (*Nemophila menziesii*), black-eyed Susan (*Rudbeckia hirta*), California poppy (*Eschscholzia californica*), Godetia (*Clarkia amoena*), mountain phlox (*Linanthus grandiflorus*, native to California), plains coreopsis (*Coreopsis tinctoria L.*).

Herbaceous Perennials

eastern purple coneflower (*Echinacea purpurea*, native to the eastern U.S.), goldsturm rudbeckia (*Rudbeckia fulgida* 'Goldsturm', native to the eastern U.S.), Indian blanketflower (*Gaillardia aristate*), lance-leaved coreopsis (*Coreopsis lanceolata*), Missouri primrose (*Oenothera missouriensis*, native to the southcentral U.S.), pink primrose (*Oenothera speciosa*, native to the southern U.S. and California), prairie coneflower (*Ratibida columnifera*, native to the central U.S.)

Succulents

broadleaf stonecrop, Cape Blanco (*Sedum spathulifolium* 'Cape Blanco'), broadleaf stonecrop, purple (*Sedum spathulifolium* 'Purpureum').

9.3.1.4 Irrigation

The subsurface drip irrigation system is used from April to October as needed during dry periods. Watering takes place twice daily at 6:00 am and 8:00 pm for about 10 min each time.

9.3.1.5 Maintenance

As a pilot project established for public viewing, the roof meadow receives regular maintenance depending upon the season. During the active growing season, visits can take place monthly or bi-weekly as needed.

9.3.1.6 Observed Wildlife

Some of the observed wildlife on the roof include grasshoppers, bees, butterflies, and a peregrine falcon.

9.3.1.7 Best Performing Native Vegetation

grasses, native sedums, lupines, blanket flower, yarrow, coreopsis, western pearly everlasting, alliums.

9.3.1.8 Post-occupancy Observations

Design Team

- We did some planting changes over the years (overseeding wildflower zones, replacing larger ornamental grasses), as well as developing and refining operations & maintenance info.

Authors' Reflections

- This project on a public building is very accessible, provides excellent views of Portland, and is likely the largest patch of a meadow habitat with some native plants in the downtown area. It is a beautiful roof garden, and with its many edits over the years, it functions as a place for the City and County to learn and explore meadow-based green roofs.
- With the green roof policy now effective in Portland, experimentation with additional plant species and watering practices could yield helpful information for building owners looking to establish grass-based green roofs to downtown Portland.

9.3.2 Gunderson's Habitat Ecoroofs Portland, Portland, Oregon

Greenbrier/Gunderson is a railcar production company that is located on the Willamette River in Portland, Oregon. The company has a long history of environmental roots and has a mission to implement green roofs on their facilities. Gunderson teamed with the Portland Green Roof Think Tank (GRiT) to come up with designs for green roofs on four of Gunderson's sites in Portland (Fig. 9.9). Senior Vice President Mark Eitzen discusses Gunderson's commitment to green, "This roof is expanding our knowledge of what we can do on-site to improve habitat and the quality of our stormwater discharge through use of habitat roofs. Our work here builds on the foundation of knowledge developed through the nationally recognized Ecoroof Program at the City of Portland's Bureau of Environmental Services (BES)." (CPRN 2012). This statement was made during the first phase of building green roofs, since then, Gunderson continued to construct four habitat roofs on open-air structures in Portland. Many concepts of biodiverse roofs were conceived and implemented into the construction of the ecoroofs (Fig. 9.10).

Fig. 9.9 Images of four Gunderson's habitat roofs located on open-air structures. (**a**) site 1, (**b**) site 3, (**c**) site 2, (**d**) site 4. (Photos courtesy of Desirae Wood, Dobro Design)

9.3.2.1 Project Team

Building Owner/Client: Greenbrier/Gunderson
Green Roof Design Team Lead: GRiT (Desirae Wood)
Landscape Architect: Dobro Design
Environmental Specialist: Dave Harvey, Soren Hill, Gunderson
Installation Contractor: Gunderson & GRiT
Maintenance Contractor: Gunderson
Project completion: spring 2011and 2014
Green roof area: 284 m^2 (3056 ft^2) of vegetated roofs on four rooftops. Two roofs are about 100 m^2 each and two roofs are about 50 m^2 each roof.

9.3.2.2 Overview and Objectives

As habitat roofs, the green roofs were designed by a multi-team effort. The GRiT organization took the design lead for the initial projects setting the tone for the roofs. Desirae Wood was a design leader in the project. The team developed a list of goals for the habitat roof. The highlights of these goals include:

- Provide food sources for pollinators from spring to fall.
- Minimum of 15–25 flower species present.

Fig. 9.10 Installation of one of the habitat roofs shows the driftwood logs, bamboo canes, coconut coir wind blanket, and mounding of substrate spring 2011. (Photo courtesy of Desirae Wood, Dobro Design)

- Minimum of three plants blooming at one time.
- Clump plants together so the insects can see them.
- For each plant, place in large swaths that overlap other swaths.
- 45% - 85% forbs to maximize the bee populations.
- Use short, warm-season grasses that can fall over to create habitat for birds.
- Install plants using a mixture of seed, cuttings/plugs, and containers (for larger plants).
- Design plant communities with a mixture of ~50 different native seed types. Die-off will be ~90% but will self-select for the site.
- Nesting habitat (snags/logs, bare ground, holes, piles of brush, tunnels, diversity of clusters), bright vs. indirect light, irregular shapes, varying depths.
- Maintenance is a once a year trim. By taking away biomass only once a year, organic matter can build on the roof.
- Vary substrate depths – include small hills, three depths (3, 4.5, and 6 inches or 7.6, 11.4, and 15.2 cm).
- Gravel can be too dry, need to add some topsoil/compost – even 4 cm for water retention is good.
- No irrigation, no fertilizer. (Wood 2011)

Several seed mixes were purchased from local seed vendors. The seed mix has a high percentage of drought-tolerant wildflowers and legumes (65%) mixed with native grasses (28%) and woodies (7%) (Wood 2018). Seeding was done during the spring and germination was successful for all four green roofs. The intention was that the green roofs would reflect the native grass habitats in the valley and go dormant during the summer (Fig. 9.11).

Site #1 Constructed in 2011, this green roof has an engineered substrate with two substrate designs. One substrate has varied depths from 2.54 cm to 15.24 cm (1 to 6 in). A local ecoroof soil company provided the growing medium and the composition was 24% sand/vermiculite, 52% loam/organic matter, and 24% clay. This roof features two raised mounds (polystyrene) with three bundles of bamboo rods buried into the berms (Wood 2018).

Site #2 Constructed in 2011, this green roof has native soil substrate with varied depth from 2.54 cm to 15.24 cm (1 to 6 in). The soil is composed of 82% sand, 14% loam matter, and 4% clay. This roof also features two raised mounds (polystyrene) with three bundles of bamboo rods buried into the berms (Wood 2018).

Site #3 Constructed in 2013, this 92 m^2 (990 ft^2) green roof has a single ridge in the middle with east and west-facing aspects. Substrate depths range from 8.9 cm to 16.5 cm (3.5–6.5 in) and was provided by a local ecoroof soil company. Soil composition was 50% sand/vermiculite, 33% loam/organic matter, and 17% clay. This roof has raised mounds made from polystyrene and driftwood logs placed on the

Fig. 9.11 A habitat roof about 6 years after planting, shown at the end of the summer dormancy. (Photo: Bruce Dvorak, September 2018)

roof. This roof was planted with sedum, native grasses, wildflowers, and bulbs. (Wood 2018).

Site #4 Constructed in 2014, this is the largest green roof at the industrial sites 94 m^2 (1015 ft^2) is constructed from metal trays, with a substrate of 12.7 cm (5 in) on several with several raised mounds made from polystyrene. This north-facing roof was seeded with Sunmark Seeds international, *Native Pollinator Mix,* and *EcoPrairie Mix*. The growing medium is 57% vermiculite/sand, 29% loam/organic matter, and 14% clay (Wood 2018).

9.3.2.3 Plant Establishment

Vegetation was established through seeding and plugging. Forms of vegetation include annuals (Table 9.1), grasses, herbaceous perennials (Table 9.2), shrubs, and succulents.

Table 9.1 Native annual wildflowers on the Gunderson's habitat roofs

Common Name	Botanical Name
American bird's-foot trefoil	*Lotus purshianus*
baby blue eyes	*Nemophila menziesii*
blackeyed Susan	*Rudbeckia hirta*
California brome	*Bromus carinatus*
California poppy	*Eschscholzia californica*
coastal tidytips	*Layia platyglossa* (native to SW USA)
common woolly sunflower	*Eriophyllum lanatum*
Douglas' meadowfoam	*Limnanthes douglasii*
elegant clarkia	*Clarkia unguiculate* (native to CA)
sicklekeel lupine	*Lupinus albicaulis*

Table 9.2 Herbaceous perennial vegetation planted on Gunderson's habitat roofs

Common Name	Botanical Name
aspen fleabane	*Erigeron speciosus*
bigleaf lupine	*Lupinus polyphyllus*
bitter root	*Lewisia rediviva*
blanketflower	*Gaillardia aristata*
common yarrow	*Achillea millefolium*
Davidson's penstemon	*Penstemon davidsonii*
Douglas aster	*Aster subspicatus*
large camas	*Camassia leichtlinii*
mosquito bills	*Dodecatheon hendersonii*
riverbank lupine	*Lupinus rivularis*
seaside fleabane	*Erigeron glaucus*
small camas	*Camassia quamash*
strawberry	*Fragaria spp.*
toughleaf iris	*Iris tenax*
western columbine	*Aquilegia formosa*

Grasses

Idaho fescue (*Festuca idahoensis*), meadow barley (*Hordeum brachyantherum*), prairie june grass (*Koeleria macrantha*), red fescue (*Festuca rubra*), squirreltail (*Elymus elymoides*), tufted hairgrass (*Deschampsia cespitosa*), western fescue (*Festuca occidentalis*).

Shrubs

kinnikinnick (*Arctostaphylos uva-ursi*), oceanspray (*Holodiscus discolor*).

Succulents

broadleaf stonecrop, purpureum (*Sedum spathulifolium* 'Purpureum'), Oregon stonecrop (*Sedum oreganum*), Pacific stonecrop (*Sedum divergens*).

9.3.2.4 Irrigation

Supplemental watering varies across the roofs. Some roofs had periodic watering during summer dry spells and some roofs received no additional watering.

9.3.2.5 Maintenance

Maintenance varies across the roofs. Some of the roofs receive annual maintenance and some roofs have not had maintenance. One plant community was overtaken by mini lupine which died out following an aphid infestation and has now transitioned to mostly stunted grasses,

9.3.2.6 Observed Wildlife

Observed wildlife includes ladybugs, bees, wasps, and evidence of geese (Wood 2018). A worker once saved a baby duck that was in danger of falling off one of the roofs.

9.3.2.7 Best Performing Native Vegetation

Across the four green roofs, prominent blooms consisted of blue button, blanket flower, alyssum, elkhorn clarkia, and Camas bulbs. Grasses dominate most of the habitat roofs and *Sedums* cover exposed areas on green roofs where grasses are not dominant.

9.3.2.8 Post-occupancy Observations

Published (Wood 2018)

- Site #2 had the greatest richness at 48 plant types, greatest bloom type richness at 26%, and bloom abundance at 133,937 blooms. It had the supplemental irrigation during seed establishment. This site also had the greatest number of insects (32) and wildlife sightings. This site also had the coolest soil temperatures and better retention of moisture.
- Site #3 with the native soil compacted from rainfall over time. The soil compaction resulted in lower diversity but greater green cover with grasses.

Authors' Reflections

- These roofs demonstrate an exemplar set up as a way to investigate ecoregional green roofs. What has been learned and reported is already valuable. Future studies could follow up with the current design, or trial new plant additions.
- The contributions and collaboration between public/private and volunteer participants are rare and demonstrate model relationships for growing ecoregional green roofs.

9.3.3 Performing Arts Center, Reed College, Portland, Oregon

Reed College has a broad mission to explore and learn about sustainability. It seeks to apply and implement sustainable development projects on and off-campus. When it came time for the construction of a new Performing Arts Center on campus, this allowed for the opportunity to include an ecoroof (Fig. 9.12). The new building had a west-facing canopy over the building entry. An outdoor seating area was made adjacent to the ecoroof so that it can also be viewed from within the building's public spaces.

9.3.3.1 Project Team

Building Owner/Client: Reed College, Portland, Oregon
Green Roof Design Team Lead: Mayer/Reed
Architect: Opsis Architecture LLP
Landscape Architect: Mayer/Reed
Installation Contractor: Teufel Landscape
Maintenance Contractor: Reed College's grounds crew
Project completion: July 2013
Green roof area: 255 m^2 (2750 sf^2)

Fig. 9.12 A mosaic of native grasses, sedums, and cactus pads (background) is shown here, as planted in 2013. The photo was taken 1 year after planting in mid-May. This green roof has a dramatic seasonal aesthetic, as the bright green and blooms fade to fall colors during the dry season (Fig. 9.13). (Courtesy of Mayer/Reed)

9.3.3.2 Overview and Objectives

Instrumental to the implementation of a living roof was the inclusion of the Biology Department's students, faculty and staff. The diverse group worked together with the design team to help select plants, to set up strategies to test plant varieties and soil, and set up measures to monitor environmental factors throughout plant life cycles. The concept was to select native vegetation or cultivars that would adapt to a low watering maintenance approach. Thus, succulents native to the Cascades, and eastern Oregon were pared together to grow in a full sun microclimate. A cultivar of Idaho fescue ('Siskiyou Blue') was selected to grow with the succulents. Idaho fescue is native to western and central Oregon and survives the dry and sometimes hot summers through dormancy. On this roof, combined with a low-watering approach, the fescue goes dormant as well (Fig. 9.13) (OA 2013).

Fig. 9.13 (**a**) *Opuntia polyacantha* and *Opuntia fragilus* grow with *Sedum divergens* and *Festuca idahoensis* on an exposed portion of the green roof. This photograph was taken 5 years after installation, near the end of summer dormancy. The dormant grasses rejuvenate their growth each year when the fall rains return. (**b**) western sword fern remains green throughout the year in a shaded corner of the roof deck. (Photos: Bruce Dvorak, September 2018)

9.3.3.3 Plant Establishment

Landscape Architect Mayer-Reed worked with the Department of Biology, students, Joy Creek Nursery, and American Hydrotech to develop an all native plant suite that would be suited for the roof's western orientation and local climate. The Biology Department participants suggested trialing some less commonly used plants on green roofs in Portland: *Lewisia* and *Opuntia*. Plants were acquired as 10-cm (4-in) pots, plugs, and cuttings. Opuntia pads were laid out onto the 15 cm-deep (6 in) growing medium for self-establishment. Western sword fern was selected for shady locations, in the shade of the building (Fig. 9.13). The 15-cm (6-in) deep blended substrate was provided by America Hydrotech.

Grasses

Idaho fescue (*Festuca idahoensis* 'Siskiyou Blue').

Herbaceous Perennials

Columbian lewisia (*Lewisia columbiana*), Siskiyou lewisia (*Lewisia cotyledon* 'Praline'), western sword fern (*Polystichum munitum*).

Succulents

brittle pricklypear (*Opuntia fragilis*), broadleaf stonecrop, Cape Blanco (*Sedum spathulifolium pruinosum* 'Cape Blanco'), broadleaf stonecrop, Carnea (*Sedum spathulifolium pruinosum* 'Carnea'), broadleaf stonecrop, purpureum (*Sedum spathulifolium pruinosum var. purpureum*), Oregon stonecrop (*Sedum oreganum*), Pacific stonecrop (*Sedum divergens*), plains pricklypear (*Opuntia polyacantha*), spearleaf stonecrop (*Sedum lanceolatum*).

9.3.3.4 Irrigation

The overhead spray irrigation system uses non-potable water from Reed Lake. Water is utilized during periods of drought. The on-campus lake is fed by Crystal Springs Creek. The irrigation system is pressurized from April/May through September. Campus-wide irrigation scheduling is modified daily using calculated evapotranspiration data, including the ecoroof.

9.3.3.5 Maintenance

The ecoroof is maintained by the ground crews at Reed College. The native grasses are cut back seasonally to maintain a more manicured appearance and to avoid thatch build-up.

9.3.3.6 Observed Wildlife

Plant diversity has become reduced through the years since it was planted. Plants that can survive the amended soils, high temperatures, have increased in population. Bees visit the Yarrow, and songbirds collect the fragments of grasses for the next building. No butterflies have been observed. There have been some bird nests and offspring sightings through the years.

9.3.3.7 Best Performing Native Vegetation

Native grasses, sedums, cactus, yarrow, western sword fern (in shaded areas).

9.3.3.8 Post-occupancy Observations

Owner

- Decisions made early in the process regarding the quality of materials paid off. The entire roofing system was provided by qualified installers American Hydrotech & Teufel Landscape.
- The design team worked with American Hydrotech and Joy Creek Nursery to specify indigenous plant species and the sizes that would be locally available based on project schedule and scale.
- The landscape architect worked alongside construction crews to customize the planting layout and made multiple site visits to ensure the plants were properly irrigated and cared for in order to maximize establishment. (OA 2013)

Maintenance Staff

- We added yarrow in the seed mix after the first planting. We had plant failures after issues with the irrigation system. We revisited the plant list and replanted it according to its preferred microclimate.
- The plant media dries out rapidly when temperatures are above about 29.4 °C (85 °F) in typical local summer humidity conditions and require up to 3 times the water of equivalent full-sun turfgrass in native soils in these conditions. Because the ecoroof irrigation zones were installed using PVC pipe with no low-point for drainage, and due to the open (unheated) area under the roof, the system regularly suffers from winter freeze damage.

Authors' Reflections

- This project exhibited a collaborative design process, where the design team, university faculty, and the green roof providers all came together. The approach to design towards microclimates on the roof, and select locally native and regionally native vegetation is commendable.
- Future updates could be warranted to revisit the irrigation system, and perhaps enhance the habitat features on the roof.

9.3.4 Columbia Building Wastewater Treatment Support Facility, Portland, Oregon

Wedged between the ecologically rich Columbia Slough habitats (Smith and Bybee Wetlands Natural Area), an active railroad line, and North Columbia Boulevard, the City of Portland had few choices for locating a needed expansion of its offices and service facilities. In addition, the site aligns with a 40-mile loop bike trail. The new services building would be in the public eye, and the design team saw an opportunity to offer a new approach to how a highly visible facility located in an industrial site could be attractive and educational to employees and visitors. Green roofs and earth embanked walls became the main features of the project to define the appearance and function of the building inside and out (Fig. 9.14).

9.3.4.1 Project Team

Building Owner/Client: Portland Bureau of Environmental Services
Green Roof Design Team Lead: 2.ink Studio Architect: Skylab Architecture

Fig. 9.14 Vegetation at the ground level slopes up to partially shelter the building walls and mitigate building temperatures and enframe the entrance. The earth dampens noise from train traffic on the adjacent railroad tracks (left of photo). The southeast sloped extensive green roofs support native grasses and a mix of other native plants. (Photo: Olyssa Starry, February 2020)

Landscape Architect: 2.ink Studio
Project completion: 2013
Green roof area: 1117 m^2 (12,030 ft^2) planted area, 1623 m^2 (15,473 ft^2) green roof total

9.3.4.2 Overview and Objectives

The landscape architects conceived of a concept for the site and building to connect with the adjacent Columbia Slough ecosystem. Native plant communities were referenced for the planning of the landscape and rooftop vegetation, which was planted/seeded into a 15-cm-deep (6-in) substrate provided by Sunmark Seeds. A rail line borders the property and the landscape was made to berm up along the rail side of the building to buffer noise, reduce heat gain, and allowing for a patch of native vegetation alongside the facility (Fig. 9.15). As the "face" of the facility, the building aesthetic needed to not only look natural but also maintained.

Fig. 9.15 Earth embanked slopes (foreground) and green roofs (background) at the offices of the Columbia Building Wastewater Treatment Support Facility. Wastewater treatment facilities are visible in the far background. (Photo: Bruce Dvorak, September 2018)

9.3.4.3 Plant Establishment

Succulents were established by cuttings, and herbaceous grasses, annuals, and forbs were established from seed.

Annuals

bicolor lupine (*Lupinus bicolor*), Oregon sunshine (*Eriophyllum lanatum*).

Grasses

bottlebrush squirreltail (*Elymus elymoides*), Indian ricegrass (*Oryzopsis hymenoides*), prairie junegrass (*Koeleria macrantha*), sandberg bluegrass (*Poa sandbergii* var. *secunda*), slender wheatgrass (*Elymus trachycaulus*).

Herbaceous Perennials

blanketflower dwarf (*Gaillardia aristata* dwarf), camas (*Camassia quamash*), nodding onion (*Allium cernuum*), prairie coneflower dwarf (*Ratibida columnifera* dwarf), western yarrow (*Achillea millefolium*).

Succulents

broadleaf stonecrop (*Sedum spathulifolium*), Oregon stonecrop (*Sedum oreganum*), Pacific stonecrop (*Sedum divergens*).

9.3.4.4 Irrigation

Overhead spray irrigation was installed and used during the establishment period for the first 2 years.

9.3.4.5 Maintenance

The green roofs and slopes are mowed once a year around June or July, occasionally twice per year. Mowing can interfere with building HVAC. It should be shut off during the process and clippings raked and removed. The grasses are mowed during the summer dormant season, as a preemptive measure to suppress any risk of fire.

9.3.4.6 Observed Wildlife

Ducks, herons, geese use the green roofs.

9.3.4.7 Best Performing Native Vegetation

Grasses and western yarrow.

9.3.4.8 Post-occupancy Observations

Owner

- The green roofs create a lot of visual interest and the location is prime for attracting wildlife. It's a little more challenging to manage compared to landscaping on the ground and developing the maintenance protocol has been a learning curve. Fortunately, there are a lot of resources in Portland, that can help troubleshoot any issues that may arise especially the Bureau of Environmental Services and Tom Liptan.

Authors' Reflections

- This project demonstrates the integration of site, building, its context to public viewing, and its relationship to the Columbia Slough. It is one of the first native grass-based green roofs in Portland.

9.3.5 Multnomah County Central Library, Portland, Oregon

Architect A. E. Doyle designed the Multnomah County Central Library. It opened on September 6, 1913, and was one of the first open-plan libraries in the United States. The roof membrane has been re-roofed many times over the years; however, in 2008 a green roof was installed to demonstrate shallow extensive green roofs for the public, and become the first library in Oregon with a green roof (Fig. 9.16). The green roof was paid for with grants from the Oregon Department of Environmental Quality and one from the City of Portland's Green Investment Fund.

9.3.5.1 Project Team

Building Owner/Client: City of Portland, Oregon
Green Roof Design Team Lead: Macdonald Environmental Planning

Fig. 9.16 Extensive green roof on the Multnomah County Central Library with massings of native and exotic *Sedum*. (Photo: Bruce Dvorak, September 2018)

Architect: Carleton Hart Architects
Landscape Architect: Macdonald Environmental Planning
Installation Contractor: Teufel Landscape
Tray system: Columbia Green Tech. Inc.
Project completion: September 15, 2008
Green roof area: 667.8 m^2 (7188 ft^2)

9.3.5.2 Overview and Objectives

The green roof was designed to showcase extensive green roof technology to the public, educate about the environmental benefits and maintenance requirements for green roofs. By constructing the green roof, the city aimed to cut energy costs by 6–8% in summer and up to 50% in winter. The vegetation and substrate will help to reduce rainwater runoff and double or triple the life of the roof (Bowie 2008). The substrate is 15 cm deep (6 in) and consists of a blend of inorganic and organic constituents. Zebra, a water retentive additive made from natural cornstarch, is included in the soil mix.

9.3.5.3 Plant Establishment

Plants were first established in Columbia Green Tech. Inc. extensive green roof trays. Additional plants (grasses, and wildflowers) were added as plugs into the trays. In addition to native sedums (Fig. 9.17) some non-native sedums and herbaceous plants were included in the plantings.

Native Succulents

broadleaf stonecrop (*Sedum spathulifolium*), Oregon stonecrop (*Sedum oreganum*), Pacific stonecrop (*Sedum divergens*).

9.3.5.4 Irrigation

Vegetation is irrigated during the summer months twice daily (morning and afternoon) for 10 minutes. There has been no research regarding irrigation on this roof.

Fig. 9.17 Native to the mid-high elevations of the Cascade Mountains, *Sedum divergens* grows naturally in scattered patches on rocky and well-drained sites. Here, it grows with other native and exotic *Sedums*. The roofs decks are sloped at about a 7% gradient, thus the roof decks provide well-drained conditions which sedums are well-adapted. (Photo: Bruce Dvorak, September 2018)

9.3.5.5 Maintenance

The green roof receives monthly maintenance, and an annual application of Biosoilds a natural fertilizer product.

9.3.5.6 Observed Wildlife

Grasshoppers, butterflies, bees, beetles.

9.3.5.7 Best Performing Native Vegetation

All of the sedums are performing very well. Herbaceous plants are still present, but not in the numbers originally planned.

9.3.5.8 Post-occupancy Observations

Authors' Reflections

- Although the native *Sedums* don't grow in large contiguous low-diversity plant communities in the wild, the *Sedums* planted on the library have adapted to the modular green roofs and are healthy growing with a moderately gentle slope.
- This project demonstrates the adaptability of the building managers and project team by integrating current resilient building techniques such as green roofs. Although the green roof is not accessible to the public, the project funded the production of onsite educational information to inform the public about ecosystem services of green roofs at a location inside that has good views to the green roof.

9.3.6 Hayden Meadows Walmart, Portland, Oregon

Previously, the site where the Hayden Meadows Walmart is now located was agricultural land and was thought to be prairie before the founding of the City of Portland. Today, the green roof on Walmart returns prairie grasses and sedums to the rooftop (Fig. 9.18). The project grew from interactions between Walmart, Portland State University, and the Portland Bureau of Environmental Services. Funders included Portland State University, the City of Portland, and Walmart. Provisions for research were initiated to study the energy savings from the green roof, water conservation, bird and insect habitats, and other environmental benefits. In 2013 when the green roof was installed, it was the largest in Portland. The substrate was provided by Roofmeadow and consists of depths of 6.5 cm and 12.7 cm depths (3 and 5 in).

Fig. 9.18 Spring wildflowers in bloom, growing with sedums on the Hayden Meadows Walmart extensive green roof. (Courtesy of Sherry Manning)

9.3.6.1 Project Team

Building Owner/Client: Wal-Mart Stores, Inc.
Green Roof Design Team Lead: Roofmeadow
Architect: Perkowitz+Ruth Architects
Structural Engineer: B&B Associates
Environmental Consultant: Terracon Consultants
Installation Contractor: SolTerra
Portland State University: Dr. David Sailor
Project completion: 2013
Green roof area: 3772 m^2 (40,600 ft^2)

9.3.6.2 Overview and Objectives

The primary goal for this green roof was to investigate energy savings, stormwater retention (Schultz et al. 2018), and the habitat provisions of a grassed and sedum green roof (Fig. 9.19). Portland's Bureau of Environmental Services (BES) continues to oversee three flumes that were installed at this building to monitor runoff from the white roof as well as the two extensive green roofs. One system is 7.6 cm deep (3 inches), and the other is 12.7 cm deep (5 inches). In collaboration with

Fig. 9.19 The Haydens Walmart Green roof combines grasses, wildflowers, and sedums, shown here at the end of the dormant season. The various forms of vegetation support habitat for wildlife. (Photo: Bruce Dvorak, September 2018)

BES, an interdisciplinary group of researchers from Portland State University (Olyssa Starry, Elliott Gall, and Todd Rosenstiel) has been continuing research on-site; since 2016, they have been investigating the influence of green roofs on indoor air quality (NSF-1605843).

9.3.6.3 Plant Establishment

The green roof was planted with a combination of native and some exotic vegetation. Native plants are reported here.

Annuals

clarkia (*Clarkia* sp.), common woolly sunflower (*Eriophyllum lanatum*), giant blue eyed Mary (*Collinsia grandiflora*), phlox (*Phlox* sp.), rusty haired popcorn flower (*Plagiobothyrs nothofulvus*).

Herbaceous Perennials

alpine yellow fleabane (*Erigeron aureus*; native to Washington and western Canada), beardtongue (*Penstemon* sp.), canary violet (*Viola praemorsa*), common lomatium (*Lomatium utriculatum*), common yarrow (*Achillea millefolium*), largeflower fame-flower (*Talinum calycinum*; native to the south-central U.S.), lewisia (*Lewisia* sp.), small camas (*Camassia quamash*).

Succulents

Pacific stonecrop (*Sedum divergens*).

9.3.6.4 Irrigation

The layered green roof system has a subgrade capillary irrigation system. The original schedule during the establishment period (24 months) was 1 h per day in three zones and no irrigation in one zone during warmer months until 80% vegetative coverage was achieved. The subgrade irrigation system is no longer operating and was decommissioned. In its place, an above ground manually-operated sprinkler system was installed in 2018.

9.3.6.5 Maintenance

The current maintenance schedule includes visits during the fall, winter, spring, and summer. Typical total number of staffed hours include fall 180 h; winter 70 h; spring 140 h; summer 200 h.

- Removal of grasses, clover, invasive weeds;
- Clear debris as necessary/use on-site gabions for disposal;
- Replace, transplant, augment plantings;
- Apply organic compost as needed;
- Monitor, record, conduct photo documentation;
- Check roof edges, roof flashings, and roof drains;
- Late summer/early fall seed collection;
- Sedum cutting and rebroadcasting;
- Turn weed debris in gabions;
- Clean and fill birdbaths;
- Maintain irrigation equipment/schedule as needed.

Plants don't seem to be re-seeding themselves, and supplemental seeding is needed. Hand pulling of weeds takes a lot of time relative to mowing, but that can help us keep unwanted invasive plants such as rat tail fescue and the shiny geranium from dominating the roof.

9.3.6.6 Observed Wildlife

Wildlife that frequents the roof includes various urban birds such as killdeer (nesting), pollinators, many geese. Some of the insects include bees, flies, praying mantis.

9.3.6.7 Best Performing Native Vegetation

Plectritis congesta, grasses (unplanted), mosses, *Sedum divergens*, and *Stellaria* species (a volunteer).

Non-native

Erodium cicutarium, *S. takesimense*, *S. rupestre*, *S. floriferum*, *Trifolium* species (unplanted), *Vicia sativa*, *geranium lucidum.*

9.3.6.8 Post-occupancy Observations

Authors' Reflections

- This project highlights a joint venture collaboration between a major corporation and leaders of green roof environmental research. The environmental data collection and reporting of research findings are important and are still needed to support the development of green roofs. This site is an ideal location for long-term research to be established.
- With the addition of a minimal budget, the collaborations could continue and new plants could be introduced for trial.

9.3.7 Native American Student & Community Center (NASCC), Portland State University, Portland, Oregon

During the 1990s at Portland State University, campus leadership began to assess the needs of students from diverse backgrounds and became inspired by the kinds of support for Native American students on campuses elsewhere. Through a decade of planning, visioning, and fundraising, the NASCC was ready for the development of a dedicated space on campus. NASCC's mission is to provide a "cultural home" where Native American, Alaskan Native, and Pacific Islander students connect to other students, faculty, staff, and community members in an inclusive and supportive environment (NASCC 2019). As such, the center was designed and programmed to provide meeting spaces for PSU students, faculty, and other members of the community to learn about native plants, and to participate in cultural traditions and ceremonies of Native Americans living in the Pacific Northwest. The design team envisioned a place where the building and site come together. As the sloped urban site has limited usable space, Nez Perce landscape architect, Brian McCormack envisioned a rooftop garden for students and faculty to learn about, maintain, and make use of historically significant native plants. Select plants are used in ceremonies, for medicinal and other uses. The rooftop also includes an important collection of Native American art (Fig. 9.20) (McCormack 2006).

9.3.7.1 Project Team

Building Owner/Client: Portland State University
Green Roof Design Team Lead: Native American Landscape Architect Brian McCormack, Nez Perce, McCormack Landscape Architecture
Architect: Stastny StastnyBrun Architects, David N. Sloan & Associates
Structural Engineer: KPFF
Landscape Architect: McCormack Landscape Architecture
Maintenance Contractor: Maintained by Native American students at the NASCC
Project completion: 2003
Green roof area: 372 m^2 (4000 ft^2)

Fig. 9.20 Roof garden on the Native American Student and Community Center at Portland State University. Native shrubs, trees, and herbaceous vegetation provide a background for a small gathering space and frame views for several sculptures by local Native American artist, Lillian Pitt (Wasco/Yakama). (Courtesy of Native American Landscape Architect: Brian McCormack, Nez Perce)

9.3.7.2 Overview and Objectives

The Native American Student and Community Center (NASCC) has semi-intensive and intensive green roof systems, with substrates 15 cm up to 76 cm (6 in up to 30 in) deep. The publicly accessible roof hosts classes on indigenous gardens and food justice. The south-facing building is oriented to maximize solar energy gain during the winter, shaded windows during the summer, and has natural daylighting throughout the building. The roof garden is integral to the functions of the building and site as the building uses natural ventilation, and the green roof prevents heat gain during the summer to help keep interior spaces cool. The green roof reduces stormwater runoff and filters rooftop runoff through cartridge filters before entering the city stormwater system.

9.3.7.3 Plant Establishment

All vegetation has been established with pre-grown vegetation installed as plugs. Like a garden, plant selection favored species that have seasonal interest summer through winter (Fig. 9.21).

Fig. 9.21 View near the edges of the green roof on the NASCC. A mix of woody and herbaceous plants cascade down several levels of the planted roof terrace. Although much of the vegetation goes dormant in the winter, ornamental features of plants (red twigs) retain a visual interest during winter. (Photo: Olyssa Starry, February 2020)

Annuals

large camas (*Camassia leichtlinii*).

Grasses

Idaho fescue (*Festuca idahoensis*), bluebunch wheatgrass (*Pseudoroegneria spicata*).

Herbaceous Perennials

showy fleabane (*Erigeron speciose*), Oregon iris (*Iris tenax*), sicklekeel lupine (*Lupinus albicauli*), Davidson's penstemon (*Penstemon davidsonii*).

Shrubs

kinnikinnick (*Arctostaphylos uva-ursi*), sagebrush (*Artemisia tridentata*), Salal (*Gaultheria shallon*), mock orange (*Philadelphus lewisii*), wax current (*Ribes cereum*).

Succulents

broadleaf stonecrop, purpureum (*Sedum spathulifolium* 'Purpureum'), Oregon stonecrop (*Sedum oreganum*), Pacific stonecrop (*Sedum divergens*).

9.3.7.4 Irrigation

Irrigation is operated weekly during the growing season, and during the summer, some plants can receive irrigation every day.

9.3.7.5 Maintenance

As a roof garden, it is intended to engage students with plants for a variety of cultural uses and functions, parts of this green roof have a high level of maintenance needs. Leadership at the NASCC works with student interns and volunteers to teach them how to maintain the roof. Some maintenance also occurs during events, where volunteers or class participants become engaged. The roof garden needed an overhaul in 2019, as there was a lapse in irrigation due to both a system malfunction and some knowledge gaps about how the system works. Some species needed to be replaced as a result of this irrigation gap, and some unwanted species, including deep-rooting trees, needed removal.

9.3.7.6 Observed Wildlife

Pollinators such as bees and butterflies frequent the roof.

9.3.7.7 Best Performing Native Vegetation

All of the vegetation is thriving.

9.3.7.8 Post-occupancy Observations

NASCC

- The ecoroof functions as intended; however, there are unplanned maintenance activities that present challenges at times, as funding for ongoing maintenance is limited. For this reason, NASCC leadership is making efforts to communicate the value that the center provides to the campus, and needs additional support from PSU.

Authors' Reflections

- This green roof reconnects a diversity of stakeholders to the enduring environmental values of a place. As such, Native Americans, Oregonians, and visitors can learn about plants that have an important place in the hearts of the tribes of the Pacific Northwest.
- As an integrated design, the green roof plays a vital role in the mission of the Center, the conservation of energy in the building, and the reduction of rooftop runoff.

9.3.8 Sacred Heart Medical Center at River Bend, Springfield, Oregon

Following advanced trends in creating healing environments and healthy hospitals, the Sacred Heart Medical Center is a 388-bed regional medical facility that was designed to bring the healing aspects of nature into its working and recovering environments (Fig. 9.22). PeaceHealth has a major commitment to sustainable

Fig. 9.22 Deer fern, western sword fern, and Oregon grape cover the ground plain of a sunny to part-shady microclimate on the hospital terrace roof. (Photo: Bruce Dvorak, September 2018)

development and incorporated multiple roof gardens to provide green space for patients and visitors, but to also manage the ecosystem services of their properties, such as stormwater and energy conservation. For example, rainfall draining from all the rooftops is directed to submerged filtration devices called "infiltration galleries" to replenish the local aquifer that feeds the adjacent McKenzie River (GR 2018). The medical center built four roof gardens, but only one is accessible. Four sunken roof gardens were built as viewing courtyards. These are visible from within recovery and waiting rooms.

9.3.8.1 Project Team

Building Owner/Client: PeaceHealth
Green Roof Design Team Lead: Rexius Forest By-Products
Architect: Anshen+Allen Architects; WATG Architects
Maintenance Contractor: Rexius
Project completion: 2008
Green roof area: 4686 m^2 (50,441 ft^2) on seven green roofs

9.3.8.2 Overview and Objectives

A main goal for the roof terraces is to provide green space for patients to enjoy, but also intercept stormwater and reduce the amount and temperature of runoff entering the city stormwater system.

Rexius designed the green roofs primarily with vegetation that is native to the Willamette Valley (Fig. 9.23). Some of the gardens are supplemented with showy drought tolerant exotic vegetation. Some of the unique challenges included the logistics of construction. The substrate was placed on the roof with a blower. Approximately 220 tons of aggregate, 120 cubic yards of mulch, and over 700 cubic yards of rooftop mix was blown and placed on the roof over 2 weeks (GR 2018).

9.3.8.3 Plant Establishment

The roof garden plantings are designed as mini vignettes, inspired by local ecosystems. Large shrubs are used as an understory layer, and shady vegetation covers the substrate below. Some of the plant selections use plants that are native to North America, but not the local forms (i.e *Arctostaphylos*). Only native vegetation used on the project is listed below.

Fig. 9.23 Native groundcovers and large shrubs grow with ornamental grasses (not native to the Pacific Northwest) on an accessible roof terrace. (Photo: Bruce Dvorak, September 2018)

Herbaceous Perennials

eastern purple coneflower (*Echinacea purpurea*; native to eastern U.S.), iris toughleaf (*Iris tenax*), western sword fern (*Polystichum munitum*), western colombine (*Aquilegia Formosa*).

Shrubs

dwarf mahonia (*Mahonia repens*), evergreen huckleberry (*Vaccinium ovatum*), Indian plum (*Oemleria cerasiformis*), low Oregon grape (*Mahonia aquifolium* 'compacta'), Massachusetts kinnikinnick (*Arctostaphylos uva-ursi* 'mass'), red flowering current (*Ribes sanguineum*), serviceberry (*Amelanchier alnifolia*), snowberry (*Symphoricarpos albus*).

Trees

shore pine (*Pinus contorata*).

9.3.8.4 Irrigation

An overhead irrigation system runs during the growing season to keep vegetation active and growing. Irrigation includes zones with three start times, 3 days a week, as each.

9.3.8.5 Maintenance

As accessible roof gardens, the green roof receives weekly maintenance. Typical activities include raking up leaf debris, thinning out grasses or other vegetation as needed, pruning of shrubs. The rhododendrons receive fertilizer once a year.

9.3.8.6 Observed Wildlife

Birds and pollinators occasionally visit the roof gardens.

9.3.8.7 Best Performing Native Vegetation

All of the vegetation is performing as expected.

9.3.8.8 Post-occupancy Observations

Authors' Reflections

- The accessible roof garden receives regular use by hospital staff and visitors. It is a popular place to hang out on a nice day.
- A majority of the roof gardens are planted with a mixture of native and non-native ornamental vegetation. However, some plant groupings have more native vegetation than others.
- The inclusion of woody shrubs and groundcovers add a colorful dimension and are visible from inside the hospital.
- Since only some of the roofs are covered with green roofs, perhaps future plans could include a fuller build-out of green roofs. A larger percentage of rooms in the towers above the green roof looks out onto traditional roofs.

9.3.9 SeQuential Biofuels Retail Station, Eugene, Oregon

SeQuential Biofuels is a Portland-based fuel company. Origins of the company begin in 2000 when co-founders Ian Hill and Tomas Endicott were home-brewing biodiesel from standard cooking oil, in a Eugene garage. Determined to find new options for low-carbon fuels, they found a break in 2006 when they were able to take advantage of the Oregon Clean Fuels Program and their efforts turned into the first all-biofuel station in the Pacific Northwest. From the beginning, they had visions to provide a more sustainable fuel and a more energy-efficient model for a fueling station.

Architect Susan Hill and landscape designers Jeff Ard and Sarah Whitney of Habitats, Inc., procured a vision for what this new type of gas station may look like and how it would function. The main building includes a green roof (Fig. 9.24), and the pumping station has 224 photovoltaic panels (Fig. 9.25). This station in Eugene was designed with "green" internal functions, education exhibits inside and outside the site, including its green roof (Fig. 9.24). The site includes bioswales and a detention basin to help clean and filter stormwater, before entering the Willamette River.

9.3.9.1 Project Team

Building Owner/Client: SeQuential
Green Roof Design Team: Jeff Ard, Sarah Whitney, and Mieko Aoki, Habitats, Inc.
Architect: Susan Hill, Tate Hill Jacobs
Civil Engineer: Doug Singer, Weber Elliott Engineers
Landscape Architect: Habitats, Inc.
Installation Contractor: Habitats, Inc.
Maintenance Contractor: Habitats, Inc. and Jeff Aryd

Fig. 9.24 This extensive green roof consists of succulents, bulbs, and forbs. The plant composition has changed over the years with persistent editing and maintenance. (Courtesy of Jeff Ard, for Habitats, Inc.)

Project completion: 2006
Green roof area: 177 m^2 (1900 ft^2) 17% slope

9.3.9.2 Overview and Objectives

The north-facing green roof was designed to manage rooftop stormwater and moderate the heat flux through the roof deck to inside the building.

9.3.9.3 Plant Establishment

The planting concept for this extensive green roof was to provide a colorful mix of drought-tolerant plants and use plants that could tolerate infrequent foot traffic. System components include gravel edging, 1.3 cm (1/2 in) geocomposite drainage layer over PVC roof membrane, a moisture retention mat, and perforated geocell filled with soil to 12.7 cm (5 in) depth. The substrate is a custom blend comprised of 50% organic planting mix, 30% coir fiber, and 20% pumice. As the green roof makes use of native and non-native vegetation, we present only the native vegetation used.

Fig. 9.25 (**a**) The extensive green roof on the SeQuential Biofuels fueling station, main building. The green roof visible here is shown in late September, near the end of the summer dormant season. (**b**) Inside the building, the green roof insulates the building above the recycled lumber used on the roof deck. (**c**) Interpretive signage inside the store informs about the sustainable features of the building and site. (**d**) Photovoltaic panels supply energy to the site and comprise the awning covering the pumping station. (Photos: Bruce Dvorak, September 2018)

Herbaceous Perennials

beach strawberry (*Fragaria chiloensis*), Cardwell's beardtongue (*Penstemon cardwellii*), moss campion (*Silene acaulis*), western yarrow (*Achillea tomentosa*).

Succulents

broadleaf stonecrop (*Sedum spathulifolium*), Camea broadleaf stonecrop (*Sedum spathulifolium* 'Camea'), cream stonecrop (*Sedum oregonense*), Oregon stonecrop (*Sedum oreganum*), Pacific stonecrop (*Sedum divergens*).

9.3.9.4 Irrigation

Subsurface drip irrigation was installed on a 30 cm by 30 cm (12 in by 12 in) grid with four soak cycles, for 3 min each (to minimize runoff), two times per week as needed during summer.

9.3.9.5 Maintenance

As needed seasonally for, approximately 4 h per month average. Activities include removal of unwanted plants by pulling and hand tools; monitoring for irrigation system, identifying invasive species and threats, plant cover and health, and soil stability.

9.3.9.6 Observed Wildlife

Pollinator insects and ladybugs frequent the green roof. Many birds also use the roof. The first year there was a pair of killdeer living on the roof with nesting young, but no nesting since then.

9.3.9.7 Best Performing Native Vegetation

A mix of native and non-native *Sedum* is now the dominant cover. Coastal strawberry (*Frageria chiloensis*) has persisted well and formed dense patches. It was overly aggressive for the first several years, with robust runners invading and over competing with other plants, but was thinned back and has become less aggressive and less problematic over time. *Lewisia* has persisted and spread. It grows on the upper few feet of the roof, apparently, a self-selecting hydro zones where the soil is the driest. Yarrow is performing well. A few volunteer western sword ferns appeared at the very lowest part of the roof and persisted for several years.

9.3.9.8 Post-occupancy Observations

Design Team (Jeff/Sarah)

- The subsurface drip irrigation system is not a good match for the green roof. It doesn't allow for the adequate spreading of the water.
- The custom-blended growing media has a natural mulching effect from the aboveground parts of the plants.
- This roof was designed and constructed in 2006. During that time, many advancements in technology and understanding of vegetated roofs have occurred. However, the custom design of SeQuential has performed beyond expectations.
- The geo-cell system is very effective and would be used again.

Authors' Reflections

- This small green roof makes a large impact on building and site functions. With no roof insulation, the green roof acts as a thermal barrier to keep the building cool during the summer, and warm during the winter. As one of the few extensive green roofs in the south end of the valley, this project could be further studied for its unique performance to the building systems.

9.4 Plants for the Willamette Valley

Across all of the ecoregional green roof case studies presented in Chap. 9, there are 79 taxa represented, 69 of which are native to the ecoregions in this chapter. Of those taxa native to the chapter ecoregions, 15 species occur more than once across the case studies. Of those occurring more than once, four are annuals, two are grasses, five are herbaceous perennials, one is a shrub, and three are succulents (Table 9.3).

With 79 native plant species trialed on the green roofs in this chapter and the potential many hundreds of native plants growing in the adjacent ecoregions, there is perhaps yet much more to learn about the adaption of native plants in the Willamette Valley. As such, many plant families that readily adapt to dry or gravely soils may be worth trialing under various green roof designs.

Plants from the following families may be worth investigating: *Asteraceae* (asters, *Antennaria* spp., *Arnicas*, *Artemisia* spp., *goldenrods*); *Aspleniaceae*, *Berberidaceae* (Oregon-grape); *Bryophyta* (mosses & liverworts); *Cactaceae* (cacti), *Carex* (sedges); *Caprifoliaceae* (snowberry); *Dryopteridaceae* (ferns); *Equisetaceae* (horsetails); *Lycopodiaceae*, *Orobanchaceae* (paintbrush); *Saxifragaceae*; *Plantaginaceae* (penstemons); *Plumbaginaceae*; *Polypodiaceae*; Pteridaceae; *Ranunculaceae* (*anemones,* columbines, buttercups); and *Rosaceae* (strawberry, non-invasive roses).

Table. 9.3 Native vegetation that occurred three or more times in the case studies in this chapter

Plant Type	Common Name	Botanical Name	A	B	C	D	E	F	G
Annual	common woolly sunflower	*Eriophyllum lanatum*	x	x					x
Herbaceous Perennial	common yarrow	*Achillea millefolium*	x	x					x
Herbaceous Perennial	small camas	*Camassia quamash*	x	x					x
Herbaceous Perennial	Indian blanketflower	*Gaillardia aristata*	x		x				x
Succulent	Pacific stonecrop	*Sedum divergens*	x	x		x	x	x	x
Succulent	Oregon stonecrop	*Sedum oreganum*	x			x	x	x	x
Succulent	broadleaf stonecrop	*Sedum spathulifolium*	x		x	x	x	x	x

Key = A (Columbia Wastewater Treatment), B (Haydens Walmart), C (Multnomah County Building), D (Multnomah County Central Library), E (Reed College, Performing Arts Center), F (SeQuential Biofuels), G (Gunderson's Habitat Roofs)

9.5 Summary

Since the early 2000's many hundreds of green roofs have been constructed in the Willamette Valley. The case studies here demonstrate that green roofs were built to:

- connect visually with the surrounding environment, e.g., Columbia Wastewater Treatment Support Facility, Multnomah County Building, and the Native American Student and Community Center at PSU;
- provide a learning tool for green roofs, e.g., Multnomah County Building, Multnomah County Central Library, SeQuential Biofuels, Haydens Walmart;
- provide an outdoor laboratory for observing nature, especially pollinators and other visitors to the green roofs, e.g., Gunderson's Habitat Roofs, Haydens Walmart, Reed College Performing Arts Center, Multnomah County Central Library;
- demonstrate ecosystem services generally associated with green roofs such as energy conservation, runoff amelioration, temperature, and noise abatement, e.g. Multnomah County Building, Haydens Walmart, Columbia Wastewater Treatment Support Facility;
- offset the effects of habitat loss, not via a mitigation process per se, but rather through the intentional introduction of habitat that increases the number of plants of a particular habitat once common to the site, e.g., Gunderson's Habitat Roofs, Haydens Walmart, Columbia Wastewater Treatment Support Facility;
- connect with students, and cultural traditions of the region, e.g., the Native American Student and Community Center at PSU, Reed College Performing Arts Center;
- increase the aesthetic appeal of buildings especially where the function of the building itself may reduce stress for workers or visitors, e.g., Sacred Heart Medical Center at River Bend, Reed College Performing Arts Center, Multnomah County Building.

Ecoregions of the Willamette Valley historically included prairie, oak woodlands, and coniferous forests. Before the mid-nineteenth century, the Chinook, Clackamas, and Kalapuya peoples burned the landscape to maintain prairie and open oak woodlands for at least the previous 4000 years. Today, many of the plants native to these ecosystems still exist; however, due to the development of the land their abundance is greatly reduced, and conservation efforts struggle to maintain the native plant communities. Several prairie preserves exist near urban regions of the Willamette Valley. Coastal temperature forest preserves abound in the region; however, plants for green roofs that are native to coastal temperate forest may be limited to ground covers, grasses, and forbs that grow in open sun pockets of forests or meadows. Some plants such as cactus that are more adapted to central Oregon or fringes of the Columbia River Gorge may be adaptable to extensive green roofs in the valley.

Urban development is expected to grow considerably throughout the valley. The 1990 Oregon census reported 1.9 million people living in the valley. By 2050, the population is expected to reach nearly four million. Based on the case studies in this chapter, ecoregional green roofs could and perhaps should play an important role in shaping new forms of development.

As of 2018, there were at least 500 green roofs in the City of Portland (Thurston 2017). The City of Portland was one of the first cities in North America to promote and conduct research on green roofs, and Portland is also one of the few cities in the United States with a mandate to require green roofs (in the City Center District) with climate-adapted vegetation, such as climate-adapted exotics or native plants. The mandate is for buildings located in the Central City District over 1858 m^2 (20,000 ft^2) in size excluding rooftop parking (LAM 2018). Because of this mandate, all green roof pilot projects that are accessible or viewable, (that have native vegetation), serve as examples of how plants from the ecoregion may adapt to green roofs in a variety of ecosystem functions. New pilot projects with native plants may be needed to present a variety of examples for ways native plants can be used on green roofs in the Willamette Valley.

The volunteer organization GRiT (Green Roof Think Tank) was organized to help promote and educate about green roofs with monthly meetings. GRiT has over 400 members and consists of design professionals, municipal members, industry leaders, and researchers from local universities. The City of Portland and collaborators have a widespread public education system with signage at public green roofs and online. Sharing of research and performance data is common between the City and County. This kind of multi-jurisdiction sharing and growing of information will likely serve as a model for municipalities new to green roofs.

The nine ecoregional green roofs were presented to cover a spectrum of ownership, management styles, and aesthetic goals. Case studies in Portland include extensive and semi-intensive green roofs that are accessible to the public. The Columbia Wastewater Treatment Plant, Multnomah County green roof and the Gunderson's habitat green roofs provide examples of grass-based green roofs that intend to be maintained not as gardens, but meadows or natural habitats. At 35 species, the Gundersons green roofs were the most diverse in the region. The meadow-based green roofs are similar to the aesthetic of a prairie, and the Gunderson's green roofs were the most biologically diverse and reflective of the prairie communities. The ecoroof at Reed College has some structure to the planting design (bed lines), which means that regular maintenance is required to achieve the aesthetic goals. The aesthetic appearance of grassed-based green roofs in Portland is dynamic as there is bright green foliage and wildflowers in the spring, and straw, brown and orange colored vegetation taking precedence during the dormant periods. New research could explore how wetland or cool-season vegetation (annuals, bulbs, grasses, sedges) can be employed on green roofs in the valley.

Extensive green roofs on the Multnomah County Library demonstrate what a green roof with native succulents can look like. This roof includes non-native sedums, however, the native sedums are performing well and sets an example of how an extensive green roof on a gentle slope can function with appropriate maintenance.

The Hayden's Walmart green roof demonstrates a hybrid-aesthetic as sedums and grasses grow together. The research and collaboration between private/public partnerships are rare. Much can be learned about how green roofs in Portland function and the maintenance requirements to achieve those benefits.

Pressures from exotic invasive species in urban areas can challenge the maintenance of prairie green roofs. One main difference between gree roofs and remnant prairie patches in the Pacific Northwest is the legacy seed banks found in prairie systems yet absent from green roofs. This is one reason they become vulnerable to invasion. In the Portland area, many green roofs with no maintenance schedule end up becoming dominated by the invasive annual rat's-tail fescue (*Vulpia myuros*), and there is some concern about fire risk in this scenario. This can be countered with maintenance that can include re-seeding regularly with desired species hand-pulling, or appropriately timed mowing, albeit potentially less effective.

Few green roofs exist in the southern end of the Willamette Valley. The Sacred Heart Medical Center roof garden makes use of native woodland groundcover vegetation, thus provide an example of how coastal temperate forest ecoregions may inspire intensive and semi-intensive green roofs. In addition to the healing benefits of the garden for patients, the inclusion of all native plants makes for a great example of regional character. Gas stations are one of the most frequent building types across urban centers. The SeQuential extensive green roof demonstrates how a hybrid plant pallet with native and exotic vegetation can bring color to green roofs with drought-tolerant vegetation on an aggressive slope.

Regarding what might be ahead for municipalities that mandate native or naturalized vegetation on green roofs (like Portland), the role of education may prove to be critical for green roof designers, building owners, policymakers, and those that maintain green roofs. For example, irrigation and maintenance were found to be important considerations for green roofs in Portland, as one study looked at owner's opinions of green roofs after they have taken on green roofs on their own buildings. Perceptions of green roofs were generally positive, but some did not have their expectations met. Of those that were uncertain about the future success of their green roof, some of those owners were told that their green roof did not need irrigation, but they found that the vegetation selected for their green roof did need irrigation. Some green roof owners that had perennial vegetation did not know that plants can become dormant during the summer or winter (Thurston 2017). What these kinds of findings reveal is that public education about basic ecological functions and their aesthetic dynamics as a garden or natural habitat be vital to a welcomed adoption of green roofs.

Regarding green roofs with native plants in the Willamette Valley, there is a need for more examples of green roofs that make use of native plants or especially those that intend to reflect a prairie or succulent habitat. The Bureau of Environmental Services website maintains a variety of different kinds of green roofs in Portland. Many of the green roofs feature exotic sedums. With the adoption of a mandate for native or naturalized vegetation on green roofs in the City Center District, more examples may be needed to demonstrate a variety of ways native plants can be used on green roofs.

Patterns of precipitation may present a real challenge for green roofs in the Willamette Valley. With scant rainfall during the summer and persistent and excessive rainfall during the fall and winter, new approaches for supplying water may be needed; or perhaps new trials and arrangements of plants and substrate and deck designs (such as Gunderson's Habitat roofs) may prove useful as well. Regarding the potential for watering or irrigating ecoregional green roofs, there is perhaps a need to demonstrate a variety of ways to water and irrigate green roofs with sustainable sources of water such as water harvested from inside buildings (greywater), harvested rainwater in tanks or other sources. Research regarding substrate depths for a variety of vegetation growing on different slope aspects and roof deck slopes is needed. Since there are no long-term green roof research sites in the valley, perhaps a goal for the region could be to secure funding to the establishment of a site where hundreds of taxa of plants could be trialed on a variety of green roof designs.

Acknowledgments We would like to thank the following individuals for taking the time, efforts, and sharing of their knowledge to help us gather information for this chapter including Tom Lipton, Brian McCormack (Nez Perce) owner McCormack Landscape Architects; Dan Sutton, with Rexius Forest By-Products; Elizabeth Hart Morris with Green Up and GRiT; Alan Proffitt with Multnomah County; Desirae Wood at Dobro Design; Johnathan Beaver at 2.ink Studio; Jeff Aryd formerly with Habitats, Inc.; Blythe Utz with Greenbelt Landtrust; staff at the Sacred Heart Medical Center; Courtney Vengarick at the Leach Botanical Garden; and the staff at Hayden Walmart.

References

Ahiablame LM, Engel BA, Chaubey I (2012) Effectiveness of low impact development practices: literature review and suggestions for future research. Water Air Soil Pollut 223(7):4253–4273

Alaback PB (1994) Plants of the Pacific Northwest Coast: Washington, Oregon, British Columbia & Alaska. Lone Pine Pub, Vancouver

Ambrose SE, Abell S (1998) Lewis & Clark: voyage of discovery. National Geographic Society, Washington, DC

Bailey RG (1997) Ecoregions of North America. U.S. Department of Agriculture, Forest Service, Washington, DC

Baker JP, Hulse DW, Gregory SV, White D, Van Sickle J, Berger PA, Dole D, Schumaker NH (2004) Alternative futures for the Willamette River basin, Oregon. Ecol Appl 14(2):313–324

Benner PA, Sedell JR (1997) Upper Willamette River landscape: a historic perspective. In: Laenen AD, Dunnette DA (eds) River quality: dynamics and restoration. Lewis Publishers, Boca Raton, pp 23–47

Bowie C (2008) The central library now sports a green roof. The Oregonian, Oct 9,

Carr GL, Bierstadt A (1997) Bierstadt's West. Gerald Peters Gallery, Santa Fe

Christy JA, Alverson ER (2011) Historical vegetation of the Willamette Valley, Oregon, circa 1850. Northwest Science 85(2):93–108

CNA (2019) Camassia Natural Area Oregon. Nature Conservancy. https://www.nature.org/en-us/get-involved/how-to-help/places-we-protect/camassia-natural-area/. Accessed 6 June 2019

Connelly, K. C. and J.B. Kauffman (1991) Ecological effects of fire in Willamette Valley wetland prairies with special emphasis on Lomatium bradshawii and Erigeron decumbens, two rare endemic plants. Oregon State University, Corvallis, OR

CPRN (2012) Gunderson dedicates second habitat roof

CRG (2020) Columbia River Gorge. http://www.columbiarivergorge.info. http://www.columbiarivergorge.info/flowers.html. Accessed 20 Feb 2020

Deur D, Turner NJ (2005) Keeping it living: traditions of plant use and cultivation on the Northwest Coast of North America. University of Washington Press, Seattle

Everett G, Lamond J, Morzillo AT, Matsler AM, Chan FKS (2018) Delivering green streets: an exploration of changing perceptions and behaviours over time around bioswales in Portland, Oregon. J Flood Risk Manage 11:S973–S985

Finley KK (1994) Hydrology and related soil features of three Willamette Valley wetland prairies. Oregon State University, Corvallis, OR

Franklin JF, Dyrness CT (1988) Natural vegetation of Oregon and Washington, vol PNW-8. Oregon State University Press, Corvallis

Gonsalves SM (2016) Green roofs and urban biodiversity: their role as invertebrate habitat and the effect of design on beetle community. Portland State University, Portland

GR (2018) Sacred Heart Medical Center at River Bend. https://www.greenroofs.com/projects/sacred-heart-medical-center-at-river-bend/. Accessed 12 June 2019

Hamrick D (2018) Portland pushes natives and green roofs for City buildings, p 1

Harvey T (2012) McCord Creek & Elowah Falls Plant list. westerncascades.com, westerncascades.com

Hauth E, Liptan T (2003) Plant survival findings in the Pacific northwest. Paper presented at the 1st North American green roof infrastructure conference: greening rooftops for sustainable communities, Chicago, IL, May 29–30

Hutchinson D, Abrams P, Retzlaff R, Liptan T (2003) Stormwater monitoring two ecoroofs in Portland, Oregon. Paper presented at the first annual greening rooftops for sustainable communities conference, awards and trade show, Chicago, IL, April 29–May

Kimmerer RW, Lake FK (2001) The role of indigenous burning in land management. J For 99(11):36–41

KPR (2019) Kingston Prairie Preserve. Greenbelt Land Trust. http://greenbeltlandtrust.org/conserving-land/kingston-prairie-preserve/. Accessed 06-07-2019

Kurtz T Flow monitoring of three ecoroofs in Portland, Oregon. In: the 2008 international low impact development conference, Seattle, WA, 2008. ASCE, pp 1–10

LAM (2018) Portland adopts a green roof requirement in the Central City 2035 Plan. Living architecture monitor, vol June 2016. Green Roofs for Healty Cities, Toronto

Liptan TW (2017) Sustainable stormwater management: a landscape-driven approach to planning and design. Timber Press, Portland

MacIvor JS, Ranalli MA, Lundholm JT (2011) Performance of dryland and wetland plant species on extensive green roofs. Ann Bot 107(4):671–679

McCormack B (2006) PSU Native American Student & Community Center. McCormack Landscape Architecture. http://www.weetes.com/index.cfm?page=psunative.cfm. Accessed 15 June 2019

NASCC (2019) Native American student and community center. Native American Student and Community Center. https://www.pdx.edu/floorplans/buildings/nascc. Accessed 16 Dec 2019

OA (2013) Reed College performing arts ecoroof. Opsis Architecture. https://www.opsisarch.com/blog/project/reed-college-performing-arts-building/. Accessed 13 June 2019

OCS (2019) Oregon Conservation Strategy: Willamette Valley. Oregon Conservation Strategy. http://www.oregonconservationstrategy.org/ecoregion/willamette-valley/. Accessed 06-10-2019

Pojar J, MacKinnon A (2004) Plants of the Pacific Northwest Coast. Partners Publishing Group, Vancouver

Reveal JL, Moulton GE, Schuyler AE (1999) The Lewis and Clark collections of vascular plants: names, types, and comments. Proc Acad Natl Sci Phila 149:1–64

Sailor DJ, Hutchinson D, Bokovoy L (2008) Thermal property measurements for ecoroof soils common in the western U.S. Energy Build 40(7):1246–1251

Schemske DW, Husband BC, Ruckelshaus MH, Goodwillie C, Parker IM, Bishop JG (1994) Evaluating approaches to the conservation of rare and endangered plants. Ecology 75(3):584–606

Scherba A, Sailor DJ, Rosenstiel TN, Wamser CC (2011) Modeling impacts of roof reflectivity, integrated photovoltaic panels and green roof systems on sensible heat flux into the urban environment. Build Environ 46(12):2542–2551. https://doi.org/10.1016/j.buildenv.2011.06.012

Schultz CB, Henry E, Carleton A, Hicks T, Thomas R, Potter A, Collins M, Linders M, Fimbel C, Black S (2011) Conservation of prairie-oak butterflies in Oregon, Washington, and British Columbia. Northwest Sci 85(2):361–389

Schultz I, Sailor DJ, Starry O (2018) Effects of substrate depth and precipitation characteristics on stormwater retention by two green roofs in Portland OR. J Hydrol Reg Stud 18:110–118

Spolek G (2008) Performance monitoring of three ecoroofs in Portland, Oregon. Urban Ecosyst 11(4):349–359

Starry O, Gonsalves S, Ksiazek-Mikenas K, MacIvor JS, Gardner M, Szallies A, Brenneisen S (2018) A global comparison of beetle community composition on green roofs and the potential for homogenization. Urban Nat 1:1–15

Thilenius JF (1968) The Quercus garryana forests of the Willamette Valley, Oregon. Ecology 49(6):1124–1133

Thurston R (2017) Defining and measuring green roof failure using a case study of incentivized industrial, commercial, and institutional vegetated roofs in Portland, Oregon. Evergreen State College, Olympia

Titus JH, Christy JA, VanderSchaaf D, Kagan JS, Alverson ER (1996) Native wetland and riparian plant communities in the Willamette Valley, Oregon. Oregon natural heritage program and the nature conservancy. Oregon Natural Heritage Program and The Nature Conservancy, Portland

Walsh MK, Pearl CA, Whitlock C, Bartlein PJ, Worona MA (2010) An 11 000-year-long record of fire and vegetation history at beaver Lake, Oregon, Central Willamette Valley. Quat Sci Rev 29(9–10):1093–1106

Williams GW (2002) Aboriginal use of fire: are there any "natural" plant communities. USDA Forest Service, Washington DC. USDA Forest Service National Office, Washington, DC

Williams RC, Steinberg AG, Gershowitz H, Bennett PH, Knowler WC, Pettitt DJ, Butler W, Baird R, Dowda-Rea L, Burch TA (1985) GM allotypes in Native Americans: evidence for three distinct migrations across the Bering land bridge. Am J Phys Anthropol 66(1):1–19

Wilson MV, Hammond PC, Schultz CB (1997) The interdependence of native plants and Fender's blue butterfly. Conservation and management of native flora and fungi. Native Plant Society of Oregon, Corvallis, pp 83–87

Wood D (2011) Gunderson habitat ecoroof: planting documentation. Dobro Design, Portland

Wood D (2018) Habitat Value of Ecoroofs-Phase 2. Paper presented at the Cities Alive: 15th annual green roof & wall conference Sept. 18–21

Zenk H (2008) Notes on Native American place-names of the Willamette Valley region. Or Hist Q 109(1):6–33

Zhou L, Shen G, Woodfin T, Chen T, Song K (2018) Ecological and economic impacts of green roofs and permeable pavements at the city level: the case of Corvallis, Oregon. J Environ Plan Manag 61(3):430–450

Chapter 10
Green Roofs in Fraser Lowland and Vancouver Island Ecoregions

Bruce Dvorak and Daniel Roehr

Abstract This chapter presents case studies of four conservation sites and seven green roofs located in western British Columbia in the ecoregions of the Fraser Lowlands and Vancouver Island. The region is geographically complex with remnants of forested, savanna, and grassland ecoregions that once populated the Fraser River delta and parts of Vancouver Island. Historically, prairie and savanna vegetation dominated the ground plain on the delta. Less than 25% of the original temperate forests remain intact, and only a few small preserves sustain the endangered native grassland habitats that were once widespread on the southeastern edge of Vancouver Island and the delta. Annual precipitation at Victoria averages about 600 mm, Nanaimo averages about 1000 mm and Vancouver 1190 mm. Much of the precipitation takes place during the fall, winter, and spring, as summers are cool and dry. This chapter highlights how 42 plant taxa native to the region have been trialed on seven ecoregional green roofs.

Keywords Summer dormancy · Rainwater harvesting · Greywater · Garry oak ecosystem · Biodiversity · Hydroseeding · First Nations · Habitat · Goats

B. Dvorak (✉)
Department of Landscape Architecture and Urban Planning, 305A Langford Architecture Center, Texas A&M University, College Station, TX, USA
e-mail: bdvorak@tamu.edu

D. Roehr
The Landscape Architecture & Environmental Design Program, The University of British Columbia, Vancouver, BC, Canada
e-mail: droehr@sala.ubc.ca

B. Dvorak (ed.), *Ecoregional Green Roofs*, Cities and Nature,
https://doi.org/10.1007/978-3-030-58395-8_10

10.1 Ecoregion Characteristics

10.1.1 Vancouver Island, British Columbia, Canada

Over 10,000 years ago, Paleo-Indians began settling the land where present-day Vancouver Island and the Fraser delta presides. Over time, the tribes of the First Nations peoples developed a culture and lifestyle where they sustained grasslands and oak savanna on the southern and eastern landscapes on Vancouver Island with anthropogenic burning of the ground-level vegetation. Research on historic vegetative patterns over the past 5000 years shows that fire-adapted landscapes were prevalent on the island (North and Dunn 1979; Clague et al. 1983). The First Nations people established grasslands, oak, and pine savanna ecosystems for hunting, diversification of the vegetation for food and other uses (Fuchs 2001; McCune et al. 2013). Today, although these grassland habitats have been reduced significantly to less than 5% across the Pacific Northwest, several small tracts of land preserve these oak ecosystems and grasslands on Vancouver Island (McCune et al. 2013). Conservation biologists consider that this ecoregion was one of the most biologically diverse on the West Coast, and is now critically endangered (Parachnowitsch and Elle 2005; Ritland et al. 2005).

Europeans first settled Vancouver Island in the mid-nineteenth century. There are only a few paintings or sketches to demonstrate the structure and quality of the Garry oak habitats that existed prior to settlement. Written descriptions also confirm ecosystems were similar to what is shown in Fig. 10.1 (Lord 1866). This painting of Victoria Harbor painted by artist Thomas J. Somerscales (1869) shows that the natural vegetation was low-growing grasses and shrubs on the rocky soils and only a scattering of trees on the far end of the harbor (Emmerson 1865). As a realist, Somerscales articulates elements of the foreground vegetation. According to historic records and images such as shown in Fig. 10.1, the southern edge of Vancouver Island was a grassland and savanna habitat (Gedalof et al. 2006). Investigations into the historic habitat support evidence that much of the southern island may have had few trees. One study at Rocky Point reveals that "Prior to 1850, the site was probably largely treeless, with only a few scattered oaks and Douglas-fir trees (Gedalof et al. 2006, p. 43)." Today on Vancouver Island, there are several nature conservation sites where these original oak habitats are preserved and can be explored. Several publicly accessible sites are covered in this chapter. These sites preserve the historic plant communities (grasslands and scrublands) with vegetation that is naturally adapted to drought, which may prove useful to future generations of green roof designers.

Fig. 10.1 *Sunset: A View Of The Entrance To Victoria Harbour, Vancouver Island, British Columbia...* by Thomas Somerscales, 1868. Grasses and shrub vegetation lie in the foreground, and the Olympic Mountains lie in the distance. (Courtesy of the Royal BC Museum and Archives)

10.1.2 Vancouver, British Columbia, Canada

Where the present-day city of Vancouver, B.C. presides, and its surrounding communities, the landscape vegetation before European settlement was a matrix of forest, savanna, open prairie, and mixed-woodlands (McCune et al. 2013). The Fraser delta, was a grassland habitat, with shrubland, bogs, moss-covered soil, woodlands, and small patches of conifers in upland locations (North and Dunn 1979; Clague et al. 1983; McCune et al. 2013). Some of the first surveys of the Fraser delta recorded a scattering of only a few small farms and the remainder of the delta was undeveloped (Fig. 10.2). Although much of the delta was grasslands, immediately upslope from the delta lied mixed woodlands and temperate coniferous forests. The forests were widespread throughout the upland hills and mountains surrounding the Fraser delta. The density and composition of the forests were likely varied, and different from many forest preserves today, as forest communities at that time were composed of mixed-age trees, with patches of open meadows where fire or other disturbances took place (Franklin et al. 1981). The even-aged stands of forests that largely represent many managed forest lands today were less common (Ford et al. 1979).

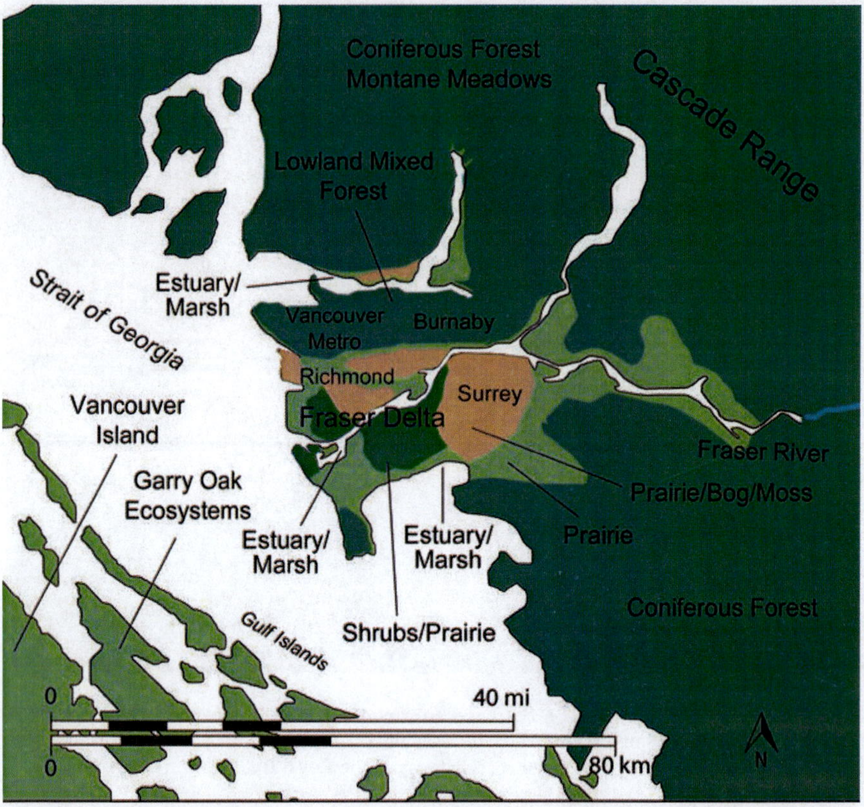

Fig. 10.2 Map of the major ecosystem types on the Fraser Delta, Fraser Lowlands, and vicinity before European settlement. The figure is inspired by the map *Vegetation of the Southwestern Fraser Lowlands, 1858–1880* (North and Dunn 1979), and research by McCune et al. (2013) and others. The figure shows the complex pattern of vegetation types that characterized the Fraser River Delta such as prairies, shrublands, peat bogs, mosslands, mixed forests, wetlands, and fresh and saltwater marshes and estuaries. Locations of current cities are provided for context. Garry oak ecosystems were dominant on Vancouver Island and the Gulf Islands. (Graphic: by B. Dvorak, June 2020)

Today, the Fraser River delta landscape is nearly entirely developed north of the Fraser River, and south of the Fraser River lies a mixture of developed and cultivated land; however, a diversity of native wildlife persists in the fringe habitats (Wang 2020). There are no preserves of the prairie landscapes and a few places where one may visit that would be similar to the Fraser delta prairies and oak ecosystems before European settlement.

Estuaries were once widespread across the delta. Estuaries are offshore habitats where freshwater and saltwater mix and interface with tidal landscapes. These transitional landscapes at the parameters of the delta are very productive and important ecosystems. A few estuaries remain protected as preserves. These serve as the last remaining habitats for a multitude of species including native and migrating birds that make their way along the Pacific Flyway each year (Robb 2014).

As green roofs continue to become more complex in their design, estuaries and wetlands may prove useful to identify plant communities that may be modeled for buildings and sites where wastewater is treated on-site. There are several case studies in this chapter where wastewater is recycled and stored short-term for use on the green roof. Making use of native vegetation from wet to moist plant communities may become more common if the construction of future buildings and communities are modeled like living systems (LBC 2017). Recapturing a diversity of habitats for birds and other habitat dependent species may be critical if the not unique function of green roofs in urban areas, especially where little of the original vegetation remains (Gedge and Kadas 2004, 2005; Brenneisen 2006; Kadas 2006; Francis and Lorimer 2011; Kowarik 2011).

10.1.3 Climate

Precipitation varies greatly across the Vancouver mainland and Island region. The city of Victoria, B.C. receives about 600 mm of annual precipitation. Nanaimo on the eastern side of Vancouver Island receives about 1000 mm annually, and Vancouver receives about 1190 mm of precipitation annually (Fig. 10.3). A majority of the annual precipitation takes place during the fall through spring. Summer in the region is dry to very dry. Snowfall is minimal near sea level, and snowfall can be significant in the adjacent mountains. Temperatures remain mild during the summer; however, the dry season typically is long enough to cause native grasses and forbs to enter a summer dormancy period that can last from July into late September or longer in drought years (Fig. 10.3).

10.1.4 Vegetation of the Fraser Lowland Ecoregions

The Fraser River is the longest river in British Columbia. The Fraser River delta was formed over hundreds of thousands of years of deposits from glaciation, upstream debris and erosion. Two major ecoregions existed prior to urban development in the Vancouver metro area and Vancouver Island. Figure 10.4 shows the extent of Coastal Temporal Forests (22) and Oak Woodlands and Grasslands (21) (ESWG 1995; Pojar and MacKinnon 2004).

10.1.4.1 Coastal Temperate Forests

Dominated by conifers, the coastal temperate forests are well adapted to summer droughts but are not fire-managed ecosystems. Ground-level fires were infrequent, and if canopy fires took place they would be damaging as they are today. Temperate forests are also known to include meadows at forest edges, disturbed sites, and high

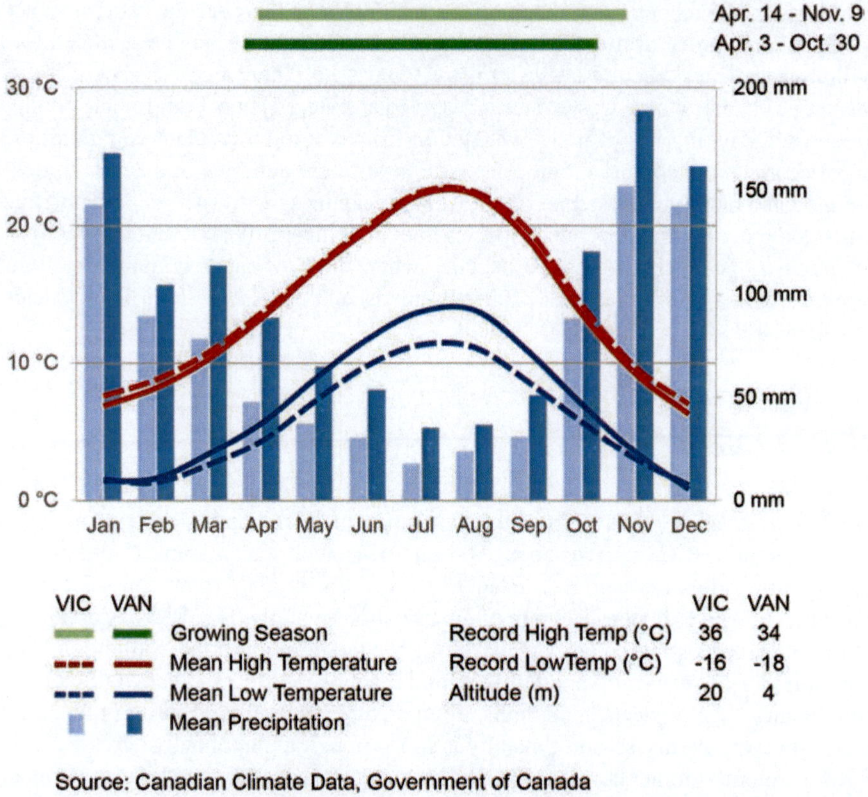

Fig. 10.3 Monthly precipitation and temperatures for Vancouver (VAN and Victoria (VIC), B.C. Precipitation is uneven throughout the year as fall, winter and spring consistently receive ample rainfall, but summers are dry and mild. Dry periods in summer have been increasing over recent years. The ecosystems in British Columbia have evolved in the Fraser Lowlands to adapt to dry summers and wet winters. The natural landscape takes on a dry and dormant appearance during late summer, where fall and spring produce vibrant green vegetation. (Graphic: Tess Menotti & Bruce Dvorak)

mountain environments. There are at least 40 tree species known to the temperate forests. Major conifer tree species include western red cedar, western hemlock, Douglas-fir, yellow cedar, lodgepole pine, ponderosa pine, and Sitka spruce. Other native conifers trees found in the ecoregion include western larch, yew, and juniper. Major deciduous trees native to the region include vine maple, Pacific dogwood, Garry oak, Pacific crab apple, birch, willow, and bigleaf maple. Madrone is native to the region and is evergreen.

Forested ecoregions are likely to have a minor role to play regarding green roofs. Certainly, intensive green roofs may make use of small native trees and shrubs; however, the ground plain vegetation may play a more important role. Plants such as western sword fern, kinnikinic, snowberry, salal, and some ferns may translate to rooftop gardens where the micro-climate and sun/shade relationships are favorable.

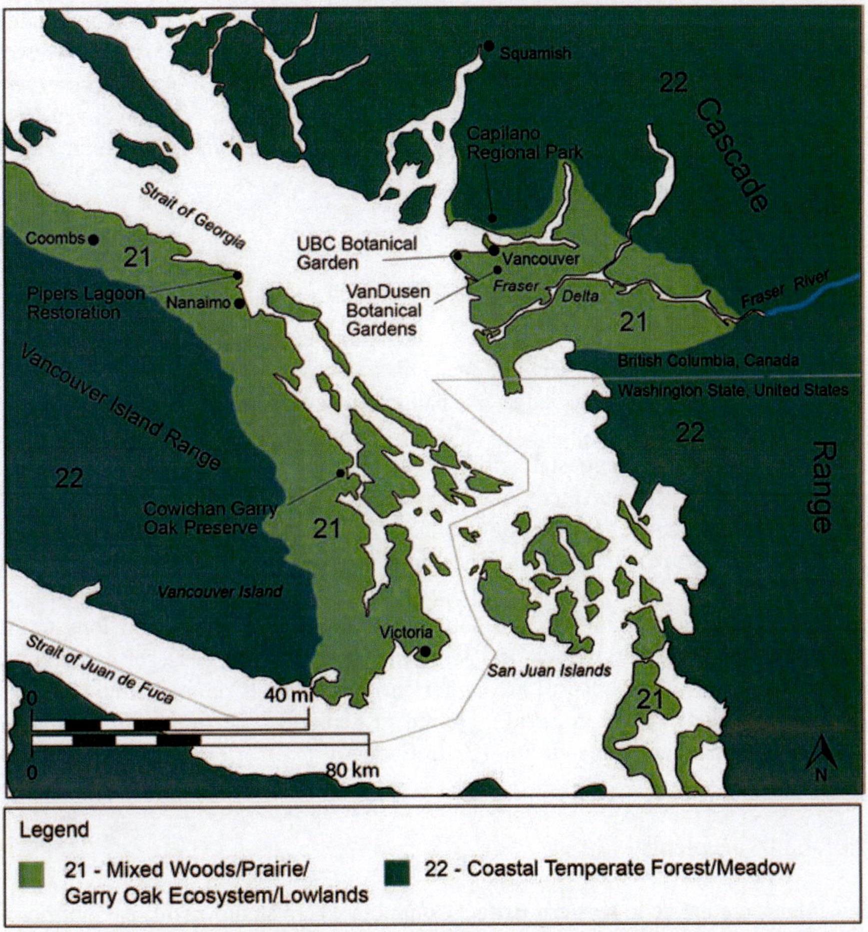

Fig. 10.4 Ecoregions map of Vancouver, B.C., Fraser River Delta, and Vancouver Island (Omernik 1995 and sources in Fig. 10.2). Dominate vegetation includes temperate coniferous forests, coastal rainforests and isolated meadows in the mountain ranges (22); in the drier rain shadow of the coastal mountains is a complex variety of ecosystems including oak savanna, prairie, mixed woods, rocky balds, forested areas, and marshes, wetlands, and bogs in the Delta lowlands (21). Oak ecosystems are still present in a scattering of a few small but critically important biological preserves. (Graphic: Trevor Maciejewski & Bruce Dvorak)

Already, western sword fern is used on roof deck structures along the west coast in both sunny and part-shade conditions. Other herbaceous plants include several species of strawberry such as wild strawberry (*Fragaria virginiana*), coastal strawberry (*Fragaria chiloensis*), and wood strawberry (*Fragaria vesca*). Native bulb species that grow throughout the region include the culturally important camas and species of allium.

Understanding of the altitude and microclimate locations of plants within the temperate forests is extremely influential as elevation, slope, and aspect influence the kind of soil and drainage characteristics, habitat, and other influencing conditions of species that thrive in those conditions. Lowland forests generally receive less precipitation than more upland and higher elevation sites. Biological preserves of native ecosystems or analogs for green roof assemblages in the Vancouver area can be found in the metro parks such as Stanley Island, Capilano Regional Park, Cypress Provincial Park, and others.

10.1.4.2 Garry Oak Ecosystems

Garry oak ecosystems were dominant on the southeast Vancouver Island, the San Juan Islands, and the Fraser River Delta. This is a broad assemblage of plants that are based on grassland and savanna-like habitats. The ecosystem is fire-adapted and fire-dependent in some cases. Garry oak ecosystems are known to have the highest diversity of vegetation in coastal British Columbia, and it supports hundreds of species of insects, reptiles, resident and migrating birds. There are over 100 types of native bees to Vancouver Island, for example. Madrone and pines may also occupy niches or be intermixed with oaks. The Nature Conservancy and other groups are working to restore Garry oak ecosystems throughout the region. Plant diseases, invasive plants, and animals make habitat restoration a complex and long-term effort. Dominant vegetation includes *Quercus garryana* (Garry oak), shrubs such as *Rosa nutkana* (Nootka Rose), *Symphoricarpos albus* (Common Snowberry), *Holodiscus discolor* (Ocean Spray). Dominant herbs include *Camassia quamash* (Common Camas) and *C. leichtlinii* (Great Camas).

10.1.4.3 Succulents

Succulents are native to western British Columbia where shallow soils exist or near edges of rocky outcrops. *Sedum spathulifolim* and *Sedum lanceolatum* var. *nesiotcium* grow on Vancouver Island; *Sedum oreganum* grows near Squamish, B.C.; *Sedum divergens* and *Sedum lanceolatum* grow in the Upper Stein Valley, B.C.; and *Sedum stenopetalum* grows near Chase, B.C. Several introduced species have naturalized on the landscape including *Sedum acre*, *Sedum forsterianum*, and *Sedum album*. These species can be found growing around the Fraser delta area on exposed sites.

10.1.5 Ecoregional Conservation Case Studies (Arranged West to East)

10.1.5.1 Pipers Lagoon Restoration, Nanaimo, B.C., Canada

Pipers Lagoon Restoration site is a small but important remnant landscape that preserves a narrow strip of Garry oak ecosystem, barrens, and rocky outcrops with vegetation native to the region and may prove useful for green roofs on the south end of the island (Fig. 10.5). Some of the native plants that can be observed include Garry oak, kinnikinic, sedums, and other drought-tolerant vegetation. Invasive species such as annual grasses, Himalayan blackberry, English ivy, and Scotch broom are also present on the preserve and difficult to remove. These invasive species compete for space and should not be mistaken as being natural to the area.

10.1.5.2 Cowichan Garry Oak Preserve, Duncan, B.C., Canada

The Cowichan Garry Oak Preserve is located on Vancouver Island near Duncan, B.C. in the Cowichan River Valley. It is known as one of the last and highest quality preserves of the globally endangered Garry oak ecosystems. The 33-hectare

Fig. 10.5 A single Garry oak tree clings to the edge of a rocky outcrop near the head of the lagoon. A variety of grassland micro-climates and various habitats are present and could be observed during various seasons of the year to better understand their productivity, function and aesthetic potential for green roofs. (Photo: Bruce Dvorak, September 2018)

(81-acre) property preserves savanna, meadow, forest, and wetland ecosystems which are supported by pollinators and other insects that perform important ecosystem services (Parachnowitsch and Elle 2005). In addition to oak and grasses, the culturally significant camas bulb is present, as well as over 100 other plant species that are linked to survival with the Garry oak ecosystems. The preserve is managed by The Nature Conservancy, and the site has guided trails and volunteer days where invasive species are removed from the landscape (MacDougall 2002). This preserve has great potential for the exploration of plants that may adapt to extensive and semi-intensive green roofs.

10.1.5.3 University of British Columbia Botanical Garden & VanDusen Botanical Gardens, Vancouver, B.C., Canada

The University of British Columbia Botanical Garden was established in 1961, and it is the oldest botanical garden at a Canadian university. The 44-hectare (110-acre) garden was established under the direction of John Davidson, British Columbia's first provincial botanist. The UBC Botanical Garden includes over 8000 different kinds of plants, has connections to UBC plant research, and one of the first native plant collections in British Columbia.

At 22 hectares (55 acres), the VanDusen Botanical Garden was named after philanthropist and lumberman Whitford Julian VanDusen. The garden is also located in Vancouver, and its collections include over 7500 plant species and varieties including a massive native plant collection. The Visitor Centre also hosts one of Vancouver's prime examples of a Garry oak ecosystem grassed green roof which is covered as a case study (Sect. 10.3.5).

10.1.5.4 Metro Vancouver Regional Parks

Metro Vancouver Regional Parks began with the formation of water and sewer districts in 1924 and 1956 respectively. Lands were set aside for biological preserves after 1967. Today, there are over 23 regional parks that preserve mature rainforests, mixed woods, and bogs. The preserves protect mainly dense forests; however, even densely wooded forests have open spaces where plants that are adapted to partial shade or sun may thrive (Fig. 10.6).

Regarding green roofs, ferns, grasses, wildflowers, groundcovers, small shrubs, and small trees may prove viable. There are over 200 species of grasses that grow in the Pacific Northwest and most of those species are native (Alaback 1994). Invasive species are present throughout; however, they are primarily on disturbed sites and urban or rural lands where early adaption species may thrive.

Fig. 10.6 Capilano Regional Park, North Vancouver, B.C., Canada. This regional park is made famous from its suspension bridge crossing the Capilano River at 140 m long and 70 m above the river. The vegetation in the park, however, is an excellent example of a temperate conifer forest. Dense second growth temperate forests dominate the steep terrain. Intermittent open canopy habitats make sun pockets, such as this, and support a more biologically diverse ground plain. Western sword fern and skunk cabbage are seen here in the foreground. (Photo: Bruce Dvorak, September 2018)

10.2 Research in the Ecoregion

British Columbia was an early adopter regarding the publication of the ecological benefits of green roofs. Early interest in Vancouver explored the possibilities of meadow-based green roofs (Pedersen 2000) and was later followed with some of the first findings regarding ecosystem services including stormwater management with green roofs, sound reduction, and climate mitigation (Connelly and Liu 2005; Connelly 2006; Roehr et al. 2008a, b, c; Roehr and Kong 2010; Connelly and Hodgson 2011; Roehr and Fassman-Beck 2015).

Canada was the first country in North America to publish peer-reviewed research regarding vegetation on green roofs. Although the study was in Quebec, it began a process that was much needed (Boivin et al. 2001). The research that Boivin and colleagues began with green roofs in 1994. Their studies looked at a mixture of six herbaceous plants growing on green roof research plots. They found that most species had adapted to the modules, but the sedum did not survive the extreme cold.

The climate and ecoregions in British Columbia are quite different from Quebec. There are several decades of green roof construction in the region; however, there is a dearth of research regarding plants on green roofs in British Columbia.

Maureen Connelly et al. pioneered green roof research at BCIT Vancouver since the 90s on rainwater monitoring, green roof assemblies and plant communities have spearheaded rigorous green roof design and experimentation in B.C. One of the significant findings was the understanding that due to dry summers, irrigation regimes at least in the early development phase of green roofs had to be employed to establish the varied vegetation and, in some cases, irrigation continued to keep the plants alive (Connelly and Liu 2005). This was a very different approach to Germanys original extensive green roofs intent were *Sedum* covered green roofs received no irrigation due to evenly distributed rainfall over the whole year. Another outcome of Connelly's collaborations a better understanding of acoustic contributions of green roofs, as they dampen sound energy through rooftops (Connelly et al. 2016).

Although there is a lack of research regarding plants on green roofs in the ecoregion, there are several sources where one can learn about, and study vegetation and plant communities that may be potentially useful for green roofs in British Columbia. Nature is the first source, in the wide range of biological preserves of the metro region. Several of those are briefly outlined in this chapter. One of the classic and most accessible publications regarding vegetation native to the region is in the book, *Plants of the Pacific Northwest Coast: Washington, Oregon, British Columbia and Alaska* (Pojar and MacKinnon 2004). This book covers 794 species of plants that are native to the Pacific Northwest. Plant profiles are arranged by plant form and family, and some diagrams of plants include images of the root systems. This may be helpful to understand how a plant may adapt to shallow substrates. Additional information regarding. Garry oak ecosystems can be found in the literature, including efforts to restore these landscapes (Erickson 1996; Fuchs 2001; Macdougall et al. 2004; Dunwiddie and Bakker 2011; McCune et al. 2013). Origins of the exploration of green roofs, their implementation, and research in North America reach back to the mid-1990s in Quebec, and Toronto, Canada (Peck et al. 1999; Boivin et al. 2001; Bass et al. 2003; Liu 2003). Some research was beginning in the United States (Miller 1998), but it was Canada, where the energy, promotion, and education of green roofs first gained traction in North America (Peck et al. 1999).

10.3 Ecoregional Case Studies (Arranged West to East)

10.3.1 Coombs Old Country Market, Coombs, B.C., Canada

Coombs, B.C. is a small community north of Nanaimo, B.C. It has a Mediterranean-like climate with little rain in the months of July and August. The community centers around the Coombs Old Country Market, which was built in the early 1970s.

The Old Country Market is a general store and restaurant that has several additions and new buildings that together make it a significant green roof cluster of old-world sustainable architecture. The green roofs on the market were originally built in 1976 by Kristian Graaten and his wife Solvieg who emigrated to Canada from Lillehammer, Norway in the 1950s (Firestone 2013). Kris learned the sod roof construction methods from his grandfather in his native land of Norway. The number of green roofs in Coombs expanded over the years with new additions to the multi-roofed building and new buildings on the grounds including additional stores, restaurants and storage sheds, and a small tree-fort condominium for the goats.

The green roof on the store has been made famous by the addition of its pygmy goats, who maintain the rooftop meadows (Fig. 10.7). Legend has it that the grass on the roof was overgrown and the Coombs County Fair was near in time. After a few glasses of wine, the family settled on an agreeable approach. The idea of maintaining the overgrown grass on the roof with goats emerged (Firestone 2013). For several decades now, the grassed roofs have kept this old-world style sod roof maintained, and famous. The green roof and goats are a top destination site on the Island. The sod has been replanted at least once since its initial installation. All of the

Fig. 10.7 View of the sod roof on the Coombs Old Country Market. All materials were locally sourced, grown, and installed by the Graaten family. The waterproofing has been replaced once since the store opened in 1976. A shallow soil substrate is edged with a log parapet. (Photo: Bruce Dvorak, September 2018)

construction was done within the family and later with the aid of new building and store management.

Construction of the green roofs originally included 15.24 cm (6 in) of native soil beneath the sod. Today only half the initial soil remains on the roof of the market store, but they have found that soil builds over time due to the fertility of the roof vegetation and natural fertilization by the goats. The original waterproofing was a rubber roof, and the most recent re-roofing in 2009 included an EPDM membrane.

10.3.1.1 Project Team

Building Owner/Client: Coombs Old Country Market
Green Roof Design Team Lead: Kristian Graaten, his wife Solveig, and their sons Svein and Andy, and son-in-law, Larry.
Maintenance Contractor: Coombs Old Country Market staff
Project completion: since 1976
Green roof area: 10,000 m^2 (107,640 ft^2) on several green roofs

10.3.1.2 Overview and Objectives

There are five green roofs on the commercial campus of buildings that total nearly 10,000 m^2 of sod grassed roofs. The old market store sod roof receives irrigation water during the dry months of the summer in order to keep appearances green along the highway and feed the goats. Other sod roofs on the campus don't receive summer irrigation, and often become dormant as the grass browns and lies flat near the end of summer. Grasses in the natural countryside are also dormant at this time, so the non-irrigated roofs take on the character of the natural landscape.

Although goats are not used to maintain any other green roofs on the island, the cultural relationship between animals and vegetated roofs could potentially prove useful in engaging the public and finding a natural solution to trimming grassed roofs and fertilizing the substrate. One issue with the goats is that they like to chew on anything including wood and trees, and they sometimes pull up the vegetation if the sod is too wet, making for bare spots. However, the managers of the roofs and goats have found a good balance, as both appear to sustain healthy relationships. There is a trend in some regions where goats are brought onto properties to maintain grass, including urban areas (Salter et al. 2013; Lovreglio et al. 2014). There is little research regarding how the presence of goats on roofs may affect the rooftop ecology.

10.3.1.3 Plant Establishment

The original soil and sod were from nearby the site. The sod for new construction and replanting of the main store that took place around 2009 came from a local landscape provider. Local grasses (fescue), clover, and meadow species.

10.3.1.4 Irrigation

The green roof on the old market store receives irrigation during periods of summer drought.

10.3.1.5 Maintenance

The goats live on the roof from April to September (Fig. 10.8). When not on the roof, the goats live on the ground on a small preserve out back. If the roof is overwatered, sometimes the goats can tear up the turf. The grass receives no artificial fertilizer, only manure from the goats. None of the other five green roofs are irrigated, thus experience summer dormancy (Fig. 10.9).

10.3.1.6 Observed Wildlife

Bees and crows are frequent visitors to the roof.

10.3.1.7 Post-occupancy Observations

Owners

- When the sod roof was first built, there were no sedum roofs in Canada. This old-world style sod grassed roof was the norm in Lillehammer, Norway. Today, after decades of development of building techniques, standards, and codes, there is some pressure to upgrade construction methods when new green roofs are proposed on the campus.
- When the sod becomes too wet, the goats can tear up the grass.
- The goats do a good job of keeping the grasses trimmed. They are well-behaved and cause little issue regarding their presence near the public.

Fig. 10.8 Nibbles (front), Minyon (middle), and Willy (under roof eve) are three of the goats that spend many of their spring and summer days grazing the rooftop meadow. In addition to the rooftop grass, the goats like to snack on maple leaves, apples, carrot tops, and just about anything. Several grasses, clover, and spring wildflowers grow with the roof sod. The main roof shown here along the roadside is watered during drought periods to keep up appearances and feed the goats. (Photo: Bruce Dvorak, September 2018)

Authors' Reflections

- This project demonstrates the simplicity and longevity of a grassed green roof.
- The maintenance of the green roof by goats is novel compared to industry standards, but could possibly be replicated on other green roofs as a viable and popular approach. This method could possibly apply to larger grassed green roofs, where mowing or trimming by maintenance crews is not feasible.

10.3.2 Shq'athut – A Gathering Place, Vancouver Island University (VIU), Nanaimo, B.C.

There are four green roofs on Vancouver Island near Nanaimo, B.C., and during their design, installation, and establishment, there was a concerted effort between the university, the city, and private industry to learn about how green roofs can be populated with native vegetation to reclaim lost habitat and to mitigate climate

Fig. 10.9 One of five buildings at the store campus with sod green roofs. Buildings such as this storage shed are not on the main highway frontage and are not irrigated. This roof photographed in late August had dormant grasses, much like the surrounding countryside. (Photo: Bruce Dvorak, September 2018)

change. Research on these roofs was largely influenced by former VIU faculty member Dr. David Gaumont-Guay. Dr. Gaumont-Guay was a member of the Department of Biology since 2009, where he established the Biometeorology Research Laboratory. Before his passing in 2016, Dr. Gaumont-Guay was the principal investigator of a green roof research lab set up at the Nanaimo campus and three other sites: the VIU Cowichan campus (VIU-COW) in Duncan, the Island West Coast Developments (IWCD) as an industrial partner, and at the transit building of the Regional District of Nanaimo (RDN), as a municipal partner.

10.3.2.1 Project Team

Building Owner/Client: Vancouver Island University
Green Roof Design Team Lead: Laura-Jean Kelly, Department of Horticulture, VIU
Architect: Alfred Waugh Architects
Installation Contractor: Vancouver Island University Horticulture
Project completion: 2011
Green roof area: 456 m^2 (4908 ft^2)

10.3.2.2 Overview and Objectives

The main objective of the Gathering Place green roof was to quantify and model the performance of extensive green roofs on Vancouver Island that were made from native plant species including three native *Sedum* species- *S. oreganum*, *S. spathulifolium* and *S. divergens* (Fig. 10.10). The extensive green roofs were studied under two different irrigation practices for their ability to sequester carbon: well-watered and drought-stressed conditions. Dr. Gaumont-Guay's mission was to learn about how extensive green roofs with native vegetation may be able to sequester carbon as a way for urban sites to reduce urban heat islands and slow global warming.

The VIC Nanaimo green roof is located on the student services building called the Shq'athut - A Gathering Place, for indigenous peoples. The building was designed by Alfred Waugh Architects, a First Nations architect, and the green roofs were designed in consolation with Dr. Gaumont-Guay and Laura-Jean Kelly an instructor in the VIC Horticulture Centre.

Fig. 10.10 The Gathering Place green roof on the Vancouver Island University student center in 2013. The newly planted roof is shown here during the plant establishment phase. (Courtesy of Rob Halsall)

10.3.2.3 Plant Establishment

The green roof research site was established with pre-grown vegetation in the forms of bulbs, herbaceous perennials, and succulents (Fig. 10.11).

Bulbs

nodding onion (*Allium cernuum*), tapertip onion (*Allium acuminatum*).

Herbaceous Perennials

crown brodiaea (*Brodiaea coronaria*).

Fig. 10.11 This image on the roof is from late August 2018 during summer dormancy and shows *Sedum spathulifolium*, *S. album* and *S. divergens* growing in the forefront of the image and allium seed heads proliferate the back of the image. (Photo: Bruce Dvorak, September 2018)

Succulents

broadleaf stonecrop (*Sedum spathulifolium*), Oregon stonecrop (*Sedum oreganum*), Pacific stonecrop (*Sedum divergens*).

10.3.2.4 Irrigation

Both green roofs received irrigation initially; however, the research demanded that portions of the roofs were not irrigated so that the role of irrigation and plant activity could be investigated. Dr. Gaumont-Guay's green roof research was not completed; however, the university maintains the vegetation to remain growing on the roofs. The roofs are not currently irrigated.

10.3.2.5 Maintenance

The roof receives only occasional weeding to remove any trees seedlings.

10.3.2.6 Observed Wildlife

Wildlife observations have not been part of the research.

10.3.2.7 Best Performing Native Vegetation

All of the original sedums still remain on the roof and some of the herbaceous vegetation remains as well. The native *Allium* are doing remarkably well as they generate blooms and seeds each year. A few grass patches exist on the roof; however, these remain small. It is believed that the grass species is *Agrostis capillaris*, an exotic species from Europe. *Sedum spathulifolium* grows at Pipers Lagoon Restoration in Nanaimo, B.C. only several kilometers from the university site. At Pipers Lagoon, the sedum grows on rocky bluffs with shallow soils on the east and partially protected sites. After visiting both locations on the same day, the patches of *Sedum spathulifolium* at the lagoon and the green roof site looked to be in similar conditions. The spread of the sedum on the VIC campus green roof appeared to be more widespread than at the natural site. Moss and lettuce leaf lichen grow in several patches on the roof.

10.3.2.8 Post-occupancy Observations

Authors' Reflections

- There is little research published regarding the performance of the native species on shallow extensive roofs such as the research site. Future research is needed to determine plant performance and maintenance of planted vegetation.
- New investigations are needed to see which additional species of the Garry oak ecosystem or coastal barrens ecosystems may be appropriate to the lowlands.
- The original carbon sequestration research begun by Dr. Gaumont-Guay is important and should be continued.

10.3.3 Island West Coast Development, Nanaimo, B.C., Canada

The Island West Coast Developments Ltd. (IWCD) is a business established to provide sustainable building solutions. Their services include design-build, construction management for commercial, multi-residential, industrial, and other building projects on Vancouver Island. The company has a mission to initiate and construct sustainable buildings including green roofs (Fig. 10.12). Their projects strive to include sustainable stormwater management, low water, and energy use, low volatile materials, and energy-efficient buildings. Green roofs are a multi-benefit investment, and so IWCD wanted to pilot test an extensive green roof with native vegetation on their headquarters building.

Based upon plant communities known to the Garry oak ecosystem, the extensive roof on their building has a variable depth substrate that ranges from 7.6 cm to 15 cm deep (3–6 in). The multi-sectioned green roofs are accessible to office workers on an observation deck. Grasses, forbs, and succulents were selected for the design. The LEED Gold-certified building was constructed in 2009, the green roof has green and flowering vegetation during the springtime and early summer. With no regular irrigation, however, the vegetation during the summer often becomes dormant, like the vegetation in the nearby Garry oak ecosystems.

10.3.3.1 Project Team

Building Owner/Client: Island West Coast Development, Nanaimo, B.C., Canada
Green Roof Design Team Lead: Victoria Drakeford, Landscape Architect
Architect: Raymond de Beeld
Project completion: 2009
Green roof area: 557 m^2 (6000 ft^2)

Fig. 10.12 This green roof is shown here when it was first planted with grasses and wildflowers native to Garry oak ecosystems. Native sedums were planted at the edges in shallow substrates. The native grasses have a 15-cm-deep (6 in) substrate at the center of the green roof (shown in 2009). (Courtesy of Island West Coast Developments, Ltd.)

10.3.3.2 Overview and Objectives

A primary objective for the green roof was to provide multiple green roof environments with vegetation native to Vancouver Island. One roof was not planted and was left as a gravel-blasted roof as a control roof to compare with the other roofs. One roof was designed as a succulent only roof and the other roofs were designed to include vegetation native to Garry oak ecosystems (Fig. 10.13). Logs and rocks were added after establishment to provide additional habitat structures. The roof is designed to attract native wildlife. Bees, butterflies, and birds are frequently seen on the one-story green roof.

Succulent vegetation is situated on the outer edges of the vegetated areas and the grasses and herbs are located throughout the interior of beds (Fig. 10.14). Growing media gently increases in depth from 7.62 cm at the edge up to 15.24 cm at the interior locations where grasses and herbs grow.

10.3.3.3 Plant Establishment

The green roof was established with pre-grown vegetation in the forms of bulbs, grasses, herbaceous perennials, and succulents (Fig. 10.15).

Fig. 10.13 Garry oak ecosystem-based green roof with mature native grasses. Shown with the addition of logs (on left) added to the roof to increase biodiversity. *Huechera micrantha* (Crevice alumroot) was observed (center) waking from summer dormancy. (Photo: Bruce Dvorak, September 2018)

Bulbs

nodding onion (*Allium cernuum*), small camas (*Camassia quamash*).

Grasses

greenleaf fescue (*Festuca ovina viridula*), Sea Urchin blue fescue (*Festuca* 'Sea Urchin').

Herbaceous Perennials

beach strawberry (*Fragaria chiloensis*), crevice alumroot (*Heuchera micrantha*), Douglas' grasswidow (*Sisyrinchium douglasii*), thrift seapink (*Armeria maritima*).

Fig. 10.14 One of the three native *Sedum* green roofs on the building, this one has a 7.6-cm-deep (3 in) substrate. Driftwood was placed on the roof post-construction to improve biodiversity. (Photo: Bruce Dvorak, September 2018)

Fig. 10.15 (**a**) Shelf fungus growing on driftwood log (behind grass), and (**b**) native alliums are in bloom, growing through a carpet of *Sedum divergens*. (Photos: Bruce Dvorak, September 2018)

Succulents

broadleaf stonecrop (*Sedum spathulifolium*), Oregon stonecrop (*Sedum oreganum*), and Pacific stonecrop (*Sedum divergens*).

Groundcovers

salal (*Gaultheria shallon*).

10.3.3.4 Irrigation

The roofs were watered initially during establishment, but have been managed without irrigation since establishment. However, in abnormally dry weather the grassed roof does receive irrigation once a week.

10.3.3.5 Maintenance

Several small compost containers are located on the roof. Periodic weeding takes place through the spring and growing season in the central protected area. The *Sedum* roofs outside the fenced areas don't receive weeding, due to lack of security.

10.3.3.6 Observed Wildlife

Bees, birds, and butterflies are frequently observed on the green roof.

10.3.3.7 Best Performing Native Vegetation

Festuca ovina viriula, Festuca sea urchin, Sedum oreganum, Sedum divergens, Huechera micrantha, Fragaria chiloensis.

10.3.3.8 Post-occupancy Observations

Owner's Observations

- The outer roofs don't need any irrigation or care. They have only been weeded three times.
- The wildflower and grassed green roof is easily accessible. This roof needs supplemental watering during the summer.

- As the vegetation is watered by hand it is not easy to remember to make trips to the roof to water.

Authors' Reflections

- This small roof is one of the few roof gardens that demonstrate the natural vegetation of the region, a Garry oak ecosystem.
- In this case, a local development company is testing green roofs as a way to explore the technology, prior to its use on future projects.

10.3.4 Burnside Gorge Community Center, Victoria, B.C., Canada

This publicly accessible green roof was funded by the City of Victoria, B.C., Canada Infrastructure Program as a way to recover lost habitat. The green roof on the Burnside Gorge Community Centre has 1066 m^2 of green roof with grasses, bulbs groundcovers, and vegetation native to Garry oak ecosystems. The building is LEED Gold-certified project, and the green roof helped attain the certification.

Fig. 10.16 Burnside Community Center roof deck during the plant establishment period. Programmable space is linked with paved and gravel paths on the roof deck that connect to a regional path seen below. Native grasses and herbs create drifts of low-profile vegetation selected for local wildlife. (Courtesy of Connect Landscape Architecture)

The previous condition of the site was an industrial use. The site had slopes falling steeply away from the street above, and posed challenges for the placement of a building. With the design of the community center, the building and green roof meet the street level access with a permeable parking lot, paths to the building entrance, and a pathway that connects to Victoria Harbor below (Fig. 10.16). The green roof habitat and paths make this green roof one of the largest publicly accessible green roofs in western Canada.

10.3.4.1 Project Team

Building Owner/Client: Burnside Gorge Community Centre, City of Victoria, B.C.
Green Roof Design Team Lead: Sharp & Diamond, Landscape Architects (Connect Landscape Architecture)
Architect: Garyali Architect Inc.,
Structural Engineer: Peterson Galloway,
Civil Engineer: Herold Engineering,
Installation Contractor: Adam Weir, Paradise Cityscapes
Project completion: 2007
Green roof area: 1066 m^2 (11,474 ft^2)

10.3.4.2 Overview and Objectives

The pre-developed project site was steeply sloped, not easily accessible, and was a dumpsite for soil and vegetation. The architectural solution included hauling away nearly 2000 loads of unusable soil to make way for a building that would cover a majority of the site (95%). The accessible green roof would connect to the street and ground-level elevation. This approach makes for the appearance of a small building from the street view, but the majority of the center is at the lower level and covers most of the site.

Randy Sharp, FCSLA, and colleagues envisioned for the green roof a connection to the historic vegetation of the region from Garry oak ecosystems (Fig. 10.17). The solution weaves waves of grass and succulent vegetation onto the roof deck, while also providing spaces for use by guests and staff. The design incorporates a combination of spaces that can be programmed and/or used, with spaces between the programmed spaces that appear to be natural.

The Garry oak ecosystem primarily supports plants that attract numerous pollinators (insects and birds). There are traditionally more than 100 small species of birds that relate to this ecosystem, including some endangered birds, so supporting the creation of new habitat was an important goal to help improve the ability for rare bird species to be maintained. The green roof has a variable depth substrate at depths of 7–15 cm (3–6 in) spanning extensive to semi-intensive systems. Where the native rose shrub is planted, soils mound to 30 cm (12 in).

Fig. 10.17 In this image (June 2008), the backside of the entrance space to the building is visible, while the community center lies below the green roof. The substrate varies in-depth with shallow substrates (7 cm) at the gravel path edges planted with native sedums, and deeper substrates (15 cm) in the mounded waves that are planted with short grasses, herbs and low shrubs, reminiscent of the Garry oak ecosystem. (Courtesy of Connect Landscape Architecture)

10.3.4.3 Plant Establishment

Vegetation was established with vegetation as plugs and container-grown plants. Twenty-four species native to the Garry oak ecosystem include sedums, cactus, grasses, bulbs, and wildflowers (Fig. 10.18). A small circular roof over the main entrance includes kinnikinic, madrone (*Arbutus menziesii*), and the rare naturally occurring hybrid: Hairy manzanita (Weir 2018).

Annuals

common woolly sunflower (*Eriophyllum lanatum*).

Bulbs

nodding onion (*Allium cernuum*), small camas (*Camassia quamash*).

Fig. 10.18 The meadow on the backside of the center has waves of grass on mounded substrate and *Sedums* planted near the gravel path in the shallow substrate. Vegetation is shown here during the fall season in late August 2018. (Photo: Bruce Dvorak, September 2018)

Grasses

California brome (*Bromus carinatus*), Idaho fescue (*Festuca idahoensis*).

Herbaceous Perennials

Columbia lily (*Lilium columbianum*), crevice alumroot (*Heuchera micrantha*), Douglas' grasswidow (*Sisyrinchium douglasii*), field pussytoes (*Antennaria neglecta*), meadow deathcamas (*Zigadenus venenosus*), rosy pussytoes (*Antennaria rosea*), thrift seapink (*Armeria maritima*).

Succulents

brittle pricklypear (*Opuntia fragilis*), broadleaf stonecrop (*Sedum spathulifolium*), Oregon stonecrop (*Sedum oreganum*), and Pacific stonecrop (*Sedum divergens*).

Low Shrubs & Groundcovers

beach strawberry (*Fragaria chiloensis*), common juniper (*Juniperus communis*), common snowberry (*Symphoricarpos albus*), dwarf rose (*Rosa gymnocarpa*), kinnikinnick (*Arctostaphylos uva-ursi*), manzanita (*Arctostaphylos* x *media*), narrowleaf sword fern (*Polystichum imbricans*), salal (*Gaultheria shallon*).

10.3.4.4 Irrigation

The irrigation system continues to be used during times of abnormal drought to prevent significant plant loss.

10.3.4.5 Maintenance

The City of Victoria maintains the green roof. Monthly visits consist of general removal of unwanted vegetation, pruning as necessary, and supplemental watering as needed.

10.3.4.6 Observed Wildlife

Pollinators including birds, bees, butterflies/moths, and beetles. They also have a colony of lizards that like to sun themselves on the concrete.

10.3.4.7 Best Performing Native Vegetation

Many of the specified plants are still present and are performing very well. However, *Sisyrinchium douglasii, Festuca ovina glauca and Festuca idahoensis* are significantly reduced on the site, and annual grasses have established.

10.3.4.8 Post-occupancy Observations

Design Team

- There are always a variety of struggles for a successful green roof. But one of the key ones for this project that we fought for and didn't win the battle completely was relating to soil depths. At the time of designing this project, the trend was to create thin roofs.
- Originally, we pushed for 30 cm (12 in) deep substrates. However, after cost estimating, project budgeting, and structural engineering, the substrate depth was

cut to half at 15 cm (6 in). Today, without continuous irrigation, the places where the soils are just too thin, the plants sometimes become stressed in the summer.

- The original plan was to try and further hide the fact that the green roof is not just part of the natural landscape, and is on a roof. So, we feel that it looks flatter than we would have liked. But especially with a public project, there was a concern about having the costs escalate for added soil loading.

Authors' Reflections

- The native grasses and low-water-use approach to irrigation allow the vegetation to go dormant during the summer. This saves water and achieves seasonal diversity. This green roof is designed to reflect the aesthetics of the natural oak ecosystems.
- It is very difficult to come back to a green roof to increase substrate depth, once a project is complete. Regardless, this project has achieved its goals to achieve a mounded aesthetic. Perhaps future projects will support deeper substrates to better support the native vegetation.

10.3.5 VanDusen Botanical Garden Visitor Center, Vancouver, B.C.

While many buildings in Vancouver have made use of succulents on extensive type green roofs, the semi-intensive green roofs on the VanDusen Botanical Garden Visitor Center are different. It makes use of a variety of vegetation that echos the ecosystems of the past—grasslands. The award-winning building and green roof have an integrated design that has attracted much attention (Fig. 10.19). The building recalls the form and shape of an orchid flower, and the vegetation recalls the Garry oak ecosystem grasses and forbs. The project represents some of the highest levels of sustainable construction integrating building, landscape, and a connection to Vancouver's natural history.

10.3.5.1 Project Team

Building Owner/Client: VanDusen Botanical Garden
Green Roof Design Team Lead: Cornelia Oberlander, with Sharp & Diamond Landscape Architects (now Connect Landscape Architecture)
Architect: Perkins + Will
Structural Engineer: Fast + EPP Engineer
Ecologist: Nick Page

Fig. 10.19 A variety of grasses, forbs, and succulents grow on the award-winning VanDusen Visitor Centre green roof. (Courtesy of Connect Landscape Architecture)

Installation Contractor: Houston Landscapes, and ZinCo Canada, Ron Schwenger, Architek
Project completion: 2011
Green roof area: 1600 m^2 (17,222 ft^2)

10.3.5.2 Overview and Objectives

Since 1975, the 22-hectare (55-acre) botanical garden has provided a setting for visitors to explore over 7500 plant species and varieties including a massive native plant collection. The visitor center is a Living Building Challenge project, which means it is categorically, one of the most sustainable buildings in the world. The building, designed by Perkins + Will, features solar hot water, photovoltaic panels, geothermal boreholes, and a prominent spire on the roof, which functions to moderate air and temperature flows inside the building (Fig. 10.20). The green roof vegetation was designed to mimic a native meadow reminiscent of Garry oak ecosystems. The pumice and organic sand-based substrate was made from local materials and are FLL compliant. The growth media was mixed with 60/20/20 percentages of pumice, sand, and organics. The media was mixed and blown up onto the roof to depths of 15–30 cm (6–12 in).

Fig. 10.20 The grassland roof transitions to a steep spire of the roof as is shown here, where many microclimates exist. Grasses, forbs, and succulents compete for space in sunny, shady, wet and dry habitats. (Courtesy of Connect Landscape Architecture, June 25, 2019)

10.3.5.3 Plant Establishment

The plant pallet includes a mix of native and some non-native plants to give the effect of a natural grassland. Due to project budget regulations, vegetation was seeded onto the roof through hydroseeding during July, instead of fall as specified. Regardless, the seed germinated 11 days after installation. Over 10,000 bulbs of allium and camas were hand planted onto the green roof during the fall of the first year.

Succulents

Native sedums were included in the mix in specific locations. The 45-degree vertical slope (Fig. 10.20) was planted with Oregon stonecrop (*Sedum oreganum*) and Pacific stonecrop (*Sedum divergens*) along with native grasses. During the establishment years, the grasses dominated the entire roof including the dramatic 45-degree central vertical slope. Over the years, drought and natural plant competition have established grasses and alliums everywhere except for the iconic vertical cone, where native sedum thrives.

Grasses

Native grasses planted on the roof include Idaho fescue (*Festuca idahoensis*). Four species of introduced grasses were included Chewing's fescue (*Festuca rubra* commutata, introduced), perennial ryegrass (*Lolium perenne*, introduced), Quatro sheep's fescue (*Festuca ovina vulgaris* 'Quatro', introduced, native to Europe), sheep fescue (*Festuca ovina* is native to Europe).

10.3.5.4 Irrigation

The green roof was irrigated for 2 weeks after hydroseeding. The roof is not currently irrigated.

10.3.5.5 Maintenance

Annual removal of tree seedlings takes place. Periodic removal of ruderal weeds is necessary. Grass is left in place each year without mowing or cutting. The lack of irrigation and lack of mowing of short stature grasses leave a thatch that helps to reduce unwanted plants.

10.3.5.6 Observed Wildlife

Bees, butterflies, and birds are observed by maintenance crews. One of the sections of the roof has a ground connection. An occasional mole and rabbit have been spotted on the roof.

10.3.5.7 Best Performing Native Vegetation

The native (and exotic) grasses are thriving on the roof. The native bulbs come and go in larger and smaller populations over the years, due to natural fluctuations in climate. Some bulbs have been replanted on the roof.

10.3.5.8 Post-occupancy Observations

Design Team

- The design team typically requests a sample mock-up (roof system and planting) especially with technically challenging design. At that time in the project, there was only one green roof supplier and we did not have the luxury of time as the project was fast-tracked to meet infrastructure funding deadlines.

- The design team typically specifies a minimum establishment watering period. For this project, the green roof did not have the luxury of reclaimed or harvested water and was prohibited from using potable water. The green roof was intended to be installed and seeded in late fall, but due to project funding requirements, planting was in July during a strong drought.
- The designers would have set up a specific management program or follow up on how it is to be managed over time - i.e. the bulb success, supplemental replanting as required. The role of the designer ended two months after construction, so they have had little influence once the project was complete.
- The designers would have pushed harder to remove the code requirement of a raised perimeter parapet which essentially hides the view of the green roof from most of the perimeter garden. It is technically feasible and common around the world to still meet roof design drainage criteria without building up a raised parapet.
- Public Access - there was an insufficient budget to provide safe access, for the public - they should consider this for later.

Authors' Reflections

- This already amazing project may have benefited from more time during the planning and design phase. Although the design team met their deadlines, complex projects such as this need more time than traditional building design to allow more time for coordination and innovation.

10.3.6 Vancouver Convention Centre, Vancouver, B.C.

In the early 2000s, when the City of Vancouver began planning to remake its convention center from the 1986 World's Fair Exposition, they set goals to make the new convention center to be a leading model for sustainable development. With the completion of the new West convention center building in April 2009, it remains as the premier example of sustainable civic construction as it is the first double LEED Platinum certified-convention facility in the world.

As the concept for the new building began to include a green roof, test plots were established beginning by 2006 to test vegetation, soils, configuration of the setup, and help sell the concept to the client. The 2.4-hectare (6 acres) meadow-roof (Fig. 10.21) was modeled after coastal meadows native to the Pacific Northwest. Since there were no remaining meadows in the Fraser delta, offsite nature preserves served as inspiration for the plant communities. Twenty-four plant species were chosen to begin the establishment phase. Over the decade that these plants have established, and just like in nature, some plants compete to thrive and others have declined. For example, a native aster has completely dominated some of the steep

Fig. 10.21 At 2.4 hectares (6 acres), the rooftop meadow is the largest patch of coastal meadow habitat in downtown Vancouver, B.C. The gently sloping roof ensures positive drainage at all locations of the roof. (Photo: Bruce Dvorak, September 2018)

slopes (56% slope) on the green roofs but fills a lesser role elsewhere on the flatter sections (3% slope) of the roof.

10.3.6.1 Project Team

Building Owner/Client: City of Vancouver
Green Roof Design Team Lead: PWL, Partnership Landscape Architects Inc., Rana Creek
Architect: CM/DA+LMN Architects
Ecologist: Paul Kephart
Installation Contractor: Holland Landscapers/NATS Nursery
Maintenance Contractor: City of Vancouver
Project completion: April 2009
Green roof area: 2.4 hectares (6 acres)

10.3.6.2 Overview and Objectives

In 2009, the City of Vancouver passed an initiative to become the "greenest city in the world" by 2020. Their goals were enforced by releasing a 10-part action plan addressing carbon, waste, and ecosystems. The green "living" roof became an iconic example of how building and green roof ecosystems can become fully integrated. The green roof in part relies on the functions of the convention center inside and below. One of the most impressive contributions of the green roof is that it recycles all water from inside the convention center as an irrigation source, and it acts as an insulator to reduce heat gain during the summer by up to 95% and heat loss in the winter by up to 26%.

A 30 cm (12 in) substrate was originally proposed for the rooftop meadow, but the depth was scaled back to a 15 cm-deep (6 in) substrate to meet structural limitations. One of the most important aspects of the meadow is how its varied slopes (3–56%) mimic natural habitats. Although flat 2% roofs dominate on standard green roofs, these roofs more reflect the natural diversity of slope and aspect found in nature. The vegetation has seemed to find its most adaptable microclimates. The growing media substrate is made from sand that is dredged from the Fraser River. The sand is part of the natural dredging process to clear routes for ships, so no new sources of sand were needed. Other substrate components include lava rock and various sources of garden waste.

10.3.6.3 Plant Establishment

The green roof vegetation was established through a combination of live plugs, bulbs, and grass seed applied as a hydromulch. Forms of vegetation include annuals, bulbs, grasses (Table 10.1), herbaceous perennials, and succulents.

Table 10.1 Grasses on the Vancouver Convention Centre green roof

Common Name	Botanical Name
bent grass	*Agrostis pallens*
slimstem reed grass	*Calamagrostis stricta*
dense sedge	*Carex densa* (native to west coast of U.S.)
chamiso sedge	*Carex pachystachya*
Pacific dune sedge	*Carex pansa*
Berkeley sedge	*Carex tumulicola*
Idaho fescue	*Festuca idahoensis*
creeping red fescue	*Festuca rubra*
prairie Junegrass	*Koeleria macrantha*
Pacific meadow sedge	*Potentilla anserina*
western blue-eyed grass	*Sisyrinchium bellum* (native to California and Oregon)

Annuals

California poppy (*Eschscholzia maritima*, native to the western U.S.)

Bulbs

common camas (*Camassia quamash*), fool's onion (*Brodiaea hyacinthina*), harvest brodiaea (*Brodiaea coronaria*), Hooker's onion (*Allium acuminatum*), nodding onion (*Allium cernuum*).

Herbaceous Perennials

beach strawberry (*Fragaria chiloensis*), common thrift (*Armeria maritima*), Douglas aster (*Aster subspicatus*), Pacifica silverweed (*Argentina pacifica*), pearly everlasting (*Anaphalis margaritacea*).

Succulents

broadleaf stonecrop (*Sedum spathulifolium*), located at edges.

10.3.6.4 Irrigation

Treated convention center black water is pre-treated and then pumped to the substrate on the roof. Drip-tubing is buried within the substrate layer and it is activated by moisture sensors (Nightingale 2010). There are 65 irrigation zones and eight moisture sensors that function on the roof to regulate when the roof needs to be watered. The roof only receives water primarily during the summer or other times of drought. The sensors are set at about a 15% volumetric threshold for minimum moisture.

10.3.6.5 Maintenance

The green roof vegetation is cut back each fall around October and material is hauled offsite for composting. Organics are fed onto the roof from scraps of food left-over from the convention center cafeteria, nearby parks, or other approved sources. Nutrients are also sustained from the irrigation water, which has some nutrients leftover from the on-site blackwater harvesting and treatment process.

Fig. 10.22 The meadow vegetation shown here (near its full hight) is mown down each fall in October. From fall through early spring the vegetation remains low in stature. From springtime and through the growing season, the meadow vegetation grows in stature can range from 60 cm, and in some areas up to 2 m. (Photo: Bruce Dvorak, September 2018)

10.3.6.6 Observed Wildlife

The lush vegetation (Fig. 10.22) creates a habitat for hummingbirds, butterflies, migratory songbirds, and insects. Six bee hive maintains over 60,000 bees, although there are also ground-nesting bees and wasps. Seagulls frequent the roof and leave behind on the roof shells from crabs and other shellfish. Two to three years into the establishment of the living roof, ants, and worms had established on the roof. Raccoons have been sited on the roof in search of mice. The grasses make for good nests for sparrows and Canada Geese. At least two Canada Geese have nested on the rooftop meadows.

10.3.6.7 Best Performing Native Vegetation

All the native grasses are thriving. Unique species growing on the roof include the native potentilla, pearly everlasting (Fig. 10.23), asters, beach strawberry, onions, sedges, fescues, poppies, and native camas bubs.

Fig. 10.23 A showy native plant pearly everlasting (*Anaphalis margaritacea*) in full bloom on the convention center. Leaves of the pearly everlasting attract the American painted lady butterfly (*Vanessa virginiensis*). (Photo: Bruce Dvorak, September 2018)

10.3.6.8 Post-occupancy Observations

Design Team (PWL Landscape Architecture)

- After 10 years of watching the roof grow, change and adapt to its surroundings, the biggest lesson learned on this project is that plants, animals, microbes, and insects are incredibly adaptable, much more than our team ever expected. Recent field reviews by an entomologist from the University of British Columbia revealed two insects, a two-spotted lady beetle (*Adalia bipunctata*) and a parasitic wasp that had not been seen in urban Vancouver since the 1930s.
- We never dreamed that the plants would perform as well as they have. Our initial suggestion for a maintenance regime was a yearly cutting that left all the biomass on the roof to decompose and provide organic material for the plants and other living organisms. This has been changed to the removal of the biomass created after the fall cutting. The plant material grows far faster and with more vigor than expected.

Authors' Reflections

- This integrated design accomplished its aggressive goals to make a resilient and low-energy and low-water-use project. The sustainable source of irrigation water keeps this biologically diverse roof meadow active throughout the summer drought.
- This project demonstrates the gains made by the City of Vancouver because it was willing to take a risk, fund, and give appropriate time for the design and development of the project. It is one of the largest and most sustainable green roofs in North America.

10.3.7 Southeast False Creek Development, Olympic and Paralympic Village, Vancouver, B.C.

In preparation for the 2010 Winter Olympics, the City of Vancouver, B.C. engaged in one of the largest sustainable urban waterfront renewal projects in North America (Fig. 10.24). The planning efforts included eight city blocks that would include mixed-use development, market spaces for sale and rental, non-market housing, senior housing, commercial development including a food store, and a Community Centre. PWL Partnership Landscape Architects, Inc. lead the site development at the public realm, PFS Landscape Architects led the Central Plaza, and DKL Landscape Architects lead private development parcels, the Community Centre, and over 1.83 hectares (197,542 ft^2) of green roofs. With the green space and green roofs, the project has nearly 48% green over the 3.8-hectare dense urban site (Kreuk 2014).

The green roofs include 7024 m^2 (75,606 ft^2) of extensive green roofs, 10,539 m^2 (113,441 ft^2) of intensive green roofs as private roof gardens, and 789 m^2 (8495 ft^2) of rooftop urban agriculture. The extensive green roofs include the iconic flowering sedum roofs depicting winter sports themes. However, these icon roofs are just a fraction of the green roofs on-site and they are no longer managed for their original Olympic themes. The many private green roofs, however, do remain and many include native trees, shrubs, and groundcovers (Fig. 10.25). As these are private roof gardens, they are not accessible to the public (Kreuk 2014).

10.3.7.1 Project Team

Building Owner/Client: the City of Vancouver and private ownership
Green Roof Design Team Lead: DKL – Landscape Architects
Architect: Nick Milkovich Architects in collaboration with Walter Francl Architecture (Creekside Community Centre)

Fig. 10.24 View from a boardwalk at the Millennium Water Front Park to apartments with many roof gardens on buildings (foreground and background). It is part of an eight-block urban redevelopment project which includes green roofs on all of the buildings. (Photo: Bruce Dvorak, September 2018)

Landscape Architect: PWL Partnership Landscape Architects, Inc. (public realm waterfront and parks)
Installation Contractor: LiveRoof, XeroFlor Canada, and others.
Project completion: 2009
Green roof area: over 1.83 hectares (4.5 acres) of green roofs.

10.3.7.2 Overview and Objectives

There were a variety of approaches used in the design of roof gardens and extensive green roofs. Native and ornamental (exotic) vegetation was used throughout, including the meadow roof on the Creekside Community Centre (Fig. 10.26). The roof gardens were designed by many different firms. The approach for the extensive *Sedum* green roofs was to make a visible impact and identify with the sporting events taking place. The green roofs on the Community Centre include a meadow roof, succulent roof, urban agriculture, and a rooftop daycare with planters.

Fig. 10.25 One of the many private roof gardens in the Southeast False Creek Olympic Village. This one includes native trees. (Photo: Bruce Dvorak, September 2018)

10.3.7.3 Plant Establishment

The vegetation for the main roof section of the center was pre-grown in LiveRoof, LLC modules. There are two designs for the community center: (1) a sedum-mix for below solar panels and (2) a meadow-mix for the main roof above the second story. The sedums have dominated on modules below the solar panels, and on the meadow roof in a few locations where microclimate favors succulents such as edges and transitions to the pavement. The modules for the solar panels are sedums only and the meadow mix includes a small proportion of sedums mixed with the grasses (Fig. 10.27). After a decade of growth, the majority of the main meadow roof is thriving.

Succulents

The native Pacific stonecrop (*Sedum divergens*) was included with nine species of exotic and naturalized sedums that were used on the extensive roofs. Sedums were planted under the photovoltaic panels.

Fig. 10.26 Image of the Creekside Community Centre meadow roof in the foreground and the solar roof in the background. Views onto the top of the Community Centre are visible from the many and taller surrounding apartment buildings. (Courtesy of Matt Collinge)

Grasses

Idaho fescue (*Festuca idahoensis*), prairie junegrass (*Koeleria macrantha*), Sandberg bluegrass (*Poa secunda*), also one exotic grass.

10.3.7.4 Irrigation

The collection of rainwater onsite and runoff from pavement and roofs allows for a massive onsite water storage and treatment system. On an annual rain year, the site receives about 3340 m^3 (117,951 ft^3) of rainfall which is collected and used for 2892 m^3 (102,130 ft^3) of water for toilet flushing, and 523 m^3 (18,469 ft^3) is used for site irrigation (Kreuk 2014). The modular green roofs receive irrigation.

10.3.7.5 Maintenance

The grasses are cut back in mid-August. This gives the grasses an opportunity for some regrowth before winter and allows more sunlight to reach the sedums. Plants are replaced as needed, in areas where they may be trampled such as around the solar panels.

Fig. 10.27 View From the top of the meadow roof on the Community Centre looking back toward apartments of the Olympic Village and the city of Vancouver in the background. (Courtesy of DKL Landscape Architecture)

10.3.7.6 Observed Wildlife

There is no formal observation of wildlife taking place on the green roof.

10.3.7.7 Best Performing Native Vegetation

All of the grasses and sedums are established and are performing well.

10.3.7.8 Post-occupancy Observations

Authors' Reflections

- This project demonstrates how native grasses can be incorporated into modular green roofs, and thrive. The context of the roof is near to water and is worthy of study regarding the use of the green roof by birds or other wildlife.
- Water harvesting is essential to making this integrated design work. In this case, all the team was brought together before the development of the project.

10.4 Plants for the Vancouver British Columbia ecoregions

Across the seven case studies in Chap. 10, there are 45 plant species in total, 42 of which are native to the ecoregions in this chapter. Of those native to chapter ecoregions, 13 species occur more than once across the case studies in this chapter. Of those occurring more than once, three are bulbs, one is a grass, one is a groundcover, one is a shrub, four are herbaceous perennials, and three are succulents. Seven species occur in three or more of the case study sites in the chapter (Table 10.2).

Many of the case studies make use of gentle or moderate slopes, varied substrate depths, mixing species in groups or communities, also mixing of plant forms such as annuals, perennials, bulbs, groundcovers as a way to make articulated habitats for the special needs of plants or wildlife that are not typical of green roofs with shallow sedum-monocultures.

In addition to plants trialed on the ecoregional green roofs, other taxa native to the region could be considered include plants of the Asteraceae family (*Aster* sp., *Antennaria* sp., *arnicas* sp., *Artemisia*, (*goldenrods*), Aspleniaceae family, Berberidaceae family (Oregon-grape), Bryophyta (mosses & liverworts), Cactaceae (cacti), *Carex* sp. (sedges), Caprifoliaceae family (snowberry), Dryopteridaceae family (ferns), Equisetaceae family (horsetails), Juncus (rushes), Lycopodiaceae family, Orobanchaceae family (paintbrush), Plantaginaceae family (*Penstemons* sp.), Polypodiaceae family, Pteridaceae family, Ranunculaceae family (*Anemones,* sp. columbines, buttercups), Rosaceae family (strawberry, shrub roses, non-invasive species).

Many of the above-mentioned species have a presence on open or disturbed sites. Making use of plants that function to revegetate after site disturbances may be useful for shallow extensive green roofs. There are also over 200 species of grasses that grow in the Pacific Northwest and most of those species are native, and few have been trialed on green roofs (Alaback 1994).

Unique to the ecoregion are mosses. There are near 700 species of mosses or liverworts listed on the *E-Flora BC: Electronic Atlas of the Flora of British*

Table 10.2 Taxa that occur in three or more of the ecoregional green roof case study sites in the chapter

Plant Type	Common Name	Botanical Name	A	B	C	D	E	F
Bulb	nodding onion	*Allium cernuum*	x	x	x	x		
Bulb	small camas	*Camassia quamash*		x	x	x		
Groundcover	beach strawberry	*Fragaria chiloensis*		x	x	x		
Herbaceous Perennial	thrift sea pink	*Armeria maritima*		x	x	x		
Succulent	Pacific stonecrop	*Sedum divergens*	x	x	x		x	x
Succulent	Oregon stonecrop	*Sedum oreganum*	x	x	x		x	
Succulent	broadleaf stonecrop	*Sedum spathulifolium*	x	x	x	x		

Key = A (Vancouver Island University), B (Island West Coast Development), C (Burnside Gorge Community Center), D (Vancouver Convention Centre), E (VanDusen Botanical Garden Visitor Center), F (Southeast False Creek Development)

Columbia site and referenced in other sources (Schofield and Drukker-Brammall 1992). Some mosses may be useful for some green roofs or to provide special functions or in unique microclimates (Pedersen 2000; Prince 2016). Given that large portions of the Fraser Lowlands were previously covered in bogs and mossy soil, there may be some function for mosses on some kinds of green roofs with shallow substrates, in shade, to absorb rainfall or make special habitats (Dunnett and Kingsbury 2004; Grant 2006; Heim et al. 2014; Prince 2016).

10.5 Summary

The seven ecoregional green roof case studies describe the breadth of experimentation that has taken place in British Columbia since the building of some of the first green roofs designed by Cornelia Hahn Oberlander and architect Artur Erickson at the Law Courts Vancouver in the late 70s. Innovations since then include the planting of native plant communities on extensive and intensive green roofs highlighted in the ecoregional case studies. The case studies here demonstrate that green roofs were built to:

- connect with the cultural and natural heritage of the place, e.g., Coombs Old Country Market, Shq'athut – A Gathering Place VIU, Island West Coast Development, Vancouver Convention Centre;
- provide a learning tool (pilot project) to demonstrate green roofs, e.g., Vancouver Convention Centre, Shq'athut – A Gathering Place VIU, Burnside Gorge Community Center;
- provide an outdoor meeting space for residents, workers, or visitors, e.g., Island West Coast Development, Southeast False Creek Development, Olympic and Paralympic Village, Burnside Gorge Community Center;
- benefit from the ecosystem services generally associated with green roofs, energy conservation, runoff amelioration, temperature and noise abatement (and this would include roofs using non-native species as well) e.g., VanDusen Botanical Garden Visitor Center, Vancouver Convention Centre, Creekside Community Centre, Burnside Gorge Community Center;
- Function as a green roof research site, e.g., Shq'athut - A Gathering Place VIU, Vancouver Convention Centre;
- increase local habitat diversity, e.g., Island West Coast Development, Burnside Gorge Community Center, Vancouver Convention Centre;
- increase the aesthetic appeal of buildings, e.g., Vancouver Convention Centre, VanDusen Botanical Garden Visitor Center, Coombs Old Country Market;
- provide seasonal attributes of vegetation that were addressed during the plant selection and ongoing maintenance, e.g., Coombs Old Country Market, Burnside Gorge Community Center, Vancouver Convention Centre.

One of the big challenges for green roofs in this region is the idea that green roofs should always be green. There is a lack of public understanding that they are not always green due to their summer dormant phase and stress caused by drought. Therefore, it is crucial to clearly communicate the design objectives of a green roof in the concept design phase, such as stormwater management or all-year-round green appearance or both. Communication of design objectives is important to make sure that recycled water, if needed, is provided for in the design early in the planning phase of the project. Another important factor is maintenance cost and the plant knowledge on how to manage desired plant communities and the removal of unwanted plants. Thus, collaborations between universities, nurseries, and municipal organizations are necessary to develop a common set of goals and expectations.

Acknowledgments We would like to thank the following individuals for dedicating their time, resources, and sharing of their knowledge of green roofs and native landscapes including John Ross, City of Vancouver; Guy Pottinger with VanDusen Botanical Garden; Bruce Hemstock, and Margot Long with PWL Partnership Landscape Architects; Dr. Caroline Josefsson and Rob Halsall with the Vancouver Island University; Peter Kreuk with Durante Kreuk; Rachel O'Neill with the Burnside Gorge Community Association; Ken Larsson and David Stoyko with Connect Landscape Architecture; Dr. Richard Hebda with the Royal BC Museum and Archives; Patricia Perrone student at the University of British Columbia; Troy-Anne Constable and Samantha Sharpe with the Island West Coast Development; Rod Nataros at NATS Nursery; Michael Wisshack with Etera; Karen Needham at the University of British Columbia; and Peter and Joy Schmidt with Vitaroofs.

References

Alaback PB (1994) Plants of the Pacific Northwest Coast: Washington, Oregon, British Columbia & Alaska. Lone Pine Pub, Vancouver

Bass B, Krayenhoff E, Martilli A, Stull RB, Auld H (2003) The impact of green roofs on Toronto's urban heat island. Proceedings of Greening Rooftops for Sustainable Communities:1–13

Boivin M-A, Lamy M-P, Gosselin A, Dansereau B (2001) Effect of artificial substrate depth on freezing injury of six herbaceous perennials grown in a green roof system. HortTechnology 11(3):409–412

Brenneisen S (2006) Space for urban wildlife: designing green roofs as habitats in Switzerland. Urban Habitats 4(1):27–36

Clague JJ, Luternauer JL, Hebda RJ (1983) Sedimentary environments and postglacial history of the Fraser Delta and lower Fraser Valley, British Columbia. Can J Earth Sci 20(8):1314–1326

Connelly M (2006) British Columbia Institute of Technology, Centre for the Advancement of Green Roof Technology, Report to Canada Mortgage and Housing Corporation (CMHC).

Connelly M, Bolbolan D, Akbarnejad M, Daneshpanah S (2016) Sound absorption and scattering properties of living walls: applications to room acoustics. Can Acoust 44(3)

Connelly M, Hodgson M (2011) Laboratory experimental investigation of the acoustical characteristics of vegetated roofs. J Acoust Soc Am 129(4):2393–2393

Connelly M, Liu K (2005) Green roof research in British Columbia-an overview. Paper presented at the third annual greening rooftops for sustainable communities conference, Washington, DC, May 4–6, 2005

Dunnett N, Kingsbury N (2004) Planting green roofs and living walls. Timber Press, Portland

Dunwiddie PW, Bakker JD (2011) The future of restoration and management of prairie-oak ecosystems in the Pacific northwest. Northwest Sci 85(2):83–93

Emmerson J (1865) British Columbia and Vancouver Island. voyages, travels, & adventures. W. Ainsley, Durham

Erickson WR (1996) Classification and interpretation of Garry oak (Quercus garryana) plant communities and ecosystems in southwestern British Columbia. University of Victoria, Victoria

ESWG (1995) A national ecological framework for Canada. Ecological Stratification Working Group, Center for Land Biological Resources Research, Canada. State of the Environment Directorate, Ottawa/Hull

Firestone BM (2013) Goats on the roof; how entrepreneurs change their communities/lessons from tiny Coombs, BC, vol 2019. Bruce M Firestone, Ottawa

Ford ED, Malcolm DC, Atterson J (1979) The ecology of even-aged forest plantations. Institute of Terrestrial Ecology, Cambridge

Francis RA, Lorimer J (2011) Urban reconciliation ecology: the potential of living roofs and walls. J Environ Manag 92(6):1429–1437. https://doi.org/10.1016/j.jenvman.2011.01.012

Franklin JF, Denison W, McKee A, Maser C, Sedell J, Swanson F, Juday G (1981) Ecological characteristics of old-growth Douglas-fir forests. Gen. Tech. Rep. PNW-GTR-118. US Department of Agriculture, Forest Service, Pacific Northwest Research Station, Portland

Fuchs MA (2001) Towards a recovery strategy for Garry oak and associated ecosystems in Canada: ecological assessment and literature review. Technical Report GBEI/EC-00-030. Environment Canada, Canadian Wildlife Service, Ladner, British Columbia

Gedalof Z, Pellatt M, Smith DJ (2006) From prairie to forest: three centuries of environmental change at Rocky Point, Vancouver Island, British Columbia. Northwest Sci 80(1):34–46

Gedge D, Kadas G (2004) Bugs, bees and spiders: green roof design for rare invertebrates. Paper presented at the second annual greening rooftops for sustainable communities conference, Portland, Oregon, June 2–4

Gedge D, Kadas G (2005) Green roofs and biodiversity. Biologist 52(3):161–169

Grant G (2006) Green roofs and façades, vol 70. IHS Bre Press, Bracknell

Heim A, Lundholm J, Philip L (2014) The impact of mosses on the growth of neighbouring vascular plants, substrate temperature and evapotranspiration on an extensive green roof. Urban Ecosyst 17(4):1119–1133

Kadas G (2006) Rare Invertebrates Colonizing Green Roofs in London. Urban Habitats 4(1):66–86

Kowarik I (2011) Novel urban ecosystems, biodiversity, and conservation. Environ Pollut 159(8–9):1974–1983. https://doi.org/10.1016/j.envpol.2011.02.022

Kreuk P (2014) Landscape on the roof: southeast False Creek the Olympic Village. Durante Kreuk Landscape Architecture, Vancouver

LBC (2017) Living building challenge. Tratto da Living Building Challenge: https://living-future.org …

Liu K (2003) Engineering performance of rooftop gardens through field evaluation. In: Proceedings of the 18th International Convention of the roof Consultants Institute, Tampa, FL, March 13, 2003. vol NRCC-46412. National Research Council, Institute of Research in Construction, pp 1–15

Lord JK (1866) The naturalist in Vancouver Island and British Columbia, vol 1. Richard Bentley, London

Lovreglio R, Meddour-Sahar O, Leone V (2014) Goat grazing as a wildfire prevention tool: a basic review. Iforest-Biogeosci For 7(4):260

MacDougall A (2002) Invasive perennial grasses in Quercus garryana meadows of southwestern British Columbia: prospects for restoration. In: Standiford RB et al (ed) Fifth symposium on oak woodlands: oaks in California's challenging landscape, Albany, CA, 2002. Pacific Southwest Research Station, Forest Service, US Department of Agriculture, pp 159–168

Macdougall AS, Beckwith BR, Maslovat CY (2004) Defining conservation strategies with historical perspectives: a case study from a degraded oak grassland ecosystem. Conserv Biol 18(2):455–465

McCune JL, Pellatt MG, Vellend M (2013) Multidisciplinary synthesis of long-term human–ecosystem interactions: a perspective from the Garry oak ecosystem of British Columbia. Biol Conserv 166:293–300

Miller C 1998 Vegetated roof covers, a new method for controlling runoff in urbanized areas. In: Proceedings from the 1998 Pennsylvania stormwater management symposium, Villanova, PA, 1998. Villanova University, 1998, October 21–22

North ME, Dunn M (1979) Vegetation of the southwestern Fraser lowland, 1858–1880 1: 50 000. Lands Directorate, Environment Canada, Ottawa

Omernik JM (1995) Ecoregions: a framework for managing ecosystems. In: The George Wright Forum. vol 1. JSTOR, pp 35–50

Parachnowitsch A, Elle E (2005) Insect visitation to wildflowers in the endangered Garry Oak, Quercus garryana, ecosystem of British Columbia. Canadian Field-Nat 119(2):245–253

Peck SW, Callaghan C, Kuhn ME, Bass B (1999) Greenbacks from green roofs: forging a new industry in Canada. Canada Mortgage & Housing Corporation, Ottawa

Pedersen KN (2000) Meadows in the sky: contemporary applications for eco-roofs in the Vancouver region. University of British Columbia, Vancouver

Pojar J, MacKinnon A (2004) Plants of the Pacific Northwest coast. Partners Publishing Group, Vancouver

Prince BA (2016) An ecological analysis of the potential for moss-based green roof design. University of Maryland, College Park

Ritland K, Meagher L, Edwards D, El-Kassaby Y (2005) Isozyme variation and the conservation genetics of Garry oak. Botany 83(11):1478–1487

Robb CK (2014) Assessing the impact of human activities on British Columbia's estuaries. PLoS One 9(6):e99578

Roehr D, Fassman-Beck E (2015) Living roofs in integrated urban water systems. Routledge, London

Roehr D, Kong Y (2010) Runoff reduction effects of green roofs in Vancouver, BC, Kelowna, BC, and Shanghai, PR China. Can Water Res J 35(1):53–68

Roehr D, Laurenz J, Kong Y (2008a) A comparison of stormwater runoff reduction by green roofs between Kelowna and Vancouver. In: Proceeding of one watershed-one water 2008 conference, pp 178–186

Roehr D, Laurenz J, Kong Y (2008b) Green envelopes. In: Proceedings of the 2008 international LID conference, pp 16–19

Roehr D, Laurenz J, Kong Y (2008c) Green envelopes: contribution of green roofs, green facades and green streets to reducing Stormwater runoff, CO2 emissions and energy demand in cities. In: She N, Clar M (eds) 2008 international low impact development conference, Seattle, WA. ASCE, p 13

Salter M, Macdonald E, Richardson Z (2013) Prescribed goat grazing in urban settings: a pilot study of the legal framework in nine us cities. Paper presented at the place, time, space, duration, CELA Austin, TX, March 27–30, 2013

Schofield WB, Drukker-Brammall P (1992) Some common mosses of British Columbia. Royal British Columbia Museum, Victoria

Wang M (2020) Biophilic community. University of British Columbia

Weir A (2018) Burnside-Gorge Community Centre. greenroofs.com. https://www.greenroofs.com/projects/burnside-gorge-community-centre/. Accessed 5 June 2018

Part III
Summary and Future Outlook

Part III is a summary discussion of concepts outlined in Parts I and II as they relate and align with key observations uncovered from information and lessons learned in the conservation site and ecoregional case studies. The chapter also discusses potential areas for future research, and potential applications of concepts presented in the book.

Chapter 11
Ecoregional Green Roofs, Infrastructure, and Future Outlook

Bruce Dvorak and Lee R. Skabelund

Abstract Chapter 11 provides a synthesis of and commentary about ecoregional green roofs informed by the case studies in Part II. Discussions address critical factors outlined in Chap. 2 such as the conservation of native plant communities, the use of native vegetation on green roofs, biodiversity, maintenance, water for irrigation, and microclimates. This chapter also discusses how ecoregional green roofs have been used as a part of therapeutic and biophilic design, how green roofs play an important role in the development of green and sustainable architecture, integrated site design, and parameters implicit to LEED, the Living Building Challenge, and SITES programs. We also discuss the use, integration, and potential expansion of ecoregional green roofs at various scales as well as landscape structures such as corridors, patches, and matrices. The chapter concludes with a look at future opportunities for green roofs. We discuss research gaps, where new knowledge and research are needed. We also note opportunities related to policies, education, industry support, and innovation.

Keywords Microclimate · Wildlife habitat · Green architecture · Ecodistrict · Landscape ecology · Corridor · Research · Integrated design · Land ethic · Biophilia · Ecological aesthetic

B. Dvorak (✉)
Department of Landscape Architecture and Urban Planning, 305A Langford Architecture Center, Texas A&M University, College Station, TX, USA
e-mail: bdvorak@tamu.edu

L. R. Skabelund
Department of Landscape Architecture and Regional & Community Planning, Kansas State University, Manhattan, KS, USA
e-mail: lskab@k-state.edu

B. Dvorak (ed.), *Ecoregional Green Roofs*, Cities and Nature,
https://doi.org/10.1007/978-3-030-58395-8_11

11.1 Ecoregional Approach to Green Roofs

The case studies in Part II of this book demonstrate many ways to plan, design, irrigate, and maintain ecoregional green roofs. The green roofs reviewed in the case studies have been designed with different goals regarding the use of native plants: all native vegetation, with a majority of the plants native, and/or partially planted with native plants. Some were planted from seeds that were harvested on the same property of the green roof, some had plants native to the ecoregion, and some had plants that were native to other ecoregions located in the western United States or Canada. Several green roofs were designed to mimic a particular plant community, while others used native plants with no particular plant community in mind—but primarily intended to create a garden-like setting. The case studies demonstrate a great variety of solutions, including examples of ecological resilience and ways that green roofs can specifically connect to local ecosystems and ecosystem functions. This section reviews some common findings from the case studies and expands on how ecoregional green roofs and native plant communities can inspire a variety of creative and sustainable design solutions for rooftop ecosystems on buildings, structures, or implement ecoregional green roofs into citywide systematic applications.

11.1.1 Native Plant Communities

Each chapter in Part II of the book begins with descriptions of the native vegetation and plant communities present in 20 major ecoregions west of the 100th meridian. Through informed interpretation by naturalists, ecologists, conservationists, landscape architects, botanists, and related fields, the native plant communities conserved in or near urban regions can teach us about which plants or plant communities may adapt to green roofs (Kephart 2005; Lundholm 2005; Brenneisen 2006; Sutton 2015). Remnant native plants located on-site where a green roof is or will be constructed can connect to the natural history of the region. When native plants are preserved on-site or on nearby conservation sites, visitors can learn about the aesthetics of native vegetation, and help build expectations for their use on green roofs. Many of the conservation sites reviewed in this book are publicly accessible. Some had green roofs on the property and included interpretative signage and discussed the use of native vegetation on the green roof. Conservation sites may prove useful for learning about how plants from one or more ecological communities may potentially adapt to green roof conditions.

For the cities covered in this book, some nearby conservation sites were selected because of the quality of the habitat, ease of accessibility, and opportunity for the public to visit these sites. Some geographic regions have many conservation sites that may provide useful when selecting vegetation for green roofs and were not included in the book. The Pacific Northwest, for example, has many conservation sites that are not included in this book but could be investigated as sources for potential green roof plants. Some locations, however, such as Salt Lake City, Utah, and Los Angeles, California have only a very limited number of conservation sites that remain. Portland, Oregon, Seattle, Washington, San Francisco, California, and Vancouver, B.C. and Nanaimo, B.C. in Canada all engender thriving conservation movements with multiple conservation sites, and most are managed with entities with specific goals to restore native landscapes (Hamman et al. 2011). These groups include volunteer organizations, universities, local and/or state agencies, and non-profit organizations (Langenfeld 2009).

When restoration activities are under the direction of knowledgeable site stewards these important places of learning and exploration can enrich and inform how native plants can be used on green roofs and other forms of green infrastructure (Ahern 2007; Kowarik 2011). Thus, environmental education can be a critical and necessary process to transfer ecological values from one generation to the next (Sauvé 1996; Bailey 2002; Hobbs and Cramer 2008). In addition, online, mobile-based citizen science programs can help build recognition of plant and animal diversity near cities, and teach the next generation about conservation of biodiversity through the use of digital field guides such as e-Bird, iNaturalist, and other user-friendly programs (Fournier et al. 2017).

As the human population increases, the protection of native landscapes near urban regions becomes more important, not just for what we can learn from them, but also for human and ecological health. Native landscapes should be protected, and green roofs should not be used as mitigation for the destruction of healthy, threatened, or culturally valuable ecosystems (Lundholm and Walker 2018). On the other hand, in developed areas where little or no native habitat remains, green roofs can reintroduce important but limited ecosystem services back into an urban location. Native landscapes in urban areas function as important reserves for the systematic study of ecoregional vegetation, and they can provide education, recreation, and enjoyment, and may be used as part of an integrated network of conservation sites.

Several studies have compared some of the native vegetation on conservation sites and green roofs. At Kansas State University, the Memorial Stadium green roofs (which collectively include over 20 native forbs that were planted or seeded) were found to support many similar butterflies an urban prairie within Manhattan, Kansas and the Konza Prairie south of Manhattan, although only generalist butterfly species were found on these two green roofs (Blackmore 2019). Ground nesting birds have made use of both nearby prairie and the green roof at the Botanical Research Institute of Texas in Fort Worth (Best et al. 2015). The California Academy of Sciences living roof in San Francisco, California has hundreds of citizen science observations of wildlife using the green roof and landscape below (CLO 2019). These examples begin to demonstrate how ecoregional green roofs can amplify conservation, research, and education efforts in urban areas.

11.1.2 Native Plants on Green Roofs

The 73 ecoregional green roof case studies present a range of successful solutions regarding the use of native plants. Across all of the ecoregions covered in Part II, more than 830 taxa of plants were found native to the ecoregions covered in the book representing 357 genera. Likewise, 36 additional taxa native to ecoregions outside of the regions covered in this book but are native to North America (e.g. Mexico, eastern or Midwestern U.S.) were found on the green roofs studied.

Nearly every geographic region has at least one or more project intended to make use of all native vegetation. In the tallgrass prairie ecoregion, nearly all of the case studies have native vegetation. The Camp Judea Young prairie roof, K-State Memorial Stadium, Trinity Unversity, and Ross Perot Museum of Nature and Science green roofs stand out as a close association and mixture of plants that are similar to the native analogs in the region (although less diverse). In the shortgrass prairie ecoregion, the Barry Biodiversity Center green roof at the University of Wyoming employs regionally native vegetation, as did a private residence near Boulder, Colorado. These projects had goals from the beginning to model the shortgrass prairie. In the Desert Southwest, the Anza-Borrego Desert State Park Visitor Center green roof mimics vegetation of the Mojave Desert living on its roof. In intermontane ecoregions, the Moda Building and Washington Fruit Growers Headquarters buildings set out to specifically use all native vegetation from the immediate environs.

Many of the grassland green roofs in multiple ecoregions used seeds of native plants to establish the green roofs. In the Pacific Northwest, many green roofs were established with all native vegetation including on the Cedar River Watershed and Gunderson eco-roofs, and at the Vancouver Convention Centre. In California, the California Academy of Sciences, the EcoCenter at Heron's Head Park, the NOAA Southwest Fisheries building, and the Slide Ranch nature center buildings provide examples of how to design and maintain green roofs assembled from local plant communities.

As it is illegal to harvest plants or seeds from conservation sites such as national or state parks and some private conservation sites, one must obtain native vegetation (both seeds and live plants) from reputable sources. Regarding the production and supply of plants for green roofs, there were green roofs that were planted from seeds harvested on the private property where the green roof was located. Green roofs were also planted with seed that was provided from local or regional and seed supply companies, and there were green roofs planted with pre-grown vegetation in pots, sod, or in modules. Some ecoregions support nurseries that actively produce, supply, and conduct investigations of native plants on green roofs. These nurseries play a vital role in supplying plants for ecoregional green roofs through the production of seed, potted plants, and cuttings.

Stronger partnerships between plant suppliers and researchers are also needed to expand upon knowledge of native vegetation on green roofs (Rowe 2019). With many regions such as the Desert Southwest, the California coastal ecoregions, and prairie ecoregions, these support hundreds (and in some areas thousands) of native plants adapted to the local climate, and many plant species native to those environments have yet to be trialed on green roofs. Only a few green roofs in each region have trialed more than 20 potential native taxa on green roofs. Across all of the ecoregions discussed, only a few green roofs have over 90 taxa of native plants: California Academy of Sciences (94 native taxa), Barry Biodiversity Center (109 native taxa), Salesforce Transit Center (110 native taxa), Denver Botanic Garden (112 native taxa), and the Mordecai Children's Garden (169 native taxa). Each ecoregion will need several or many such highly diverse green roofs. This can be accomplished with many potential design variations, different substrate compositions, depths of substrates, watering regimes, substrate dynamics and evolution, and habitat maintenance practices. More diverse green roofs provide excellent opportunities for private or public entities to team up with universities to conduct research and to highlight plant knowledge. With the diversity of native vegetation in North America, there are potentially hundreds or thousands of taxa that have yet to be explored at conservation sites and used on green roofs.

If constructed wetland green roofs prove to be a valid method for treating greywater from inside buildings (Chaps. 7, 9 and 10), then new research is needed within associated ecoregions since little is known about which native wetland vegetation from a region can function and effectively clean water and successfully grow on green roofs.

Regarding taxa of plants used on the case study green roofs that are not native to North America, there were many exotic plants used for their ornamental character; however, there are native plants (and or cultivars) that have very similar ornamental character and could be used. One example is *Festuca glauca* 'Elijah Blue' which is native to Europe. *Festuca idahoensis*, however, is native to much of the western U.S., has been trialed on green roofs in several case studies (Chaps. 6, 7, 8, 9, 10), is very similar in ornamental character to 'Elijah Blue', and supports habitat for the native Sonora Skipper (*Polites sonora*) and other butterflies (Beyer and Schultz 2010). Many ornamental fountain grasses (*Pennisetum*) were used on green roofs, however, many grasses native to the ecoregions also exhibit ornamental character and have proved to perform well on green roofs (*Bouteloua*, *Schizachyrium*, *Sporobolus*) with appropriate substrate depth, drainage, and irrigation practices. Other research supports the successful use of native vegetation that exhibits ornamental characteristics and their ability to compete and thrive on green roofs in the western ecoregions (Dewey et al. 2004).

11.1.3 Biodiversity

The case studies demonstrated a wide range of biodiversity on the green roofs. Some projects, such as the California Academy of Sciences (CAS) green roof initially installed only a few plant species (9) but were later diversified to explore and expand the palette of native plants. Staff at CAS added many native plants to the green roof and now have trialed at least 94 species of plants. Some of the different plant forms include annuals, ferns, grasses, herbaceous perennials, shrubs, and succulents (Silver 2019). The exploration of plant diversity at CAS is intentional and decided upon during annual maintenance meetings. New plants were installed directly, while selective retention of non-aggressive volunteer species takes place. Regarding insects, birds, and butterflies, CAS has an exceptional, mobile-based citizen science system in place. Here volunteers record observations of wildlife on the green roof to online sources such as iNaturalist, e-Bird, and similar sites (CLO 2019).

Monitoring programs can help build knowledge about the biodiversity of green roofs and is readily accessible to anyone with internet access. Most of the green roofs examined did not have a formal process in place to collect information about organisms living in substrates or insects and birds using green roof vegetation. Most of the green roofs likely have been visited by wildlife, but there is a lack of identification or recorded observations of fauna. Maintenance personnel could be trained and tasked with systematically noting observations while they are on a green roof. Cameras could be installed to record bird visits, however, each of these approaches will generally require extra time and additional funding or volunteers. These kinds of activities could become part of research projects at local universities.

Some of the most diverse green roofs include different forms of plants, with additions of rocks, gravel, varied substrate depths, water, decayed wood, or other objects to attract a range of wildlife (Brenneisen 2006). As one example, hummingbird sage (*Salvia spathacea*) is planted on many green roofs in California. Hummingbirds have been observed using many of those green roofs by visitors, building occupants, and maintenance crews. At the coastal meadow-based green roofs on the Vancouver Convention Center, two insects, a two-spotted lady beetle (*Adalia bipunctata*), and a parasitic wasp that had not been seen in urban Vancouver since the 1930s were each observed on the Convention Center green roofs during a field visit by a biologist at the University of British Columbia (Hemstock 2019).

Where green roofs connect to the ground, mammals easily inhabit rooftop ecosystems. In August 2019, following the observation of many burrows, traps for mice were set on the K-State East Memorial Stadium green roof. Eight cotton rats, a native species, were captured, and rabbit scat observed (Andrew Hope, K-State Biologist, personal communications, Aug. 2019). At the green roof at the VanDusen Botanical Garden Visitor Center in Vancouver, B.C., moles, and rabbits have been observed on the green roof. None of the maintenance staff reported any serious damage from visits from mammals, however, crows and seagulls have pecked vegetation off of several green roofs in Seattle, Washington, and Vancouver, B.C.

Biodiversity on green roofs typically comes about from an intentional effort by designers or maintenance staff, and by deftly matching a building owner committed to maintaining biodiversity with an effective, knowledgeable design and maintenance program. However, increased biodiversity can also occur incidentally when additional species become introduced on the roof after being carried to a rooftop via wind, birds, mammals, or people, and perceived as part of the desired aesthetic.

Only a few projects, such as some of the meadow-based green roofs in Jackson, Wyoming (Sect. 6.3.2), the design team directed the selection of vegetation to target specific species such as migrating butterflies (Monarchs). How migrating species use green roofs is only beginning to be studied in North America (Coffman and Waite 2011; Partridge and Clark 2018). Graduate student research at Kansas State University systematically studied butterfly use of two large green roofs, including visits by monarch butterflies, which migrate through Manhattan, Kansas each spring and fall (Blackmore 2019). Monarch butterflies were observed on other green roofs mentioned in case studies, but there is little formal monitoring, data collection, or publication in the peer-reviewed literature in the West, or elsewhere regarding migrating species (Gedge and Kadas 2004; Partridge and Clark 2018).

11.1.4 Maintaining Ecoregional Green Roofs

The appearance and aesthetic expectations for green roofs go hand-in-hand, though maintenance practices vary greatly. Many of the ecoregional roofs in Part II prescribe maintenance expectations, funds, staffing, and accurately executing maintenance. Yet many other case studies found that green roofs were poorly maintained because either the maintenance plan was not well-articulated, well-understood, or the owner's finances were limited, and thus maintenance crews were understaffed. In reality, the intended aesthetic often determines the frequency and kind of maintenance recommended. Besides institutional capacity (personnel, equipment, and funding), how often and how well green roofs are maintained is frequently associated with the knowledge of those doing the work, and the ease of access and maneuverability in managing and removing materials (Cantor 2008; Snodgrass and McIntyre 2010) from the rooftop (Figs. 11.1 and 11.2). Several green roofs in the case studies had no provisions for easy removal of unwanted vegetation from the roof. At several green roofs on hospitals, maintenance workers had to arrive very early in the morning so they could walk through hallways with bags of rooftop refuse to avoid visiting hours.

The spectrum of maintenance needs on a green roof varies from active and frequent maintenance to more passive management where visits take place during a limited number but at strategic times during the growing season such as those discussed in the case studies (Sects. 3.33, 4.3.1, 7.3.1, 7.3.2 and 7.4.4) and others (Nagase et al. 2013). High maintenance green roofs may have formal plantings or geometric patterns where plants are located in planting beds and drifts of similar species are grouped together such as those in the case studies (Sects. 3.3.3, 4.3.9, 7.3.5, 7.3.6, 8.3.3 and 8.3.5) and others (Cantor 2008; Snodgrass and McIntyre 2010).

Green roofs such as the Bill and Melinda Gates Foundation in Seattle, the SRM (Google) campus in Kirkland, Washington, the Salesforce Transit Center in San Francisco, and the Denver Botanic Garden all have high expectations for a well-kept appearance. These projects established a sufficient budget to hire skilled workers. They are roof gardens, and as such, require high levels of maintenance. At the lower frequency and intensity of maintenance, projects such as the Slide Ranch (Muir Beach, California), Cedar River Education Center (within the greater Seattle

Fig. 11.1 Hauling materials on and off of a green roof may require various roof-specific methods. Access for paths or accommodations for hauling devices as shown in (**a**) may need designated paths for workers to achieve access. The maintenance shed shown in (**b**) was built post-construction on the California Academy of Sciences, as the staff determined that composting organics and managing soil nutrients from on-site materials was desirable, and they needed a protected location to store tools and materials. (Photos: Bruce Dvorak)

Fig. 11.2 A worker maintains the TWA Headquarters green roof in Kansas City by removing unwanted vegetation. This is a common maintenance activity for garden-like green roofs, and intervention is sometimes required to prevent invasive plants from distributing its seed. Controlling seed production and dispersal greatly reduces outbreaks of invasive and/or undesirable vegetation. (Photo: Bruce Dvorak, June 2018)

Fig. 11.3 This modular green roof system was installed in a major city park on the U.S. west coast. Lack of maintenance and no provisions for supplemental watering has led to a decline in the vegetation during an extended drought. It can be expensive and disruptive to remove or replant a green roof. In instances where vegetation declines or dies out, collecting and adding native seed, or cuttings (succulents), that may adapt to the substrate depth and microclimate conditions can be a very useful method to revegetate. (Photo: Bruce Dvorak, October 2018)

metropolitan area), and Camp Young Judea prairie roof (Austin, Texas) used all native plants where the vegetation is free to adapt to its preferred microclimate. For these types of green roofs, invasive species are sometimes removed or reduced, but there are no formal bed lines and plants are free to spread and compete. Volunteer staff can certainly be a great help in maintaining green roof vegetation, nevertheless, volunteers and inexperienced staff typically require close supervision until they are well trained. Knowledgeable staff and volunteers can make mistakes if they are in too much of a hurry to attend to their work because they are inattentive or are overburdened and need to move quickly to their next task.

Green roof failures can take place when green roofs are not maintained (Fig. 11.3) or lose funding for maintenance (Snodgrass and McIntyre 2010). It is probably better for a potential builder or owner of a green roof to postpone the design, construction, and installation of a green roof until appropriate measures are in place to ensure that the green roof has provisions in place to maintain the green roof in perpetuity.

New, hybrid, or emerging aesthetic approaches for maintenance are appearing on some ecoregional green roofs. Green roofs at the Moda Building (Bend, Oregon), the Berry Biodiversity Center (University of Wyoming), NOAA Southwest Fisheries Science Center (La Jolla, California), and the EcoCenter at Heron's Head (San Francisco, California), function to provide an informal appearance that is appreciated with an understanding of the local plant communities. This aesthetic can be embraced and understood with specific knowledge of ecosystems, plants, beneficial associates with wildlife, a diminished presence of pests, or coincide with seasonal expectations within the ecoregion (Sutton 2014, 2020). For example, without knowledge of the summer dormant season, a summertime visitor to the Scripps Coastal Reserve near San Diego may believe that the grey and leafless shrubs are dead, not dormant (Sect. 7.1.2). The NOAA green roof was inspired by the vegetation native to the Scripps Coastal Reserve, and that vegetation consists of annuals, grasses, herbaceous perennials that go through seasonal changes. Armed with knowledge about how these plants survive by avoiding growth during the dry period of the summer, one would not expect these plants to be green year-round. Thus, appreciation can be acquired through knowledge of the ecosystem and its processes. If the observer understands the seasonal cycle of vegetation, knows their typical ecological associations in the landscape, and sees their similar associations in a garden, he or she can develop an informed ecological aesthetic (Sutton 2014).

The Moda Building located in Bend, Oregon, for example, makes use of local manzanita, sagebrush, and rabbitbrush, plants typically associated with scrub habitat, and not typically associated with green roofs or gardens. The distribution of these plants on the green roof into articulated groupings with their natural associated plant community members makes ecological connections that those familiar with the natural vegetation might appreciate. However, their skilled placement into groupings also takes advantage of the colors, forms, textures, and aromas of the plants to create a garden-like setting (Sect. 6.3.10). It is important to note that deeper substrates (semi-intensive) are typically needed to support shrubs on green roofs.

Another acquired attribute of aesthetic appreciation is the potential role of individual plant species within different plant communities. Blue grama grass (*Bouteloua gracilis*) for example, is native across some parts of the West in many different habitat types and has common and varied associations across ecoregions. In the Great Plains, it grows with other grasses, prickly-pear (*Opuntia*), big sagebrush (*Artemisia*), and rubber rabbitbrush (*Ericameria*). In ponderosa pine (*Pinus ponderosa*) communities it grows with oaks, juniper, manzanita, and many grasses different from those found in the Great Plains. A local ecologist, landscape architect, botanist, or naturalist may know these companion plant associations and recommend them for a green roof. Therefore, a particular species such as *Bouteloua gracilis* may become appreciated across many ecoregions, even though it may perform various ecological functions. In the case studies, *Bouteloua gracilis* was integrated into five ecoregional green roofs in the Tallgrass Prairie ecoregion, five ecoregional green roofs in the Shortgrass Prairie ecoregion, and three ecoregional green roofs in the Desert Southwest ecoregions. *Bouteloua gracilis* is native to southeastern California but was not used on green roofs in the case studies examined for this region. It is not native to Oregon or Washington and was not used there. Western sword fern (*Polystichum munitum*), native to the Pacific Northwest, was used on green roofs in California, Oregon, and Washington, but not in ecoregions east of the Cascade Mountains. Thus, a particular species may become appreciated in one ecoregion, but may not be seen as ecologically important in another ecoregion. The appreciation for the local flora may become an important part of ecoregional green roofs for them to become locally adopted.

Green roofs with native vegetation growing with their familiar land-based ecological associations are emerging as a specialized kind of green roof as demonstrated in many of the ecoregional case studies in this book. Building owners with plant community-based ecoregional green roofs must be informed, and be prepared to pass this specialized knowledge along to future building owners. This approach is radically different from the concept of a green roof as a uniformly green, *Sedum* monoculture. Ecoregional green roofs can feature unique plant associations that are present in local plant communities, and perhaps become more visually dynamic than monocultures when observed throughout the growing seasons, but may require specialized plant knowledge and care.

As opposed to simple maintenance procedures where only one or several species are maintained, ecoregional green roofs that use multiple members of plant communities may take specialized care for them to be appreciated. To an untrained eye; however, nothing special may be taking place. These ecological associations on green roofs can demonstrate new expectations for green roofs since they can provide important ecological functions and high biodiversity. For example, green roofs on the TWA Headquarters, Barry Biodiversity Center (Fig. 11.4), Moda building, EcoCenter (Heron's Head), California Academy of Sciences, Sonoma Academy, TWA Headquarters, K-State Memorial Stadium, and others all receive maintenance, but without formal bed lines, and sometimes dynamic or changing appearances through the seasons make for visually interesting green roofs that support ecosystem services for local insects and fauna. The Memorial Stadium green roofs in Manhattan, Kansas display large and complex prairie-like ecosystems, dynamic in their seasonal changes, with some movement of species over time expected (Skabelund et al. 2017; Blackmore 2019). These kinds of green roofs are much more dynamic and diverse than a low diversity sedum green roof (Cook-Patton and Bauerle 2012).

Some green roofs receive very little precipitation or maintenance and their appearance is, perhaps, more predictable with little change in appearance through the seasons. The green roof on the Anza-Borrego Visitor Center (Figs. 5.26 and 5.27) receives very little precipitation in the Mojave Desert, and thus the evergreen presence of succulent desert plants provides a stable habitat and structures its visual experience. Parts of the NOAA Southwest Fisheries green roofs have a similar low-precipitation climatic context, where large native succulents (e.g. agave, cacti) sustain an evergreen appearance year-round.

One of the lessons learned from many of the post-occupancy case study observations is that prior to the beginning of a project, clients must be realistic regarding their needs, wants, and capabilities to maintain ecoregional green roofs. Designers of green roof ecosystems must work to meet those needs and should provide a maintenance handbook that can be understood (bi-lingual) and followed. Some of the best handbooks had laminated pages, a listing of common and botanical names of plants, photographs of plants installed on the green roof, and photographs and names of common invasive plants. This approach allows for changes in maintenance staff, and almost anyone become familiar with the vegetation and expected maintenance activities.

Fig. 11.4 The dynamic seasonal characteristics of the Berry Prairie green roof on the campus of the University of Wyoming (**a**) entering dormancy, (**b**) winter character, (**c**) early spring, (**d**, **e**) summer, and (**f**) fall. The visual and ecological dynamics of this green roof change throughout the year. This ecoregional green roof represents three local habitats (found within 160 km/100 mi) and maintains over 115 native taxa including 9 species of *Penstemon* (**e**), one native *Sedum* (yellow bloom in (**d**) *Sedum lanceolatum*), and ten native grasses (**f**), including blue grama. (Photos: courtesy of Dorothy Tuthill a, b, c, f, Bruce Dvorak d, e)

Generally, the more dependent a green roof ecosystem is on frequent human intervention and skilled maintenance procedures, the more likely the green roof may fail if there is not a clear and effective transfer of ownership or transfer of maintenance providers (paid and/or volunteer) or a consistent stream of maintenance funding or support including equipment and knowledgeable coordination. Highly visible green roofs such as roof gardens should be designed with expectations for frequent maintenance (weekly or several times a month) to take place by skilled workers trained to maintain the desired green roof ecosystem characteristics. Conventional landscape-level maintenance providers may have little or no knowledge of how to maintain an ecoregional green roof. If this is true, training or some other strategy will be needed to ensure that maintenance is done well.

Along the west coast, examples exist where a number of ecologists were actively involved with the design, implementation, and management of green roofs. Some of the green roofs that closely resemble the natural landscape plant communities had caretakers that understood the local ecosystems. For these kinds of green roofs, untrained workers can easily damage or harm the green roof ecosystem. For example, one green roof in the Seattle area contained a large bed of thriving western sword fern. Yet during a visit by untrained maintenance staff, all of the sword ferns were deadheaded and pruned to the root balls during the summer. With porous growing media, infrequent irrigation, and a few days of warm temperatures, deadheading quickly killed the ferns they and they needed to be replaced. Untrained staff knew how to care for cultivated landscapes on the ground, and made an uninformed decision on the roof garden.

One of the challenges of ecoregional green roofs is that the green roof industry may have a limited workforce that is not properly trained to care for ecologically-based green roofs. In such cases, it is better to retain higher fee maintenance contracts to fund well-informed work, over less-expensive, untrained workers. While some ecoregional green roofs did require specialized maintenance, many ecoregional green roofs did not require much care. The important point here is that the level of required maintenance should match the expectations and financial capacity of the building manager responsible for the green roof.

11.1.5 Water for Ecoregional Green Roofs

Perhaps there is no issue more important for green roofs west of the 100th meridian than the availability of water. Indeed, precipitation is limited in most of the urbanized regions of the West, so there is a presumed priority for the need and use of potable water inside buildings. Thus, while it would seem wasteful to designate potable water exclusively to expand a green roof industry in arid or semi-arid climates, many, if not most, of the green roof case studies demonstrate valid alternatives to the use of potable water for green roofs. And, where the use of potable water is deemed essential, it can be used wisely and sparingly.

The harvesting of rooftop rainwater for future use is an old idea that once again gaining traction. Many of the green roofs in this book made use of rainwater harvesting systems. Cisterns were used to retain water in external locations above the ground (Fig. 11.5a) or at internal locations (basements) integrated into the structure of the building (Fig. 11.5b). A few of the projects with water harvesting systems experienced setbacks, as pumps sometimes failed, or they were undersized or underfunded (because the owner could not afford the appropriate size tank), or

Fig. 11.5 Water harvesting cisterns located external (**a**) and internal (**b**) to projects. The external cistern (**a**) located in Flagstaff, Arizona has a small shed to enclose the irrigation pump system. A pump room located on the basement floor of a hospital (Sect. 7.3.5) in San Francisco (**b**) houses valves and monitoring and filtering systems for a 1400 m^3 (49,440 ft^3) rainwater harvesting system located behind the wall. (Photos: Bruce Dvorak)

insufficient water was stored to irrigate a green roof when needed during dry periods. Thus, backup or secondary solutions were designed and implemented for most projects to maintain an adequate, supplemental water supply. However, based upon discussions (by Bruce Dvorak) with facilities managers, most of the large-scale rainwater harvesting systems (Fig. 11.5b) observed seem to be operating well as integrated features of both building and green roof systems.

One of the more innovative and successful ways water is secured for green roofs comes from the re-use of harvested rainwater (from rooftops or pavement) and the use of greywater from inside buildings. For greywater systems, water draining from sinks, showers, dishwashers, and floors is collected, pre-filtered, and held in a tank or cistern. In a final step of cleansing, the greywater can be filtered through a rooftop wastewater wetland before its exit into a rain garden located at grade (Bullitt Center 2016). With this approach, there is typically a one-time cost for the installation of additional pipe systems (compared to single-use water systems) to achieve a steady, reliable source of water as long as the building is occupied and water is being used.

Several buildings with green roofs were located off-grid from public utilities (including the Bullitt Center in Seattle, and the Heron's Head Environmental Learning Center in San Francisco). This means that water inputs and outputs need specialized engineering, permitting, and regulation. These non-potable water solutions may require special regulation, permitting, and attention from local governing authorities. The cities in the case studies where rainwater harvesting or greywater was being used on green roofs in 2019 (e.g. San Francisco, Seattle, Portland, Salt Lake City, San Diego, and others) made accommodations for special water uses for green roof irrigation. If procedures for such measures are not in place locally, special measures may need additional project time to receive approval from local authorities.

When committed to this approach at the planning stages of a building with a green roof, using some alternative sources of supplemental water opens up possibilities for plant communities requiring a moderate or high need for water. Wetlands or mesic grasslands may be possible on green roofs including dry and hot regions if greywater systems are used. Xeric or drought-tolerant vegetation on green roofs will likely need some source of water for irrigation during prolonged dry spells, and non-potable water collected on-site may be the best choice. Building owners should be prepared to invest in the best system for the building and project to sustain long-term operations. The type of vegetation and plant communities and the desired coverage of a green roof by healthy vegetation are vital considerations when determining how much water will be needed for plant survival and growth year-to-year. On steeply sloped green roofs, more supplemental water will typically be needed compared to flat or gently sloping green roofs. Monitoring systems typically coincide with harvested or greywater systems to ensure efficiency and conservation of water.

Water conservation was a common goal for most green roofs in case studies. The green roof and irrigation designers set out to efficiently manage water on green roofs deploying several concepts. Selecting vegetation for drought tolerance (or avoidance) was the main strategy for most of the green roofs in the case studies. Regardless, supplemental water was needed to maintain even drought-tolerant vegetation. Moisture sensors were used in many of the green roofs to regulate irrigation run times and prevent waste. Weather-based irrigation systems were also used to prevent irrigation before, during, or after substantive rain events. Multiple measures were used for many projects to conserve water and avoid wasteful use of potable or harvested water. Extra vigilance is required to monitor substrate moisture levels on steep-sloped green roofs.

Zoning of irrigation systems for green roofs based upon the water needs of the vegetation was also a common design feature. Most projects with advanced irrigation systems (i.e., moisture sensors, weather-smart devices) also had vegetation designed in common moisture zones (hydro zones) so that irrigation zone run times could correspond to the water needs of the selected plant species. Some projects experienced problems establishing vegetation where plant selection, irrigation zones, and run times were not well-matched. Planting designs where succulents and grasses shared the same irrigation zone often experienced maintenance issues as either the grasses were underwatered or succulents or other very drought-tolerant species were overwatered.

Many landscapes west of the 100th meridian lack contiguous vegetative cover, and in some ecoregions, vegetation is dispersed with bare ground between plants. Landscapes with intermittent plant cover limits available moisture in the soils. Thus, green roofs can emulate this approach, however, their appearance may be quite different from a fully-covered sedum-based green roof typical of moist climates. For example, green roofs on the Anza-Borrego Desert State Park Visitor Center in the Mojave Desert, and the NOAA Southwest Fisheries Science Center, in La Jolla, California intentionally used this approach. Many ecoregional green roofs in the more arid and hot climates of the West aid moisture retention by further mulching the substrate with light-colored gravel, or by adding biomass clippings.

11.1.6 Microclimate

Buildings make contextual settings for green roofs. From the case studies in Part II, a variety of microclimatic conditions affected the green roofs. The most common negative effect of a building on green roofs was damage to vegetation from reflected sunlight (Sects. 5.3.2 and 8.3.3). In some cases, at locations where south or west-facing windows adjacent to green roofs, reflected sunlight caused significant damage to vegetation. At a building in Lincoln, Nebraska (not covered in the case studies) a green roof was situated along a continuous south-facing façade of windows (Fig. 11.6). Sunlight reflected directly onto the adjacent vegetation during the late afternoon. Over time, the vegetation was burned and stressed immediately adjacent to the windows and the green roof. The building façade was made from dark gray-brown materials, which also absorbed and reflected heat, contributing to plant stress. Prevention of these conditions should be considered at the planning stages of the building, and if needed during the design of a green roof. When irrigation is

Fig. 11.6 Heat and solar reflection from the south-facing windows and a dark-colored tile facade in combination with lack of a functional automated irrigation system led to the decline of the vegetation closest to the facade. In this case, building owners changed and new ownership was not interested in maintaining the green roof. Portions of the green roof that are furthest from the facade show full coverage of live vegetation, even though the irrigation was not operational at the time. Previously, the green roof was fully vegetated with a functional irrigation system (Sutton 2013). (Photo: Bruce Dvorak, June 2018)

eliminated, an extremely drought-tolerant plant community will be required to retain living vegetation in areas receiving intense or concentrated solar radiation.

Regarding design solutions from the case studies, some preventative measures might include the following: provide eves or overhangs to shade windows and prevent reflection, set back windows from a façade to make self-shading conditions (Sects. 5.3.3 and 5.3.6), avoid high-reflective glass adjacent to green roofs, locate green roofs with views looking north or east so that there is no direct sunlight on glass (Sect. 4.3.1), place pavement adjacent to or nearby reflective windows instead of vegetation (Sect. 4.3.6), select plant materials for the green roof that are tolerant of high reflectivity (Sect. 5.3.2), or provide zoned irrigation along the west or south-facing facade to increase supplemental watering to prevent stress to plants.

The reflective glass on the north or east-facing façades may prevent reflective damage to vegetation, but some designs where a large unbroken plane of glass can cause other issues such as high reflectivity of the sky, trees, or landscape. This condition can confuse birds and lead to collisions and injury or death to birds that fly into windows (Klem Jr et al. 2009; Parkins et al. 2015). Learning from this research regarding how bird collisions can be reduced on and adjacent to buildings is important if green roof designers and managers are to reduce impacts to birds (Arkles 2007; TCL 2017).

Locations of exhaust vents or air intake vents present another common building condition providing stress for vegetation. If air velocities from exhaust vents are high and persistent, accelerated evapotranspiration ensues in adjacent vegetation on green roofs. The green roof on the One Van Ness Avenue building (Fig. 7.14) in San Francisco, was added on after the building was constructed. Mechanical exhaust vents are located near green roofs and cause some stress for vegetation directly near them. In addition, green roofs that are located near exhaust intakes should not generally include vegetation that has fine plant parts (e.g. seeds, blooms, pollen) which could impede air flows or clog intake valves.

Air temperatures and humidity on roofs vary from the ground and create different microclimate conditions for green roofs. For green roofs located near the ground level, this difference may be negligible, but green roof conditions located multiple stories above ground level or where adjacent building materials heat up or reflect sunlight, green roof designers should anticipate drier or more stressful conditions for plants.

Roof deck slope can create potentially beneficial or stressful conditions for vegetation. Flat-sloped roof decks might make for persistently damp conditions where roots of drought-tolerant vegetation may rot, mold may develop, or other problems occur, if irrigation is run too frequently. Some flat roof decks were modified with stratified layers of polystyrene to gently elevate and slope the green roof substrate to make the substrate conditions more xeric, and function similar to conditions where plants grow on the ground (Sects. 4.3.1 and 4.3.7). Other projects took advantage of a flat roof deck and selected moisture-loving vegetation (Sects. 7.3.6, 8.3.3, and 8.3.4).

Although most of the green roofs in the case studies represented flat roof decks, several case studies were designed with dramatic slopes at some location of the roof (Fig. 11.7). In many of the sloped green roofs with south-facing or west-facing

Fig. 11.7 Examples that demonstrate the effects of slope and microclimate. Image (**a**) shows a sparsely vegetated steep south-facing slope on a roof garden in Salt Lake City, Utah. The vegetation present is healthy, however, at some previous time, about half of the plants died, and were not replaced. (**b**) A portion of a steep and variable slope on a green roof in Los Angeles, California is being revegetated with the assistance of an erosion control blanket. (**c**) A green roof in Vancouver, B.C. has a steep south-facing steep slope and was fully established with grasses and sedums. The sedums are established, but the grasses have waned in the south-facing aspect during a dry year shown in August 2018. However, by May of 2019, all the vegetation fully recovered (Fig. 10.20). (**d**) South-facing slopes at the California Academy of Sciences living roof are successfully being replanted and managed as unique microclimates on the roof. (Photos: Bruce Dvorak).

aspects, the steeply-sloped sections had less than 100 percent coverage of vegetation. Meanwhile the east, north, and northwest locations on the same roof had healthy vegetation. In these cases, the apparent cause of stress for plants in these specific locations was poor plant selection, ineffective or inconsistent methods of irrigation zoning or delivery, or perhaps accelerated wind or some other unanticipated environmental factor. These conditions require special measures such as a more frequent delivery of supplemental irrigation, appropriate plant selection (for the microclimate), or a combination of measures to address these conditions.

Some of the mounds on the California Academy of Sciences living roof experienced plant stress on south-facing and steep aspects of the roof. However, because there is now an active annual review of vegetation on the green roof, there has been an increased diversity of plants arrayed around the mounds to address the rooftop microclimates. Special measures include the use of drought-tolerant grasses and forbs and succulents as a resilient matrix of above surface biomass and network of roots (Hauser 2013). This suggests that special measures may be needed to maintain vegetation on steep and south-facing zones on green roofs in the Northern Hemisphere. In addition, a periodic review of vegetation on such roofs may be necessary to keep them healthy.

Another approach to address green roofs with steep slopes is to select a wide variety of vegetation that can adapt to the local condition. The K-State Memorial Stadium green roofs were designed with species composition to adapt to east-facing and west-facing slopes. Over time, vegetation adapted to its preferred microclimate and substrate moisture levels (Sect. 3.3.4). In contrast, The Ross Perot Museum of Science and Nature has steep slopes, but its solar orientation is facing east-northeast. There has been little or no plant stress on this steep, irrigated, and very visible location of the green roof (Sect. 3.3.5).

11.2 Green Architecture and Integrated Design

Green architecture is a broad term that encompasses many aspects of sustainable, regenerative, low-energy, passive, renewable, healthy, and ecologically-sound buildings (Wines and Jodidio 2000). It can reference buildings constructed from local materials, renewable materials, and materials intended to last for hundreds of years. Green architecture refers to buildings that are designed to maintain healthy air, light, or the handling of waste (McDonough and Braungart 2010; GhaffarianHoseini et al. 2013; Loftness 2019). Many organizations exist that encourage different methods to inspire the design, construction, and maintenance of buildings that make use of green approaches.

When buildings, green roofs, and sites are designed together, they can more efficiently engage sustainable design parameters and become efficient users or producers of energy (Wines and Jodidio 2000; Fowler et al. 2010). Perhaps one common goal is that green architecture and green roof designs must become more dynamic and attentive to ecological functions in their building and mechanical systems (Ragheb et al. 2016). Such approaches are implicit to programs such as Living Building Challenge (Fig. 11.8), LEED, and SITES.

These systems demand articulated performance goals, that often result in integrated buildings and sites, and include green roofs with native plants (Fowler and Rauch 2006). With the development of integrated technologies such as green roofs, rainwater harvesting systems, solar energy systems, and other integrated technologies, rooftops need no longer be considered only an overhead cover to protect the insides from the elements or a method to redirect rainwater to storm sewers (Das et al. 2015; Ragheb et al. 2016). Multiple case studies in this book demonstrate that when rooftops are considered for their total potential functions and benefits to the building systems (e.g. energy, water, waste) from their integrated fabrication, it becomes difficult to plan a future where rooftops remain only single-purpose

Fig. 11.8 View of the leachate storage tanks and greywater system built into the basement of the Bullitt Center building in Seattle, Washington. Water from sinks, showers, and kitchens is directed to these holding tanks, before their cleansing in the constructed wetland green roof over the first floor of the building. (Photo: Bruce Dvorak)

surfaces for buildings, especially municipal, commercial, industrial, institutional buildings (Sects. 3.3, 7.3, 8.3 and 10.3).

When clients and designers set building projects goals to become more self-sustaining, or almost completely self-sustaining, then rooftops must be appropriated to perform multiple functions (Ragheb et al. 2016; Schindler et al. 2016). With the development of active and passive solar energy, rooftops can house photovoltaic panels (PVE) for the generation of energy for building use (Cubi et al. 2016). Indeed, an enhanced PVE generation efficiency can be expected when integrated with plantings (Schindler et al. 2016). A couple of projects that were off the electrical grid, could not achieve their goals without making use of rooftop space for the generation of solar energy, a place to collect rainwater, space for green roofs to reduce heat flux through roof decks, and habitat for native ecosystems (Sects. 7.3.7, 8.3.4).

Because of the negative environmental effects of the widespread use of single-purpose rooftops in cities, stresses such urban heat islands, urban flooding, and drought, need to be mitigated to make cities more resilient (Carter and Jackson 2007; Akbari and Kolokotsa 2016; Sharma et al. 2016; Park et al. 2018). Well-designed and implemented green roofs can play an important role in regards to promoting sustainable, resilient, and ecological development, and impacts of climate change (Hobbie and Grimm 2020).

11.2.1 Site Integration and Multi-Purpose Green Roofs

When rooftops are viewed as a part of the ecosystem of the entire site, they can be easily integrated into functions of the building and site more effectively than a single-purpose roof that is connected to a municipal stormwater system (Ragheb et al. 2016). Rooftops can be incorporated as usable space for recreation, healing gardens, food production, or outdoor dining (Cantor 2008). Rooftops can be linked together with bridges to connect rooftop parks (Sect. 7.3.6), and they can be used to make multi-block ecosystem patches to serve as stepping stones for certain types of wildlife (for example, birds and pollinators) similar to Olympic Village in Vancouver, B.C. (Sect. 10.3.7). Rooftop runoff can be drained into cisterns or percolated into rain gardens or other water infiltration techniques (Sects. 7.3.1 and 8.3.4). When a development considers the entire site for its potential contribution to resilient urban ecology, then conventional single-functioned rooftops become much less affordable (Osmundson 1999; Cantor 2008; Weiler and Scholz-Barth 2009) and a less efficient use of urban space.

11.2.2 Biophilia and Ecoregional Green Infrastructure

Biophilia is the concept that people and ecosystems have co-evolved, and therefore humans need to be in contact with nature to sustain a sense of well-being (Kellert and Wilson 1995). In urban centers, this means that vegetation may need to be pervasive so that buildings and vegetated landscapes coexist throughout as people need contact with plants and natural materials (Beatley 2011). This could also mean the inclusion of plants on structures, or could be included inside buildings or at least designed with provisions for people to have frequent views to vegetation from inside buildings (Peck 2012).

Biophilic design includes buildings that make use of green roofs or living walls. Because humans have a long history of living in close contact with plants, we need vegetation or frequent views to vegetation to balance our emotional and psychological health and well-being, especially when we spend time inside buildings (Fig. 11.9) (Suppakittpaisarn et al. 2017). Regarding green roofs, many case studies in this book demonstrate that ecoregional green roofs are also biophilic designs.

Any kind of contact with nature may be beneficial for humans (Reese and Myers 2012). This may be true for human behavior, but it may not be true for native flora and fauna. If only exotic vegetation is used on green roofs, human psychological or emotional needs may be met (biophilic), but the needs of the remnant and urban ecosystems may not benefit. In some cases, local biodiversity could be diminished

Fig. 11.9 View from inside a patient recovery room at the Palomar Hospital in Escondido, California. Coastal meadows inspired the green roof seen here and is planted with a high percentage of native vegetation. A green parking lot is visible beyond. The green roof is visible from inside recovery room beds, visiting areas inside recovery rooms, outdoor spaces for visitors, and from skylights located in surgery rooms. (Photo: Bruce Dvorak, October 2018)

if invasive exotic vegetation is used on green roofs, or if plants that specialist wildlife species need are cleared from the landscape and replaced with exotic vegetation. This pattern is largely the popular method of land development in the United States, and green roofs with exotic vegetation may also diminish the presence of wildlife in urban areas (Cook-Patton and Bauerle 2012).

If one is promoting and making use of biophilic design for cities, we suggest that it be accomplished with native vegetation when feasible (Beatley and Newman 2013). A green roof with exclusively exotic vegetation makes for a generic habitat in a similar way that buildings of the International Style can diminish the regional character of a place. If repeated extensively across an urban center, the International Style of architecture and use of zoning and transportation systems that are placeless, contribute to loss of regional identity because they reduce and ignore a region's history, economy, culture, and environment (Arefi 1999). The same is true for the repetition of sedum-based green roofs. The mixing of non-native sedums with native sedums can be ecologically beneficial, but the use of all native species helps to better connect people to the aesthetics of the ecoregion and can communicate greater clarity (Nassauer 1995; Sutton 2020).

With green roofs, the same degradation as a place is possible. Where green roofs were historically built to give back to the prior vegetation of a site where a building or structure is constructed, an exotic palette of plants does less to contribute to the environmental history of a place or region. When native plants are included, human contact with native plants and their community members can take place, even though vegetation on the ground is removed or reduced. For example, one user of the NOAA Southwest Fisheries green roof noted that when the rains come, the colors and aroma of sage on the rooftops are pervasive, recalls the chaparral, and draws in hummingbirds (Sect. 7.3.10). Is this not the fullness of the concept of biophilia—a connection to the plants of the regional landscape (Wilson 2017), and as a learned form of aesthetic appreciation that connects one to native landscapes (Hadavi and Sullivan 2018; Sutton 2020)?

The native vegetation supports the native and migrating biodiversity of and through an ecoregion. When native plants are included, the biophilic design becomes supportive of the nature of the place. Biophilic design with only exotic vegetation is perhaps a more neutral kind of contact with nature, and there is less contribution to the biodiversity dependent upon the ecoregion. The case studies in this book make example of many ways that green roofs are already making connections to native places and contribute to conservation (Wilson 2017).

11.3 Green Roofs and Roles of Municipal Planning Tools

Ecoregional green roofs will likely make only minor contributions to cities without the substantial support of a triad of local municipalities, corporate leaders in the industry, and academic research. When policies are supported by dialog from the research and industry communities, ecoregional green roofs can begin to make some improvements to energy conservation, reduction of flooding, reduced urban heat islands, and the reintroduction of native habitat in urban areas, especially when backed by academic research and industry support (Carter and Fowler 2008; Francis and Lorimer 2011).

In addition to the several green roof bylaws in North America (Toronto, Portland, Denver, and San Francisco), other forms of support may be needed. Haphazard or uncoordinated placement of ecoregional green roofs may begin to develop an integrated and informed industry, but with regional planning tools implemented, more meaningful and planned aggregations of green roofs and green infrastructure could be achieved in cities (Francis and Lorimer 2011; Lehmann 2014). Ecodistricts are one tool already in use in many cities around the world, and they could be edited to include ecoregional green roofs (Lehmann 2014; Bottero et al. 2019).

Ecodistricts have been developing in popularity internationally including China and Europe (Hult 2015; Jing 2008). Malmo, Sweden for example, has created two ecodistricts since 1998. Examples include establishing a carbon footprint to be used daily as a way to conserve energy and promote renewable energy, manage stormwater, recycle waste, and make use of green roofs and living walls (Juvera 2015).

The City of Metz, France implemented ecodistricts to manage energy and encourage renewable energy sources. The challenge in Metz was to decide how to complete the plan with an array of inherited systems lacking current standards (Gabillet 2015).

In the United States, ecodistricts have been adopted in Portland, Oregon; Seattle, Washington; Austin, Texas; Los Angeles, California; Washington, D.C., and several other communities. The Target Cities program in 2014 additionally included Ontario, Atlanta, Boston, Cambridge, Denver, and Ottawa (Juvera 2015). EcoDistricts are being tested as a way for multiple interests to come together to achieve sustainability locally (Agbannawag 2015). The City of Portland, Oregon has been a leader in developing and implementing these concepts in the United States. Portland has five ecodistricts including The Lloyd Crossing, Lentes, Gateway, South Waterfront, and South of Market. Their focus has largely been the sourcing and managing of energy and waste reduction, but also have expanded to include pedestrian-friendly streets, water neutral, urban agriculture, diverse neighborhoods, and other non-energy related goals (G. Brown & Dykema; Juvera 2015; Seltzer et al. 2010).

As of 2018, the City of Portland, Oregon established a downtown district requiring green roofs. Native vegetation is not required, but climate-adapted vegetation is required. In China, the government has invested in massive greening, reforestation, and replanting programs to restore ecoregional vegetation in targeted districts in many of its largest cities (Jie et al. 2001). The Bullitt Center office building with its wetland green roof (Sect. 8.3.4) is located in the Capitol Hill Ecodistrict, in Seattle, Washington (Reynolds 2019). Such examples should be leveraged to accomplish more inclusive views of green roofs as an integral part of ecodistricts.

Green roofs are already included in some planning tools such as ecodistricts and could be included in overlay districts. An overlay district is a planning tool that can apply a particular policy or concept across multiple neighborhoods or other political boundaries. An example could be a historic overlay district such as the one in New Orleans, Louisiana (Haughey and Basolo 2000). New developments and building restorations in the New Orleans Historic District are subject to full and partial controls according to designated zones. They strictly regulate building façade designs, materials, and such. An example of how this concept can influence green roofs is the addition of the roof garden to the TWA Headquarters building (a historic building) in downtown Kansas City, Missouri. The roof garden was required to have a set back from the edge of the roof to retain the visual integrity of this historic building (Sect. 3.3.3).

Overlay districts and ecodistricts may have the potential to set up a framework or planning tool to implement and manage widespread ecoregional green infrastructure tools such as green roofs specifically targeting locally vulnerable species or habitat preservation. In the case of prairies that once covered North America from Canada to the southern United States near Mexico, they are one of the most endangered biomes in North America and worldwide (Eidson and Smeins 2017; Ricketts et al. 1999). Ecoregional green roof districts could be used to help rebuild prairie habitats where they are currently endangered.

11.4 Landscape Ecology and Ecoregional Green Infrastructure

Landscape ecology studies landscape structure function and change. It includes landscape features such as mosaics, corridors, patches, edges, and networks (Forman and Godron 1986). Of special interest are its regional spatial configurations, processes, and modes of disturbance for plants, animals (including humans), and their habitats. The roles of these features are vital to the interplay of species dynamics including population expansion, decline, and extinction (Forman 2014). Large contiguous landscapes are characterized as mosaics. As mosaics become parceled or fragmented, they can become patches. Patches then form networks where species can subsist because of the proximity of patches to each other and nearby mosaics. As patch size becomes smaller, the viability of interior dwelling species becomes vulnerable. Climate events, predation, and other stressors such as disease or decline of host vegetation can contribute to the decline and extinction of some species (Forman and Godron 1986; Dramstad et al. 1996).

In urban landscapes, the amalgamation of green spaces, water features, buildings, hardscapes, and/or brownfield sites constitute the landscape (Forman 2016). Some urban spaces have set aside open space for various functions as green infrastructure. Some green roofs more than others can contribute to green infrastructure, and improved ecological functions (Ahern 2007; Gill et al. 2007). In short, green roofs can serve as important ecological stepping stones within the larger ecoregions.

11.4.1 Green Roofs as Patches

Any green roof acts as a patch and most any kind of green roof vegetation provides some habitat value (Oberndorfer et al. 2007). However, as demonstrated in the case studies of this book and elsewhere, ecoregional green roofs can be designed to be biologically diverse and provide articulated habitats targeted for particular species or a particular plant community (Gedge 2003; Brenneisen 2006; Blackmore 2019). It is the composition and arrangement of vegetation and substrate that can determine the ecological usefulness of a green roof as a patch (Lovell and Johnston 2009).

11.4.2 Green Infrastructure Networks and Corridors

Green infrastructure in urban regions includes parks, wildlife habitats, conservation sites, and green roofs connected according to the needs of target species of wildlife (fauna). Ecoregional green roofs could become part of a green infrastructure network if they are planned to form rooftop clusters that achieve connectivity or form networks for the fauna of interest. The example proposed in Fig. (11.10) is a

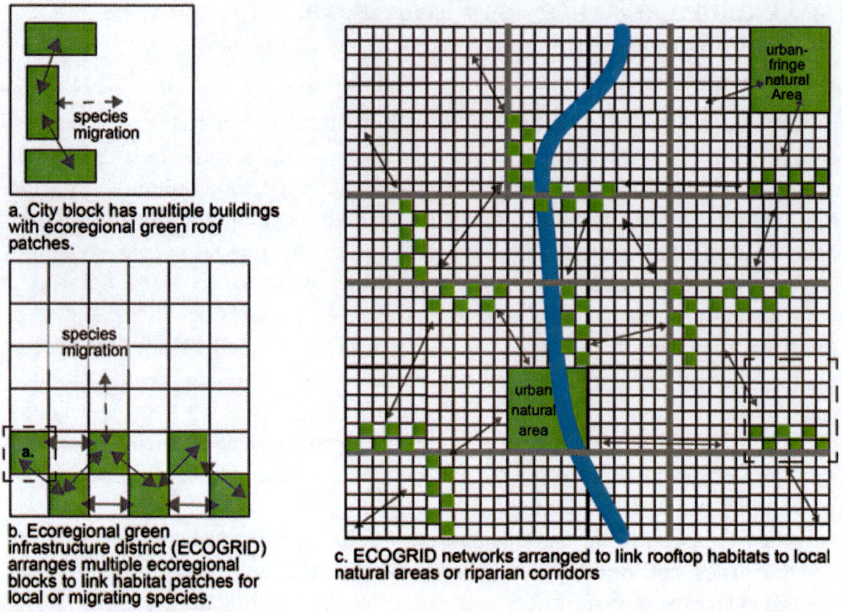

Fig. 11.10 (**a**) Conceptual arrangement of ecoregional green roofs on multiple buildings on a site or city block to form an ecoregional block or eco-block. (**b**) Eco-blocks can be clustered to form ecoregional green infrastructure districts (ECOGRID) to support local or migrating species. (**c**) Corridors can be formed from the alignment of eco-districts that connect to natural sites or riparian corridors across an entire urban region. Planning tools such as ecodistricts or overlay districts may be necessary to accomplish these conservation goals. (Graphic: Bruce Dvorak)

hypothetical multi-scaler arrangement used to demonstrate how the clustering of ecoregional green roof patches within and between city blocks could be arranged to create rooftop habitat patches, networks, and corridors in cities where limited conservation is taking place on the ground (e.g. mega-cities, sprawl).

Green spaces preserved on the ground can be linked to city blocks that in turn, form linked patches of habitat in networks. Several of the case studies begin to demonstrate this effect. In Vancouver, B.C., Canada, the Southeast False Creek Olympic Village (Fig. 11.11) was planned with green roofs and roof gardens on multiple rooftops across eight city blocks. This creates a district of green roof patches. If planned to maintain native plant communities associated with the local ecoregion, such an amalgamation of green roof patches could support particular habitats, especially in cities where little or no native habitat remains. Since green roofs may not support complete ecosystem functions, the preservation of quality native habitat on the land should be considered as a first priority.

The concepts presented here are articulated as an example of how ecodistricts or overlay districts might be conceived at the city scale to implement habitat patches and corridors. The City of San Antonio, Texas is located in the Blackland Prairie ecoregion in the southern part of Texas. Bexar County, where San Antonio lies, has

Fig. 11.11 Many extensive and intensive green roof patches were constructed at the Olympic Village (Southeast False Creek) in Vancouver, B.C. Several blocks were built in 2010, and now nearly a dozen city blocks at False Creek Harbour demonstrate how concepts in 11.10 have already been implemented in Canada. As there were no particular habitat goals for this installment, many exotic plants were used. However, native plants were included on the semi-intensive green roofs (Sect. 10.3.7) seen in the upper right, and throughout private roof gardens (Courtesy Flightpath Aerial Photography–Vitaroofs International Inc.)

little original prairie habitat remaining. Some plant species are endemic and are threatened due to a lack of habitat. San Antonio began to rapidly grow in land area by the 1950s, and the landscape composition and species abundance were already significantly altered. One insect species, the manfreda giant-skipper butterfly (*Stallingsia maculosus*) was locally abundant in San Antonio in 1951. Just several years later in 1955 this butterfly was already in decline. The Naturalist H. A. Freeman made the following observation: "…in 1955, I started visiting the locations where I had first collected it and found that several habitats had been destroyed by progress such as new roads, buildings, overgrazing and other methods that had destroyed the larval food plant (*Manfred maculosa*)" (Quinn 2007). *Manfreda maculosa* was and still is in decline and is currently listed as a threatened species. This plant is available in the nursery trade and has also been trialed on green roof test plots. *Manfreda maculosa* grows in Texas on green roofs with and without irrigation (Sect. 3.3.7, and Dvorak and Volder 2013).

Bexar County is also in the path of the central flyway where hundreds of species of birds migrate through and around Bexar County each year. Monarch butterflies also migrate through Bexar County each year and species of *Asclepius* are a native host plant used by monarchs for reproduction. An Ecoregional Green Roof Overlay District could be adopted to help protect threatened and valued ecosystems, and plant and animal species (Fig. 11.12).

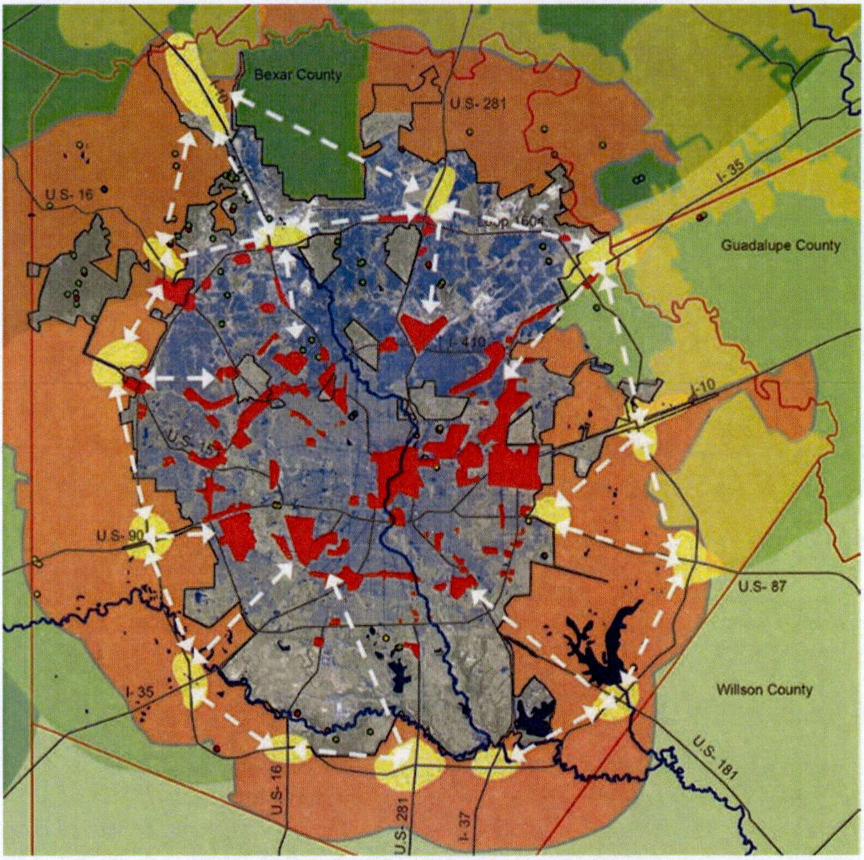

Fig. 11.12 Hypothetical use of overlay districts as an example for the City of San Antonio, Texas. Blue colors represent current impervious surfaces in the city; grey represents undeveloped land. Potential ecoregional green roof overlay districts within the city are shown in red (prairie ecoregion) and magenta (Cross Timbers ecoregion). These zones are determined by the concentration of flat-roofed buildings located in commercial, institutional, and industry zoned properties. Each district would require or incentivize the use of ecoregional green roofs on municipal, institutional, commercial, and industrial buildings. Habitat patches, clusters, or networks (white arrows) could be formed at the rooftop level. Outer belt networks show potential expansion areas located in extraterritorial jurisdiction zones (yellow, orange, and gold). This overlay is based upon the City of San Antonio online impervious cover maps and various historic and current city maps. (Drawn by Xin Zhu and Bruce Dvorak)

11.5 Future Outlook

The development of a vibrant, long-term green roof industry is largely supported by active research in the region, a diverse industry and market, an educated workforce to build and maintain green roofs, and policy to support the growth and quality of work. Unlike Europe, where the concept of green roofs with native vegetation emerged, the green roof industry in North America evolved without a particular focus or objective to design green roofs with native vegetation (Dvorak and Volder 2010; Magill et al. 2011). As demonstrated by the case studies in this book, an available support industry already exists in most regions of North America west of the 100th meridian. However, some metropolitan areas have advanced further in this process than others. At the time of this assessment, only a few of the cities were actively encouraging native vegetation on green roofs, and no municipality required native vegetation on green roofs.

While it is important to encourage the use of native plants on green roofs, it is also very important to emphasize that one cannot simply use native plants on green roofs and expect that they will thrive over the long-term without close attention to the particular needs of selected native species, some supplemental water during dry periods, and other appropriate maintenance practices (including removal or control of invasive species and plants that undermine the integrity of the green roof system). A future outlook for the development of green roofs that are planted and maintained with vegetation from local ecoregions is further expanded upon below.

11.5.1 Research

Most of the green roofs discussed in this book are located near urban centers that have universities or colleges. Where green roof research is taking place, the content of much of the green roof research regards ecosystem services of green roofs. A recent study of publication activity on green roofs across the world, suggests that much of the green roof research relates to the architectural or engineered benefits of green roofs. The authors concluded that there is much potential for growth regarding the study of biodiversity on green roofs (Blank et al. 2013).

According to the research summaries provided in the case study chapters in this book, much of the research focuses on stormwater benefits and building heating or cooling benefits of green roofs. Some research includes vegetation, but little attention has been given to biodiversity, native plants, or plant communities (Rowe 2019). Perhaps this is because little or no perceived benefits or economic incentives accrue to the growth or preservation of biodiversity. Thus, much potential exists for the future study of the ecological functions, performance and qualities of native plants, and the function of plant communities on green roofs. Some western green roofs with native plants support active research (Bousselot et al. 2009; Drennan et al. 2011; Dvorak and Volder 2012; Sutton et al. 2012; Liu et al. 2019), but on a number of research sites dissemination of what is being learned into peer-reviewed outlets is lacking.

Observational and experimental studies need to be published and learned from (Rowe 2019). Long-term studies of vegetation on green roofs are extremely valuable for the growth of the industry but are not yet common in North America (Dvorak and Volder 2010; Rowe 2015). Research regarding plant viability, plant associations (i.e. plant communities), and various functions for native plants on green roofs are needed in each ecoregion (Oberndorfer et al. 2007; MacIvor et al. 2011; Lundholm et al. 2015).

Regarding vegetation dynamics on green roofs in the western ecoregions, no long-term studies have been published, and few studies of different kinds of green roofs with various maintenance or irrigation practices and depths of substrates exist (Dvorak and Volder 2010). Aside from the case studies, in Part II of this book, no peer-reviewed studies are published of vegetation for green roofs that clean or treat greywater as a part of the building system functions. More than a few built projects with greywater or blackwater irrigating green roofs exist, but there are no peer-reviewed publications regarding the design and ongoing performance of these systems.

If city administrators begin to encourage or require a more sustainable use of natural resources such as water, energy, and air, ecoregional green roofs can play a significant role concerning integrated building and green roof designs. In many instances, the green roof industry seems to be ahead of the research community in terms of documenting what is taking place or predicting the benefits of integrated design in various ecoregions, and then presenting that information for educators, students, and other researchers. Examples of the green roof industry leading the way include the Bullitt Center in Seattle, Washington, and the Vancouver Convention Center in British Columbia, Canada. Projects such as the Sonoma Academy incorporate greywater and photovoltaic systems onto the rooftop. Many of these projects have monitoring systems in place, however, there is little connection to the research community to investigate the performance of these innovative systems.

11.5.2 Policy and Education of Ecoregional Green Roofs

Growth in the green roof industry can be stimulated with an informed policy that is supported by local research and outlets for education. Several cities discussed in the case studies have green roof policies and outlets for education. Portland (Oregon) and San Francisco are taking a leading role in the environmental education of children. Portland, San Francisco, Denver, and Austin are cities in the study area that have incentives, bonus credits, or requirements for green roofs.

Green Roofs for Healthy Cities supports the development of policies for green roofs across North America. Their education programs work to inform and prepare a region for green roofs—including all aspects of design, maintenance, and policy. Such outside expertise and support are critical for growth to take place. The sustainable evolution of both policy and education may be dependent upon local organizations such as the Green Roof Information Think Tanks (GRiT) and similar organizations in cities such as in Portland, San Francisco, Denver, and Vancouver, B.C. The formation and ongoing support of these groups are vital to a region because they are venues for local collaboration, sharing of knowledge, and integrated thinking between professionals, researchers, and those that develop policy and a maintenance workforce. When there are representatives across the spectrum of activities, then growth and stability can take place, with innovations passed along to the next generation.

Pilot projects are essential elements to grow the green roof industry. Multiple pilot projects are needed for the many kinds of green roofs, many options for vegetation, and a range of functional roles for ecoregion-based green roofs. The cities of Portland, San Francisco, Seattle, and Austin have pilot projects where the public and industry can visit and learn about green roof plants and plant communities. Many other locations in the case study ecoregions have green roofs in publicly accessible areas such as nature centers, transit centers, public utility buildings, and university campuses, while some are located on private buildings where tours can be arranged.

The most effective pilot projects are those where native vegetation is included, monitoring and research are active, and the maintenance and upkeep of the green roof is part of the pilot project. Since green roof plant communities evolve, and seasonal or cyclical climate variations like El Niño and La Niña occur with some consistency, much can be learned about how owners and managers of green roofs anticipate and adapt to these and other environmental changes and dynamics. Pilot projects are important places where this kind of learning can take place and then be disseminated for open discussion and constructive critique. The Bullitt Center in Seattle is an outstanding example of a private organization that is making known the monitoring and outcomes of its resilient design (Bullitt Center 2016).

11.5.3 Industry Support

The production of native plants, seeds, and bulbs for green roofs is a vital part of the local function of a green roof industry. There are important relationships that exist between the designers of green roofs, their knowledge, and experiences, the availability of native bulbs, seeds, and plants, and their collective knowledge of green roof ecosystems. Growers and producers of native plants, seeds, and bulbs need the demand for native vegetation on green roofs if they are to invest in supplying these materials. Pilot projects, research, and sharing of information at local or regional conferences can help grow and inform that demand. Collaborations between plant growers, researchers, and industry are vitally important to the implementation and success of ecoregional green roofs.

Many of the case studies demonstrated a variety of approaches to the maintenance of ecoregional green roofs. Some institutions and municipalities had a very informed staff and leadership that understood local ecosystems and how to maintain native plants on green roofs. These entities had funding available for adequate maintenance of the green roofs. Some public institutions had multiple green roofs but were underfunded to provide adequate maintenance. The maintenance staff and leaders had knowledge of green roofs, but their workforce was understaffed. Other entities lacked maintenance staff with knowledge of green roof and native plant maintenance practices and did not have adequate funding to support ongoing maintenance.

We recommend that informed decision-making includes discussions of available funding, aesthetic expectations, and integrated functions of green roofs with sites so that green roofs are not neglected and socio-cultural and ecological needs are met. When green roofs were integrated into building system functions—such as the Bullitt Center, Vancouver Convention Center, Sonoma Academy, and the EcoCenter (at Heron's Head)—the green roofs are not just another surface for someone to maintain, they become part of an integrated system where regular maintenance is integrated into the site and building maintenance. When monitoring and research are taking place, multiple perspectives and observations can support the regular or periodic review of system dynamics and performance.

11.5.4 Innovation

It is said that necessity is the mother of invention. When people make personal connections with nature, when people have knowledge of local ecosystems, then appropriate care of those places is more likely to take place (Nassauer 1995). If a culture is disconnected from a place, does not know its native plants or native ecosystems, or how to care for them, then the demise or loss of an ecosystem may go unchecked and unnoticed. It is the naturalist, botanist, ecologist, landscape architect, and others who may understand local ecological functions that can make significant contributions to the innovations of ecoregional green roofs. Scientists and owners or managers of buildings who observe the nature of a place and respect its existence can make innovations using native plants and plant communities a reality.

Landscapes west of the 100th meridian lack the reliability of water that people are used to east of the 100th meridian. John Wesley Powell argued against the rapid population of the West because there was not enough precipitation to sustain rapid growth. With the innovation of reservoirs, canals, environmental engineering solutions, computers, and a technotronic way of life, human populations have continued to grow in arid and semi-arid regions. However, with the scarcity and unreliability of precipitation west of the 100th meridian, new innovations, or a new openness to existing technologies (ecoregional green roofs) may be needed to sustain human developments for well into the future.

In-depth development and research are needed to better understand and improve integrated systems—where buildings, water, and plants function and work together to maintain natural cycles on-site while contributing to the evolution of beautiful and greatly appreciated energy-efficient building systems. The variety of the 20,000 or more native plants on the North American continent should be sufficient to inspire and invigorate a more creative and enriching future for nature in our cities, small towns, and rural places. Ecoregional green roofs can play an important role in future cities and human communities of all sizes. This will happen if people work together effectively, give back to the community, and learn to work with local ecosystems and biodiversity.

Settlers and residents of the western U.S.A. and Canada have a long history of environmental destruction and abuse, and for its planned recovery and rejuvenation (Dobson et al. 1997; Goudie 2018). Development patterns that result in ecological resilience *should* point to the future (Kowarik 2011; Forman 2014). We have much to learn from the native land. Native American and First Nations populations lived off the land for the past ten thousand years without electricity, automobiles, airplanes, smartphones, computers, or (in many instances) permanent buildings. They formed, shaped, and co-evolved with the native plants and plant communities for millennia because they taught their children how to grow and live with the plants and ecosystems of a place. As Native American author N. Scott Momaday wrote, we ought to learn deeply from each particular landscape in our experience—"to look at it from as many angles as [we] can, to wonder about it, to dwell upon it" (Momaday 1969 p. 83).

If people living in urban centers need more contact with nature (Van den Berg et al. 2007), then the nature that has co-evolved with the people living there seems to be worth preserving, learning from and making use of (Velarde et al. 2007) on ecoregional green roofs (Suppakittpaisarn et al. 2017). With innovation, exploration, and dissemination of new knowledge, followed by attentive monitoring, management, and adaption, ecoregional green roofs can become an integral element in cities that work with nature.

11.6 Conclusions

This book set out to introduce the concept of ecoregions as a source of inspiration for green roofs. More than 20 dominant ecoregions of the western United States and Canada were explored through observation of 26 conservation sites near or in urban areas. Seventy-three (73) ecoregional green roof case studies demonstrated how over 830 taxa represented by 357 genera of native plants are maintained on over 122 rooftops. It turns out that native sedums, which were represented by 13 species, played only a minor role in the composition of ecoregional green roofs. Alternatively, grasses, wildflowers, bulbs, cacti, low-growing woody vegetation, and some wetland vegetation all thrived in deep-extensive and semi-intensive green roofs west of the 100th meridian. The case studies represent varied sizes of public and privately owned green roofs.

Green roofs using native plants were demonstrated in various climates, and with varied maintenance and cultural practices. In doing so, the book provides many diverse examples of how native vegetation can be managed on green roofs. This book also discusses how ecoregional green roofs provide essential ecosystem services for local and migrating wildlife including threatened species, some not seen in urban areas since the 1930s (Sect. 10.3.6). Many common plant species were used, as well as some endemic or rare species. This evidence provides support for the growth of the green roof industry by demonstrating how native plants can inform integrated designs, and lead the way to form new possibilities for urban ecology at a local or regional scale. Building owners, designers, installers, providers of materials, maintenance personnel, scientists, and educators all played a vital role in the success of ecoregional green roofs. Thus, interdisciplinary planning, design, and management may be necessary for ecoregional green roofs to become effectively employed, and economically viable.

Conservationist Aldo Leopold argued that a diverse flora and fauna—including native plants and animals—keep the energy circuit open and retain the health, stability, and beauty of ecological systems (Leopold 1966). He also reasoned that human alterations on the land ought to be done with minimal violence or systemic harm. He indicated that landscapes are fountains of energy "flowing through soils, plants, and animals" (Leopold 1966, p. 253). As such, Leopold said, we ought to retain and help to recreate these flows and complex interactions "for the health of the land" (p. 258) as we interact with natural systems and processes to meet our economic needs and other interests. Aldo Leopold's land ethic was perhaps intended for the conservation of ecosystems in exurban environments. However, the findings presented in this book and chapter demonstrate that ecosystem services from green roofs with native vegetation can help bring the energy circuit, improved ecological health, and an informed land ethic into urban environments where buildings tend to dominate the landscape.

The interdisciplinary activities that led to the design, maintenance, and continuation of ecoregional green roofs covered in this book suggest that remnant wild places can inspire ecological thinking across professions. Policy and incentives may be needed to encourage energy and water conservation in buildings through the use of ecoregional green roofs. Also, ecoregional green roofs could be used to form habitat corridors or networks across built-up regions. The ecoregional green roof market, political landscape, and supportive research network, however, may require many years or decades to mature to achieve such meaningful goals such as conservation in urban areas. Regardless, this book serves as a beginning point to reference progress in these important efforts.

References

Ahern J (2007) Green infrastructure for cities: the spatial dimension. In: Cities of the future: towards integrated sustainable water and landscape management. IWA Publishing, London, pp 267–283

Agbannawag C (2015) EcoDistricts: Bridging the gap between plans and projects. University of Arizona, Tucson, AZ

Akbari H, Kolokotsa D (2016) Three decades of urban heat islands and mitigation technologies research. Energ Buildings 133:834–842

Arefi M (1999) Non-place and placelessness as narratives of loss: rethinking the notion of place. J Urban Des 4(2):179–193

Arkles B (2007) Bird-deterrent glass coatings. US Patent Application 11/702,779. Google Patents, USA

Bailey RG (2002) Ecoregion-based design for sustainability. Springer, New York

Beatley T (2011) Biophilic cities: integrating nature into urban design and planning. Island Press, Washington, DC

Beatley T, Newman P (2013) Biophilic cities are sustainable, resilient cities. Sustainability 5(8):3328–3345

Best BB, Swadek RK, Burgess TL (2015) Soil-based green roofs. In: Green roof ecosystems. Springer, Cham, pp 139–174

Beyer LJ, Schultz CB (2010) Oviposition selection by a rare grass skipper Polites mardon in montane habitats: advancing ecological understanding to develop conservation strategies. Biol Conserv 143(4):862–872

Blackmore P (2019) Butterflies, Tallgrass Prairie, and Green Roofs. K-State, Manhattan, KS

Blank L, Vasl A, Levy S, Grant G, Kadas G, Dafni A, Blaustein L (2013) Directions in green roof research: a bibliometric study. Build Environ 66:23–28

Bottero M, Caprioli C, Cotella G, Santangelo M (2019) Sustainable cities: a reflection on potentialities and limits based on existing eco-districts in Europe. Sustainability 11(20):5794

Bousselot JM., Klett JE., Koski RD. (2009) High elevation semi-arid taxa: evaluations on an extensive green roof. Paper presented at the seventh annual greening rooftops for sustainable communities conference, awards and trade show, Atlanta, June 3–5

Brenneisen S (2006) Space for urban wildlife: designing green roofs as habitats in Switzerland. Urban Habitats 4(1):27–36

Bullitt Center (2016) Case Studies: Bullitt Center

Cantor S (2008) Green roofs in sustainable landscape design. Norton, New York

Carter T, Fowler L (2008) Establishing green roof infrastructure through environmental policy instruments. Environ Manag 42(1):151–164

Carter T, Jackson CR (2007) Vegetated roofs for stormwater management at multiple spatial scales. Landsc Urban Plan 80(1–2):84–94

CLO (2019) California academy of science–living roof cornell lab of ornithology: eBird. https://ebird.org/hotspot/L2721356. Accessed 10 Sept 2019

Coffman RR, Waite T (2011) Vegetated roofs as reconciled habitats: rapid assays beyond mere species counts. Urban Habitats 6(1):1–10

Cook-Patton SC, Bauerle TL (2012) Potential benefits of plant diversity on vegetated roofs: a literature review. J Environ Manag 106:85–92

Cubi E, Zibin NF, Thompson SJ, Bergerson J (2016) Sustainability of rooftop technologies in cold climates: comparative life cycle assessment of white roofs, green roofs, and photovoltaic panels. J Ind Ecol 20(2):249–262

Das O, Bera P, Moulick S (2015) Water conservation aspects of green buildings. Int J Res Eng Technol 4:75–79

Dewey DW, Johnsone PG, Kjelgren PK (2004) Species composition changes in a rooftop grass and wildflower meadow implications for designing successful mixtures. Native Plants Journal 5(1):56–65

Dobson AP, Bradshaw A, Baker A (1997) Hopes for the future: restoration ecology and conservation biology. Science 277(5325):515–522

Dramstad W, Olson JD, Forman RT (1996) Landscape ecology principles in landscape architecture and land-use planning. Island Press, Washington, DC

Drennan PM, Leon A, Tyrrell C (2011) Green roofs: developing a native plant palette for Southern California. Proc Water Environ Fed 2011(10):5972–5978

Dvorak B, Volder A (2010) Green roof vegetation for north American ecoregions: a literature review. Landsc Urban Plan 96(4):197–213

Dvorak BD, Volder A (2012) Plant establishment on unirrigated green roof modules in a subtropical climate. AoB Plants. https://doi.org/10.1093/aobpla/pls049

Dvorak BD, Volder A (2013) Plant establishment on unirrigated green roof modules in a subtropical climate. AoB Plants 5:10

Eidson J, Smeins FE (2017) Texas Blackland Prairies World Wildlife Fund. Accessed 06-09-17 https://www.worldwildlife.org/ecoregions/na0814

Forman RT (2014) Urban ecology: science of cities. Cambridge University Press, Cambridge, UK
Forman RT (2016) Urban ecology principles: are urban ecology and natural area ecology really different? Landsc Ecol 31(8):1653–1662
Forman RT, Godron M (1986) Landscape ecology. Wiley, New York
Fournier AM, Sullivan AR, Bump JK, Perkins M, Shieldcastle MC, King SL (2017) Combining citizen science species distribution models and stable isotopes reveals migratory connectivity in the secretive Virginia rail. J Appl Ecol 54(2):618–627
Fowler KM, Rauch EM (2006) Sustainable building rating systems summary. Pacific Northwest National Lab. (PNNL), Richland
Fowler KM, Rauch EM, Henderson JW, Kora AR (2010) Re-assessing green building performance: a post occupancy evaluation of 22 GSA buildings. Pacific Northwest National Lab (PNNL), Richland
Francis RA, Lorimer J (2011) Urban reconciliation ecology: the potential of living roofs and walls. J Environ Manag 92(6):1429–1437
Gabillet P (2015) Energy supply and urban planning projects: Analysing tensions around district heating provision in a French eco-district. Energ Policy 78:189–197. https://doi.org/10.1016/j.enpol.2014.11.006
Gedge D (2003) From rubble to redstarts. Paper presented at the 1st Annual Greening Rooftops for Sustainable Communities, Chicago, pp 29–30
Gedge D, Kadas G (2004) Bugs, bees, and spiders: Green roof design for rare invertebrates. In: Second annual greening rooftops for sustainable communities conference, awards and trade show, 2004. pp. 2–4
GhaffarianHoseini A, Dahlan ND, Berardi U, GhaffarianHoseini A, Makaremi N, GhaffarianHoseini M (2013) Sustainable energy performances of green buildings: a review of current theories, implementations and challenges. Renew Sust Energ Rev 25:1–17
Gill SE, Handley JF, Ennos AR, Pauleit S (2007) Adapting cities for climate change: the role of the green infrastructure. Built Environ 33(1):115–133
Goudie AS (2018) Human impact on the natural environment. John Wiley & Sons, Hoboken
Hadavi S, Sullivan WC (2018) Environmental aesthetics. In: Handbook of behavioral and cognitive geography. Edward Elgar Publishing, Cheltenham, pp 307–321
Hamman ST, Dunwiddie PW, Nuckols JL, McKinley M (2011) Fire as a restoration tool in Pacific northwest prairies and oak woodlands: challenges, successes, and future directions. Northwest Science 85(2):317–328
Haughey P, Basolo V (2000) The effect of dual local and national register historic district designations on single-family housing prices in New Orleans. Apprais J 68(3):283
Hauser K (2013) An evolving landscape: the living roof at California Academy of Sciences. Pacific Horticulture Society, Berkeley
Hemstock B (2019) Personal communication: vancouver convention center living roof field visit
Hobbie SE, Grimm NB (2020) Nature-based approaches to managing climate change impacts in cities. Philos Trans R Soc B 375(1794):20190124
Hobbs RJ, Cramer VA (2008) Restoration ecology: interventionist approaches for restoring and maintaining ecosystem function in the face of rapid environmental change. Annu Rev Environ Resour 33:39–61
Hult A (2015) The circulation of Swedish urban sustainability practices: to China and back. Environ Plan A 47(3):537–553
Jie FB, Hua LG, Ding CL, Ming MK, Ran LJ (2001) Scheme of ecological regionalization in China [J]. Acta Ecologica Sinica 1
Jing Z (2008) Study on the ecological regionalization in Qinhuangdao City based on GIS Graticule Method. Journal of Anhui Agricultural Sciences 35(28):9088
Juvera EM (2015) Ecodistricts in San Francisco: The Implementation of Neighborhood Regional Planning and Its Potential Effects on Environmental Resilience. University of San Francisco, San Francisco, CA
Kellert SR, Wilson EO (1995) The biophilia hypothesis. Island Press, Washington, DC
Kephart P (2005) Living architecture – an ecological approach. Paper presented at the third annual greening rooftops for sustainable communities conference, Washington, DC, 4–6 May 2005

Klem D Jr, Farmer CJ, Delacretaz N, Gelb Y, Saenger PG (2009) Architectural and landscape risk factors associated with bird—glass collisions in an urban environment. The Wilson Journal of Ornithology:126–134

Kowarik I (2011) Novel urban ecosystems, biodiversity, and conservation. Environ Pollut 159(8–9):1974–1983. https://doi.org/10.1016/j.envpol.2011.02.022

Langenfeld C (2009) The value of ecological restoration volunteer programs: a case study in Western Washington state. Evergreen State College, Olympia

Lehmann S (2014) Low carbon districts: mitigating the urban heat island with green roof infrastructure. City Cult Soc 5(1):1–8

Leopold A (1966) A Sand County almanac: with essays on conservation from Round River. Oxford University Press, Inc., New York

Liu J, Shrestha P, Skabelund LR, Todd T, Decker A, Kirkham M (2019) Growth of prairie plants and sedums in different substrates on an experimental green roof in Mid-Continental USA. Sci Total Environ 697:134089

Loftness V (2019) Environmental surfing: design for human health and environmental sustainability. In: Paper presented at the ARCC: Bridging Research & Practice, Ryerson University, Toronto. ARCC

Lovell S, Johnston D (2009) Designing landscapes for performance based on emerging principles in landscape ecology. Ecol Soc 14(1)

Lundholm J (2005) A habitat template approach to green building surfaces. Paper presented at the third annual greening rooftops for sustainable communities conference, Washington, D.C, 4–6 May 2005

Lundholm JT, Walker W (2018) Evaluating the habitat template approach applied to green roofs. Urban Naturalist S 1:39–51

Lundholm J, Tran S, Gebert L (2015) Plant functional traits predict green roof ecosystem services. Environ Sci Technol 49(4):2366–2374

MacIvor JS, Ranalli MA, Lundholm JT (2011) Performance of dryland and wetland plant species on extensive green roofs. Ann Bot 107(4):671–679

Magill JD, Midden K, Groninger J, Therrell M (2011) A history and definition of green roof technology with recommendations for future research. Master's Thesis Carbondale Southern Illinois University Carbondale

McDonough W, Braungart M (2010) Cradle to cradle: remaking the way we make things. North point press, New York

Momaday NS (1969) The way to rainy mountain. University of New Mexico Press, Albuquerque

Nagase A, Dunnett N, Choi M-S (2013) Investigation of weed phenology in an establishing semi-extensive green roof. Ecol Eng 58:156–164

Nassauer JI (1995) Messy ecosystems, orderly frames. Landsc J 14(2):161–170

Oberndorfer E, Lundholm J, Bass B, Coffman R, Doshi H, Dunnett N, Gaffin S, Köhler M, Lui K, Rowe B (2007) Green roofs as urban ecosystems: ecological structures, functions, and services. Bioscience 57(10):823–833

Osmundson T (1999) Roof gardens – history, design and construction. W.W. Norton & Company, New York, NY

Park J, Kim J-H, Dvorak B, Lee DK (2018) The role of green roofs on microclimate mitigation effect to local climates in summer. International Journal of Environmental Research 12(5):671–679

Parkins KL, Elbin SB, Barnes E (2015) Light, glass, and bird—building collisions in an urban park. Northeast Nat 22(1):84–95

Partridge DR, Clark JA (2018) Urban green roofs provide habitat for migrating and breeding birds and their arthropod prey. PLoS One 13(8):e0202298

Peck SW (2012) The rise of living architecture. GreenRoofsforHealthy Cities, Toronto

Quinn M (2007) Manfreda Giant-Skipper Stallingsia maculosus (H. A. Freeman, 1955). Texas Entomology. Accessed May 17 2012 http://texasento.net/Stallingsia.htm

Ragheb A, El-Shimy H, Ragheb G (2016) Green architecture: a concept of sustainability. Procedia Soc Behav Sci 216:778–787

Reese RF, Myers JE (2012) EcoWellness: the missing factor in holistic wellness models. J Couns Dev 90(4):400–406

Reynolds J (2019) Ecodistricts: models for sustainable urban development at the neighborhood scale. Portland State University, Portland

Ricketts TH, Dinerstein E, Olson DM, Eichbaum W, Loucks CJ, Kavanaugh K, Hedao P, Hurley P, DellaSalla D, Abell R (1999) Terrestrial ecoregions of North America: a conservation assessment, vol 1. Island Press, Washington, DC

Rowe B (2015) Long-term rooftop plant communities. In: Sutton RK (ed) Green roof ecosystems. Springer International Publishing, Cham, pp 311–332. https://doi.org/10.1007/978-3-319-14983-7_13

Rowe B (2019) Reflections on 20 years of Green Roof Research. Living Architecture Monitor, vol 21. Green Roofs for Healthy Cities, Toronto

Sauvé L (1996) Environmental education and sustainable development: a further appraisal. Can J Environ Educ 1:7–34

Seltzer E, Smith TW, Cortright J, Bassett EM, Shandas V (2010) Making EcoDistricts Concepts and Methods for Advancing Sustainability in Neighborhoods. Portland State University, Portland, OR

Schindler BY, Blank L, Levy S, Kadas G, Pearlmutter D, Blaustein L (2016) Integration of photovoltaic panels and green roofs: review and predictions of effects on electricity production and plant communities. Israel Journal of Ecology & Evolution 62(1–2):68–73

Sharma A, Conry P, Fernando H, Hamlet AF, Hellmann J, Chen F (2016) Green and cool roofs to mitigate urban heat island effects in the Chicago metropolitan area: evaluation with a regional climate model. Environ Res Lett 11(6):064004

Silver W (2019) A guide to the plant species that inhabit the Osher Living Roof of Cal Academy in Golden Gate Park, San Francisco, California

Skabelund LR, Decker A, Moore T, Shrestha P, Bruce JL (2017)Monitoring two large-scale prairie-like green roofs in Manhattan, Kansas. In: CitiesAlive: 15th Annual Green Roof and Wall Conference 2017 Conference Proceedings, Seattle, WA, September 18–21 2017. The Cardinal Group, p 30

Snodgrass EC, McIntyre L (2010) The green roof manual: a professional guide to design, installation, and maintenance. Timber Press, London

Suppakittpaisarn P, Jiang X, Sullivan WC (2017) Green infrastructure, green stormwater infrastructure, and human health: a review. Current Landscape Ecology Reports 2(4):96–110

Sutton RK (2013) Seeding green roofs with native grasses

Sutton RK (2014) Aesthetics for green roofs and green walls

Sutton RK (2015) Green roof ecosystems, vol 223. Springer, Cham

Sutton RK (2020) Reading the Nebraska Landscape: An Ecological Aesthetic. In. CreateSpace Independent Publishing Platform, p 16

Sutton RK, Harrington JA, Skabelund L, MacDonagh P, Coffman RR, Koch G (2012) Prairie-based green roofs: literature, templates, and analogs. Journal of Green Building 7(1):143–172. https://doi.org/10.3992/jgb.7.1.143

TCL (2017) Why birds hit windows—and How You Can Help Prevent It. Cornell University. https://www.allaboutbirds.org/news/why-birds-hit-windows-and-how-you-can-help-prevent-it/. Accessed 23 Oct 2019

Van den Berg AE, Hartig T, Staats H (2007) Preference for nature in urbanized societies: stress, restoration, and the pursuit of sustainability. J Soc Issues 63(1):79–96

Velarde MD, Fry G, Tveit M (2007) Health effects of viewing landscapes–landscape types in environmental psychology. Urban For Urban Green 6(4):199–212

Weiler S, Scholz-Barth K (2009) Green roof systems: a guide to the planning, design, and construction of landscapes over structure. John Wiley & Sons, Hoboken

Wilson EO (2017) Biophilia and the conservation ethic. In: Evolutionary perspectives on environmental problems. Routledge, pp 263–272

Wines J, Jodidio P (2000) Green architecture, vol 240. Taschen Köln, Cologne

Afterword

Bruce Dvorak

On June first, 2018, I set out from College Station, Texas, with my wife Erin, and our two sons Andrew and Mark, on a 20,000-mile five-month journey across the western U.S.A. and Canada. With our trailer in tow, we set out a route where we would visit ecoregional green roofs and conservation sites that preserve some of the last remnant native landscapes near urban centers. During that time, I visited 140 green roofs, 14 national parks, and 40 conservation sites. I met with the co-authors of this book, architects, landscape architects, ecologists, building owners, green roof builders, suppliers, municipal representatives, and others. Upon return, beginning in November 2018, until the end of February 2020, *Ecoregional Green Roofs* was written with the aid of the eight co-authors of this book. Blind peer-review (by Springer) occurred from March to May 2020, and final revisions to the manuscript took place from June to July 2020.

This book articulates a background discussion, theory, and application of ecoregional green roofs located in North America west of the 100th meridian. This Afterword provides a discussion regarding why the book was written, an abbreviated summary of the main points of the book, and a discussion of topics or issues that are not conclusive and warrant further exploration. Although different chapters of the book address these issues, this afterword provides a concise discussion of these topics within a broader context.

B. Dvorak
Department of Landscape Architecture and Urban Planning, 305A Langford Architecture Center, Texas A&M University, College Station, Texas, USA
e-mail: bdvorak@tamu.edu

B. Dvorak (ed.), *Ecoregional Green Roofs*, Cities and Nature,
https://doi.org/10.1007/978-3-030-58395-8

Why This Book Was Written

Ecoregional Green Roofs was written for several reasons: (1) there is a general need for discussion regarding how green roofs can be employed in new locations and the role of research in that process (Rowe 2019); (2) there is a lack of publications on green roofs in hot and dry climates (Williams et al. 2010); (3) there is a need for reporting of emerging ideas for green roofs (Sutton and Lambrinos 2015); (4) there is a need to expand discussions on native plants on green roofs and biologically diverse green roofs (Butler et al. 2012; Cook-Patton and Bauerle 2012); and (5) due to the nature of experimental setups, green roof research often consists of small-scale simulations (Dvorak and Volder 2010). This book was written to address these needs, and more.

First, an extensive or semi-intensive green roof—as initially conceived—is typically planted with native or naturalized vegetation (FLL 2008). The concept that a green roof consists of native vegetation was developed in Europe and other parts of the world and has been practiced that way for many decades (Köhler and Poll 2010). Green roofs in Germany and Switzerland, where modern green roofs emerged, are typically planted with vegetation from the region, or vegetation native to the alpine ecosystems (Landolt 2001; Brenneisen 2006; Köhler 2006). Green roofs along the Mediterranean regions, and in China, have also made use of local vegetation (Papafotiou et al. 2013; Van Mechelen et al. 2014; Brandão et al. 2017; Deng and Jim 2017). Although there was some direction in North America regarding the application of native plants on green roofs (Monterusso et al. 2005; Lundholm 2006; Snodgrass and Snodgrass 2006), the emergence and dissemination of green roof technology in North America initially began without the benefit of long-term research and lacked funding for in-depth research on vegetation for green roofs (Rowe 2015, 2019). Green roof designers and industry providers, in North America, often turned to traditions of gardening as a metaphor for green roofs, including extensive green roofs (Osmundson 1999; Peck 2008; Luckett 2009). Green roof vegetation for extensive green roofs (east of the 100th meridian) has been largely selected from a range of exotic sedums from Europe, Asia, and South Africa and is planted as a kind of low-diverse exotic habitat (Luckett 2009; Cook-Patton and Bauerle 2012). However, many of these exotic sedums don't survive on green roofs in many parts of the western U.S. and Canada, where high daytime temperatures and prolonged drought can limit their use. This book was written to provide some background, theory, and diverse examples of how plants native to hot and dry climates can be used on green roofs.

Main Points of the Book

The book includes three parts: Part I Background and Theory, Part II Application: Ecoregional Green Roof Case Studies, Part III Summary, and Future Outlook. The main points of the book are summarized below.

Part I Background and Theory

1. Ecoregions are delineated boundaries of existing or potential native plant communities that have adapted to a particular place based upon environmental factors such as climate, altitude, soils, fire, biota, etc. (Bailey 1983; Omernik 1987).
2. The ecoregions of North America covered in this book generally began forming over 10,000 years ago (Holocene Epoch) as the Northern Hemisphere saw the warming of the climate and retreat of continental ice. About 4000 years ago the climate stabilized, and the pre-Columbian ecoregions were formed.
3. During that time, the widespread population of Native American tribes and nations developed cultural practices that resulted in sustainable food sources, plants for medicinal and cultural uses, and habitat for the hunting of fish and wild game. The ecosystems were largely self-sustaining, but experienced significant anthropomorphic activities such as the intentional burning of the ground vegetation, hunting of wildlife, harvesting of plants and plant parts, planting, seeding of native and adapted food crops.
4. The native terrestrial vegetation of the western U.S.A. and Canada includes varied taxa of trees, shrubs, grasses, wildflowers, bulbs, sedges, rushes, succulents, and low-growing drought-adapted woody vegetation, and others.
5. By the beginning of the 1900s, most of the native ecosystems in North America had been significantly altered by a growing North American population and expansion into the West.
6. Currently, the habitat quality of conservation sites near urban centers varies greatly. Some remnant conservation sites represent their pre-Columbian conditions, but many have experienced heavy anthropogenic influences or exists only as small isolated patches.
7. Prairie ecosystems were once a dominant ecosystem in North America, but with about 1% remaining, they are now one of the most critically endangered ecosystems due to loss of habitat.
8. Vegetation from prairie ecosystems can thrive on green roofs in North America (with necessary provisions); however, current green roof planting favors exotic succulents.
9. There are at least 61 species of *Sedum* native to North America; however, few of the native *Sedum* has been trialed on green roofs.
10. Sedums growing in their natural habitats (covered in this book) do not typically form large patches of contiguous groundcover. Many grow in isolated patches and are members of a plant community. They are frequently found growing on sloped, steeply sloped, and well-drained conditions.
11. The historic use of native vegetation on green roofs is rich in Europe; however, there are only a limited number of case studies where native plant communities have been trialed on green roofs west of the 100th meridian.
12. The first wave of green roofs in North America made use of native vegetation on pioneer houses (1862–1890s), and recreational roof gardens in urban centers (1880s–1940s) during the late Nineteenth Century and early Twentieth Century.

13. Contemporary research and development of the science of ecoregions have begun to translate to green roofs; however, the depth and exploration of the concepts are just emerging in North America.
14. Based on the case studies in this book, ecoregional green roofs belong to one of four categories:
    1) A green roof where all of the intended vegetation is native to the site, or nearby;
    2) A green roof where the intended vegetation is native to one or several adjacent.
       ecoregions (i.e. vegetation from an alpine ecosystem that is planted on a green roof in a valley below);
    3) A green roof where the intended vegetation is native to various, but similar ecoregions of North America (west of the 100th meridian).
    4) A green roof with vegetation from a combination of native (west of the 100th meridian) and non-native (introduced/exotic) vegetation. These climate-adapted green roofs feature native species and also include non-native plants.
15. The methods used to collect information for this book included on-site visits, interviews, written correspondence, online sources such as websites, reports, and peer-reviewed publications.
16. Each case study followed a pre-determined format and set of criteria guided by the objectives of this book.

Part II Application: Ecoregional Green Roof Case Studies

Case Studies

1. All of the urban regions investigated in this book had remnant conservation sites where native plants and plant communities are preserved. The book covers 26 conservation sites in 20 ecoregions west of the 100th meridian.
2. At least 830 taxa of native plants were explored by green roof designers in the 73 ecoregional case studies presented in this book.
3. During the early adoption phase of green roofs in North America, some argued that there is nothing native about rooftops, and therefore native plants don't belong on green roofs. Although rooftops are indeed manufactured environments, the evidence as demonstrated in the pioneering work of the building owners, designers, and providers of the 73 ecoregional green roof ecosystems featured in this book suggests that there is a role for native plants on green roofs. The case studies demonstrate multiple ways how native plants can adapt to green roofs, even in hot, dry or, semi-arid climates.
4. Grasses, herbaceous perennials, succulents, ferns, sedges, rushes, bulbs, woody groundcovers, small shrubs, and trees were used on ecoregional green roofs.
5. Ecoregional green roofs were adapted to various roof deck configurations: flat roof decks, moderately sloped roof decks, varied slopes and aspects, various

depths of growing media, various methods used to irrigate, and many different plant assemblages.

6. Nearly all of the ecoregional green roofs had access to supplemental irrigation. A majority of the irrigation systems used harvested rainwater, and a few used greywater as their irrigation water source.
7. Most of the green roofs that were initially installed without irrigation systems had irrigation systems installed post-construction after a drought or heat event stressed or killed the vegetation. Some of the replanted green roofs had original vegetation reinstalled, and some were replanted with vegetation that was not part of the original green roof design. In some cases, local landscapers (who lacked knowledge of green roofs) were hired to select plants and replant the green roof. Green roof design is a specialized area of knowledge, and building owners should contact the original designer for post-installation consulting.
8. Many green roofs were functioning with one irrigation run cycle throughout the growing season, with no process in place for periodical inspection. Many of the green roofs with heavy weed infestations were overwatered.
9. When there was dieback of vegetation on some of the case study green roofs, it appeared that the primary causes of dead plants were due to either no irrigation system present, a broken or non-functioning irrigation system, overwatering, or poor plant selection such as where drought-tolerant plants were located in a poorly drained substrate or there was reflected sunlight from glass/buildings that burned the vegetation.
10. Successful ecoregional green roofs had multiple irrigation zones to allow for different watering rates and were articulated for different microclimates on the roof. Many also used moisture sensors with smart watering systems and had resources for a periodic site visit and review of plant performance by a trained professional.
11. Most of the green roofs had monthly maintenance of vegetation. Even natural-appearing green roofs had some form of maintenance, at least once or twice during a growing season.
12. Case studies that featured green building practices (e.g. energy efficient, water efficient) had design teams that implemented an integrated design process where all the stakeholders were present and engaged from the beginning of the project: owners, architects, landscape architects, engineers, building maintenance and facilities staff, etc.
13. Most of the integrated designs in the case studies reported significant benefits to building owners including reduced energy use, natural daylighting, reduced demand for potable water, reduced stormwater runoff, biophilic environments for building users and visitors, and conservation of native vegetation and wildlife.
14. Wetland vegetation is being used on green roofs as part of a constructed wetland system to treat greywater or blackwater from inside buildings. Specialized permitting and regulation review was required. Ongoing monitoring and maintenance are required for constructed wetland green roofs.
15. Native local and migrating wildlife made use of the native vegetation on green roofs. Some of the larger ecoregional green roofs described in this book had

hundreds of sightings of native birds, butterflies, bees, and other forms of wildlife.

16. Some wildlife observed on ecoregional green roofs have not been seen in the region for decades. The habitat value of including native vegetation on green roofs could prove vital to the conservation of native biodiversity in cities.
17. Some cities have projects where green roofs are being used on every building in a city block or multiple blocks. This approach begins to employ concepts of landscape ecology where multiple blocks can form multiscale rooftop habitat patches or corridors. Multiple blocks could be planned to create rooftop habitat corridors for local and migrating wildlife.
18. Some cities such as Portland, Oregon have a venue for meaningful and ongoing dialogue about green roofs. The Green Roof info Think-tank (GRiT) is a model system to support green roofs locally. The Portland GRiT is comprised of local business owners, designers, researchers, naturalists, guests, and municipal representatives. Monthly meetings support current topics and help advise policy and exploration of new ideas.

Lessons Learned

1. In cities that had no previous study of green roofs, pre-testing and plant trials were an important part of the successful use of plants on green roofs. Not all native vegetation will adapt to any green roof design or maintenance program.
2. Ecoregional green roofs can be designed to be low maintenance (2–3 visits annually) and look natural in appearance.
3. Green roofs should not be neglected (no annual visits) but should be designed as if they will be neglected. This means that green roof ecosystems should be designed for the second or third owners, anticipating that a future owner may know little or nothing about green roof ecosystems.
4. Green roof maintenance is a specialized task not appropriate for untrained staff. Damage can be done to ecoregional green roof ecosystems if caretakers lack knowledge of how to care for them.
5. The best performing ecoregional green roofs had a "champion", someone that was knowledgeable about the local ecosystem and green roof ecosystem dynamics, anticipating potential outbreaks of unwanted volunteer or invasive exotic vegetation. It is much easier to manage a green roof proactively than to recover from neglect.
6. The best performing ecoregional green roofs had a nominal budget for minor replanting of zones (annually), and to respond to minor system maintenance. For these projects, the presence of a nominal annual budget allows for critical updates as needed. The absence of funding for ongoing maintenance or replanting often resulted in neglect and eventually a major disturbance to the green roof ecosystem.
7. Ecoregional green roofs west of the 100th meridian may need deep extensive or semi-intensive substrates to thrive. For new building construction, this may impose little additional cost (12.7 or 15 cm minimum depth versus 10 cm) for

extensive green roofs. The substrate depths of most of the green roofs in this book ranged from 12.7 cm to 30 cm (5 in-12 in).

8. The best performing ecoregional green roofs have an active soil (growth media) inspection and nutrient management system in place. Vegetation on ecoregional green roofs may require micronutrients or organisms inoculated into the soil, or periodically added to the substrate to establish and maintain a healthy green roof ecosystem.
9. Successful managers of green roofs in the case studies had easy access to a laminated maintenance manual (provided by the design team) that included photographs of desirable plants on the green roof, and common unwanted plants. Plant photos had captions with common and botanical names. Maintenance procedures were articulated in clear language with seasonal instructions (bi-lingual recommended).
10. Steep south or west-facing sections of green roofs (in the Northern Hemisphere) may need to be irrigated more frequently than low slope sections or north or east-facing sections. Sections of green roofs under these conditions should be provided with separate irrigation zones.
11. Many of the green roofs that were retrofitted onto an existing building did not have an adequate or designated maintenance access route. Maintenance access should be designed to be safe with appropriate measures in place.
12. Some green roofs adjacent to west or south facing reflective glass experienced die-back of vegetation. Special measures or design features may be needed to prevent reflected sunlight or heat from dark-colored facades from damaging vegetation. These considerations should be resolved during the design phase of the building, with an experienced green roof consultant.
13. Integrated designs require a process where multiple perspectives and skills are considered early in the planning stages of a project. Case study projects that *could* have taken advantage of a more integrated design process were limited because major project decisions were made without a variety of inputs. Those interested in gaining the benefits of integrated designs should plan accordingly.
14. Green roofs with native vegetation may require some interpretation and education for users to appreciate their aesthetic appearance. For example, in southern California, some of the native vegetation goes dormant during the dry summers and can appear 'dead'. However, with some diversity of plant forms and colors, visual interest can be maintained. Interpretive signage regarding the seasonal dynamics of native vegetation may be appropriate for green roofs that are visible to the public.
15. Due to the inherent differences in ecoregional vegetation across the western U.S. and Canada, the aesthetic appearance of the ecoregional green roofs varied greatly. These differences should be appreciated by local industry leaders who are invested in resilient and sustainable green roofs that reflect the local vegetation of their region. Thus, monitoring, research, education, and dissemination are integral to the successful adoption of ecoregional green roofs.

Part III Summary and Future Outlook

Many U.S. cities are still primarily influenced by land development tactics that largely exclude ecological values and replace natural systems with gray infrastructure and ornamental landscapes. This approach has contributed to a significant decline in native ecosystem functions and native biodiversity in urban regions. As the re-emergence of green building and ecological planning ideas continues to be explored and developed, green roofs could prove to become a vital component to the conservation of native vegetation in urban areas. However, more knowledge and dissemination about the performance, design parameters, and maintenance requirements of a variety of kinds of ecoregional green roofs are needed to guide designers and decision-makers.

There are many beautiful and biologically diverse green roofs in the western U.S.A. and Canada, but little is known about their composition, maintenance requirements, ecosystem services, their contributions towards mitigating climate change, and their potential conservation of native biodiversity. Research is needed regarding the application of these many integrated designs, with their exploratory methods for developing soils for green roofs, trialing of many hundreds of plants and plant communities, exploring alternative sources of water for irrigation, treating wastewater, improving energy savings for buildings, economic and life-cycle benefits, and their part in reducing carbon footprints.

Several ecoregional green roofs in each chapter of this book were intentionally inspired by local native plant communities, while many of the plant assemblages didn't intend to accurately represent local ecosystems. More design exploration and research on native plant assemblages and their design requirements are needed. For example, more research is needed on grass-based and succulent-based green roofs modeled after local analogs (e.g. coastal or alpine meadows, prairies, scrub habitats, rocky outcrops) as well as low-maintenance designs, floristically diverse green roofs, green roofs for native pollinators, green roofs for migrating wildlife, and constructed wetland green roofs to treat grey or black water from inside buildings. There has been little peer-reviewed research west of the 100th meridian regarding which plant assemblages best manage stormwater, how green roofs perform during monsoons, or can function during long cool and wet winters.

The green roof market needs research to inform political decisions and ongoing support; however, it may require many years or decades to mature to a point where meaningful conservation goals are conceived of and realized. Our hope for this book is that it may motivate others to learn from and appreciate how native ecosystems can be a source of inspiration for green roofs.

What is Inconclusive (and Research Gaps)

1. Although there has been some research and education on green roofs in the western U.S. and Canada, most urban centers currently have little or no research taking place on green roofs. Much of the existing research is on green roof ecosystem services (e.g. energy, water), and there is little or no research (in most ecoregions) on the application of plants and plant assemblages, or their

application to varied substrate compositions, or different roof deck slopes or microclimates for plant success.

2. There is a lack of research by developers regarding the successes and challenges of integrating green roofs into master-planned communities, mixed-use developments, or the ecosystem service benefits of green roofs with large scale applications.
3. There is a lack of research by maintenance providers, regarding feedback and challenges of maintaining green roof ecosystems.
4. Some plant nurseries are well-adapted to explore and trial the use of native plants on green roofs. However, few nurseries in the western U.S.A. and Canada are taking advantage of their potential to lead and team up with local universities to study some of the thousands of plants native to North America.
5. Much of the monitoring and data collection that is taking place on green roofs from the case studies are not being disseminated. Many owners of ecoregional green roofs lack connections to the industry or research community.
6. Some regions have strong support from local governments, and other regions are just beginning to integrate green roof technology into municipal codes, ordinances, policies, and incentives. There is a lack of research by urban planners on how green roofs can be integrated into municipal policies, regional planning efforts, address resiliency, habitat loss, or climate change.
7. There is a need to better understand how current green roof policies are serving their intended goals, economic structures, insurance investments, and performance of public utilities.
8. There is a need to understand the public perception of ecoregional green roofs and aesthetic understanding of ecoregional green roofs.
9. Climate change can be addressed locally through the use of planning tools such as ecodistricts or overlay zones to implement ecoregional green infrastructure zones where low-energy use, low water use, wastewater recycling, and native habitat are achieved through ecoregional green roofs.
10. There is little research taking place on the wider application of green roofs as part of green architecture and green infrastructure.

References

Bailey RG (1983) Delineation of ecosystem regions. Environ Manag 7(4):365–373

Brandão C, do Rosário Cameira M, Valente F, de Carvalho RC, Paço TA (2017) Wet season hydrological performance of green roofs using native species under Mediterranean climate. Ecol Eng 102:596–611

Brenneisen S (2006) Space for urban wildlife: designing green roofs as habitats in Switzerland. URBAN Habitats 4(1):27–36

Butler C, Butler E, Orians CM (2012) Native plant enthusiasm reaches new heights: perceptions, evidence, and the future of green roofs. Urban For Urban Green 11(1):1–10

Cook-Patton SC, Bauerle TL (2012) Potential benefits of plant diversity on vegetated roofs: a literature review. J Environ Manag 106:85–92

Deng H, Jim CY (2017) Spontaneous plant colonization and bird visits of tropical extensive green roof. Urban Ecosyst 20(2):337–352
Dvorak B, Volder A (2010) Green roof vegetation for North American ecoregions: a literature review. Landsc Urban Plan 96(4):197–213
FLL (2008) Guidelines for the planning, construction and maintenance of green roofing, English edn. Forschungsgesellschaft Landschaftsentwicklung Landschaftsbau e. V, Bonn
Köhler M (2006) Long-term vegetation research on two extensive green roofs in Berlin. Urban Habitats 4(1):3–26
Köhler M, Poll PH (2010) Long-term performance of selected old Berlin greenroofs in comparison to younger extensive greenroofs in Berlin. Ecol Eng 36(5):722–729
Landolt E (2001) Orchideen-Wiesen in Wollishofen (Zürich)-ein erstaunliches Relikt aus dem Anfang des 20. Jahrhunderts Vierteljahrsschrift der Naturforschenden Gesellschaft in Zürich 146(2/3):41–51
Luckett K (2009) Green roof construction and maintenance. McGraw-Hill, New York
Lundholm JT (2006) Green roofs and facades: a habitat template approach. Urban habitats 4(1):87–101
Monterusso MA, Rowe BD, Rugh CL (2005) Establishment and persistence of *Sedum* spp. and native taxa for green roof applications. HortScience 40(2):391–396
Omernik JM (1987) Ecoregions of the conterminous United States. Ann Assoc Am Geogr 77(1):118–125
Osmundson T (1999) Roof gardens – history, design and construction. W.W. Norton & Company, New York
Papafotiou M, Pergialioti N, Tassoula L, Massas I, Kargas G (2013) Growth of native aromatic xerophytes in an extensive Mediterranean green roof as affected by substrate type and depth and irrigation frequency. HortScience 48(10):1327–1333
Peck S (2008) Award winning green roof designs. Schiffer, Atglen
Rowe B (2015) Long-term rooftop plant communities. In: Sutton RK (ed) Green roof ecosystems, vol 223. Springer, Cham, pp 311–332
Rowe B (2019) Reflections on 20 years of green roof research. living architecture monitor, vol 21. Green Roofs for Healthy Cities, Toronto
Snodgrass E, Snodgrass L (2006) Green roof plants. Timber Press, Portland
Sutton RK, Lambrinos J (2015) Green roof ecosystems: summary and synthesis. In: Green roof ecosystems. Springer, pp 423–440
Van Mechelen C, Dutoit T, Hermy M (2014) Mediterranean open habitat vegetation offers great potential for extensive green roof design. Landsc Urban Plan 121:81–91
Williams NSG, Rayner JP, Raynor KJ (2010) Green roofs for a wide brown land: opportunities and barriers for rooftop greening in Australia. Urban For Urban Green 9(3):245–251. https://doi.org/10.1016/j.ufug.2010.01.005

Index

B. Dvorak (ed.), *Ecoregional Green Roofs*, Cities and Nature,
https://doi.org/10.1007/978-3-030-58395-8

F